RESPIRATION AND METABOLISM OF EMBRYONIC VERTEBRATES

Perspectives in vertebrate science

Volume 3

Series Editor
EUGENE K. BALON

Respiration and metabolism of embryonic vertebrates

Satellite Symposium of the 29th International Congress of Physiological Sciences,
Sydney, Australia, 1983.

Organized and edited by

ROGER S. SEYMOUR

Department of Zoology
University of Adelaide

1984 **Dr W. JUNK PUBLISHERS**
A MEMBER OF THE KLUWER ACADEMIC PUBLISHERS GROUP
DORDRECHT / BOSTON / LANCASTER

Distributors

for the United States and Canada: Kluwer Boston, Inc., 190 Old Derby Street, Hingham, MA 02043, USA
for all other countries: Kluwer Academic Publishers Group, Distribution Center, P.O.Box 322, 3300 AH Dordrecht, The Netherlands

Library of Congress Cataloging in Publication Data

Library of Congress Cataloging in Publication Data
Main entry under title:

Respiration amd metabolism of embryonic vertebrates.

   (Perspectives in vertebrate science ; v. 3)
   Includes index.
   1. Embryology--Vertebrates--Congresses. 2. Respira-
tion--Congresses. 3. Metabolism--Congresses. I. Seymour,
Roger S. II. International Congress of Physiological
Sciences (29th : 1983 : Sydney, N.S.W.) III. Series.
QL959.R39 1984   596'.0333   84-9751
ISBN-13:978-94-009-6538-6

ISBN-13:978-94-009-6538-6   e-ISBN-13:978-94-009-6536-2
DOI: 10.1007/978-94-009-6536-2

Copyright

Preface

The papers in this volume were presented at an international symposium, held in South Australia on September 8–10, 1983. The purpose of the meeting was to present the comparative physiology of gas exchange, water balance and energetics of developing vertebrate embryos. contributions were invited from leading research workers in an attempt to represent the forefront of investigation of all vertebrate classes and to promote a broadly comparative approach to the study of embryonic physiology.

These proceedings therefore reflect the current level of research activity focusing on each group of vertebrates. While considerable expansion and specialization has occurred in the area of avian embryos over the last decade, work on reptilian embryos is less developed and that on fish and amphibians is still in its infancy. Although a great deal is known about respiration and metabolism in embryos of placental mammals, the physiology associated with the curious mode of development of monotreme and marsupial embryos has not been examined until recently. In this symposium, the well-studied vertebrate classes are represented primarily by specific research papers that document original work. These are balanced by more extensive reviews of the lesser known classes.

Studying vertebrate embryos is fascinating because the growing organism is often remarkably circumscribed yet its physiology is wonderfully dynamic. Unlike adult vertebrates that have materials continually passing through them at more or less constant rates, the embryos are never in the steady state, but regularly change the rates of metabolic processes as they grow. The cleidoic embryos of many reptiles and most birds develop within shells that permit the exchange of gas only; therefore they carry with them nearly all materials and energy required for development. Non-cleidoic embryos can take up various materials from the environment and may control exchange according to the ever-changing requirements of the growing embryo. Regulated instability is also demonstrated in the dramatic transitions in gas exchange organs during development or at birth, shifting from diffusional exchange in the early stages of development to convective systems involving skin, gills, extraembryonic membranes or lungs. These transitions are correlated with increasing metabolic rate and are accompanied by suites of regulatory changes in blood flow rate and distribution, in ventilation and in the blood properties associated with gas transport and acid-base balance. The examination of these changes represents one theme of the symposium.

Another theme is adaptation of the embryos to their immediate environments. Increasing metabolic rate tends to increase diffusion gradients between the embryo and its surroundings unless there are adaptive changes in the diffusion paths. In cases such as the avian eggshell, the path for gas diffusion appears to be genetically determined by the parent and is often remarkably appropriate for the incubation environment, therefore providing tolerable conditions inside the egg. There may also be adaptive changes in the gas exchange organs, for example, changes in the avian chorioallantois or the amphibian larval gill in response to hypoxic environments.

This symposium has its roots in two previous meetings. In 1977 Johannes Piiper organized a satellite symposium in Göttingen, F.R.G., associated with the 27th International Union of Physiological Sciences Congress in Paris that year. The topic of the symposium was avian respiration, and Hermann Rahn was instrumental in the inclusion of a session on embryonic birds. Hence the final title of that meeting became 'Respiratory Function in Birds: Adult and Embryonic' and the proceedings were published by Springer-Verlag in 1978. The success of this symposium was repeated in 1979 by Cynthia Carey who organized a symposium for the American Society of Zoologists meeting in Tampa, Florida, U.S.A. Entitled 'Physiology of the Avian Egg' and published in *American Zoologist* in 1980, the symposium was devoted to birds but also included some recent work on reptiles. On that occasion, Dr Piiper suggested that it was my turn to plan a symposium for the 'egg people', as Dr Carey termed us, as a satellite of the I.U.P.S. 29th International Congress, to be held in Sydney, Australia, in 1983. I felt that it might be useful to expand the coverage even further and include not only the oviparous birds and reptiles, but all vertebrates. Hermann Rahn, Knut Schmidt-Nielsen and Pierre Dejours sustained this view and the symposium was approved by the I.U.P.S. Organizing Committee.

The symposium occurred aboard the 'Proud Mary', a riverboat on the Murray River. This unusual venue isolated the 35 participants from distraction and provided intensive contact during which we shared ideas and approaches far beyond the formal presentations. The papers were presented over two and a half days and discussions continued into the nights. Although these discussions were sometimes as interesting as the presentations, they could not be included in this volume.

All papers published here were reviewed by anonymous referees to whom I offer my appreciation. Some authors subsequently resubmitted revised manuscripts. Although an effort was made to resolve and eliminate controversy, the authors were allowed full expression of their conclusions and some issues must remain contentious. I have attempted to edit the papers in a uniform style and a useful presentation. In general, the units have been standardized according to the

International System of Units (SI). Unfortunately, the SI unit for gas partial pressure, the Pascal (Pa), is not yet fully assimilated by most respiratory physiologists. In the interest of effective communication, therefore, the familiar unit, the torr, is used here. The conversion is 1 torr = 1 mm Hg = 133 Pa. This usage, although incorrect, is practical because it clearly distinguishes gas partial pressure (torr) from hydrostatic pressure or tension (kPa).

This symposium was supported wholly by the participants and the granting agencies that funded their travel. While recognising the attractiveness of the I.U.P.S. Congress, I am nevertheless delighted that the symposium was important enough to justify the enormous travel expenses incurred. We greatly appreciate the courtesy of Robert Foord who permitted us to use the 'Proud Mary' and we thank the captain and crew for warm hospitality. David Booth, David Bradford, Cathie Buddle, Phil Kempster, Sandi Poland, Greg Powell, Michael Thompson and David Williams assisted the organization in many ways. Ruth Altmann drafted almost all of the figures and Sandra Lawson and Heather Kimber retyped all of the manuscripts. To these and the other forbearing members of my Department, I offer my sincere thanks.

Roger S. Seymour,
Adelaide, December 1983

Contents

1. Facilitation of maternal-fetal oxygen transfer in fishes: Anatomical and molecular specializations

Rolf L. Ingermann[1] and Robert C. Terwilliger[2]

Abstract

Oxygen transfer in viviparous vertebrates is facilitated by the close proximity of fetal and maternal circulatory systems. Among the fishes, facilitation is accomplished by a remarkable diversity of maternal-fetal anatomical specializations. Maternal-fetal O_2 transfer is further enhanced by a fetal blood which has a higher O_2 affinity than that of the adult. The diversity of anatomical specializations suggests that there may be a similar variety of molecular specializations as well to ensure a relatively high fetal blood O_2 affinity. Study of adult and fetal bloods of the seaperch, *Embiotoca lateralis* have shown that fetal blood has a higher O_2 affinity than that of the adult. It appears that the *E. lateralis* fetus utilizes a different hemoglobin and a lower organic phosphate concentration to ensure this higher blood O_2 affinity; a lower mean corpuscular hemoglobin concentration may contribute to the higher O_2 affinity as well. A higher fetal than adult O_2 affinity of whole blood exists in at least some sharks. The molecular basis for such affinity differences also appears to involve different fetal and adult hemoglobins; different intraerythrocytic organic phosphate concentrations may contribute as well. However, further studies are required to clarify the molecular basis of the facilitation of maternal-fetal O_2 transfer in the fishes.

Introduction

Viviparity has evolved repeatedly and independently in all the vertebrate classes except the birds. Although viviparity is most commonly studied and the details well known in mammals, it is widespread in the fishes which were the first vertebrates to exhibit this mode of reproduction. It is amongst the fishes that the

[1] *Obstetrics and Gynecology Research, L458, Oregon Health Sciences University, Portland, Oregon 97201, U.S.A.*
[2] *Oregon Institute of Marine Biology, University of Oregon, Charleston, Oregon 97420, U.S.A.*

Seymour, R.S., (ed.) Respiration and metabolism of embryonic vertebrates.
© *1984, Dr W. Junk Publishers, Dordrecht/Boston/London. ISBN-13: 978-94-009-6538-6*

known anatomical adaptations for maternal-fetal interactions (yolk sac placenta, various placental analogs and specialized epithelial absorptive and secretory structures) first appear (Wourms, 1981). Because of the remarkable spectrum of anatomical structures in fishes as well as the lower taxonomic position of fishes, an understanding of fish maternal-fetal interactions may provide important insights into functional aspects and evolution of viviparity.

Facilitation of maternal-fetal exchanges by anatomical specializations

Efficient and rapid maternal-fetal exchange of molecules is facilitated by the close proximity of maternal and fetal circulatory systems. In most mammals, the yolk sac and/or chorioallantoic placenta provides a direct juxtaposition of the two circulatory systems (Ramsey, 1982). Among the fishes, placentae as well as other fetal and maternal anatomical specializations can be found with wide structural variations existing within families (see reviews by Amoroso, 1960; Hoar, 1969; Wourms, 1981). Placentae, which are found only in some sharks, are all modifications of the yolk sac. Shark placentae consist of the distal portion of the yolk sac lying in contact or interdigitating with a highly vascular area of the uterus. This apposition is characterized by a reduction in the epithelia of the yolk sac and the uterine wall (Schlernitzauer and Gilbert, 1966; Gilbert and Schlernitzauer, 1966). However, the presence of finger-like extensions of the umbilical stalk (appendiculae) in certain species suggest that some maternal-fetal exchanges occur by an intrauterine route as well. Among the chondrichthyans, a different placental analog is found in some bats and rays. In these fishes, long, thin, glandular villi (trophonemata) extend from the uterine wall and, in at least one species, enter the esophagus of the fetus via the spiracles (Wood-Mason and Alcock, 1891). Among other viviparous cartilaginous fishes, such as *Squalus acanthias* (= *S. suckleyi*, Hart, 1973), there are no structures equivalent to a placenta; instead, there is a vascular plexus covering the fetal body, tail and yolk sac (Manwell, 1958a).

Different kinds of maternal and fetal anatomical specializations, related to nutritional and respiratory exchanges, exist in the viviparous osteichthyan fishes. In the families, Poeciliidae and Anablepidae, where the fetus remains within the ovarian follicle until hatching (followed immediately by birth), maternal-fetal exchanges occur by follicular pseudoplacenta (Turner, 1940a, b). The follicular pseudoplacenta is a complex structure, composed of a villous follicular wall, a follicular space and the adjacent vessels of a portal system which cover a greatly expanded yolk or belly sac. In the poeciliids, the belly sac is actually an enlarged pericardial sac; in the anablepids, the portal system lies over both expanded pericardial and peritoneal sacs. In members of some families, gestation occurs within the ovarian cavity and other maternal-fetal anatomical specializations are employed. For example, in the family Jenynsiidae, vascular pericardial and

2

peritoneal sacs, characteristics of follicular gestation, are present only temporarily. Eventually, primary exchanges appear to involve a 'branchial placenta' where extensions of the ovarian wall are in close association with the gills and mouth cavities of the fetus (Turner, 1940d). Fetuses of most of the goodeid and at least some of the parabrotulid and ophidioid fishes are characterized by highly vascularized, long extensions of the hind gut (Turner, 1936, 1937, 1940c; Wourms and Cohen, 1975; Cohen and Wourms, 1976). These extensions (trophotaeniae) probably function in both nutrient and respiratory exchanges. In the Embiotocidae, highly vascularized ovarian folds form compartments around the developing fetus (Blake, 1867; Eigenmann, 1892; Turner, 1938; Webb and Brett, 1972a, b; McMenamin, 1979). In at least one species of embiotocid, an ovarian fold enters the fetal operculum and rests against the gill tissue, a maternal-fetal structural relationship similar to the branchial placenta of the jenynsiids (Turner, 1952). In addition, embiotocid fetuses have highly vascularized hypertrophied fins terminating in fin extensions or fin spatulates. An example, *Embiotoca lateralis,* is shown in Fig. 1. The apposition of fetal fins and spatulates with the

Fig. 1. Embiotoca lateralis mid- (upper) and late-gestation (lower) fetuses obtained in April and late-June/July, respectively. Scale in upper right is 2 cm.

3

vascularized ovarian folds of the mother forms another type of pseudoplacenta. Finally, in other fishes including representatives of the scorpaenids (reviewed by Wourms, 1981), the clinids (Veith, 1980) and the zoarcids (e.g. *Zoarces viviparus*, Kristofferson et al., 1973), fetuses lack specialized trophic structures; they probably supplement nutrient ingestion with epidermal uptake as in some chondrichthyans.

Thus, as in viviparous mammals, many viviparous fishes utilize either a placenta of a placenta analog, formed by fetal and/or maternal anatomical specializations, to bring fetal and maternal circulations into juxtaposition. The close proximity or fetal and maternal bloods therefore facilitates the exchange of both soluble nutrients and gases. However, in addition to such anatomical mechanisms, facilitation of maternal-fetal molecular transfers in fishes is likely to involve molecular mechanisms as well. One vital transfer is that of O_2 from mother to fetus. Although the molecular basis for the facilitation of O_2 uptake by the mammalian fetus has been studied extensively, little is known of the molecular aspects of such transfer in fishes.

Facilitation of maternal-fetal oxygen transfer by molecular specializations

General

In most viviparous animals examined, fetal blood has a higher affinity for O_2 than does adult blood. This is thought to facilitate the maternal-fetal O_2 transfer (Meschia, 1978; Bauer et al., 1981). Among the mammals, the molecular basis underlying this facilitation falls into two general categories. Fetal blood may have a relatively high O_2 affinity due to the presence of a structurally unique fetal hemoglobin with a higher O_2 affinity than adult hemoglobin. For example, different adult and fetal hemoglobins appear responsible for the relatively high O_2 affinity of fetal blood in the ruminants (Blunt et al., 1971; Baumann et al., 1972). In other mammals, such as the dog, horse and pig, the hemoglobins of the adult and fetus are structurally identical; however, the adult and fetal concentrations of the intraerythrocytic organic phosphate, 2,3-disphosphoglycerate (2,3-DPG), are different (Dhindsa et al., 1972; Bunn and Kitchen, 1973; Tweeddale, 1973). Organic phosphates lower the oxygen-binding affinity of hemoglobins by binding to and stabilizing the deoxygenated, rather than the oxygenated hemoglobin (Benesch and Benesch, 1967, 1969; Chanutin and Curnish, 1967; Perutz, 1970). Consequently, low intraerythrocytic concentrations of 2,3-DPG in the fetus can account for the relatively high O_2 affinity of its whole blood.

A modification of these strategies is found in the primates. Distinct adult and fetal hemoglobins exist in man as well as in the baboon and japanese monkey (Bauer et al., 1968; De Verdier and Garby, 1969; Tyuma and Shimizu, 1969; Takenaka and Morimoto, 1976; De Simone and Mueller, 1979). In contrast to the

4

ruminants, these adult and fetal hemoglobins have very similar intrinsic O_2 affinities and are sensitive to allosteric modulation by organic phosphates. However, adult and fetal hemoglobins have different affinities for organic phosphates; thus the O_2 affinity of the fetal hemoglobin in the presence of an equal concentration of organic phosphate is lowered less than that of adult hemoglobin, accounting for the relatively higher O_2 affinity of fetal blood. Regardless of the molecular mechanism employed, the higher O_2 affinity of fetal versus adult blood is a widespread occurrence in viviparous animals and suggests that it has an important role in ensuring an adequate O_2 supply to the fetus. A difference in O_2 affinities between fetal and maternal bloods does not, however, appear necessary for fetal development; no such affinity difference exists in the cat (Novy and Parer, 1969).

Teleostean fishes

In contrast to a detailed understanding of maternal-fetal O_2 transfer which exists in mammals, much less is known about the molecular bases of this process in fishes. From measurements of ovarian fluid O_2 tensions in *Rhacochilus vacca* (a close relative of *E. lateralis*), Webb and Brett (1972b) have estimated that the O_2 tension of fluid in the ovarian sac during the latter half of gestation is substantially less than that of the maternal blood. This is likely to be true in *E. lateralis* pregnancy as well. It is thus reasonable to suspect that the O_2 affinity of fetal blood is higher than that of adult blood to ensure adequate O_2 uptake by the fetus. Consequently, we have measured adult and fetal whole blood O_2 affinities by a thin-layer spectrophotometric method and have found that, as in most viviparous vertebrates, fetal *E. lateralis* blood has a higher O_2 affinity than does adult blood (Ingermann et al., 1984) (Fig. 2). Furthermore, blood from a mid-gestation fetus has a higher O_2 affinity than blood from late-gestation fetus. This implies that the O_2 affinity of fetal blood becomes lower as fetal development approaches birth.

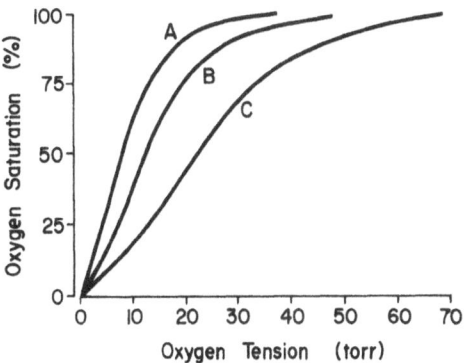

Fig. 2. Oxygen equilibrium curves of *E. lateralis* mid- (A) and late-gestation fetal (B) and adult (C) whole bloods generated in the absence of exogenous CO_2.

Oxygen binding studies of mid- and late-gestation fetal and adult hemoglobins, previously stripped of allosteric modifiers, show a pattern of O_2 affinities similar to that observed for whole blood (Ingermann and Terwilliger, 1981a) (Fig. 3). The intrinsic O_2 affinity of the hemoglobin of the mid-gestation fetus is greater than that of the late-gestation fetus which in turn is greater than that of the adult over a wide range of pH. The O_2 binding data suggest structural differences among hemoglobins from the different developmental stages.

Fetal and adult hemoglobins were examined for structural differences in several ways (Ingermann and Terwilliger, 1981a). The molecular weights of the intact hemoglobins and subunits were determined by gel chromatography and sodium dodecyl sulfate electrophoresis, respectively. These molecular weights are comparable to those of other tetrameric hemoglobins but reveal no adult-fetal differences. Furthermore, no differences could be detected with polyacrylamide disc gel electrophoresis or ion exchange chromatography of the native molecules. Structural differences were detected, however, by electrophoresis in the presence of a high urea concentration and low pH. The electrophoretic patterns suggest differences in the subunit composition of the mid-gestation fetal and adult hemoglobins (Fig. 4). It is likely that late-gestation fetal hemoglobin is a mixture of unique fetal and adult hemoglobins; this probably accounts for the intermediate O_2 affinity of late-gestation fetal hemoglobin. In animals such as sheep, man and baboon, adult and fetal hemoglobins share a common globin chain; the other chain is uniquely fetal or adult (Kitchen and Brett, 1974; De Simone and Mueller, 1979). Based on the electrophoretic patterns (Fig. 4), it appears likely that *E. lateralis* fetal and adult hemoglobin also share a common subunit. Mid-gestation fetal hemoglobin can also be distinguished from that of the adult by enzymatic digestion and analysis of the resulting peptide fragments by separation on thin-layer silica sheets using chromatography in the first dimension and electro-

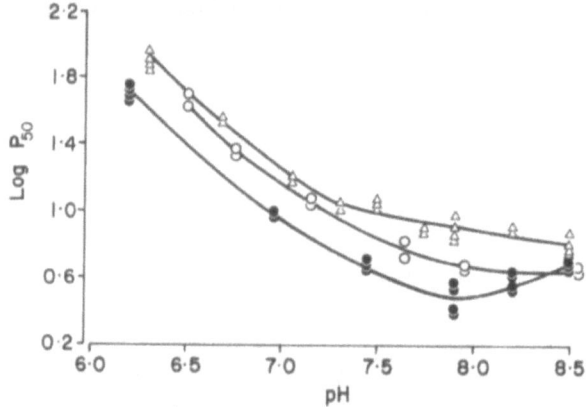

Fig. 3. Oxygen binding studies of *E. lateralis* hemoglobins stripped of organic phosphates from the mid- (●) and late-gestation (○) fetus and adult (△).

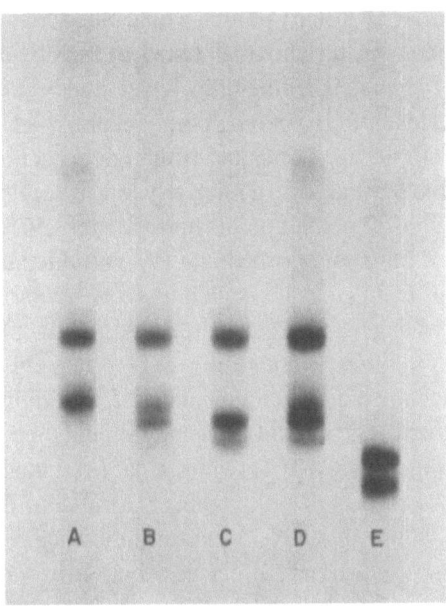

Fig. 4. Low pH, urea disc gel electrophoresis of *E. lateralis* purified globins from the mid- (A) and late-gestation (B) fetus, adult (C), mid-gestation fetus plus adult (D) and human (E).

phoresis in the second. Analysis of amino acid composition of purified globin from the three stages shows differences which are consistent with this interpretation (Ingermann and Terwilliger, 1981a). It is concluded that *E. lateralis* fetus and adult possess structurally and functionally distinct hemoglobins.

Oxygen binding by fish hemoglobins is generally subject to allosteric control by organic phosphates (Gillen and Riggs, 1971; Wood and Johansen, 1972; Coates, 1975). The O_2 affinity of mammalian and fish hemoglobins is reduced by organic phosphates in a similar manner (Perutz and Brunori, 1982). However, whereas 2,3-DPG is the primary organic phosphate of mammalian erythrocytes, adenosine triphosphate (ATP) and/or guanosine triphosphate (GTP) are the principle erythrocyte organic phosphates of most fishes (Coates, 1975; Bartlett, 1978, 1980; Leray, 1979). Examination of mid- and late-gestation fetal and adult *E. lateralis* hemoglobins has revealed that their O_2 affinities are lowered by ATP and that the effect of ATP is comparable for all three hemoglobins (Ingermann and Terwilliger, 1981a, b). Qualitative analysis of late-gestation fetal and adult erythrocytes has shown that with respect to total concentrations, ATP and GTP are the primary and secondary organic phosphates, respectively. Total nucleoside triphosphate (NTP) concentrations are not significantly different between erythrocytes of mid- and late-gestation fetuses, however, NTP concentrations of fetal cells are appreciably lower than adult concentrations (Ingermann and Terwilliger, 1981b). Lower fetal erythrocytic concentrations of NTP are likely to contribute to the higher O_2 affinity of fetal whole blood for several reasons. First, since organic phosphates reduce hemoglobin O_2 affinity by direct binding to, and stabilization of, deoxygenated hemoglobin, lower fetal NTP concentrations probably result in higher levels of oxyhemoglobin. Second, high intracellular organic phosphate concentrations are associated with high concentrations of protons in the cell relative to the plasma; i.e., a relatively low intracellular pH exists in this situation via the Gibbs-Donnan equilibrium (Duhm, 1971, 1976). Thus, lower fetal NTP concentrations should contribute to an elevated intraerythrocytic pH. Since fetal and adult *E. lateralis* hemoglobins show similar normal Bohr effects (Ingermann and Terwilliger, 1981a), elevated fetal cellular pH is likely to result in higher hemoglobin O_2 affinity. In addition, ATP is less effective in reducing the O_2 affinity of *E. lateralis* hemoglobin as pH rises and approaches a value of 8; consequently, relatively less organic phosphate should bind to fetal hemoglobin, raising O_2 affinity (Ingermann and Terwilliger, 1981a). Finally, a reduction in cell ATP concentrations can lead to cell swelling (Hoffmann, 1977; MacKnight and Leaf, 1977). Thus, lower fetal NTP concentrations may be associated with swollen cells. *E. lateralis* fetal erythrocytes are approximately 60% larger and have lower mean corpuscular hemoglobin concentrations (MCHC) than the adult cells; hemoglobin in the fetal cells is thus more dilute than that in adult cells (Ingermann and Terwilliger, 1982). A decrease in intracellular hemoglobin concentration has been correlated with an increase in hemoglobin O_2 affinity (Bellingham et al., 1971; Forster, 1972; Sinet et al., 1976; Laver et al., 1977). Although the molecular basis of this increased affinity has not yet been clearly established, it is possible that the lower MCHC in the *E. lateralis* fetal erythrocytes, in contrast to adult cells, may contribute to the differences in O_2 affinity between maternal and fetal whole bloods (Ingermann and Terwilliger, 1982).

It thus appears that *E. lateralis* may utilize several molecular mechanisms to ensure that the fetal blood has a high O_2 affinity, or at least that it does not have a lower O_2 affinity than the maternal blood. Clearly, the structurally distinct fetal versus adult hemoglobin contributes to the higher affinity of the fetal blood. However, the O_2 binding characteristics of the hemoglobins themselves cannot account entirely for the differences in affinities of the whole bloods; this suggests an important role for organic phosphate (Ingermann et al., 1984). It is likely that maternal-fetal blood and erythrocyte pH differences, as well as the MCHC differences, are involved in fetal respiratory physiology, however, their roles remain to be established. Nonetheless, it appears that the two primary mammalian strategies for facilitating fetal O_2 uptake at the level of the erythrocyte, namely distinct adult-fetal hemoglobins and different intraerythrocytic organic phosphate concentrations, are evident in a fish, *E. lateralis*.

A number of perplexing questions remain regarding the respiratory physiology of the intact animals. For example, during the mid- to late-gestation period, the ovarian fluid P_{O_2} decreases and fetal O_2 consumption per unit mass increases in the seaperch, *Rhacochilus vacca* (Webb and Brett, 1972b). This is very likely to occur in *E. lateralis* as well. During this period, the relative surface area of the highly vascularized fins versus the body size decreases (Fig. 1), and the body becomes covered with scales, thereby reducing the ability of these surfaces to function in respiratory transfer. Therefore, it seems surprising that the O_2 affinity of the fetal blood decreases during this period (Fig. 2). A similar decrease in O_2 affinity of blood with development has been reported in mouse, man and the viviparous lizard, *Sphenomorphus quoyii* (Wells, 1979; Bard and Teasdale, 1979; Grigg and Harlow, 1981). The physiological significance of prenatal lowering of blood O_2 affinity with increasing development is not clear. However, since the decrease in fetal blood in *E. lateralis* affinity is due to changes in blood hemoglobin molecules (NTP levels do not change during this period) and since NTP levels rise rapidly after birth (Ingermann and Terwilliger, 1981b), this physiological adaptation may ensure quick acquisition of adult blood properties by the neonate.

With respect to teleosts other than *E. lateralis*, little is known. The hemoglobins of fetal and adult *Zoarces viviparus* are electrophoretically distinguishable (Hjorth, 1974); on this basis, they presumably have different functions as well. M. Hartvig and R.E. Weber present a thorough study of the molecular basis for maternal-fetal O_2 transfer in *Z. viviparus* in this volume.

Chondrichthyan fishes

Fetuses of most common viviparous teleosts are very small and thus technically difficult to study; on the other hand, shark and ray fetuses are generally much larger and therefore yield sufficient volumes of blood for experimentation. Thus, examination of adult-fetal blood characteristics of other fishes has been almost exclusively restricted to the chondrichthyans.

McCutcheon (1947) examined the O_2 binding characteristics of dilute hemoglobin solutions of fetuses and adults from three viviparous ray species and found no maternal-fetal differences in O_2 affinity. He pointed out, however, that all fetuses examined were in late stages of gestation. Manwell (1958b) found ontogenetic changes in hemoglobin-O_2 binding properties in an oviparous ray where advanced embryonic hemoglobin was functionally very similar to that of adult hemoglobin. The possibility thus remains that in the species examined by McCutcheon, a fetal to adult switch in hemoglobins occurs, similar to the hemoglobin transition in man (Bard and Teasdale, 1979) and in *E. lateralis*, and that the switch had occurred prior to his examination.

Distinct fetal hemoglobins occur in several species of shark. The bloods of fetal, neonate and adult dogfish, *Squalus acanthias*, have been studied by several groups, the neonate presumably being one transition state with respect to blood properties. Suspensions of fetal erythrocytes of dogfish have higher O_2 affinities than do those of adult cells (Manwell, 1958a, 1963). Structurally and functionally different hemoglobins appear to account for at least part of the affinity difference between the fetal and adult erythrocyte suspensions. Consistently different fetal-adult hemoglobin patterns are shown by starch gel electrophoresis (Manwell and Baker, 1970). Hemoglobins from neonatal and adult dogfish are electrophoretically similar but can be distinguished on the basis of polymerization by disulfide bridging (Fyhn and Sullivan, 1975). Hemoglobins of neonates have higher O_2 affinities than those of the adult (Weber et al., 1983a) and the same is true for neonate whole blood versus adult blood (Wells and Weber, 1983). Finally, the O_2 binding characteristics of *Carcharhinus milberti* fetal and adult hemoglobins (stripped of organic phosphates), examined in the presence and absence of ATP, are interpreted as being different proteins due to the difference in their intrinsic O_2 affinities; fetal hemoglobin has the higher O_2 affinity (Pennelly et al., 1975). Both hemoglobins are sensitive to ATP. Thus, distinct fetal and adult hemoglobins exist in at least some sharks and appear to have a functional role in fetal respiratory physiology. These sharks therefore appear to share at least one of the molecular mechanisms which facilitates fetal O_2 uptake in some mammals and *E. lateralis*.

Since organic phosphates are important allosteric regulators of hemoglobin function and are involved in maternal-fetal O_2 transfer in many animals, it is important to consider their role in fetal physiology of the cartilaginous fishes. Dilute solutions of both fetal and adult *S. acanthias* hemoglobins have higher O_2 affinities than their erythrocyte suspensions (Manwell, 1958a). This suggests the possibility of different allosteric regulation and/or different MCHCs, between fetal and adult erythrocytes. Unfortunately, relatively little is known of organic phosphate concentrations in fetal and adult shark erythrocytes. However, mid-gestation fetal and adult shark erythrocyte ATP and GTP concentrations have been measured in *S. acanthias* and *Mustelus canis* by Bartlett (1982). Fetal and adult *S. acanthias* erythrocytes contain ATP as their primary organic phosphate. In contrast, fetal *M. canis* erythrocytes contain ATP as the primary organic phosphate (at a concentration higher than that of adult cells) while GTP concentrations are higher in the adult cells. However, adult erythrocytes in these two species contain higher total concentrations of organic phosphates than do their fetal counterparts. Erythrocytes of *S. acanthias* neonates also contain lower organic phosphate concentrations than do adult cells (Wells and Weber, 1983). However, two considerations suggest that the organic phosphate concentration differences might not be major factors in facilitating fetal O_2 uptake in these sharks. First, the difference in [ATP + GTP] concentrations between adult and fetal erythrocytes is low, about 1.5 and 2.8 mM for *S. acanthias* and *M. canis*, respectively, as compared to a 7 mM difference in *E. lateralis*. Second, sharks and rays have high concentrations of urea in their bloods and urea apparently blocks organic phosphate binding to hemoglobins (Coates, 1975; Weber et al., 1983a, b). In addition, the Bohr effect for *S. acanthias* whole blood is small (Wells and Weber, 1983); thus, relatively small differences in erythrocyte adult-fetal NTP concentrations should have little influence on ensuring a higher fetal blood O_2 affinity via an effect on intracellular pH. The data therefore suggest that fetal sharks, and possibly rays, have bloods with higher O_2 affinities than those of their mothers. Structurally and functionally different fetal and adult hemoglobins probably make a major contribution to the blood O_2 affinity differences. Additional factors such as differences in organic phosphate concentration and/or MCHC may also be involved, although the available data do not confirm their role.

Summary

Most mammals utilize the yolk sac and/or chorioallantoic placenta to facilitate fetal-maternal exchanges of important molecules by bringing fetal and maternal circulations close to one another. The widespread distribution of a placenta among mammals is testimony to its functional importance. A chorioallantoic placenta is absent and the yolk sac placenta is rare among the fishes; however, viviparous fishes show a wide variety of placenta-like structures which facilitate the efficient transfer of nutrients and gases, including oxygen.

To further facilitate the uptake of O_2 uptake by the fetus, almost all viviparous animals examined have a fetal blood with a higher O_2 affinity than adult blood. Among the mammals, this maternal-fetal difference in blood O_2 affinity is due either to structurally and functionally distinct fetal hemoglobins or to lower concentrations of important allosteric regulators of hemoglobin function within fetal versus adult erythrocytes. The teleost, *Embiotoca lateralis,* uses both these molecular strategies, and possibly a third, a lower fetal erythrocyte hemoglobin concentration, to ensure a higher (or prevent a lower) fetal than maternal whole blood O_2 affinity. In other fishes, fetal blood O_2 affinity is also higher than that of the adult; this difference probably involves functionally different fetal and adult hemoglobins. Whether the different erythrocytic concentrations of organic phosphates seen in *E. lateralis* contribute to maternal-fetal blood O_2 affinity differences in other chondrichthyan and osteichthyan fishes remains to be determined. However, this is a reasonable possibility based on the analysis of high fetal blood O_2 affinity of *E. lateralis,* and of other 'lower' vertebrates, e.g., the viviparous caecilian, *Typhlonectes compressicauda* and the lizard, *Sphenomorphus quoyii* (Garlick et al., 1979; Grigg and Harlow, 1981).

The remarkable diversity of maternal-fetal anatomical specializations found in fishes suggests that fishes may also exhibit a variety of molecular adaptations as well. Further studies of maternal-fetal O_2 transport mechanisms in other species could clarify the nature of such molecular diversity and perhaps shed light on how O_2 transfer mechanisms have evolved in viviparous vertebrates.

Acknowledgement

R.L.I. was supported by NIH Training Grant # HD07084; this work was also supported in part by NSF Grant # PCM8207548 to R.C.T.

References

Amoroso, E.C. (1960). Viviparity in Fishes. *Symp. Zool. Soc. (Lond.)* 1: 153–181.

Bard, H. and Teasdale, F. (1979). Red cell oxygen affinity, hemoglobin type, 2,3-diphosphoglycerate, and pH as a function of fetal development. *Pediatrics* 64: 483–487.

Bartlett, G.R. (1978). Water-soluble phosphates of fish red cells. *Can. J. Zool.* 56: 870–877.

Bartlett, G.R. (1980). Phosphate compounds in vertebrate red blood cells. *Am. Zool.* 20: 103–114.

Bartlett, G.R. (1982). Phosphate compounds in red cells of two dogfish sharks: *Squalus acanthias* and
- *Mustelus canis. Comp. Biochem. Physiol.* 73A: 135–140.

Bauer, C., Ludwig, I. and Ludwig, M. (1968). Different effects of 2,3-diphosphoglycerate and adenosine triphosphate on the oxygen affinity of adult and fetal human haemoglobin. *Life Sci.* 7: 1339–1343.

Bauer, C., Jelkmann, W. and Moll, W. (1981). High oxygen affinity of maternal blood reduces fetal weight in rats. *Respir. Physiol.* 43: 169–178.

Baumann, R., Bauer, C. and Ratschlag-Schaefer, A.M. (1972). Causes of the postnatal decrease of blood oxygen affinity in lambs. *Respir. Physiol.* 15: 151–158.

Bellingham, A.J., Detter, J.C. and Lenfant, C. (1971). Regulatory mechanisms of hemoglobin oxygen affinity in acidosis and alkalosis. *J. Clin. Invest.* 50: 700–706.

Benesch, R. and Benesch, R.E. (1967). The effect of organic phosphates from the human erythrocyte on the allosteric properties of hemoglobin. *Biochem. Biophys. Res. Comm.* 26: 162–167.

Benesch, R. and Benesch, R.E. (1969). Intracellular organic phosphates as regulators of oxygen release by haemoglobin. *Nature* 221: 618–622.

Blake, J. (1867). On the nourishment of the foetus in embiotocid fishes. *Proc. Calif. Acad. Nat. Sci.* 3: 314–317.

Blunt, M.H., Kitchen, J.L., Mayson, S.M. and Huisman, T.H.J. (1971). Red cell 2,3-diphosphoglycerate and oxygen affinity in newborn goats and sheep. *Proc. Soc. Exp. Biol. Med.* 138: 800–803.

Bunn, H.F. and Kitchen, H. (1973). Hemoglobin function in the horse: the role of 2,3-diphosphoglycerate in modifying the oxygen affinity of maternal and fetal blood. *Blood* 42: 471–479.

Chanutin, A. and Curnish, R.R. (1967). Effect of organic and inorganic phosphates on the oxygen equilibrium of human erythrocytes. *Arch. Biochem. Biophys.* 121: 96–102.

Coates, M.L. (1975). Hemoglobin function in the vertebrates: an evolutionary model *J. Mol. Evol.* 6: 285–307.

Cohen, D. and Wourms, J.P. (1976). *Microbrotula randalli,* a new viviparous Ophidioid fish from Samoa and New Hebrides, whose embryos bear trophotaeniae *Proc. Biol. Soc. Wash.* 86: 339–350.

De Simone, J and Mueller, A.L. (1979). Fetal hemoglobin (HbF) synthesis in baboons, *Papio cynocephalus.* Analysis of fetal and adult hemoglobin synthesis during fetal development, *Blood,* 53: 19–27.

De Verdier, C.H. and Garby, L. (1969). Low binding of 2,3-diphosphoglycerate to haemoglobin F. A contribution to the knowledge of the binding site and an explanation for the high oxygen affinity of foetal blood. *Scand. J. Clin. Lab. Invest.* 23: 149–151.

Dhindsa, D.S., Hoversland, A.S. and Templeton, J.W. (1972). Postnatal changes in oxygen affinity and concentration of 2,3-diphosphoglycerate in dog blood. *Biol. Neonate* 20: 226–235.

Duhm, J. (1971). Effects of 2,3-diphosphoglycerate and other organic phosphate compounds on oxygen affinity and intracellular pH of human erythrocytes. *Pflügers Arch.* 326: 341–356.

Duhm, J. (1976). Dual effect of 2.3-diphosphoglycerate on the Bohr effects of human blood. *Pflügers Arch.* 363: 55–60.

Eigenmann, C.H. (1982). *Cymatogaster aggregatus* Gibbons. A contribution to the ontogeny of viviparous fishes. *Bull. U.S. Fish. Comm.* 12: 401–478.

Forster, R.E. (1972). The effects of dilution in saline on the oxygen affinity of human hemoglobin. In: Oxygen Affinity of Hemoglobin and Red Cell Acid Base Status, M. Rorth and P. Astrup, eds., Alfred Benzon Symposium IV-Copenhagen, 1971. Academic Press, New York, pp. 162–165.

Fyhn, U.E.H. and Sullivan. B. (1975). Elasmobranch hemoglobins: dimerization and polymerization in various species. *Comp. Biochem. Physiol.* 50B: 119–129.

Garlick, R.L., Davis, B.J., Farmer, F., Fyhn, H.J., Fyhn, U.E.H., Noble, R.W. Powers D.A., Riggs, A. and Weber, R.E. (1979). A fetal-maternal shift in the oxygen equilibrium of hemoglobin from the viviparous caecilian. *Typhlonectes compressicauda. Comp. Biochem. Physiol.* 62A: 239–244.

Gilbert, P.W. and Schlernitzauer, D.A. (1966). The placenta and gravid uterus of *Carcharhinus falciformis. Copeia* 1966: 451–457.

Gillen, R.G. and Riggs, A. (1971). The hemoglobins of a freshwater teleost *Cichlasoma cyanoquttatum:* The effects of phosphorylated organic compounds upon oxygen equilibria. *Comp. Biochem. Physiol.* 38B: 585–595.

Grigg, G.C. and Harlow, P. (1981). A fetal-maternal shift of blood oxygen affinity in an Australian viviparous lizard. *Sphenomorphus quoyii* (Reptilia, Scincidae). *J. Comp. Physiol.* 142: 495–499.

Hart, J.L. (1973). Pacific Fishes of Canada. Bulletin 180, Fisheries Research Board of Canada, Ottawa.

Hjorth, J.P. (1974). Genetics of *Zoarces* populations. VII. Fetal and adult hemoglobins and a polymorphism common to both. *Hereditas* 78: 69–72.

Hoar, W.S. (1969). Reproduction. In: Fish Physiology, Vol. 3. W.S. Hoar and D.J. Randall, eds., Academic Press, New York, pp. 1–72.

Hoffmann, E.K. (1977). Control of cell volume. In: Transport of Ions and Water in Animals. B.L. Gupta, R.B. Moreton, J.L. Oxchman and B.J. Wall, eds., Academic Press, London, pp. 285–332.

Ingermann, R.L. and Terwilliger, R.C. (1981a). Oxygen affinities of maternal and fetal hemoglobins of the viviparous seaperch, *Embiotoca lateralis. J. Comp. Physiol.* 142: 523–531.

Ingermann, R.L. and Terwilliger, R.C. (1981b). Intraerythrocytic organic phosphates of fetal and adult seaperch *(Embiotoca lateralis):* their role in maternal-fetal oxygen transport. *J. Comp. Physiol.* 144: 253–259.

Ingermann, R.L. and Terwilliger, R.C. (1982). Blood parameters and facilitation of maternal-fetal oxygen transfer in a viviparous fish *(Embiotoca lateralis). Comp. Biochem. Physiol.* 73A: 497–501.

Ingermann, R.L., Terwilliger, R.C. and Roberts, M.S. (1984). Foetal and adult blood oxygen affinities of the viviparous seaperch, *Embiotoca lateralis. J. Exp. Biol.* In press.

Kitchen, H. and Brett, I. (1974). Embryonic and fetal hemoglobin in animals. *Ann. N.Y. Acad. Sci.* 241: 653–671.

Kristofferson, R., Broberg, S. and Pekkarinen, M. (1973). Histology and physiology of embryotrophe formation, embryonic nutrition and growth in the eel pout, *Zoarces viviparus* (L.). *Ann. Zool. Fennici* 10: 467–477.

Laver, M.B., Jackson, E., Scherperel, M., Tung, C., Tung, W. and Radford, E.P. (1977). Hemoglobin-O_2 affinity regulation: DPG, monovalent anions, and hemoglobin concentration. *J. Appl. Physiol.: Respir. Environ. Exercise Physiol.* 43: 632–642.

Leray, C. (1979). Patterns of purine nucleotides in fish erythrocytes. *Comp. Biochem. Physiol.* 64B: 77–82.

MacKnight, A.D.C. and Leaf, A. (1977). Regulation of cellular volume. *Physiol. Rev.,* 57: 510–573.

Manwell, C. (1958a). A 'fetal-maternal shift' in the ovoviviparous spiny dogfish *Squalus suckleyi* (Girard). *Physiol. Zool.* 31: 93–100.

13

Manwell, C. (1958b). Ontogeny of hemoglobin in the skate *Raja binoculata*. *Science* 128: 419–420.

Manwell, C. (1963). Fetal and adult hemoglobins of the spiny dogfish *Squalus suckleyi*. *Arch. Biochem. Biophys.* 101: 504–511.

Manwell, C. and Baker, C.M.A. (1970). Molecular Biology and the Origin of Species. University of Washington Press, Seattle.

McCutcheon, F.H. (1947). Specific oxygen affinity of hemoglobin in elasmobranchs and turtles. *J. Cell. Comp. Physiol.* 29: 333–344.

McMenamin, J.W. (1979). Functional morphology of sheathed blood vessels in the ovarian curtains of embiotocid fishes. *Am. Zool.* 19: 947 (abst).

Meschia, G. (1978). Evolution of thinking in fetal respiratory physiology. *Am. J. Obstet. Gynecol.* 132: 806–813.

Novy, M.J. and Parer, J.T. (1969). Absence of high blood oxygen affinity in the fetal cat. *Respir. Physiol.* 6: 144–150.

Pennelly, R.R., Noble, R.W. and Riggs, A. (1975). Equilibria and ligand binding kinetics of hemoglobins from the sharks. *Prionace glauca* and *Carcharhinus milberti*. *Comp. Biochem. Physiol.* 52B: 83–87.

Perutz, M.F. (1970). Stereochemistry of cooperative effects in haemoglobin. *Nature* 228: 726–739.

Perutz, M.F. and Brunori, M. (1982). Stereochemistry of cooperative effects in fish and amphibian haemoglobins. *Nature* 299: 421–426.

Ramsey, E.M. (1982). The Placenta: Human and Animal. Praeger Publi., New York.

Schlernitzauer, D.A. and Gilbert, P.W. (1966). Placentation and associated aspects of gestation in the bonnethead shark, *Sphyrna tiburo*. *J. Morphol.* 120: 219–232.

Sinet, M., Joubin, C., Lachia, L. and Pocidalo, J.J. (1976). Effect of osmotic changes on intracellular pH and haemoglobin oxygen affinity of human erythrocytes. *Biomedicine* 25: 66–69.

Takenaka, O. and Morimoto, H. (1976). Oxygen equilibrium characteristics of adult and fetal hemoglobin of japanese monkey *(Macaca fuscata)*. *Biochim. Biophys. Acta* 446: 457–462.

Turner, C.L. (1936). The absorptive processes in the embryos of *Parabrotula dentiens*, a viviparous, deep-sea brotulid fish. *J. Morphol.* 59: 313–325.

Turner, C.L. (1937). The trophotaeniae of the Goodeidae, a family of viviparous cyprinodont fishes. *J. Morphol.* 61: 495–523.

Turner, C.L. (1938). Histological and cytological changes in the ovary of *Cymatogaster aggregatus* during gestation. *J. Morphol.* 62: 351–373.

Turner, C.L. (1940a). Pseudoamnion, pseudochorion, and follicular pseudoplacenta of poeciliid fishes. *J. Morphol.* 67: 59–89.

Turner, C.L. (1940b). Follicular pseudoplacenta and gut modifications in anablepid fishes. *J. Morphol.* 67: 91–105.

Turner, C.L. (1940c). Pericardial sac, trophotaeniae, and alimentary tract in embryos of goodeid fishes. *J. Morphol.* 67: 271–289.

Turner, C.L. (1940d). Adaptations for viviparity in jenynsiid fishes. *J. Morphol* 67: 291–297.

Turner, C.L. (1952). An accessory respiratory device in embryos of the embiotocid fish, *Cymatogaser aggregata*, during gestation. *Copeia* 1952: 146–147.

Tweeddale, P.M. (1973). DPG and the oxygen affinity of maternal and foetal pigblood and hemoglobins. *Respir. Physiol.* 19: 12–18.

Tyuma, I. and Shimizu, K. (1969). Different response to organic phosphates of human fetal blood and adult hemoglobins. *Arch. Biochem. Biophys.* 129: 404–405.

Veith, W.J. (1980). Viviparity and embryonic adaptations in the teleost *Clinus superciliosus*. *Can. J. Zool.* 58: 1–12.

Webb, P.W. and Brett, J.R. (1972a). Respiratory adaptations of prenatal young in the ovary of two species of viviparous seaperch, *Rhacochilus vacca* and *Embiotoca lateralis*. *J. Fish. Res. Board Can.* 29: 1525–1542.

Webb, P.W. and Brett, J.R. (1972b). Oxygen consumption of embryos and parents, and oxygen transfer characteristics within the ovary of two species of viviparous seaperch, *Rhacochilus vacca* and *Embiotoca lateralis*. *J.Fish. Res. Board Can.* 29: 1543–1553.

Weber, R.E., Wells, R.M.G. and Rossetti, J.E. (1983a). Allosteric interactions governing oxygen equilibria in the haemoglobin system of the spiny dogfish, *Squalus acanthias. J. Exp. Biol.* 103: 109–120.

Weber, R.E., Wells, R.M.G. and Tougaard, S. (1983b). Antagonistic effect of urea on oxygenation-linked binding of ATP in an elasmobranch hemoglobin. *Life Sci.* 32: 2157–2161.

Wells, R.M.G. (1979). Haemoglobin-oxygen affinity in developing embryonic cells of the mouse. *J. Comp. Physiol.* 129: 333–338.

Wells, R.M.G. and Weber, R.E. (1983). Oxygenation properties and phosphorylated metabolic intermediates in blood and erythrocytes of the dogfish, *Squalus acanthias. J. Exp. Biol.* 103: 95–108.

Wood, S.C. and Johansen, K. (1972). Adaptation to hypoxia by increased HbO_2 affinity and decreased red cell ATP concentration. *Nature* 237: 278–279.

Wood-Mason, J. and Alcock, A. (1891). On the uterine villiform papillae of *Pteroplatea micrura*, and their relation to the embryo. *Proc. R. Soc. (London)* 49: 359–367.

Wourms, J.P. (1981). Viviparity: the maternal-fetal relationship in fishes. *Am. Zool.* 21: 473–515.

Wourms, J.P. and Cohen, D. (1975). Trophotaeniae, embryonic adaptations in the viviparous ophidioid fish, *Oligopus longhursti*: A study of museum specimens. *J. Morphol.* 147: 385–401.

2. Blood adaptations for maternal-fetal oxygen transfer in the viviparous teleost, *Zoarces viviparus* L.

Marianne Hartvig and Roy E. Weber
Institute of Biology, University of Odense, Denmark

Abstract

Oxygen affinities of the whole blood are considerably higher in late gestational fetuses than in the maternal specimens of the viviparous, ovary-breeding teleost, *Zoarces viviparus* (half-saturation O_2 tensions, P_{50}, approximate 9 and 23.5 torr, respectively, at pH 7.5 and 10 °C). The maternal blood P_{50} values moreover show a greater sensitivity to pH and a lower sensitivity to temperature than the fetal blood. We also report on the P_{CO_2}/pH/bicarbonate relationships in the fetal and adult blood in vitro and the P_{O_2} and pH values in the ovarian fluids.

The higher O_2 affinity of the fetal blood is mainly due to higher intrinsic O_2 affinities of the cofactor-free fetal hemoglobin compared to the adult hemoglobin, and is not attributable to significantly lower concentrations of the nucleoside triphosphates. ATP and GTP, in the fetal erythrocytes, or to lower sensitivities of the fetal hemoglobin to these allosteric modulators of O_2 affinity. The large fetal-maternal shift in blood O_2 affinity appears to compensate for a low 'structural capacity' for maternal-fetal O_2 transfer in the blood system of zoarcid teleost fish.

Introduction

Although viviparity (giving birth to live young) is rare among teleosts, it shows very diverse mechanisms and no correlation with habitat, geographical region or phylogeny, suggesting that it evolved separately on more than one occasion in fishes (Hoar, 1969).

Very little is known about the respiration and metabolism of embryonic stages of viviparous fish. In mammals, maternal-fetal O_2 transfer is facilitated by a higher O_2 affinity in the fetal than in the maternal blood. The molecular and cellular mechanisms underlying this 'fetal-maternal shift' may be (a) differences in the erythrocytic concentrations of 2,3 diphosphoglycerate (DPG) which depresses hemoglobin-oxygen affinity by allosteric interaction (e.g. dog, horse and seal – Dhindsa et al., 1972; Bunn and Kitchen, 1973; Qvist et al., 1981) and/or (b)

Seymour, R.S., (ed.) Respiration and metabolism of embryonic vertebrates.
© *1984, Dr W. Junk Publishers, Dordrecht/Boston/London. ISBN-13: 978-94-009-6538-6*

structural differences between fetal and maternal hemoglobins, resulting in different intrinsic O_2 affinities (e.g. sheep – Baumann et al., 1972) or different sensitivities of the hemoglobin to cofactors (e.g. man – de Verdier and Garby, 1969; Tomita, 1981). The only exception is the domestic cat where the absence of an O_2 affinity shift appears to be compensated by a greater O_2 carrying capacity in the fetus and a counter-current arrangement of blood flow in the placenta (Novy and Parer, 1969).

Fetal-maternal shifts in blood O_2 affinity have recently been recorded in viviparous representatives of reptiles and amphibians, i.e. the lizard *Spheno-morphus quoyii* and the caecilian *Typhlonectes compressicauda* (Grigg and Har-low, 1981; Toews and Macintyre, 1977; Garlick et al., 1979). In both these species the fetal and maternal hemoglobins appears to be electrophoretically identical. In the caecilian the maternal red cells contain higher concentrations of the nu-cleoside triphosphates (NTP), ATP and guanosine triphosphate (GTP) (Garlick et al., 1979; Weber et al., 1975), which decrease hemoglobin-oxygen affinity of the hemoglobin of ectotherm vertebrates as does DPG in mammals. Corre-sponding O_2 affinity differences do not appear to have been reported for the whole blood of fish. In the elasmobranch, *Squalus acanthias*, however, the isolated fetal hemoglobin has a higher affinity than the maternal hemoglobin (Manwell, 1958, 1963). This also applies to the teleost *Embiotoca lateralis*, where a higher fetal blood O_2 affinity furthermore is predicted by lower concentrations of nucleoside triphosphates and hemoglobin in fetal than in adult erythrocytes (Ingermann and Terwilliger, 1981a, b; 1982).

The marine fish *Zoarces viviparus* is an ovary breeder and appears to be exceptional among viviparous teleosts in that ovulation precedes fertilization (Hoar, 1969). The embryos lack anatomical connections with the mother and occur freely in the ovarian cavity during a four month pregnancy period (Kors-gaard and Petersen, 1979). The fetal and adult hemoglobins exhibit heterogeneity and polymorphism; in Danish coastal waters where each stage shows one of three hemoglobin patterns, the ontogenetic change from the fetal to adult hemoglobin is completed within 6 months (Hjorth, 1974, 1975). We examined the oxygenation and acid-base characteristics of the whole blood and the cofactor-free hemo-globins from maternal and late-gestational fetal stages of *Z. viviparus*, with the view of contributing to the understanding the mechanisms responsible for mater-nal-fetal O_2 transfer in teleosts.

Material and methods

Material

Adult specimens of *Zoarces viviparus* L., 25–32 cm long and weighing 90–200 g were collected in Vejle Fjord, Denmark, and kept in aerated sea water at 10 °C for at least one week before the experiments.

Adult blood was drawn from the caudal vessels or the heart of fish slightly anesthetized with 0.5% MS-222 (Santos) using 1-ml hypodermic syringes where the dead space had been filled with heparin in saline (5000 IU/ml) and was used immediately for P_{O_2} and pH measurement with Radiometer Copenhagen equipment (BMS-3 and PHM 71). These determinations were commenced within 30 sec of removing the fish from the water.

Ovarian fluid samples were drawn into a syringe by piercing the body wall of the mothers via a small incision in the skin of the belly, and used directly for P_{O_2} and pH measurement (as above). Late gestational fetuses were obtained by 'caesarian' section. The ovary was exposed and incised and the fetuses transferred to a beaker of aerated sea water. The mean number obtained per adult female was 110.

Fetal blood was collected by placing individual specimens ventral side up under a dissecting microscope and piercing the heart with thinly drawn out heparanized glass capillaries. A blood sample of 2–5 μl was obtained per specimen. Blood samples from one brood were pooled and kept on ice.

Whole blood studies

Oxygen equilibria of whole blood were determined at 10 °C using a Hem-O-Scan (American Instruments Co., Silver Springs) coupled to Wösthoff (Bochum, FRG) pumps for mixing O_2, CO_2, and N_2 (as earlier described by Wells and Weber, 1983). For each determination separate 50 μl aliquants of the blood samples were equilibrated with gas mixtures of the same CO_2 tension in Radiometer BMS-2 tonometers for pH measurement.

Blood hemoglobin concentrations were measured spectrophotometrically after hemolysing the cells in 0.01 M Tris buffer pH 7.5, using molar extinction coefficients of 14600 and 14200 at 576 and 542 nm, respectively. ATP and GTP were assayed by thin layer chromatography, and NTP enzymatically (Weber et al., 1976a; Lykkeboe and Weber, 1978). Intracellular concentrations were calculated from hematocrit values assuming all hemoglobin and NTP are intracellular.

Acid-base status of the blood was investigated by equilibrating 70 μl samples to varying CO_2 tensions in the absence and presence of O_2 (~150 torr) for at least 20 min at 10 °C and then measuring pH as above. Bicarbonate concentration was calculated from the Henderson-Hasselbalch equation:

$$[HCO_3^-] = 10^{(pH-pK_1')} \cdot \alpha \cdot P_{CO_2}$$

using values of Severinghaus (1971) for CO_2 solubility (α) and pK_1' (interpolating where necessary).

Hemoglobin solutions: preparation and measurements

Hemoglobin solutions were obtained from erythrocytes washed in 2.5% NaCl as previously described (Weber et al., 1976b) except that the hemoglobin was 'stripped' of cofactors and electrolytes by filtration through a 30 cm column of Sephadex G25 superfine gel (Berman et al., 1971). Samples showing spectrophotometric evidence for methemoglobin formation were reduced by adding a trace of solid sodium dithionite before gel filtration. Preliminary experiments showed that CO-treatment resulted in strong decreases in the α absorption band of the fetal hemoglobin, without significantly affecting the spectrum of the adult hemoglobin. This treatment was thus omitted for hemoglobins from both stages.

Measurement of O_2 equilibria of hemoglobin solutions, using a diffusion chamber, was carried out as earlier described (Weber et al., 1976b).

Results

Oxygenation properties of the blood and hemoglobin

The whole blood showed much higher affinities in the maternal than in the fetal specimens (Fig. 1). At pH 7.5 and 10 °C the maternal and fetal P_{50} values were 23.5

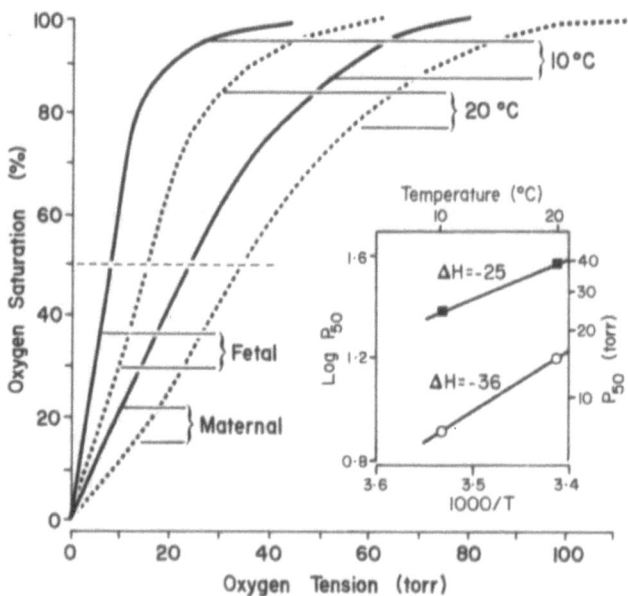

Fig. 1. Oxygen equilibrium curves of whole blood of maternal (broken curves) and fetal (continuous curves) *Zoarces viviparus*, measured at pH 7.5 and 10 °C and 20 °C. Inset: The relationship between log P_{50} and absolute temperature (T), showing apparent heats of oxygenation (ΔH, in kJ · mol[-1]), in the maternal (■) and fetal (○) whole blood at pH 7.5.

and 9 torr, respectively; at 20 °C these values were 34 and 15 torr, respectively.

The apparent enthalpy of oxygenation calculated from these values using the van't Hoff isochore (Wyman, 1964):

$$\Delta H = -2.303 \cdot R \cdot \Delta \log P_{50} \cdot \{\Delta(1/T)\}^{-1}$$

(where R is the gas constant) was -25 and -36 kJ $\cdot$ mol^{-1} for maternal and fetal blood, respectively, showing a greater temperature sensitivity in the latter (Fig. 1, inset).

The Bohr factors ($\varphi = \Delta \log p_{50}/\Delta pH$) were -0.95 and -0.55 in maternal and fetal blood, respectively, at pH 7.6 to 8.0. The fetal-maternal O_2 affinity difference thus increased with decreasing pH (Fig. 2). Below pH 7.6 the blood from both stages lost pH dependence. The hemoglobin in fetal blood exhibited greater cooperativity in O_2 binding than that in the maternal animals (Hill's cooperativity coefficient, $n \sim 2.4$ and 1.9 respectively, at pH 7.1 to 7.8).

In 'stripped' (cofactor-free) state the fetal hemoglobin consistently displayed higher O_2 affinities than the adult pigment; at pH 7.5 and 10 °C the P_{50} values were 2.8 and 7.4 torr, respectively (Fig. 3). We also observed a much greater Bohr factor in the maternal hemolysate (φ at pH 7.5 = -1.2 compared to -0.5 in the fetal hemoglobin).

We investigated the pH dependence of the temperature effect in the adult hemolysate by measuring the Bohr effect at two temperatures (12 and 22 °C, see Fig. 4A) calculating the ΔH values from the P_{50} shifts observed at individual pH values. As shown (Fig. 4B) the temperature effect is small at low, and large at

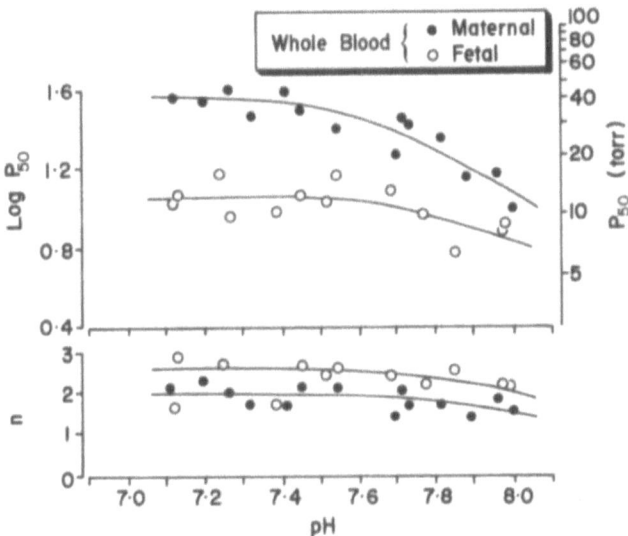

Fig. 2. Half-saturation oxygen tensions, P_{50}, cooperativity coefficients, n, and their variation with pH in maternal (●) and fetal (○) *Zoarces* blood, measured at 10 °C. Blood pH was varied by changing the carbon dioxide tension.

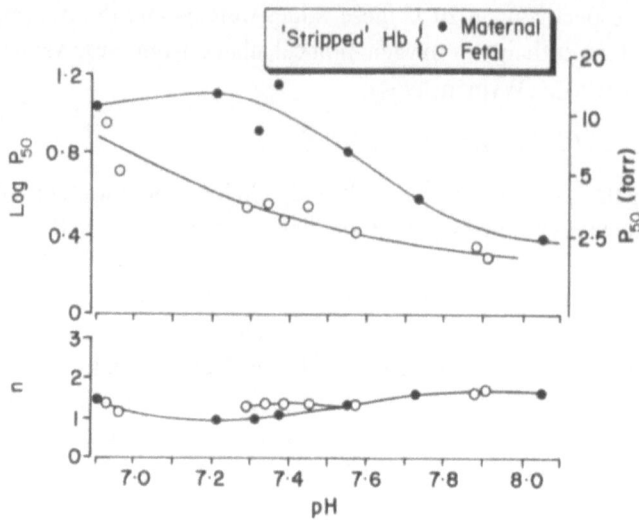

Fig. 3. Influence of pH on P_{50} and *n* values of 'stripped' (cofactor-free) hemoglobins from maternal (●) and fetal (○) specimens of *Zoarces*, suspended in 0.05 mol·l⁻¹ Na Hepes buffer at 10 °C. Tetrameric Hb concentration, 0.14–0.17 mmol·l⁻¹.

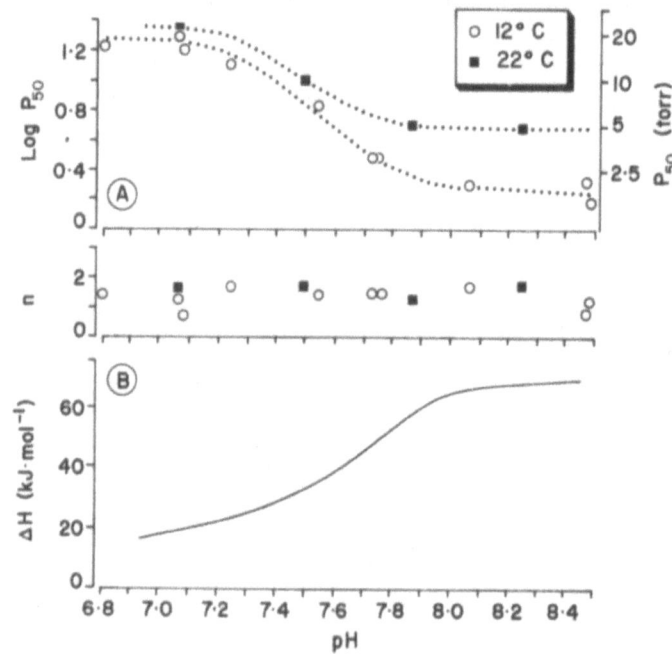

Fig. 4. (A) pH dependence of P_{50} (Bohr effect) and cooperativity coefficient, *n*, in the adult hemoglobin of *Zoarces*, measured at 12 °C (○) and 22 °C (■) in 0.05 mmol·l⁻¹ Na Hepes buffer. Tetrameric Hb concentration, 0.43 mmol·l⁻¹. (B) pH dependence of the apparent heat of oxygenation, ΔH, derived as described in the text from the difference in P_{50} between 12 and 22 °C (Fig. A).

high pH, ΔH approximating $-18\,kJ \cdot mol^{-1}$ at pH 7.0 and $-68\,kJ \cdot mol^{-1}$ at pH 8.2. The latter value agrees well with those reported for other fish hemoglobins at alkaline pH (Powers, 1980).

Allosteric factors

The main erythrocytic phosphate cofactor was found to be ATP, which was 5 to 7 times more abundant than GTP (Table 1). The [NTP] : [Hb] ratios were 1.2 and 1.0 in adult and fetal *Zoarces* but the difference was not statistically significant. Although the concentrations measured in the fetal blood samples may be slightly low due to contamination with adherent water or body fluids during blood sampling, this will not have affected the [NTP] : [Hb] ratio. Again, we found no significant differences between mothers and fetuses as regards hematocrit, the blood and erythrocytic concentrations of ATP, GTP and hemoglobin, and blood pH (Table 1).

The effects of ATP and GTP on O_2 affinity of the stripped fetal and adult hemolysates are shown in Fig. 5. In accordance with previous findings on adult fish hemoglobins (Lykkeboe et al., 1975; Weber et al., 1975, 1976a; Kaloustian and Poluhowich, 1976; Weber and Lykkeboe, 1978), GTP depressed the O_2 affinity of both hemolysates more than ATP at the same concentration. Unexpectedly, the stripped fetal hemoglobin exhibited greater sensitivity to both co-factors. The hyperbolic nature of these curves (Fig. 5) shows that the hemoglobin-oxygen affinities are highly sensitive to variation in NTP concentrations at the low [NTP] : [Hb] ratios, which we found in the fetal and adult red cells (Table 1).

Table 1. Some hematological properties of the blood of maternal and fetal *Zoarces viviparus*.

	Maternal	Fetal	Sign if [*]
Hb (mmol/l blood)	0.52 ± 0.32 (6)	0.45 ± 0.12 (6)	N.S.
Hct (%)	13.3 ± 7.0 (4)	12.4 ± 1.3 (3)	N.S.
ATP (mmol/l blood)	0.509 ± 0.12 (5)	0.403 ± 0.15 (5)	N.S.
GTP (mmol/l blood)	0.071 ± 0.052 (4)	0.076 ± 0.05 (5)	N.S.
ATP/Hb	1.14 ± 0.42 (5)	0.84 ± 0.27 (5)	N.S.
(ATP + GTP)/Hb	1.21 ± 0.41 (5)	1.01 ± 0.35 (5)	N.S.
Hb (mmol/l RBC)	3.95 ± 1.8 (4)	3.44 ± 0.8 (3)	N.S.
NTP (mmol/l RBC)	5.49 ± 2.3 (3)	5.13 ± 0.62 (3)	N.S.
pH	7.52 ± 0.14 (8)	7.48 ± 0.23 (8)[**]	N.S.

[*] N.S. = not significant
[**] measured in the ovarian fluid that bathes the fetuses

The mean pH value (±s.d.) measured in freshly drawn maternal blood was 7.52 ± 0.14 (N = 8). In view of the minute quantities of fetal blood it was not possible to measure its pH without exposing it to air. For the ovarian fluid collected as described above, however, we found a value of 7.48 ± 0.23 (N = 8), which is not significantly different from that in the adult blood (Table 1). It is likely that the in vivo blood pH in the adult was higher than the values here obtained for blood drawn by venepuncture. In rainbow trout, the pH values of blood taken from undisturbed fish via ventral aortic cannulae were 0.25 pH units higher than those taken by cardiac puncture (Nikinmaa, 1981).

The small size of *Zoarces* precluded implantation of catheters for measurement of arterial and venous O_2 tensions. In blood rapidly sampled by cardiac puncture we found O_2 tensions of 3.3 ± 1.4 torr (N = 5). Blood taken from the caudal artery and vein gave higher values (13.7 ± 3.8 torr; N = 5) presumably due to admixture of venous and arterial blood. Significantly, much higher tensions (38.3 ± 7.5 torr, N = 7) were seen in the ovarian fluid, indicating the presence of even higher O_2 tensions in the ovarian afferents and that an efficient transfer of O_2 occurs to the medium bathing the fetuses.

We used blood from non-pregnant adults to determine the in vitro CO_2/pH/bicarbonate relationships. The blood exhibited a Haldane effect; at the same P_{CO_2} the pH in the deoxygenated blood was higher ($\Delta pH \sim 0.05$) than in the oxygenated blood (Fig. 6A). The log P_{CO_2}/pH relationship allows estimation of in vivo CO_2 tensions, indicating values were near 2 and 5 torr at pH 7.7 and 7.5, respectively.

The P_{CO_2}/pH values measured in whole fetal and maternal blood during O_2 equilibrium studies (Fig. 6B) show good agrement with the blood of non-preg-

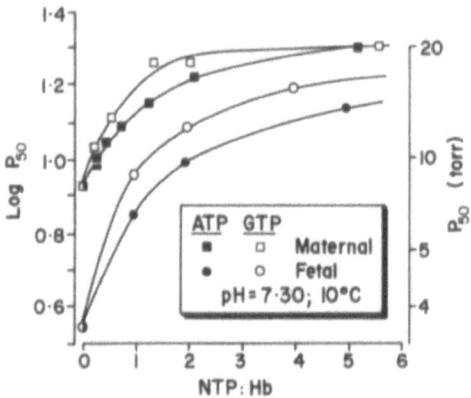

Fig. 5. Dependence of P_{50} values of the stripped maternal (■,□) and fetal (○,●) hemoglobins of *Zoarces*, measured at 10 °C and pH 7.3 in 0.08 mol · l⁻¹ Na Hepes buffer, on ATP (closed symbols) and GTP (open symbols). Tetrameric hemoglobin concentration, 0.11 mmol · l⁻¹.

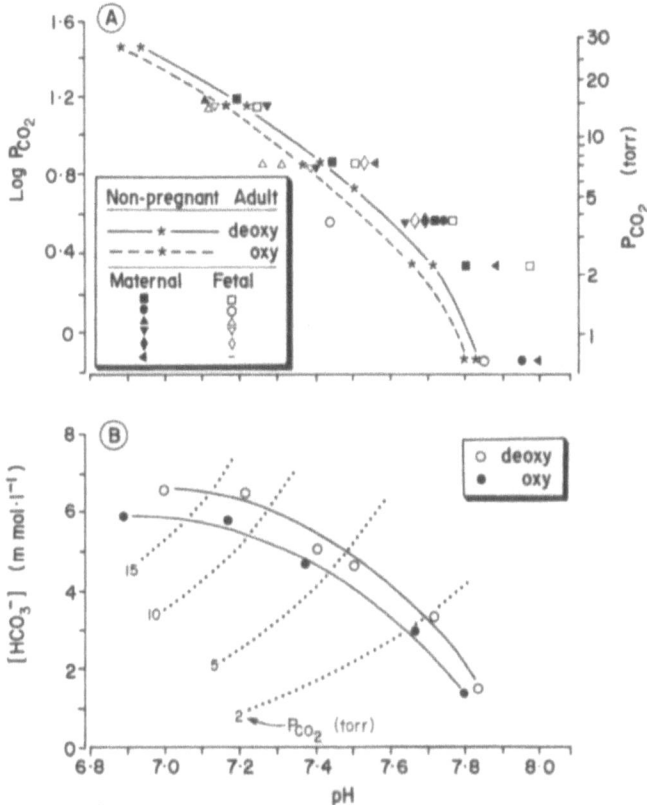

Fig. 6. A. Relationship between CO_2 tension and pH in oxygenated (broken curve) and deoxygenated (continuous curve) blood from non-pregnant adult *Zoarces* at 10 °C. Also included are the values for maternal (closed symbols) and fetal (open symbols) blood, obtained during whole blood oxygen equilibrium determinations (cf. Fig. 2). Closed and open symbols of the same shape refer to individual mother-brood pairs. B. Bicarbonate vs pH diagram with P_{CO_2} isopleths for adult *Zoarces* blood at 10 °C.

nant adults at CO_2 tensions above 5 torr. At lower values, however, blood pH was higher in the maternal and fetal specimens reflecting an altered capacity to buffer CO_2 compared to the non-pregnant adults. The buffering capacity ($B = \Delta HCO_3^-/\Delta pH$) decreased with falling pH but was similar for oxygenated and deoxygenated states (Fig. 6B). At pH 7.7 and 7.5, B amounted to 9.1 and 6.7 mmol $\cdot$ l^{-1} $\cdot$ (pH unit)$^{-1}$.

Discussion

Zoarces viviparus shows a large difference in O_2 affinities of fetal and maternal whole blood (P_{50} at 10 °C = 9 and 23 torr, respectively). The higher affinity in the fetuses aligns teleosts with previously studied viviparous animals from the other classes of vertebrates. That the P_{50} shift is much greater than those encountered in mammals (cf. Bunn and Kitchen, 1973) may be adaptive to the lesser efficiency of the brood ovary system for gas exchange than the mammalian placenta. It should, however, be borne in mind that the in vivo shift may be slightly smaller because the pH in the fetal blood may be lower than that in the mother (although we observed no significant difference between maternal venous pH and ovarian fluid pH).

The O_2 tension found in the ovarian fluid ($\bar{x} = 38$ torr) is high compared to the P_{50} value of the blood. This fact, together with the large fetal-maternal differences in O_2 affinity and cooperatively for hemoglobin-oxygen binding will favour O_2 saturation of the fetal blood and thus a maternal-fetal O_2 transfer, compensating for the lack of special anatomical structures to facilitate exchanges between the mother and fetuses (i.e. the low 'structural transfer capacity', see Webb and Brett, 1972). Accordingly, the O_2 tensions in the intra-ovarian fluid of the seaperch *Rhacochilis vacca* (fam. Embiotocidae) are much lower (decrease from about 36 torr to 14 torr during pregnancy - Webb and Brett, 1972) correlating with the presence of highly vascularized, spatulate respiratory structures on the embryonic fins in the embiotocids, whereas supplementary respiratory organs are lacking in Zoarcidae (Hoar, 1969).

The greater Bohr factor in the maternal than in the fetal blood predicts an increased fetal-maternal difference as pH falls (cf. Fig. 2), further favoring O_2 transfer to the fetal organisms. This difference in pH sensitivity could be advantageous to a benthic fish like *Zoarces*, where burst activity alternates with periods of rest. Decreases in maternal blood pH may thus be expected to enhance O_2 transfer, safeguarding fetal O_2 supply during maternal activity.

Since the mean ATP + GTP concentration in the fetal and adult erythrocytes are not significantly different, these cofactors do not account for the observed P_{50} difference in the whole blood. The blood shift is, however, reflected in a pronounced difference in the intrinsic O_2 affinity of the cofactor-free fetal and maternal hemolysates (Figs. 3, 5). This difference accords with electrophoretic evidence for different hemoglobin components in the two stages (Hjorth, 1974). Since the fetal hemoglobins are retained for up to 6 months after birth, the young stages of free living *Zoarces* may be expected to have a higher blood O_2 affinity than adults in the presence of equal effects of intraerythrocytic modifiers of O_2 affinity. Studies on the effects of hypoxia on the properties of the fetal blood will give insight into its dependence upon environmental O_2 tensions, given the observations that hypoxia reduces erythrocytic NTP concentrations and increases O_2 affinity in adult teleosts (Wood and Johansen, 1972; Weber et al., 1976a; Weber and Lykkeboe, 1978; Soivio et al., 1980), and raises cooperativity of O_2 binding in blood of adult *Zoarces* (Hartvig and Weber, unpublished).

The P_{50} values observed in the fetal and maternal hemoglobins in solution are lower than corresponding values in the whole bloods even when ATP had been added to raise [NTP] : [Hb] ratios to in vivo levels (cf. Figs. 1, 5). These blood-hemoglobin differences may be explained partly by the markedly higher hemo-globin and NTP concentrations in the erythrocytes (Table 1) than in the hemo-globin solutions, and partly by the fact that pH values inside fish erythrocytes are lower than those measured in the blood (cf. Wood and Johansen, 1973; Jensen and Weber, 1982). Also CO_2, which was present only in the whole blood studies lowers affinity of fish hemoglobin at high pH (Weber and Lykkeboe, 1978; Weber and Johansen, 1979).

The main organic phosphates in the erythrocytes of adult and fetal *Zoarces* are ATP, the most common modulator of O_2 affinity in fish, and GTP, which occurs in some fish species, particularly those which periodically are exposed to hypoxic conditions (Bartlett, 1980; Weber, 1982). As previously established for hemo-globins from adult fish, the hemoglobin from fetal *Zoarces* is more sensitive to GTP than to ATP. An unexpected finding is the greater sensitivity of the fetal hemoglobin to both cofactors. This is opposite to man in which adult hemoglobin binds more DPG and shows a greater DPG effect than fetal hemoglobin (Bauer et al. 1968; de Verdier and Garby, 1969; Tomita, 1981).

The temperature sensitivities of the oxygenation reactions of the fetal and maternal bloods of *Zoarces* reflect apparent ΔH values of -36 and $-25\,kJ \cdot mol^{-1}$, respectively (Fig. 1). The ΔH value in the stripped adult hemoglobin varies from about $-18\,kJ \cdot mol^{-1}$ at low pH to about $-68\,kJ \cdot mol^{-1}$ at high pH. The overall ΔH value consists of several components: (a) the intrinsic heat of oxygenation, ΔH_O, (b) about $-13\,kJ \cdot mol^{-1}$ attributable to the heat of solution of oxygen, and (c) contributions from other ligand-linked reactions. The latter include the binding of organic phosphates (which were removed by 'stripping') and protons, which subtract from the value for the intrinsic heat of reaction (rendering it less negative). This explains the pH dependence of ΔH at pH values where the Bohr effect is operative (cf. Fig. 4), and is consistent with the lower apparent heat of oxygenation in the adult whole blood (Fig. 1), which has a larger Bohr factor than the fetal blood (Fig. 2).

The pH induced change in ΔH of about $50\,kJ \cdot mol^{-1}$ (from about -68 to $-18\,kJ \cdot mol^{-1}$; Fig. 4) can thus be aligned with the Bohr factor (cf. Fig. 3) which indicates that 1.2 protons are bound per O_2 molecule released. This suggests that the heat of reaction per proton is about $42\,kJ \cdot mol^{-1}$. The latter value is inter-mediate to the heats of ionization of histidine imidazole groups and α amino groups, 29 and $46\,kJ \cdot mol^{-1}$, respectively (Reeves and Rahn, 1979), which tallies with the fact that the histidine residues in position 146 of the β chains and the α amino groups of the α chains of human hemoglobin are responsible for about 40 and 30 % respectively, of the alkaline Bohr effect (Kilmartin, 1977).

The present findings for *Zoarces* call for comparison with the seaperch *Embio-toca lateralis* which seems to be the only other teleost where the fetal and adult hemoglobins have been studied (Ingermann and Terwilliger, 1981a,b, 1982).

As in *Zoarces* the fetal hemoglobin of *Embiotoca* is structurally different from the adult hemoglobin and has a higher intrinsic O_2 affinity. In contrast to the larger Bohr factor in adult than in fetal *Zarces* hemoglobin, the hemoglobins from early as well as late gestational stages of *Embiotoca*, show the same pH sensitivity as in the adult (Ingermann and Terwillinger, 1981a).

Erythrocytes from adult *Embiotoca* furthermore contain significantly higher NTP concentrations than fetal erythrocytes, thus compounding the affinity difference between fetal and maternal hemoglobin (Ingermann and Terwilliger, 1981b). Finally, the fetal stages of *Embiotoca* show much higher hematocrit values than neonate and adult forms (37 to 46% compared to 26 to 27%, respectively – Ingermann and Terwilliger, 1982), and lower intraerythrocytic hemoglobin levels, which predictably will contribute to a fetal-maternal difference in blood O_2 affinity. Curiously the situation in *Zarces* appears to be markedly different, because hematocrit is low in pregnant females and increases again after parturition, presumably as a result of minor bleedings during pregnancy (Kristofferson et al. 1974). No information, however, appears to be available concerning the whole blood O_2 equilibria, temperature effects and acid-base status in *Embiotoca* mothers and fetuses.

The striking differences in the fetal-maternal blood adaptations between *Zarces* and *Embiotoca* witness diverse adaptive strategies to ensure fetal O_2 supply in zoarcid and embiotocid teleosts and support the view of a polyphyletic origin of viviparity in teleosts.

Note added in proof:

Ingermann et al. (1984) recently recorded higher O_2 affinities in mid- and late gestational fetal blood than in adult blood of *Embiotoca lateralis*, although CO_2 was either absent or present at very high tension (about 41 mmHg) in their measurements.

Acknowledgement

We thank Dr Rufus Wells, Auckland, New Zealand, for advice in the use of the Hem-O-Scan.

References

Barlett, G.R. (1980). Phosphate compounds in vertebrate red cells. *Am. Zool.* 20: 103–115.
Bauer, C., Ludwig, J. and Ludwig, M. (1968). Different effects of 2,3-diphosphoglycerate and adenosine triphosphate on the oxygen affinity of adult and fetal human hemoglobin. *Life Sci.* 7: 1339–1343.
Baumann, R., Bauer, C. and Ratschlag-Schaefer, A.M. (1972). Causes of the postnatal decrease of oxygen affinity in lambs. *Respir. Physiol.* 15: 151–158.

Bermann, M., Benesch R. and Benesch, R.E. (1971). The removal of organic phosphates from hemoglobin. *Arch. Biochem. Biophys.* 145: 236–239.

Bunn, H.F. and Kitchen, H. (1973). Hemoglobin function in the horse: The role of 2,3-diphosphoglycerate in modifying the oxygen affinity of maternal and fetal blood. *Blood* 42: 471–479.

de Verdier, C.-H. and Garby, L. (1969). Low binding of 2,3-diphosphoglycerate to hemoglobin F. *Scand. J. Clin. Lab. Invest.* 23: 149–151.

Dhindsa, D.S., Hoversland, A.S. and Templeton, J.W. (1972). Post-natal changes in oxygen affinity and concentration of 2,3-DPG in dog blood. *Biol. Neonate* 20: 226–235.

Garlick, R.L., Davies, B.J., Farmer, M., Fyhn, U.E.H., Noble, R.W., Powers, D.A., Riggs, A. and Weber, R.E. (1979). A fetal-maternal shift in the oxygen equilibrium of hemoglobin from the viviparous caecilian *Typhlonectes compressicauda. Comp. Biochem. Physiol.* 62A: 239–244.

Grigg, G.C. and Harlow, P. (1981). A fetal-maternal shift of blood oxygen affinity in an australian viviparous lizard *Spenomorphus quoyii* (Reptilia, Scindidae). *J. Comp. Physiol.* 142, 495–499.

Hoar, W.S. (1969). Reproduction. In: Fish Physiology. Vol. 3. W.S. Hoar and D.J. Randall, eds., Academic Press, New York, pp. 1–72.

Hjorth, P. (1974). Genetics of *Zoarces* populations. VII. Fetal and adult hemoglobins and a polymorphism common to both. *Hereditas* 78: 69–72.

Hjorth, J.P. (1975). Molecular and genetic structure of multiple hemoglobins in the eelpout, *Zoarces viviparus* L. *Biochem. Genet.* 13: 379–391.

Ingermann, R.L. and Terwilliger, R.C. (1981a). Oxygen affinities of maternal and fetal hemoglobins of the viviparous seaperch, *Embiotoca lateralis. J. Comp. Physiol.* 142: 523–531.

Ingermann, R.L. and Terwilliger, R.C. (1981b). Intraerythrocytic organic phosphates of fetal and adult seaperch *Embiotoca lateralis:* Their role in maternal-fetal oxygen transport. *J. Comp. Physiol.* 144: 253–259.

Ingermann, R.L. (1982). Blood parameters and facilitation of maternal-fetal oxygen transfer in a viviparous fish *(Embiotoca lateralis). Comp. Biochem. Physiol.* 73A: 497–501.

Ingermann, R.L., Terwilliger, R.C. and Roberts, M.S. (1984). Foetal and adult blood oxygen affinities of the viviparous seaperch, *Embiotoca lateralis. J. Exp. Biol.* 108: 453–457.

Jensen, F.B. and Weber, R.E. (1982). Respiratory properties of tench blood and hemoglobin. Adaption to hypoxic-hypercapnic water. *Molec. Physiol.* 2: 235–250.

Kaloustian, K.V. and Poluhowich, J.J. (1976). The role of organic phosphates in modulating the oxygenation behaviour of eel haemoglobins. *Comp. Biochem. Physiol.* 53A: 245–248.

Kilmartin, J.V. (1977). The Bohr effect of human hemoglobin. TIBS, Nov. 1977, 247–249.

Korsgaard, B. and Petersen, I. (1979). Vitellogenin, lipid and carbohydrate metabolism during vitellogenesis and pregnancy, and after hormonal induction in the blenny, *Zoarces viviparus. Comp. Biochem. Physiol.* 63B: 245–251.

Kristoffersen, R., Broberg, S. and Oikari, A. (1974). Annual changes in some blood constituents of female brackish water *Zoarces viviparus,* Teleostei, with special reference to the reproductive cycle and the embryotrophe. *Hydrobiol. Bull.* 8: 117–123.

Lykkeboe, G., Johansen, K. and Maloiy, G.M.O. (1975). Functional properties of hemoglobins in the teleost, *Tilipia grahami. J. Comp. Physiol.* 104: 1–11.

Lykkeboe, G. and Weber, R.E. (1978). Changes in the respiratory properties of the blood of carp induced by diurnal variation in ambient oxygen tension. *J. Comp. Physiol.* 128: 117–125.

Manwell, C. (1958). A fetal-maternal shift in the ovoviviparous spiny dogfish, *Squalus suckleyi. Physiol. Zool.* 31: 93–100.

Manwell, C. (1963). Fetal and adult hemoglobins of the spiny dogfish, *Squalus suckleyi. Arch. Biochem. Biophys.* 101: 504–511.

Nikinmaa, M. (1981). Respiratory adjustments of rainbow trout (*Salmo gairdneri*, Richardson) to changes in environmental temperature and oxygen availability. Ph.D. Thesis, University of Helsinki.

Novy, M.J. and Parer, J.T. (1969). Absence of high blood oxygen affinity in the fetal cat. *Respir. Physiol.* 6: 144–150.

Powers, D.A. (1980). Molecular ecology of teleost fish hemoglobins: Strategies for adapting to changing environments. *Am. Zool.* 20: 139–163.

Qvist, J., Weber, R.E. and Zapol, W.M. (1981), Oxygen equilibrium properties of blood and hemoglobin of fetal, newborn and adult Weddell seals. *J Appl. Physiol.* 50: 999–1005.

Reeves, R.B. and Rahn, H. (1979). Patterns in vertebrate acid-base regulation. In: Evolution of Respiratory Processes. A comparative approach. Series: Lung Biology in Health and Disease, Vol. 13. S.C. Wood and C. Lenfant, eds., Marcel Dekker, New York, pp. 225–252.

Severinghaus, J.W. (1971). Carbon dioxide solubility and first dissociation constant (pK') of carbonic acid in plasma and cerebrospinal fluid: Man. In: Handbook of Respiration and Circulation. P.L. Altman and Dittmer, eds., *Fed. Am. Soc. Exp. Biol., Bethesda*, pp. 218–219.

Soivio, A., Nikinmaa, M. and Westman, K. (1980). The blood oxygen binding properties of hypoxic *Salmo gairdneri. J. Comp. Physiol.* 136: 83–87.

Toews, D. and Macintyre, D. (1977). Blood respiratory properties of a viviparous amphibian. *Nature* 266: 464.

Tomita, S. (1981). Modulation of the oxygen equilibrium of human fetal and adult hemoglobins by 2,3 diphosphoglyceric acid. *J. Biol. Chem.* 256: 9495–9500.

Webb, P.W. and Brett, J.R. (1972). Oxygen consumption of embryos and parents, and oxygen transfer characteristics within the ovary of two species of viviparous seaperch, *Rhacochilus vacca* and *Embiotoca lateralis. J. Fish. Res. Bd. Can.* 29: 1543–1553.

Weber, R.E. (1982). Intraspecific adaptation of hemoglobin function in fish to oxygen availability. In: Exogenous and Endogenous Influences on Metabolic and Neural Control. A.D.F. Addink and N. Spronk, eds., Pergamon Press, Oxford, pp. 87–102.

Weber, R.E. and Johansen, K. (1979). Oxygenation-linked binding of carbon dioxide and allosteric phosphate cofactors by lungfish hemoglobin. In: Animals and Environmental Fitness. R. Gilles. ed., Pergamon Press. Oxford, pp. 49–50.

Weber, R.E. and Lykkeboe, G. (1978). Respiratory adaptations in carp blood. Influence of hypoxia, red cell organic phosphates, divalent cations and CO_2. *J. Comp. Physiol.* 128: 127–137.

Weber, R.E., Lykkeboe, G. and Johansen, K. (1975). Biochemical adaptation of hemoglobin-oxygen affinity of eel to hypoxia. *Life Sci.* 17: 1345–1350.

Weber, R.E. (1976a). Physiological properties of eel haemoglobin: Hypoxic acclimation, phosphate effects and multiplicity. *J. Exp. Biol.* 64: 75–88.

Weber, R.E., Wood, S.C. and Lomholt, J.P. (1976b). Temperature acclimation and oxygen-binding properties of blood and multiple haemoglobins of rainbow trout. *J. Exp. Biol.* 65: 333–345.

Wells, R.M.G. and Weber, R.E. (1983). Oxygenational properties and phosphorylated metabolic intermediates in blood and erythrocytes of the dogfish. *Squalus acanthias. J. Exp. Biol.* 103: 95–108.

Wood, S.C. and Johansen, K. (1972). Adaptation to hypoxia by increased HbO_2 affinity and decreased red cell ATP concentration. *Nature (London) New Biol.* 237: 278–279.

Wood. S.C. and Johansen, K. (1973). Organic phosphate metabolism in nucleated red cells: Influence of hypoxia on eel $Hb-O_2$ affinity. *Neth. J. Sea Res.* 7: 328–338.

Wyman, J. (1964). Linked functions and reciprocal effects in hemoglobin: A second look. *Adv. Protein Chem.* 19: 223–286.

3. Transition of respiratory processes during amphibian metamorphosis: from egg to adult

Warren Burggren

Department of Zoology, University of Massachusetts, Amherst, Massachusetts 01003, U.S.A.

Abstract

Transitions in respiratory and circulatory functions during development in amphibians are reviewed. Gas exchange in the amphibian egg occurs primarily by diffusion. The heart beat and consequent internal convection of body fluid through the primordia of the larval respiratory organs occurs quite late in embryonic development, and is probably not necessary for gas exchange given the very small size of amphibian eggs and embryos. Post-hatch larvae respire predominantly with the well-perfused skin, with supplemental respiration provided by internal or external gills. As larval development proceeds, gill arch VI provides blood to the developing pulmonary circulation. Most amphibian larvae rely progressively on gas exchange with the paired unicameral lungs, but aquatic gas exchange normally predominates until complete metamorphosis. Neural reflexes involving mechano- and chemo-receptors coordinate gill and lung ventilatory movements and heart beat in the larva. Metamorphosis of the larva involves regression of the gills and further simplifications of the circulation involving the loss of entire gill arches or their distal segments and the proximal conjunction of systemic and pulmocutaneous arches. The skin of the adult remains a major site of CO_2 elimination, but O_2 uptake is achieved primarily via the lungs. The physiological transitions, particularly of regulatory mechanisms, which occur during amphibian metamorphosis remain only poorly understood, and serve as a challenging and rewarding area for future research.

Introduction

As a condensed study in water-to-land transitions, morphological transformations during metamorphosis in amphibians have quite correctly received extensive scrutiny. However, recent investigations have begun to reveal parallel and equally momentous transformations in the physiological systems of amphibians undergoing metamorphosis.

Many of the physiological 'problems' associated with amphibian metamor-

Seymour, R.S., (ed.) Respiration and metabolism of embryonic vertebrates.
© *1984, Dr W. Junk Publishers, Dordrecht/Boston/London. ISBN-13:978-94-009-6538-6*

phosis, and the consequent adaptations which serve as 'solutions', revolve directly or indirectly around profound transitions in modes of gas exchange. Strictly fluid 'breathing' in the egg and earliest larval stages of many amphibians is replaced by a pattern in the adult of either combined water/air breathing or completely aerial respiration. Lest such a transition be considered unique to amphibians, consider that similar ontogenetic developments associated with an air-to-water respiratory transition are to be found in air breathing fishes, in the embryos of reptiles and birds, and as vestiges in the mammalian fetus. The study of these developmental processes is thus fundamental to our understanding of vertebrate evolution and diversity.

The intent of this paper is to review current knowledge of the transition of respiratory and supporting circulatory processes during amphibian metamorphosis and, perhaps more importantly, to indicate the considerable gaps in our understanding of this fascinating developmental process. The following pages contain somewhat categorical statements describing the 'typical' anuran or urodele condition. I ask the reader familiar with the tremendous diversity in amphibian development and life cycles (see Feder, present volume) to withold criticism, and offer by way of excuse the fact that much of the published physiological data on respiration is for amphibian species with similar lifestyles occupying temperate Europe and North America.

Respiration in amphibian eggs: diffusion in small embryos

The requirement in amphibians for direct gas exchange with the environment begins with egg extrusion, and intensifies as cell division in the embryo progresses. Like the eggs of fishes, there are no extra-embryonic organs analagous to the chorioallantoic vascular network of the avian egg for convection of gases between egg surface and embryonic tissues. Hence, the amphibian embryo with its attached yolk sac can be viewed as a self-contained entity surrounded by jelly envelopes or membranes and covered by a non-adhesive, elastic surface coat more or less continuous with the intercellular matrix (Holtfreter, 1943; Bell, 1960). Convection of blood begins comparatively late in the development of amphibian embryos (see below), so gas exchange between the embryo's surface and the innermost embryonic tissues must occur entirely by diffusion for much of embryonic development. However, diffusion distances are comparatively very short, since amphibian eggs are very small relative to those of reptiles or birds. A chicken egg, for example, has 30,000–40,000 times the volume of a bullfrog egg (though the eggs of other anurans may be slightly larger). Quite clearly, small size is advantageous in an egg completely lacking extra-embryonic convective flow and is a feature of eggs of almost all aquatically-breeding vertebrates.

Interestingly, the smallest amphibian eggs have both the greatest O_2 uptake and the greatest rate of development (Wills, 1936). It is tempting to speculate that the even shorter diffusion distances of these smallest eggs allow greater bulk flow of O_2 and CO_2 between embryo and environment, and consequently support an elevated metabolic rate.

Although in theory gas exchange by diffusion alone can supply adequate bulk flow of O_2 and CO_2 when egg and embryo are very small, many amphibians lay their eggs in large gelatinous aggregations up to 20 cm in diameter. Almost nothing is known of the diffusive properties for O_2 and CO_2 exhibited by this jelly, but it obviously presents a significant barrier to gas exchange. Partial pressures of O_2 in the center of spawn masses of *Rana temporaria* may be as low as 4 torr, with pH as much as 0.8 units lower than ambient water (Savage, 1935). Lactic acid concentrations in the eggs of *Rana pipiens* in the center of the jelly mass are nearly 200 times greater than in individual eggs previously separated from the jelly mass (Barth, 1946). Clearly, some degree of asphyxia is a natural feature of embryonic development in amphibians laying egg aggregations, and appears to be countered by an extraordinary resistance to hypoxia. While the early embryos of many vertebrates are less sensitive to O_2 limited situations than at later developmental stages, amphibian embryos and even post-hatch larvae survive several hours to one week of anoxia, depending on developmental stage and temperature (Detwiler and Copenhaver. 1940; Gregg, 1960; Rose et al., 1971; Weigmann and Altig, 1975; Adolph, 1979). Such anoxic tolerance is probably due in part to the large glycolytic potential for anaerobic utilization of yolk substrate (Gregg, 1962; Weigmann and Altig, 1975). In addition, the O_2 consumption of amphibian embryos is extremely low up until the folding of the neural crest (Bialaszewicz and Bledowski, 1915; Parnas and Kraskinska, 1921; Brachet, 1934; Wills, 1936; Hopkins and Handford, 1943; Gregg, 1960; J. Roberts, pers. com.). Fig. 1 presents the changes in O_2 uptake which occur during development in *Rana pipiens*. This profile is quite representative of the amphibians which have been

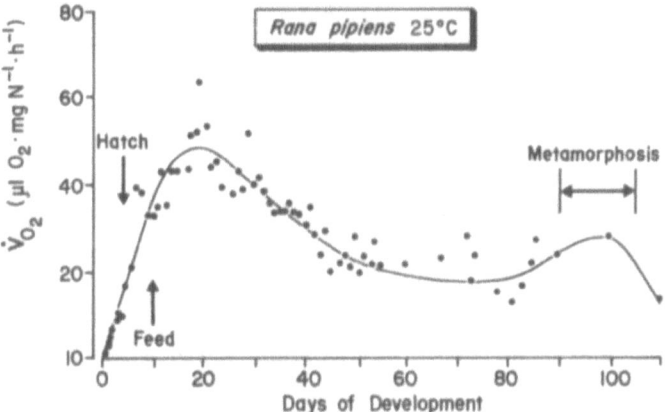

Fig. 1. Rates of oxygen uptake at 25 °C from the time of fertilization until completion of metamorphosis in the anuran *Rana pipiens* (after Wills, 1936).

examined, and reveals a comparatively low metabolic rate for much of embryonic development.

Gas exchange by diffusion has the potential to be enhanced by the convective flow of body fluids in the last third of the period of embryonic development as the elements of the circulation differentiate and grow. Unfortunately, observations on the ontogeny of the amphibian circulation are limited to morphological aspects of the embryonic and larval circulation and scattered observations of changing heart function. Morphological changes in the embryonic circulation are probably best described for ranid frogs and ambystomid salamanders, primarily due to their extensive use in embryological studies of chordate development (Manner, 1975; Lehman, 1977). The primordium for the circulation is the splanchnic mesoderm underlying the pharynx. This tissue folds to form a hollow tube, ultimately to become the pericardial sac. Isolated endothelial cells within this space organize themselves into an endothelial tube which posteriorly forms the paired amphalomesenteric veins growing directly into the yolk sac, and anteriorly forms paired ventral aortae. The central, unpaired region of this endothelial tube becomes the endothelial lining of the heart, while the myocardium is formed from surrounding splanchnic mesoderm. The heart continues to elongate in the pericardial space by folding into an S-shaped tube, whereupon the sinus venosus, paired atria, ventricle and conus and bulbus arteriosus become differentiated. Concomitantly the embryonic hematoblasts differentiate into erythrocytes and the capillary walls of the developing circulation, while the larger vessels are formed from isolated mesenchymal cells which round up to form hollow, elongate tubes.

Originally, six aortic arches develop within the visceral arches, which are connected ventrally to the truncus arteriosus of the heart by the paired aortae. The anterior most arches (I and II) degenerate early in embryonic development, but the anterior extensions of the ventral aortae which perfused them are retained as the external (ventral) carotid arteries (Fig. 2). Aortic arches III and IV supply their respective gill arches, which appear on and then grow out from the body wall of the embryo before hatching. Aortic arches V and VI continue to develop, but these gill arches remain internalized until hatching. Efferent blood from all four arches collects dorsally and is distributed between the dorsal aorta and the internal (dorsal) carotid arteries (Fig. 2). In the later embryonic stages before hatching, the myogenic heart begins to beat in a rhythmic fashion and propels blood through the circulation as it grows and elaborates (Manner, 1975; Lehman, 1977). Apart from observations of electrical properties of the embryonic heart of *Ambystoma* (see Justus, 1978), extremely little is known of the cardiovascular physiology of amphibian embryos. Convection of blood between gill arches III and IV and the deep body tissues of the embryo no doubt facilitates gas exchange with the outer regions of the egg and the environment. In addition, a high degree of vascularization and the large surface area of the tail, which is the primary site of growth and elongation after formation of the cephalic structures and the alimen-

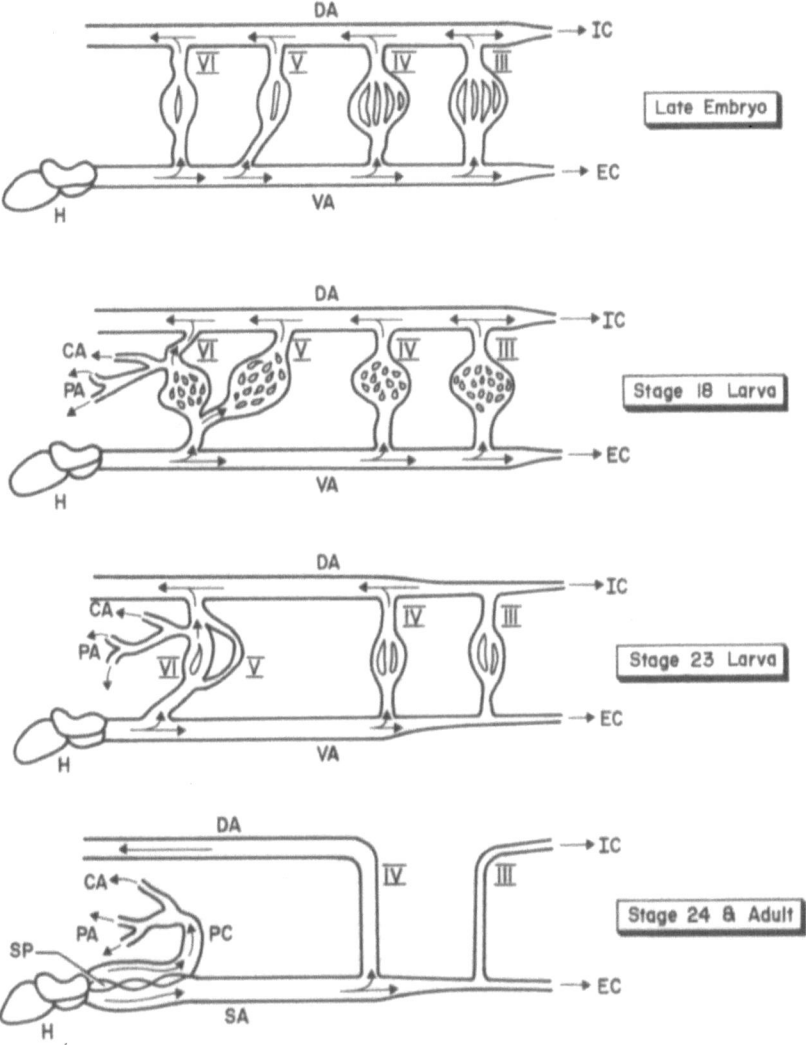

Fig. 2. A highly diagrammatic representation of changes in the central circulation of *Rana catesbeiana* during metamorphosis. Arabic numerals indicate developmental stages of larvae according to Taylor and Kollros (1946). Roman numerals indicate the aortic arches associated with the gills. CA, cutaneous artery; DA, dorsal artery; EC, external (ventral) carotid artery; H, heart; IC, internal (dorsal) carotid artery; PA, pulmonary artery; PC, pulmocutaneous arch; SA, systemic arch; SP, spiral valve; VA, ventral aorta (after Witschi, 1956, and Just et al., 1973).

tary tract, probably contributes greatly to gas exchange by the skin. Certainly, a major gas exchange function has been suggested for the tail of post-hatch larvae of *Rana* (Medvedev, 1937; Strawinski, 1956; Branch and Taylor, 1977; Burggren and West, 1982).

While the role of blood convection in gas exchange of the amphibian embryo has yet to be demonstrated, an experimental model exists in which the role of the cardiovascular system in embryonic respiration could be tested. A mutant strain of *Ambystoma*, designated 'cardiac lethal', in which the heart develops but induction of heartbeat fails to occur, has been identified (see review by Justus, 1978). Amazingly, the affected embryos nonetheless hatch and even swim. They don't feed, however, and usually die within two weeks of hatching. Clearly, blood convection is not necessary in the embryo or very young larva. It would be extremely interesting to compare aspects of gas exchange in 'cardiac lethal' mutant embryos and larvae, where all gas exchange is by diffusion, with that in the normal unaffected strain. Unfortunately, our current understanding of the ontogeny of gas exchange and circulatory processes in both individual amphibian eggs and particularly in naturally laid egg clusters, is extremely rudimentary, and lags far behind what is known for avian and reptilian eggs. Respiratory physiologists have treated the amphibian egg largely as a black box which consumes O_2 and eliminates CO_2. Use of recent advances in miniaturization of electrodes for measurement of respiratory gases, pH and blood pressure, together with micro-analysis techniques for hemoglobin function, acid-base balance, and metabolic substrates and end-products, for example, will hopefully remedy this situation.

Post-hatch larvae: respiration with gills and skin

Branchial gas exchange

Immediately after hatching, most amphibian larvae undergo a brief sessile stage, anchored to a suitable substrate by mucous glands. In anurans visceral gill arches V and VI form at this time (Fig. 2) and gill filaments or 'tufts' develop on the four pairs of external gills (see Burggren and Mwalukoma, 1983; McIndoe and Smith, present volume; and Fig. 5 and 6 for details of branchial stucture). Interestingly, recent studies on the microcirculation of the gill filaments of larval urodeles and anurans have revealed a complex system of vasoactive branchial shunt vessels which may constitute a non-respiratory by-pass through the gills (Malvin and Wood, 1982; McIndoe and Smith, present volume).

Larval or neotenous urodeles have only three pairs of external gill arches. Aortic arch VI remains relatively undifferentiated within the body, but the pulmonary arch divides into a labyrinth-like network of vessels that subsequently coalesces back into a major artery before entering the lung (Malvin and Wood, 1982); this network may represent a vestige of an external branchial arch. Ligation of the external gills of urodeles does not fatally interrupt circulation (Shield and Bentley, 1973; Heath, 1976), because blood flow through the internal arch VI apparently operates as a collateral pathway of sufficient size between the heart and systemic circulation.

The gills, though structurally complete and perfused, receive no form of active ventilation in early post-hatch larvae. However, diffusion across a body wall even devoid of specialized respiratory structures could probably still serve a significant component of the larva's gas exchange needs. In fact, even in hypoxic water (P_{O_2} = 80 torr), the 'short radius' for O_2 penetration into tissue by diffusion alone exceeds the 'short radius' of the body of post-hatch *Ambystoma* larvae (see Adolph, 1979). Adolph (1979) quite aptly describes the aquatic larval amphibian immediately post-hatching as '... too small to lack oxygen even without oxyen transport.' Certainly, short-term survival of 'cardiac lethal' larvae in which no convective blood flow can occur (see above) offers ample support for this hypothesis. However, as the larva begins to forage and development progresses, body mass increases many fold and diffusion distances consequently lengthen. Metabolic rates of larvae rise well above those of the embryos in *Ambystoma, Taricha, Triturus, Bufo* and *Rana* (Fig. 1), at least in the early larval stages (Wills, 1936; Bialaszewics and Bledowski, 1915; Parnas and Kraskinska, 1921; Gayda, 1921; Brachet, 1934; Hopkins and Handford, 1943; Gregg, 1960; Kaplan 1980). Clearly new demands are made upon gas exchange processes as growth proceeds after hatching. It is at this point in development that a major dichotomy in the morphology of the gas exchange organs between amphibian orders becomes evident. In anuran larvae, opercular membranes develop and cover the gills, internal branchial chambers are formed, and the buccal musculature differentiates and elaborates (DeJongh, 1968; McIndoe and Smith, present volume). The buccal floor develops rhythmic dorso-ventral movements, which draw a unidirectional stream of water into the mouth, through the gills, and out the single opercular spout or 'spiracle' on the left side of the body wall (DeJongh, 1968; Kenny, 1969; Severtzov, 1969; Gradwell, 1972a and 1972b; Burggren and West, 1982). In urodeles, which retain external gills throughout larval development, no comparable internal chambers or spiracles are formed.

The mechanics of ventilation of the internal branchial chambers of anuran larvae are relatively well understood for ranid larvea. At room temperature, buccal pumping at 70–100 cycles · min⁻¹ produces an oscillating pressure gradient from buccal to opercular cavity of approximately 1–2 cm H_2O (Fig. 3), which in turn drives a total branchial water flow of approximately 0.15–0.30 ml · g⁻¹ · min⁻¹

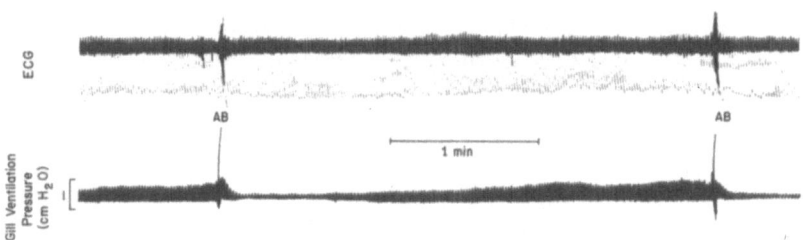

Fig. 3. The suppression of gill ventilation cycles following spontaneous air breaths (AB) in an unrestrained larva of *Rana catesbeiana*. Upper trace is the ECG; lower is buccal pressure recorded via a narial cannula (from West and Burggren, 1983).

(Gradwell and Pasztor, 1968; Gradwell, 1970; Wassersug and Hoff, 1979; Burggren and West, 1982; West and Burggren, 1982). Oxygen utilization from this ventilatory stream is high, about 65% in air saturated water (Burggren and West, 1982). Thus, the branchial performance of anuran larvae is quite comparable to that of a variety of fishes (see Shelton, 1970; Itazawa and Takeda, 1978; Burggren and Cameron, 1980 for references). Unlike fishes, however, branchial gas exchange in ranid larvae does not appear to be highly diffusion limited, for substantial increases in the P_{O_2} of exhaled water, reflecting an increase in the P_{O_2} gradient from water to blood in the gills, have no significant effect on branchial O_2 uptake (Burggren and West, 1982). Acute aquatic hypoxia stimulates gill ventilation in anuran larvae (Wassersug and Seibert, 1975; Wassersug and Feder, 1979; Wassersug et. al, 1981; West and Burggren, 1982), and the increased flow of branchial water maintains O_2 uptake by the gills in all but severely hypoxic water (West and Burggren, 1982).

It is tempting to assign great respiratory importance to the development in anuran larvae of a large, unidirectional flow of water over internal, as opposed to external, branchial structures. However, even with the involvement of a buccal pump to ventilate the internal gills, they only contribute 40% of total O_2 uptake and CO_2 elimination in the early, strictly aquatic stage of *Rana catesbeiana* (Fig.4). Moreover, almost all larval urodeles retain large external gills which are not 'actively' ventilated in the sense of receiving a flow of water generated by a muscular pump, although, together with the skin, the gills can still provide ample respiratory exchange for large and active larvae. Interestingly, in the neotenous salamander *Necturus* the action of branchiomeric levator and depressor muscles results in small rhythmic movements of the gills, a motion that is amplified considerably when in hypoxic water (Guimond and Hutchison, 1972, 1976). The movement of the external gills in this urodele probably serves to disturb boundary layers of O_2-depleted water next to the gill epithelium.

Chronic hypoxia serves as a potent stimulus for growth of branchial tissue in the larvae of frogs or salamanders. Gill filaments enlarge and in some instances the water-blood diffusion distance decreases (Preyer, 1885; Babak, 1907; Drastich, 1925; Bond, 1960; Guimond and Hutchison, 1976; Burggren and Mwalukoma, 1983). Hypertrophy of the gills and other changes in skin and lungs in response to chronic hypoxic exposure in *Rana catesbeiana* are indicated in Fig. 5. These long-term morphological changes in gill structure, together with short-term increases in gill ventilation, constitute a broadly flexible set of responses to facilitate branchial gas exchange in O_2-limited situations.

Finally, it should be emphasized that the gills of many amphibian larvae serve not only for respiration, but also for filter feeding and probably for osmoregulation (Feder, present volume). Consequently, any morphological and physiological adaptations of the gills for respiration must be compatible with these functions.

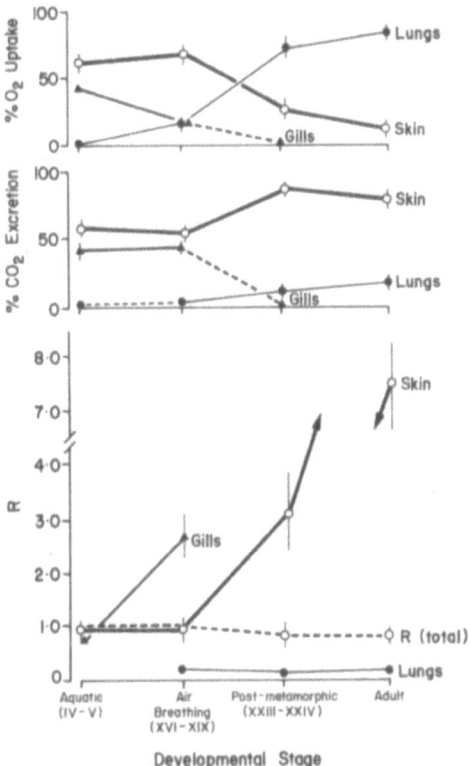

Fig. 4. Changes in the gas exchange ratios (R) and in gas exchange partitioning between gills, lungs and skin during metamorphosis at 20 °C in the bullfrog, *Rana catesbeiana*. Mean values ± 1 standard error are presented (from Burggren and West, 1982).

Cutaneous gas exchange

Regardless of the extent of gill development in the larval amphibian, branchial gas exchange is supplemented to varying degrees by gas exchange across additional respiratory structures, the skin and the lungs (if developed and ventilated). In the larvae of most frogs and salamanders a stage of strictly aquatic respiration occurs for some time after hatching. If lungs are present in this 'aquatic' stage, they are often rudimentary and, in any event, are not ventilated. Hence, the skin shares a major respiratory role with the gills, a feature long appreciated (see Krogh, 1904; Noble, 1924, 1931; Dolk and Postma, 1927 for early references). More recent investigations have quantified cutaneous gas exchange under a variety of experimental conditions and in various developmental stages of amphibians (see Vinegar and Hutchison, 1965; Whitford and Hutchison, 1965, 1966; Whitford and Sherman, 1968; Hutchison et al., 1968; Guimond, 1970; Guimond and Hutchison, 1972, 1976; Shield and Bentley, 1973; Gottlieb and Jackson, 1976; Wakeman and Ultsch, 1976; MacKenzie and Jackson, 1978; Feder,

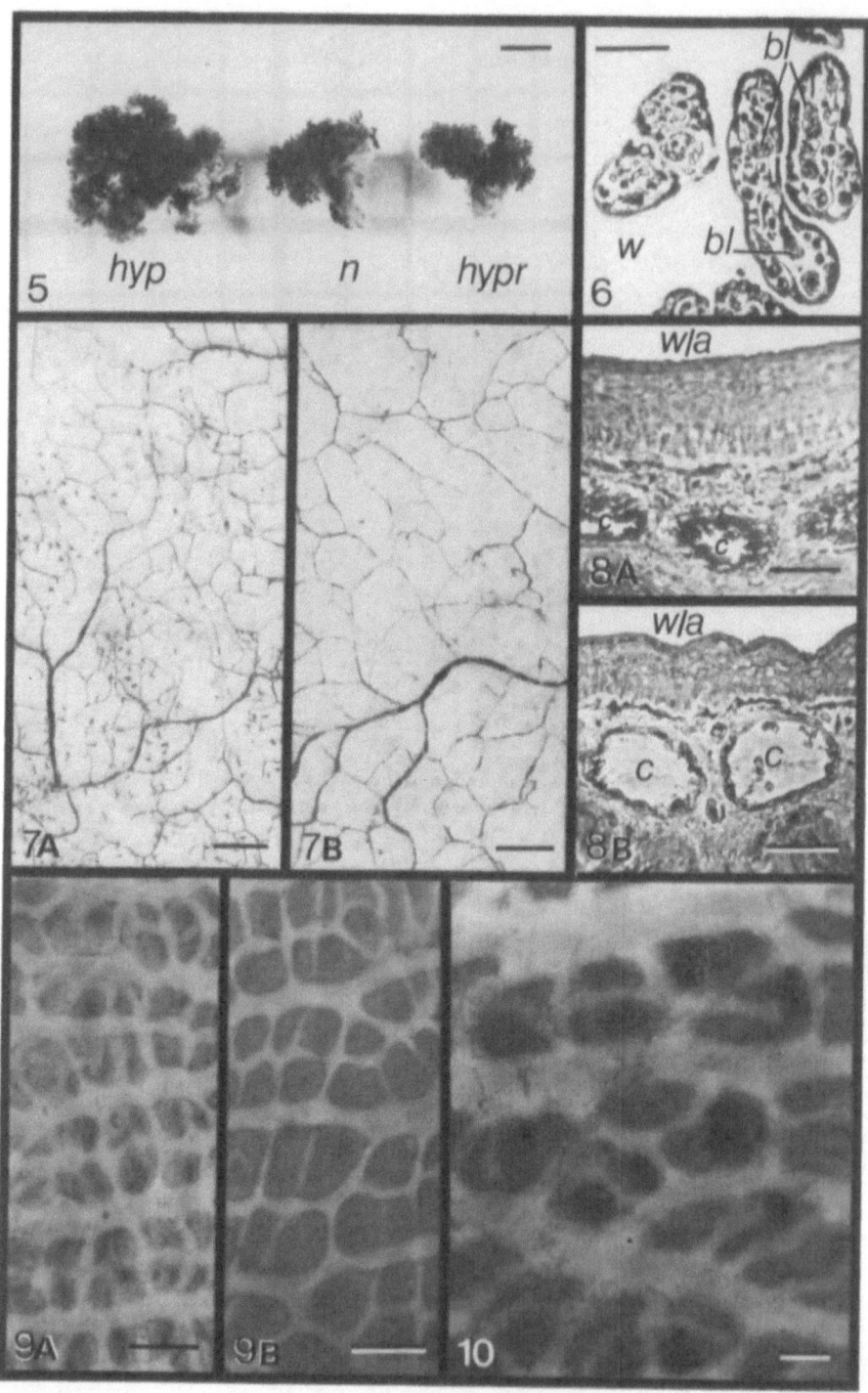

1982; Feder and Wassersug, 1983; Burggren and West, 1982; Burggren et al., 1983, among many others). Considerable interspecific variation emerges, particularly because body temperature and variable pulmonary contributions may have a major influence (see below). Unfortunately, few data exist on gas exchange partitioning for amphibian larvae in the early stage of strictly aquatic respiration. Ligation of the gills of larval or neotonic urodeles at 22 °C reduces O_2 uptake by 40% (*Necturus*) or causes no decrease (*Siren, Ambystoma*) (Shield and Bentley, 1973; Heath, 1976). While these data suggest a considerable cutaneous involvement in gas exchange, such experimental manipulations affect normal gas partial pressure gradients across the skin. In less invasive experiments, intact *Necturus* and *Siren* showing minimal air breathing activity relied upon the skin for 30–40% of both O_2 uptake and CO_2 excretion at 25 °C (Guimond and Hutchison, 1972, 1976). Data from large neotenic (paedomorphic, paedogenic) amphibians may not necessarily be typical of amphibians which undergo metamorphosis to an adult form, and should be regarded with some caution when studying developmental changes.

Prior to lung development in early anuran larvae, the skin is the major organ for both O_2 uptake and CO_2 elimination (Fig. 4), in spite of the ventilation of well developed branchial filaments. The effectiveness in gas exchange of the skin of amphibian larvae is largely due to the dense capillary network of the skin, which lies within 40 μm of the skin surface (Medvedev, 1937; Poczopko, 1957; Czopek, 1955, 1957, 1965; Strawinski, 1956; Burggren and Mwalukoma, 1983; Figs. 7, 8). Cutaneous blood flow in amphibians is a mosaic derived from both aortic arterial blood originating primarily from the left atrium and pulmocutaneous arterial blood originating primarily from the right atrium (see Moalli et al., 1980). Thus,

Fig. 5. The center gill filaments removed from the left gill arch V of three 15 g larval *Rana catesbeiana* exposed to chronic normoxia (n), hypoxia (hyp) and hypcroxia (hypr). Black bar indicates 1.0 mm.

Fig. 6. Cross section through the gill filament of a larval *Rana catesbeiana* exposed to chronic normoxia. Black bar indicates 10 μm; bl, blood channel; w, branchial water.

Fig. 7. Tail skin perfused with India ink and removed from larval *Rana catesbeiana* exposed to chronic hypoxia (A) and normoxia (B). Black bars indicate 250 μm on both figures.

Fig. 8. Thin sections through the tail of larval *Rana catesbeiana* exposed to chronic normoxia (A) and hypoxia (B). Bars indicate 20 μm in both figures; w/a, water or air; c, capillary.

Fig. 9. Internal surface of the lung walls of larval *Rana catesbeiana* exposed to chronic hypoxia (A) and normoxia (B). Bars indicate 500 μm.

Fig. 10. Internal surface of the lung wall of an adult bullfrog exposed to chronic normoxia. Bar indicates 500 μm.

Fig. 5–10 from Burggren and Mwalukoma, 1983.

significant partial pressure gradients between water and cutaneous blood should prevail when the animal is in well aerated water, particularly across those capillaries carrying comparatively deoxygenated right atrial blood. In anuran larvae the large, laterally compressed tail has not only an extensive subdermal capillary network but also a surface-to-mass ratio twice that of the trunk (Medvedev, 1936; Strawinski, 1957; Burggren and West, 1982; Burggren and Mwalukoma, 1983), and is probably a major site of skin gas exchange.

Use of the skin as a respiratory structure would clearly be most effective if gas exchange across the skin could be regulated. Clearly, regulatory changes must be mainly in blood perfusion, since the skin is not ventilated per se. Early studies have suggested that capillary recruitment in the skin of amphibians varies with exposure to experimental hypoxia and hypercapnia (see Poczopko, 1957). However, the prevailing view has been that changes in skin perfusion are not effective in regulating cutaneous respiration, and that cutaneous gas exchange varies mainly with the transcutaneous gas partial pressure (Gatz et al., 1975; Piiper et al., 1976; Gottlieb and Jackson, 1976; MacKenzie and Jackson, 1978). Recent experiments in our lab have shown that a brief air exposure of a few hours in unrestrained, adult *Rana catesbeiana* results in a highly significant and reversible decrease in capillary recruitment by the skin (Fig. 11). This active neural reflex, eliminated by the alpha-blocker phenoxybenzamine, results in a large fall in cutaneous CO_2 elimination (Burggren and Moalli, 1984; A. Caplan, unpublished). It would appear that descriptions of cutaneous gas exchange in amphibi-

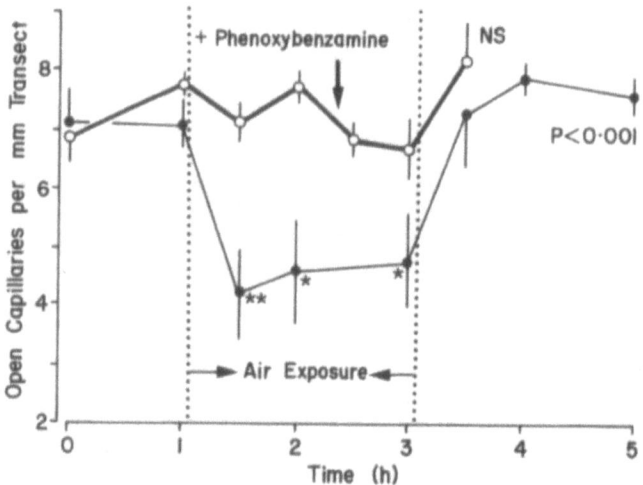

Fig. 11. Numbers of perfused capillaries in the hind foot web of unanaesthetised *Rana catesbeiana* before, during, and after complete air exposure. Data are also shown before and after the injection of phenoxybenzamine, an alpha-adrenergic blocker. Means ±1 standard error (N = 5) and significance levels of ANOVA are given. Asteriks beside data points indicate significant differences (1*, P<0.05; 2*, P<0.001) by independent t-test from control values before air exposure. NS, not significant. (from Burggren and Moalli, 1984.)

ans as 'passive' and 'poorly regulated' may now need revision. It is not yet known to what extent larvae of anurans or urodeles can regulate CO_2 elimination and O_2 uptake by altering skin perfusion, but observations of 'blushing' of the ventral trunk surface of larval anurans during hypoxic stress and exercise suggests that capillary recruitment does occur (independent unpublished observations of R. Wassersug, M. Feder, D. Quinn).

Changes in skin capillarization as a function of developmental stage in larval amphibians have been claimed. Czopek (1957) reported that capillary density in *Ambystoma* decreased following metamorphosis. However, either no change or an increase in skin capillary density following metamorphosis has been reported for anuran larvae (Strawinski, 1957; Burggren and Mwalukoma, 1983). Inter-specific or developmental variations in skin capillarization may relate as much to other skin functions (i.e. water and ion balance, protection) as to gas exchange (Feder and Burggren, 1984).

As reported for branchial structures, chronic hypoxic exposure in anuran larvae induces profound morphological changes in the skin. In larval *Rana catesbeiana* four weeks of exposure to a water and air P_{O_2} of 70 torr causes the cutaneous capillary density to double, and the diffusion distance between water and blood in skin capillaries to be halved (Burggren and Mwalukoma, 1983; Figs. 7, 8). These morphological adjustments facilitating cutaneous gas exchange indicate an important ability for the skin of larvae to accommodate to changing respiratory conditions.

Ventilation of the lungs: the transition to air breathing begins

Considerable interspecific variation exists in the onset of lung ventilation in larval amphibians. In some species, particularly those living in flowing water where inflated lungs could present buoyancy problems, the lungs develop only upon metamorphosis (see Feder, present volume). Some urodeles (e.g. Plethodontidae) fail altogether to develop lungs at any developmental stage. In the great majority of amphibian larvae, however, primary septation and further elaboration of the inner surfaces of the paired, unicameral lungs proceeds soon after hatching, and infrequent ventilation of the lungs begins to occur. It should be emphasized that the onset of lung ventilation depends not only on developmental stage of the larva, but also on more insidious factors such as degree of water hypoxia, ambient temperature (and therefore metabolic rate), and experimental handling. Increases in all of these factors will stimulate lung ventilation, and have doubtlessly led to some of the discrepancies in the literature describing the onset of air breathing in larval amphibians (Noble, 1931; Wassersug and Seibert, 1975; Burggren and West, 1982; West and Burggren, 1982; Burggren et al., 1983; Feder, present volume).

As in all other tetrapods, the blood supply to the lung derives from gill arch VI, and in larval amphibians consists of a posterior pulmonary arterial branch. The non-vertebral cutaneous circulation is also derived from arch VI (Fig. 2). Blood drains the lungs of larval and adult amphibians via distinct pulmonary veins returning directly to the left atrium. This veno-atrial development is one of the major circulatory improvements over the circulatory patterns of air breathing teleosts (see Johansen and Burggren, 1980; Randall et al. 1981).

How important a respiratory function can be ascribed to the lungs of larval amphibians? Amphibian larvae that normally ventilate their lungs apparently tolerate forced submergence (Shield and Bentley, 1973; Guimond and Hutchison, 1976; Feder, 1983b; Feder and Wassersug, 1983), but such experiments only demonstrate that ventilation of the lungs is not absolutely essential to survival. At low temperatures in air saturated waters, a common situation during the spring or fall in temperate climates, intervals between air breaths in unrestrained anuran and urodele larvae or neotenous adults may be 30–60 min or longer (Burggren and Wood, 1981; Burggren et al., 1983). At 15 °C, the contribution of the lungs to total O_2 uptake or CO_2 elimination is usually less than 20% (Guimond and Hutchison, 1972; Burggren et al., 1983). The major sites of gas exchange at low ambient temperatures thus remain the skin and gills.

At warmer temperatures the metabolic rate rises, and pulmonary gas exchange increased in partial compensation in larval amphibians. Increases in the percentage contribution of the lungs to gas exchange with increasing temperature occur in urodele larvae (Whitford and Sherman, 1968; Lenfant et al., 1970; Guimond and Hutchison, 1976), and reflects a pattern common to adult amphibians and air breathing fishes (Das, 1927; Hutchison and Dady, 1964; Hutchison et al., 1968; Johansen et al., 1970; Rahn et al., 1971; Whitford, 1973; Burggren, 1979; Hutchison and Miller, 1979). Nevertheless, in two species of anuran larvae, partitioning of O_2 uptake between skin (ca. 70%), gills (ca. 15%) and lungs (ca. 15%) is surprisingly constant over a temperature range from 15–33 °C (Burggren et al., 1983), even though total O_2 uptake increases nearly 3 times. Apparently neither the metabolic costs of increased gill ventilation nor the lower O_2 concentration of water at high environmental temperatures are sufficiently restrictive to require a disproportionate increase in pulmonary ventilation to support elevated metabolic rates (see Feder, this volume, for elaboration of these arguments). Indeed, transit costs of movement up and down the water column to breathe at the surface, together with ecological considerations such as increased risk of aerial predation at the air-water interface, may represent selection pressures against enhanced air breathing in air breathing fishes and larval amphibians (Kramer and Graham, 1976; Kramer and McClure, 1981; Wassersug et al., 1981; Burggren et al., 1983; Feder, 1983a).

In natural habitats both anuran and urodele larvae avoid hypoxic water where possible (Savage, 1935; Costa, 1967; Whitford and Massey, 1970; Wassersug and Seibert, 1975; Feder, 1983b). However, if the water becomes sufficiently warm and hypoxic and cannot be avoided, pulmonary ventilation rate and the percentage contribution of the lung to gas exchange do ultimately increase to help maintain O_2 uptake (Rose et al., 1971; Wassersug and Seibert, 1975; Heath, 1976; Wassersug and Feder, 1979, 1983; Feder, 1981, 1983b, 1984; Burggren and Wood, 1981; West and Burggren, 1982). Exposure to chronic hypoxia induces morphological changes in the lungs of larval anurans, as it does to the skin and gills. In *Rana catesbeiana* both lung volume and the degree of primary septation of the lung wall increases markedly (Burggren and Mwalukoma, 1983; Fig. 9). Blood-gas diffusion distances, already very small at $1-2\,\mu m$, remain unchanged.

Regardless of developmental or interspecific variation in the degree of dependence upon pulmonary respiration, it remains a common feature among larval amphibians that the lungs serve primarily as organs of O_2 uptake with the majority of CO_2 elimination occurring via the skin or gills. Thus, pulmonary gas exchange ratios of 0.2 or lower occur in larval or neotenic amphibians (Guimond and Hutchison, 1976; Burggren and West, 1982; Fig. 4).

Little is known of the transitions in hypobranchial muscle structure, innervation and function as buccal pumping is modified for lung ventilation. The basic mechanism for lung ventilation in anuran larvae is thought to be similar to that described for adults (DeJongh and Gans, 1969; West and Jones, 1975; MacIntyre and Toews, 1976; Brett and Shelton, 1979; West and Burggren, 1982), and consists of modification of the buccal movements normally used by the aquatic larvae to propel water through the branchial chambers. In anuran larvae, air is drawn into the mouth rather than through the nares as in adults. This inhaled air (and possibly some exhaled gas remaining in the buccal cavity – see Gans, 1970) is compressed to a pressure of $4-6\,cm\ H_2O$ by an exaggerated upward movement of the buccal cavity. Lung filling subsequently occurs during the very brief opening of the glottis. Exhalation in anurans appears to be largely passive, powered by pulmonary elastic recoil and by hydrostatic pressure on the body wall. The volume of air trapped in the buccal cavity upon inspiration must be relatively large, for a single pumping of the buccal cavity suffices to fill the lungs.

Ventilatory mechanisms have also been examined in the neotenous urodele *Siren*, where buccal movements serve for either olfaction or buccal/pharyngeal respiration rather than for ventilation of the external gills (Guimond and Hutchison, 1972, 1976). *Siren* first exhales gas at the surface, with 'active lung contraction' purportedly facilitating exhalation. On inspiration the buccal cavity dilates and draws air in through the mouth. The nares close, the buccal floor raises, and air is forced into the lungs. Buccal volume is small relative to lung volume, and several buccal pumping cycles are required to fill the lung after exhalation. More than half of the lung gas is renewed after a series of buccal pumping movements (Guimond and Hutchison, 1976). Whether lung ventilation mechanisms differ in urodele larvae that lose their gills and metamorphose to an adult body form is not known.

A reflex coordination of gill and lung ventilation exists in larval amphibians, as occurs in many vertebrates breathing both air and water through multiple gas exchange sites (see Randall et al., 1981). In anuran larvae, an air breath is followed by a 15–20 sec cessation of the small amplitude buccal movements producing water flow over the gills (Fig. 3). After this brief hiatus, buccal movements of progressively greater pressure amplitude return, until their inhibition by the next air breath. Experimental lung inflations with humidified air, N_2 or O_2 in unrestrained, unanaesthetised larvae of *Rana catesbeiana* all result in a transitory reduction in the frequency and pressure amplitude of buccal pumping (West and Burggren, 1983; Fig. 12). However, inhibition of gill ventilation is most pronounced with the injection of O_2-rich gas. Complex integration of afferent input, both from internally located chemoreceptors sensitive to P_{O_2} and from pulmonary mechanoreceptors sensitive to lung wall stretch, appear to mediate this oscillation between gill and lung ventilation in anuran larvae (West and Burggren, 1983). Larvae in hypoxic water risk losing blood-borne O_2 derived from lung gas to the flow of hypoxic water over the gills (West and Burggren, 1982; Feder, 1984). Reflex inhibition of gill ventilation following an air breath

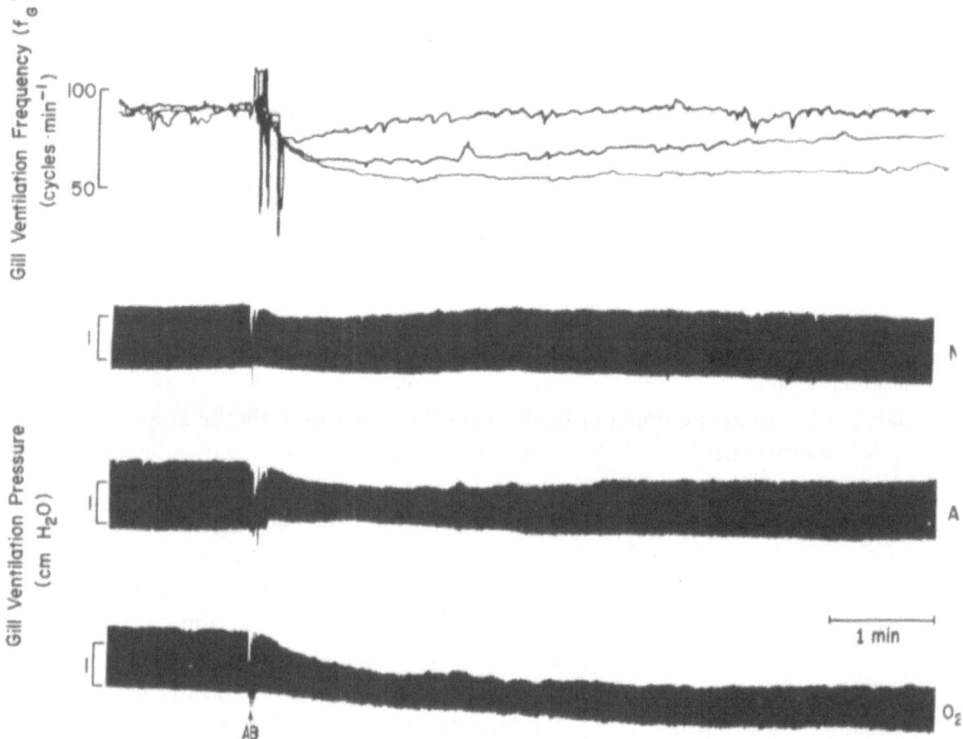

Fig. 12. Effects of experimental lung inflation with humidified N_2, air and O_2 to an estimated 100% of lung volume in a larval *Rana catesbeiana*. Upper traces, frequency of gill ventilation; lower traces, ventilation cycle pressure amplitude. The upward deflection (AB) indicates an air breath. (from West and Burggren, 1983.)

may well reduce this risk. In addition, blood vessels serving to shunt afferent branchial blood through non-respiratory pathways in the gills have been recently identified in larvae of both anurans and urodeles (Malvin and Wood, 1982; McIndoe and Smith, this volume). A partial circulatory by-pass of the gills would further reduce risk of O_2 loss back to hypoxic water following an air breath, particularly if coupled with transient decreases in skin perfusion.

Cardiac reflexes associated with lung ventilation have also been described for larval amphibians. A mild tachycardia occurs subsequent to an air breath in 'older' larvae of *Ambystoma tigrinum* (Heath, 1980), though chronotropic heart responses associated with spontaneous lung ventilation have been looked for but not found in the larvae of several anurans (Wassersug et al., 1981, West and Burggren, 1982, 1983; Fig. 3). Bradycardia associated with hypoxia as well as coupling of heart rate and gill ventilation rate have been reported for both anuran and urodele larvae (Heath, 1980; Wassersug et al., 1981), but is certainly not universal among these animals (West and Burggren, 1982). Matching of perfusion to ventilation is fundamental to the efficacy of a gas exchange organ, and is observed repeatedly in lower vertebrates with multiple gas exchange organs (see Johansen and Burggren, 1980; Randall et al., 1981). Noteworthy, then, is the 'metronome-like' quality of the heart beat of early larval amphibians compared with the more labile heart rates of later developmental stages (W. Burggren and N. West, unpublished; Wassersug et al., 1981). Perhaps complete functional development of cardiovascular reflexes may occur only after a period of post-hatch development. Such developmental immaturity of the nervous system could explain many of the inconsistencies in the literature. Our knowledge of the ontogeny of neural reflexes in amphibians remains rudimentary at best (see Kollros, 1981, for discussion). The reflex control of both ventilation and circulation and how they change during the transition to air breathing in amphibian larvae is an extremely challenging and rewarding area for further attention, particularly because of our limited understanding of the ontogeny of ventilatory and circulatory control in vertebrates generally.

Metamorphosis complete: respiration with lungs and skin

As metamorphosis proceeds, a rapid regression of the gills occurs, gill arch V simplifies to a non-differentiated shunt vessel and progressively becomes incorporated into arch VI, and the dorsal connection between arches III and IV disappears (Fig. 2). Upon completion of metamorphosis, the vestiges of arch III remain as the internal carotid artery, and arch IV serves as the major systemic arterial trunk. Arch VI forms the pulmocutaneous trunk supplying discrete pulmonary and cutaneous trunks. Interestingly, an anastomosis between the pulmonary artery and the efferent branchial artery of arch V occurs in *Ambystoma tigrinum*, and this places the lungs both in parallel and in series with the gills (Malvin and Wood, 1982). The pulmocutaneous and systemic trunks of amphibians are actually joined proximally along their length as they emanate from the single ventricle. This conjunction constitutes a potentially large route for admixture of blood between these two vessels. However, a spiral valve arising from the arterial arch wall separates the lumens of the two arches, and its complex valving action helps maintain separation of the blood streams in the two arterial trunks (see Shelton, 1976).

With regression of the gills and the profound changes in patterns of circulation outlined above, gas exchange in the adult becomes redistributed between skin and lungs. In post-metamorphic and adult *Ambystoma tigrinum* and *Rana catesbeiana* the lungs account for approximately 70% of the O_2 uptake at 20–25 °C (Whitford and Sherman, 1968; Gottlieb and Jackson, 1976, Burggren and West, 1982; Fig. 4). CO_2 eliminaton typically still occurs primarily via the skin in adult salamanders and ranid frogs whether they are in water or air, and the cutaneous gas exchange ratio is consequently very high (Gottlieb and Jackson, 1976; Guimond and Hutchison, 1976; MacKenzie and Jackson, 1978; Jackson and Braun, 1979; Burggren and West, 1982; Burggren and Moalli, 1984; Fig. 4). This consistent elimination of CO_2 by the skin throughout the amphibian life cycle may reflect the important differences in P_{CO_2} gradients across capillaries of the skin compared with those of the lungs. The P_{CO_2} of lung gas of any intermittently breathing vertebrate that has been examined is always substantially higher than ambient air, even immediately after inhalation (see Randall et al., 1981; Shelton and Boutilier, 1982, for references). Generally, the lungs of tetrapods are never fully emptied at expiration and significant anatomical dead space persists even in respiratory trees with very short airways. Thus, tidal pulmonary ventilation invariably results in admixture of inspired air with residual gas in the lungs and airways. In adult amphibians the P_{CO_2} of lung gas even immediately after inspiration is 10–20 torr (Toews et al., 1981; Boutilier, 1981). When the animal is in air or air-saturated water, the P_{CO_2} gradient across the skin will thus be considerably larger than across the lungs, and this probably accounts for the efficacy of cutaneous CO_2 elimination. (Although similar differences between lungs and skin in P_{O_2} gradient also exist, the much lower cutaneous O_2 uptake typical of amphibians may result from the considerably smaller diffusivity of O_2 through the relatively thick tissue barrier presented by the skin.)

As described earlier, our recent experiments have in fact linked skin capillary recruitment with large changes in CO_2 elimination across the skin of post-metamorphic and adult bullfrogs (Burggren and Moalli, 1984; A. Caplan, unpublished; Fig. 11). Much more remains to be determined about the regulation of cutaneous gas exchange in amphibians of all developmental stages.

Conclusion

Penetrating the morphological, physiological, biochemical and behavioral complexities that constitute amphibian metamorphosis still represents one of the more significant challenges of developmental vertebrate zoology. Too little is understood of the development of regulatory mechanisms for respiration and circulation. Too little is known of regulatory function once development is completed. Finally, too little information has been gathered on all of these processes in the myriad of amphibians living, developing, and reproducing in specialized ways in specialized environments. Hopefully the recent technological and conceptual advances required to remedy this paucity of information will be brought to bear on these and other mysteries of amphibian metamorphosis.

Acknowledgements

I am grateful to Alan Pinder for criticisms of this manuscript during its preparation, and to Martin Feder, Juan Markin, Rich Moalli, Anganille Mwalukoma, Alan Pinder, Nigel West and Steve Wood for sharing the frustrations and joys of experimentation on larval amphibians. Financial support has been provided by the NSF of the United States, and by the University of Massachusetts.

References

Adolph, E.F. (1979). Development of dependence on oxygen in embryo salamanders. *Am. J Physiol.* 236(5): R282–R291.

Babák, E. (1907). Über die funktionelle Anpassung der äusseren Kiemen bei Sauerstoffmangel. *Zentralbl. Physiol.* 21: 97.

Barth, L.G. (1946). Studies on the metabolism of development. *J. Exp. Zool.* 463–486.

Bell, E. (1960). Some observations on the surface coat and intercellular matrix material of the amphibian ectoderm. *Exp. Cell Res.* 20: 378–383.

Bialaszewicz, K. and Bledowski. R. (1915). Wplyw zaplodnienia na oddyehanie jaj. *Comptes Rendus des Séances de la Societe des Sciences de Varsovie* 8: 1.

Bond, A.N. (1960) An analysis of the response of salamander gills to changes in the oxygen concentration of the medium. *Dev. Biol.* 2: 1–20.

Boutilier, R.G. (1981). Gas exchange and transport during intermittent ventilation in the aquatic amphibian, *Xenopus laevis.* Ph.D. Dissertation. University of East Anglia, Norwich, England.

Brachet, J. (1934). Étude du métabolisme de l'oeuf de Grenouille *Rana fusca* au cours de developpement. I La respiration et la glycolyse, de la segmentation a l'eclosion. *Arch. Biol. T.* 45: 611.

Branch, L.C. and Taylor. D.H. (1977). Physiological and behavioural responses of larval spotted salamanders *(Ambystoma maculatum)* to various concentrations of oxygen. *Comp. Biochem. Physiol.* 58A: 269–274.

Brett, S.S. and Shelton. G. (1979). Ventilatory mechanisms of the amphibian, *Xenopus laevis*; the role of the buccal force pump. *J. Exp. Biol.* 80: 251–270.

Burggren, W.W. (1979). Bimodal gas exchange during variation in environmental oxygen and carbon dioxide in the air breathing fish *Trichogaster trichopterus. J. Exp. Biol.* 82: 197–214.

Burggren, W. and Cameron, J.N. (1980). Anaerobic metabolism, gas exchange and acid-base balance during hypoxic exposure in the channel catfish, *Ictalurus punctatus. J. Exp. Zool.* 213: 405–416.

Burggren, W.W., Feder, M.E. and Pinder, A.W. (1983). Temperature and the balance between aerial and aquatic respiration in larvae of *Rana berlandieri* and *Rana catesbeiana. Physiol. Zool.* 56(2): 263–273.

Burggren, W. and Moalli, R. (1984). 'Active' regulation of cutaneous gas exchange by capillary recruitment in amphibians: Experimental evidence and a revised model for skin respiration. *Respir. Physiol.* In press.

Burggren, W. and Mwalukoma, A. (1983). Influence of chronic hypoxia and hyperoxia on gas exchange processes in larval and adult bullfrogs (*Rana catesbeiana*): I. Morphological responses of lungs, skin and gills. *J. Exp. Biol.* 105: 191–203.

Burggren, W.W. and West, N.H. (1982). Changing respiratory importance of gills, lungs and skin during metamorphosis in the bullfrog *Rana catesbeiana. Respir. Physiol.* 47: 151–164.

Burggren, W.W. and Wood, S.C. (1981). Respiration and acid-base balance in the tiger salamander *Ambystoma tigrinum:* Influence of temperature acclimation and metamorphosis. *J. Comp. Physiol.* 144(B): 241–246.

Costa, H.H. (1967). Avoidance of anoxic water by tadpoles of *Rana temporaria. Hydrobiologia* 30: 374–384.

Czopek, J. (1955). The vascularization of the respiratory surfaces of some *Salientia. Zool. Pol.* 6: 101–134.

Czopek, J. (1957). The vascularization of respiratory surfaces in *Ambystoma mexicanum* (Cope) in ontogeny. *Zool. Pol.* 8: 131–149.

Czopek, J. (1965). Quantitative studies on the morphology of respiratory surfaces in amphibians. *Acta Anat.* 62: 296–323.

Das, B.K. (1927). III. The bionomics of certain air-breathing fishes of India, together with an account of the development of their air-breathing organs. *Phil. Tran. R. Soc. B.* 216: 183–218.

DeJongh, H.P. (1968). Functional morphology of the jaw apparatus of larval and metamorphosing *Rana temporaria. Neth. J. Zool.* 18: 1–103.

DeJongh, H.J. and Gans, C. (1969). On the mechanisms of respiration in the bullfrog *Rana catesbeiana:* A reassessment. *J. Morph.* 123: 259–290.

Detwiler, S.R. and Copenhaver, W.M. (1940). The developmental behaviour of *Ambystoma* eggs subjected to atmospheres of low oxygen and high carbon dioxide. *Am. J. Anat.* 66: 393–410.

Dolk, H.E. and Postma, N. (1927). Uber die Haut-und die Lungenatmung von *Rana temporaria. Z. vergl. Physiol.* 5: 417–444.

Drastich, L. (1925). Ueber das Leben der Salamandra-Larven bei hohem und niedrigem Sauerstoffpartialdruk. *Z. Vgl. Physiol.* 2: 632–657

Feder, M.E. (1981). Effect of body size, trophic state, time of day, and experimental stress on oxygen consumption of anuran larvae: an experimental assessment and evaluation of the literature. *Comp. Biochem. Physiol.* 70A: 497–508.

Feder, M.E. (1982). Effect of developmental stage and body size on oxygen consumption of anuran larvae: a reappraisal. *J. Exp. Zool.* 220: 33–42.

Feder, M.E. (1983a). The relation of air breathing and locomotion to predation on tadpoles, *Rana berlandieri*, by turtles. *Physiol. Zool.* 56: 522–531.

Feder, M.E. (1983b). Responses to acute aquatic hypoxia in larvae of the frog *Rana berlandieri. J. Exp. Biol.* 104: 79–96.

Feder, M.E. (1984). Consequences of aerial respiration for amphibian larvae. Present Volume.

Feder, M.E. and Burggren, W. (1984) Cutaneous gas exchange in the vertebrates: Design, patterns, control and implications. *Biol. Rev.* In press.

Feder, M.E. and Wassersug, R.J. (1983). Aerial versus aquatic oxygen consumption in larvae of the clawed frog. *Xenopus laevis. J. Exp. Biol.* 108: 231–245.

Gans, C. (1970). Strategy and sequence in the evolution of the external gas exchangers of ectothermal vertebrates. *Forma et Functio* 3: 61–104.

Gatz, R.N. Crawford, E.C. and Piiper, J. (1975). Kinetics of inert gas equilibration in an exclusively skin-breathing salamander, *Desmognathus fuscus. Respir. Physiol.* 24: 15–29.

Gayda, T. (1921). Ricerche di calorimetria. Nota II. La produzione di calore nello svolgimento onteogenico del *Bufo vulgaris. Arch. Fisiol.* 19: 211.

Gottlieb, G. and Jackson, D.C. (1976). Importance of pulmonary ventilation in respiratory control in the bullfrog. *Am. J. Physiol.* 230: 608–613.

Gradwell, N. (1970). The function of the ventral velum during gill irrigation in *Rana catesbeiana. Can. J. Zool.* 48: 1179–1186.

Gradwell, N. (1972a). Gill irrigation in *Rana catesbeiana.* I. On the anatomical basis. *Can. J. Zool.* 59:481–499.

Gradwell, N. (1972b). Gill irrigation in *Rana catesbeiana.* II.On the musculoskeletal mechanism. Can. J. Zool. 50: 501–521.

Gradwell, N. and Pasztor, V.M. (1968). Hydrostatic pressures during normal ventilation in the bullfrog tadpole. *Can. J. Zool.* 46: 1169–1174.

Gregg, J.R. (1960). Respiratory regulation in amphibian development. *Biol. Bull.* 119: 428–439.

Gregg, J.R. (1962). Anaerobic glycolysis in amphibian development. *Biol. Bull.* 123: 555–561.

Guimond, R.W. (1970). Aerial and aquatic respiration in four species of paedomorphic salamanders: *Amphiuma means means, Cryptobranchus alleganiensis alleganiensis, Necturus maculosus maculosus* and *Siren lacertinam.* Ph.D. Dissertation, University of Rhode Island.

Guimond, R.W. and Hutchison, V.H. (1972). Pulmonary, branchial and cutaneous gas exchange in the mud puppy, *Necturus maculosus maculosus*(Rafinesque). *Comp. Biochem. Physiol.* 42A: 367–392.

Guimond, R.W. and Hutchison, V.H. (1976). Gas exchange of the giant salamanders of North America. In: Respiration of Amphibious Vertebrates. G.M. Hughes, ed., Academic Press, New York, pp. 313–338.

Heath, A.G. (1976). Respiratory responses to hypoxia by *Ambystoma tigrinum* larvae, paedomorphs, and metamorphosed adults. *Comp. Biochem. Physiol.* 55A: 45–49.

Heath, A.G. (1980). Cardiac responses of larval and adult tiger salamanders to submergence and emergence. *Comp. Biochem. Physiol.* 65A: 439–444.

Holtfreter, J. (1943). Properties and functions of the surface coat in amphibian embryos. *J. Exp. Zool.* 93: 251–323.

Hopkins, H.S. and Handford, S.W. (1943). Respiratory metabolism during development in two species of *Ambystoma. J. Exp. Zool.* 93: 403–414.

Hutchison, V.H. and Dady, M.J. (1964). The viability of *Rana pipiens* and *Bufo terrestris* submerged at different temperatures. *Herpetologica* 20: 149–162.

Hutchison, V.H., Whitford, W.G. and Kohl, M. (1968). Relation of body size and surface area to gas exchange in anurans. *Physiol. Zool.* 41: 65–85.

Hutchison, V.H. and Miller, K. (1979). Aerobic and anaerobic contributions to sustained activity in *Xenopus laevis. Resp. Physiol.* 38: 93–103.

Itazawa, Y and Takeda, T. (1978). Gas exchange in the carp gills in normoxic and hypoxic conditions. *Respir. Physiol.* 35: 263–269.

Jackson, D.D. and Braun, B.A. (1979). Respiratory control in bullfrogs: cutaneous versus pulmonary response to selective CO_2 exposure. *J. Comp. Physiol.* 129: 339–342.

Johansen, K. and Burggren, W.W. (1980). Cardiovascular function in the lower vertebrates. In: Hearts and Heart-like Organs. G.H. Bourne, ed. Academic Press, New York, Vol. 1.

Johansen, K., Hanson, D. and Lenfant, C. (1970). Respiration in a primitive air breather, *Amia calva. Respir. Physiol.* 9: 162–174.

Just, J.J., Gatz, R.N. and Crawford, E.C. (1973). Changes in respiratory functions during metamorphosis of the bullfrog, *Rana catesbeiana. Respir. Physiol.* 17: 276–282.

Justus, J.T. (1978). The Cardiac Mutant: An Overview. *Am. Zool.* 18: 321–326.

51

Kaplan, R.H. (1980). Ontogenetic energetics in *Ambystoma*. *Physiol. Zool.* 53: 43–56.

Kenny, J.S. (1969). Feeding mechanisms in anuran larvae. *J. Zool.* 157: 225–246.

Kollros, J.J. (1981). Transitions in the nervous system during amphibian metamorphosis. In: Metamorphosis: A Problem in Developmental Biology, 2nd ed. L.I. Gilbert and E. Frieden, eds., Plenum Press, New York, pp. 445–459.

Kramer, D.L. and Graham, J.F. (1976). Synchronous air breathing, a social component of respiration in fishes. *Copeia* 1976: 689–697.

Kramer, D.L. and McClure, M. (1981). The transit cost of aerial respiration in the catfish *Corydoras aeneus* (Callichthyidae). *Physiol. Zool.* 54(2): 189–194.

Krogh, A. (1904). On the cutaneous and pulmonary respiration of the frog. *Skand. Arch. Physiol.* 15: 328–419.

Lehman, H.E. (1977). Chordate Development. Hunter, Winston-Salem, 369 p.

Lenfant, C., Johansen, K. and Hanson, D. (1970). Bimodal gas exchange and ventilation-perfusion relationship in lower vertebrates. *Fed. Proc. Fed Am. Soc. Exp. Biol.* 29(3): 1124–1129.

MacIntyre, D.H. and Toews, D.P. (1976). The mechanics of lung ventilation and the effects of hypercapnia on respiration in *Bufo marinus. Can J. Zool.* 54(8): 1364–1374.

MacKenzie, J.A. and Jackson, D.C. (1978). The effect of temperature on cutaneous CO_2 loss and conductance in the bullfrog. *Respir. Physiol.* 32: 313–323.

Malvin, G. and Wood, S.C. (1982). Morphology and control of blood flow in the gills of the salamander. *Ambystoma tigrinum.* Symposium on Gas Exchange, Gas Transport and Acid-Base Regulation, Max-Planck-Institut fur experimentelle Medizin. Gottingen, F.R.G., August, 1982.

Manner, H.W. (1975). Vertebrate Development. Kendall/Hunt, Dubuque, 242 p.

McIndoe, R. and Smith, D.G. (1984). Internal gills of the tadpole, *Litoria ewingi*: morphology and vascular anatomy with reference to the transition from water to air breathing. Present Volume.

Medvedev, L. (1936). The vessels of the caudal fin in amphibian larvae and their respiratory function. *Zool. J.* 16: 393–403.

Moalli, R., Meyers, R.S., Jackson, D.C. and Millard, R.W. (1980). Skin circulation in the frog, *Rana catesbeiana*: Distribution and dynamics. *Respir. Physiol.* 40: 137–148.

Noble, G.K. (1925). The integumentary, pulmonary, and cardiac modification correlated with increased cutaneous respiration in the Amphibia: A solution of the 'hairy frog' problem. *J. Morphol. Physiol.* 40: 341–416.

Noble, G.K. (1931). The Biology of the Amphibia. MacGraw Hill, New York.

Parnas, J.K. and Krasinska, Z. (1921). Uber den Stoffwechsel der Amphibienlarven. *Biochem. Z.* 116: 108.

Piiper, J., Gatz, R.N. and Crawford, E.C. (1976). Gas transport characteristics in an exclusively skin-breathing salamander, *Desmognathus fuscus* (Plethodontidae). In: Respiration of the amphibious vertebrates. G.M. Hughes, ed., Academic Press, New York.

Poczopko, P. (1957). Further investigations on the cutaneous vasomotor reflexes in the edible frog in connexion with the problem of regulation of the cutaneous respiration in frogs. *Zool. Pol.* 8: 162–173.

Preyer, W. (1885). Specielle Physiologie des Embryo. Grieben, Leipzig, p. 1.

Rahn, H., Rahn, K.B., Howell, B.J., Gans, C. and Tenney, S.M. (1971). Air breathing of the garfish (*Lepisosteus osseus*). *Respir. Physiol.* 11(3): 285–307.

Randall, D.J., Burggren, W.W., Farrell, A.P. and Haswell, M.S. (1981). The Evolution of Air Breathing in Vertebrates. Cambridge University Press, New York, 133 p.

Rose, F.L., Armentrout, D and Roper, P. (1971). Physiological responses of paedogenic *Ambystoma tigrinum* to acute anoxia. *Herpetologica* 27: 101–107.

Savage, R.M. (1935). The ecology of young tadpoles, with species reference to some adaptations to the habit of mass spawning in *Rana temporaria temporaria* Linn. *Proc. Zool. Soc. London:* 605–610.

Severtzov, A.S. (1969). Food seizing mechanisms of Anura larvae. *Dokl. Akad. Nauk. SSSR* 187: 211–214 (Trans).

Shield, J.W. and Bentley, P.J. (1973). Respiration of some urodele and anuran Amphibia I. In water, role of the skin and gills. *Comp. Biochem. Physiol.* 46A: 17–28.

Shelton, G. (1970). The regulation of breathing. In: Fish Physiology, Vol. 4. W.S. Hoar and D.J. Randall, eds., Academic Press, New York, pp. 293–359.

Shelton, G. (1976). Gas exchange, pulmonary blood supply, and the partially divided amphibian heart. In: Perspectives in Experimental Biology, Vol. 1. P. Spencer Davies, ed., Pergamon Press, Oxford, pp. 247–259.

Shelton, G. and Boutilier, R.G. (1982). Apnoea in amphibians and reptiles. *J. Exp. Biol.* 100: 245–273.

Strawinski, S. (1957). Vascularization of respiratory surfaces in ontogeny of the edible frog *Rana esculenta* L. *Zool. Pol.* 7: 327–365.

Toews, D.P., Shelton, G. and Randall, D.J. (1971). Gas tensions in the lungs and major blood vessels of the urodele amphibian, *Amphiuma tridactylum. J. Exp. Biol.* 55: 47–61.

Vinegar, A. and Hutchison, V.H. (1965). Pulmonary and cutaneous gas exchange in the green frog *Rana clamitans. Zoologica* N.Y. 50: 47–53.

Wakeman, J.M. and Ultsch, G.R. (1976). The effects of dissolved O_2 and CO_2 on metabolism and gas-exchange partitioning in aquatic salamanders. *Physiol. Zool.* 48(4): 348–359.

Wassersug, R.J. and Feder, M.E. (1979). Respiratory behaviour of *Xenopus laevis* larvae. *Am. Zool.* 19: 863.

Wassersug, R.J. and Feder, M.E. (1983). The effects of aquatic oxygen concentration, body size, and respiratory behaviours on the stamina of obligate aquatic (*Bufo americanus*) and facultative air-breathing (*Xenopous laevis* and *Rana berlandieri*) anuran larvae. *J. Exp. Biol.* 105: 173–190.

Wassersug, R.J. and Hoff, K. (1979). A comparative study of the buccal pumping mechanism of tadpoles. *Biol. J. Linn. Soc.* 12:225–259.

Wassersug, R.J., Paul, RD. and Feder, M.E. (1981). Cardio-respiratory synchrony in anuran larvae (*Xenopus laevis Pachymedusa dacnicolor*. and *Rana berlandieri*). *Comp. Biochem. Physiol.* 70A: 329–334.

Wassersug, R.J. and Seibert, E.A. (1975). Behavioural responses of amphibian larvae to variation in dissolved O_2. *Copeia* 1975: 86–103.

Weigman, D.L. and Altig, R. (1975). Anaerobic glycolysis in two larval amphibians. *J. Herpetol.* 9: 355–357.

West, N.H. and Burggren, W.W. (1982). Gill and lung ventilatory responses to steady-state aquatic hypoxia and hyperoxia in the bullfrog tadpole (*Rana catesbeiana*). *Respir. Physiol.* 47: 165–176.

West, N.H. and Burggren, W.W. (1983). Reflex interactions between aerial and aquatic gas exchange organs in the larval bullfrog. *Am. J. Physiol.* 244(6): R770–R777.

West, N.H. and Jones, D.R. (1975). Breathing movements in the frog *Rana pipiens*. I. The mechanical events associated with lung and buccal ventilation. *Can. J. Zool.* 53: 332–344.

Whitford, W.G. (1973). The effects of temperature on respiration in the Amphibia. *Am. Zool.* 13: 505–512.

Whitford, W.G. and Hutchison, V.H. (1965). Gas exchange in salamanders. *Physiol. Zool.* 25: 228–242.

Whitford, W.G. and Hutchison, V.H. (1966). Cutaneous and pulmonary gas exchange in ambystomatid salamanders. *Copeia* 1966: 573–577.

Whitford, W.G. and Massey, M. (1970). Responses of a population of *Ambystoma tigrinum* to thermal and oxygen gradients. *Herpetologica* 26(3): 372–376.

Whitford, W.G. and Sherman, R.E. (1968). Aerial and aquatic respiration in axolotl and transformed *Ambystoma tigrinum. Herpetologica* 24(3): 233–237.

Wills, I.A. (1936). The respiratory rate of developing amphibian with special reference to sex differentiation. *J. Exp. Zool.* 73: 481–510.

Witschi, E. (1956). Development of Vertebrates. W.B. Saunders, Philadelphia.

4. Functional morphology of gills in larval amphibians

Rosemary McIndoe[1] and D.G. Smith[2]

Department of Zoology, University of Melbourne, Parkville, Victoria, 3052, Australia.

Abstract

Scanning electron microscopy and vascular casting were used to study the morphology and vascular anatomy of the fully developed internal gills of *Litoria ewingi* tadpoles and the external gills of mexican axolotls, *Ambystoma mexicanum*.

The exchange units in *L. ewingi* consist of simple capillary loops in which blood flow should be well described by the Poiseuille relation. The random orientation of loops implies that there is no special con- or contra-flow arrangement of blood and water flows. In *A. mexicanum* gills blood flows through exchange sites comprising a double sheet of capillary space where the mathematics of sheet flow probably apply.

In both species, the gills are regarded as being important sites of aquatic gas exchange.

Branchial shunts linking afferent and efferent branchial arteries are present in both species.

Introduction

Larval and neotenous amphibians are bimodal breathers, using various combinations of gills, skin, lungs and bucco-pharyngeal mucosa as respiratory organs. Despite numerous thorough studies of the partitioning of total gas exchange between these various sites in adult amphibians (e.g. Hutchison, Whitford and Kohl, 1968), relatively little effort has been directed to larval forms. In particular, detailed anatomical information, such as that provided for fish gills by the

Present address:
[1] *Department of Zoology, Monash University, Clayton, Victoria, 3168, Australia.*
[2] *Natural History Unit, Australian Broadcasting Corporation, P.O. Box 9994 G.P.O., Melbourne, Victoria, 3001, Australia.*

Seymour, R.S., (ed.) Respiration and metabolism of embryonic vertebrates.
© *1984, Dr W. Junk Publishers, Dordrecht/Boston/London. ISBN-13:978-94-009-6538-6*

application of modern methods of vascular casting and electron microscopy (e.g. Farrell, 1980) is conspicuously lacking for anuran and urodele larvae.

A recent study by the present authors has attempted to address this problem for a single species of anuran (*Litoria ewingi*) at a particular stage in its larval development (Stage 41: Gosner, 1960). An overview of those results is presented here, together with some results from an unpublished study of the axolotl, *Ambystoma mexicanum*.

General morphology of amphibian gills

The external and internal gills of tadpoles, and the external gills of urodeles, are strikingly different from each other and both differ markedly from fish gills. In the many comprehensive accounts of anuran development there have been only incidental references to the structure of the internal gills (see, for example, Schulze, 1889; Marshall, 1893, 1932; Gradwell, 1972a, b; Kenny, 1969; Wassersug, 1980). In general terms these are usually arranged as four arches lying in each of two branchial baskets within the bucco-pharyngeal cavity. In *L. ewingi* there are three gill clefts: the second, third and fourth, the first being absent in this species (Figs. 1, 2). When present the first cleft forms a direct connection between the buccal cavity and opercular chamber, thus by-passing the pharynx (Gradwell, 1969).

Water enters the branchial system through the buccal cavity, passing over sensory papillae before crossing the ventral velum to enter the filter chambers (Fig. 3). These three chambers contain filter plates which support the ruffled epithelium forming the surface of the gill filters. Water enters the opercular chambers through the gill clefts. Since the water outlet or spiracle is sinistral, water from the right opercular chamber passes to the opposite side in a broad channel ventral to the pericardium. The opening of the left opercular chamber communicates directly with the spiracle.

The ventilatory mechanics have been described in great detail in a series of papers by Gradwell (1969, 1972a, b, 1975). Water flow is pulsatile, being achieved by the alternating action of buccal and pharyngeal pumps (Gradwell, 1972b). Although not described here external gills, where present, are very much simpler in form and are merely protruded into the surrounding water. Thus, the ventilatory efficiency achieved will probably be much lower than that occurring in the internal gills. Wassersug et al. (1981) have observed that, except in disturbed animals, cardio-respiratory synchrony is normally seen in many species of anuran tadpoles.

Figge (1936) provided the first comprehensive physiological and anatomical account of the external gills of the Colorado axolotl (*Ambystoma tigrinum*). It is disappointing to note that almost 50 years later very little has been added to that perceptive and imaginative study. In axolotls, there are three gill arches on each

56

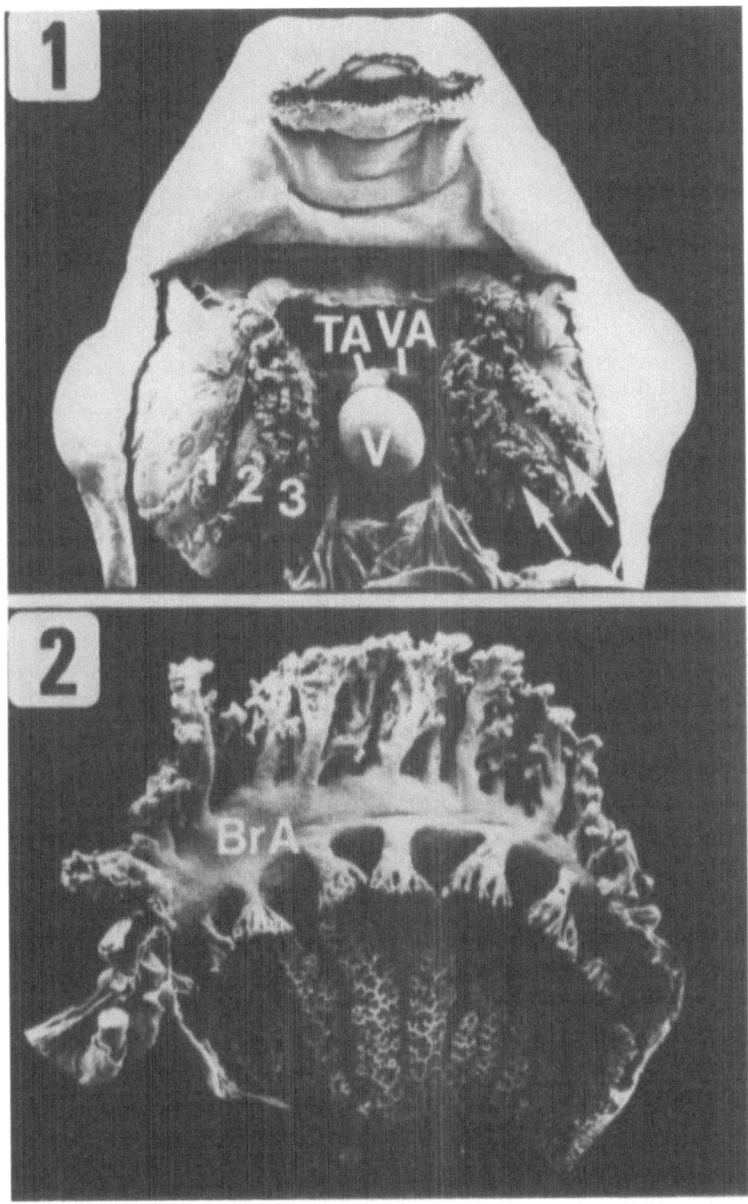

Fig. 1. Ventral view of anterior of tadpole. Note shape of branchial baskets and location of ventricle (V) between them. First three gills (Gl-3) are shown at left (G4 obscured) and two of the three gill clefts are shown with arrows. Critical point dried (CPD) preparation. ×20

Fig. 2. Lateral side of 3rd gill showing gill tufts above (ventral) and gill filters below (dorsal) to prominent branchial arch (BrA). CPD preparation. ×30

57

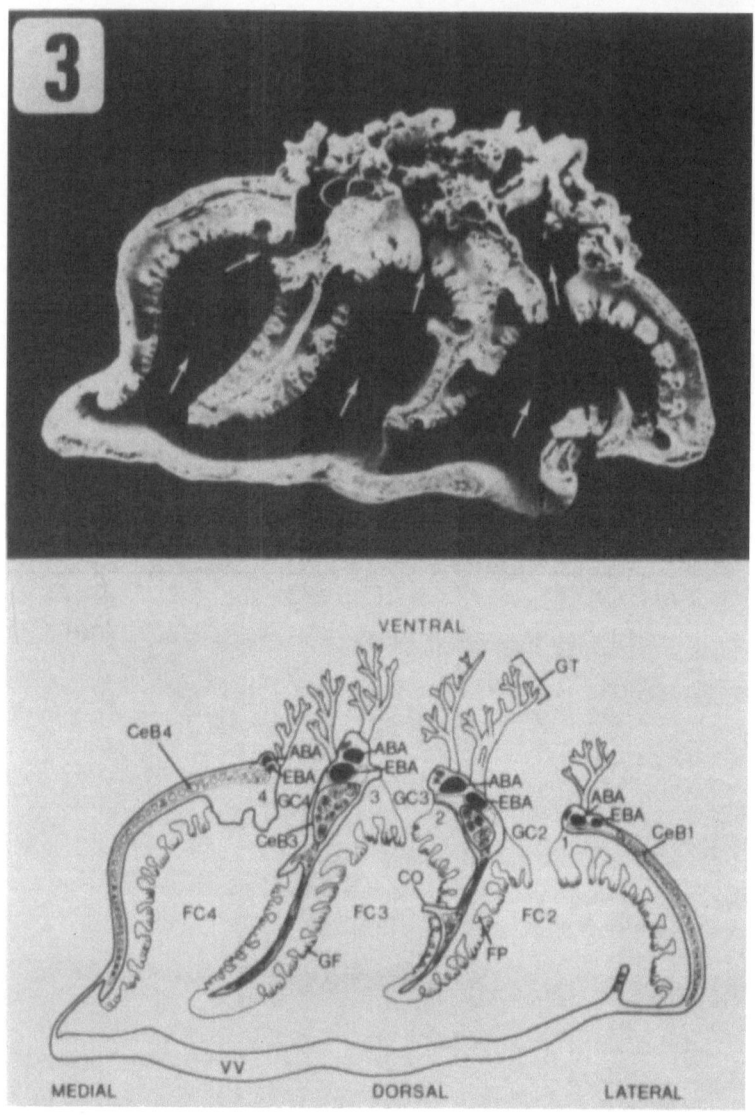

Fig. 3. Transverse razor-section of branchial basket (upper panel) with accompanying diagram (lower panel). Path of water through filter chambers (FC 2–4) to gill clefts (GC 2–4) is shown with arrows. Note extensive cartilage of 1st and 4th gills supporting branchial basket and rod-like cartilages of gills 2 and 3.

ABA: afferent branchial artery;
CO: crescentic organ;
CeB: ceratobranchial cartilage;
EBA: efferent branchial artery;
FF: filter plate;
GF: gill filter
GT: gill tuft;
VV: ventral velum. ×35

side, each provided with numerous lamellae (Fig. 4). Ventilation is achieved by waving these gills gently in the water, or when more vigorous ventilation is required, snapping them back against the body. The ventilatory frequency and intensity are under chemical control, both increasing if the animal is exposed to hypoxic water (Figge, 1936).

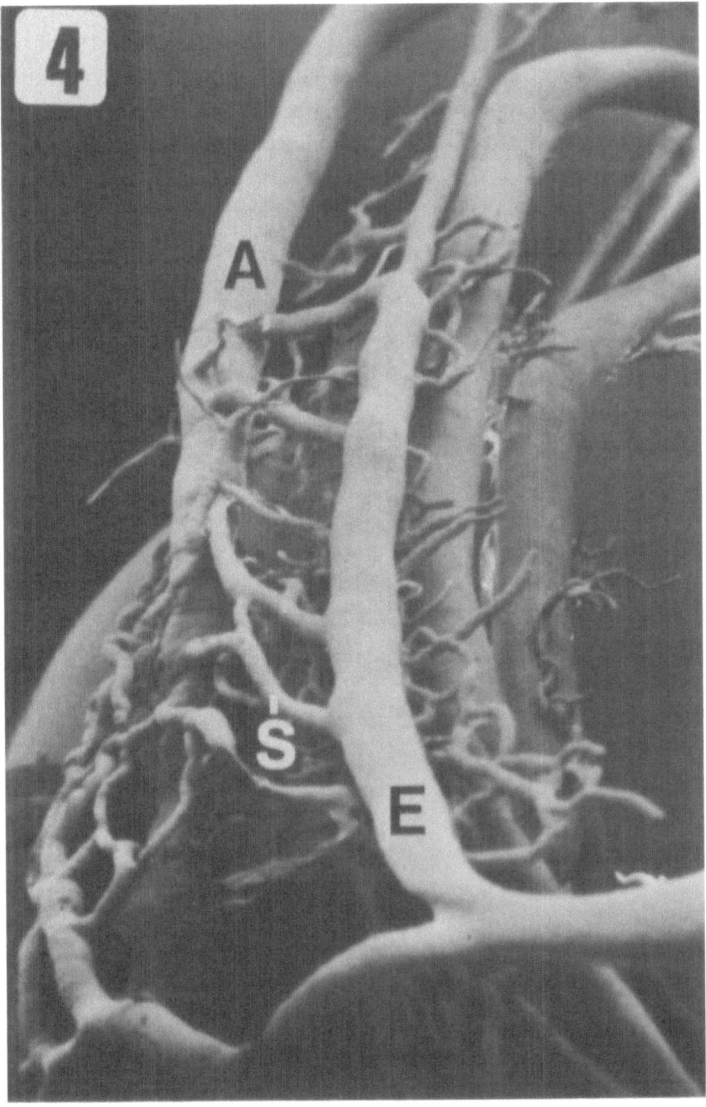

Fig. 4. L.P. view of methacrylate cast of arch vasculature in *A. mexicanum* gill. Note afferent (A) and efferent (E) arch arteries, and shunt vessels (S). ×12

Branchial vasculature

Litoria ewingi

Gill tufts. – The gills receive blood from the paired ventral aortae via the afferent branchial arteries (ABAs) and blood enters the dorsal aorta via the efferent branchial arteries (EBAs) and paired lateral aortae. The ventral aorta divides on each side to supply the first, second and third ABAs, the third ABA dividing again to supply the fourth ABA, which is much smaller in diameter (Fig. 5).

Numerous subdivisions of the ABAs on each arch supply blood to the exchange units, the capillary loops (Figs. 6, 7). These loops are projected ventrally into the ventilatory current and typically have a thin tissue barrier separating blood and water (Figs. 8, 9, 10). Blood leaving the capillary loops is collected by a system of efferent vessels which join the four efferent branchial arteries (EBAs) leading to the paired lateral dorsal aortae (Fig. 6). The approximate diameters are: ABA and EBA, $100\,\mu$m; afferent and efferent tuft vessels, $40\,\mu$m; capillary loops, $12\,\mu$m.

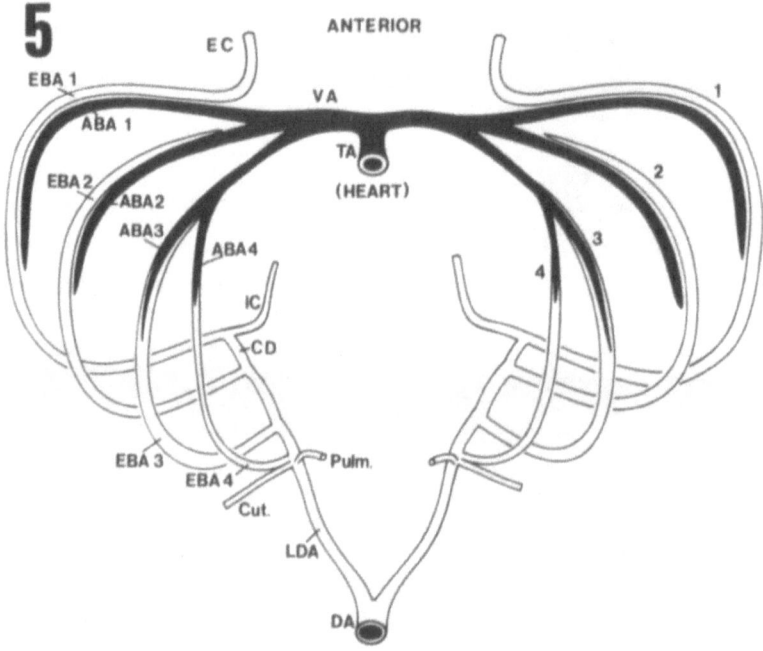

Fig. 5 Diagram of arterial arches in *L. ewingi* tadpole based on Microfil preparations. Blood is distributed from truncus ateriosus (TA) to paired ventral aortae (VA) from which arise four afferent branchial arteries (ABA 1–4) on each side. After passing through gill tufts or shunts (not shown) blood collects into four efferent branchial arteries (EBA 1–4) which give rise anteriorly to the internal carotid arteries (IC) and posteriorly to paired lateral dorsal aortae (LDA). The LDAs fuse to form the dorsal aorta (DA). Note persistent carotid duct (CD) and origin of pulmonary (Pulm) and cutaneous (Cut) arteries. Note also charactersitic relationship of ABA with EBA in each arch.

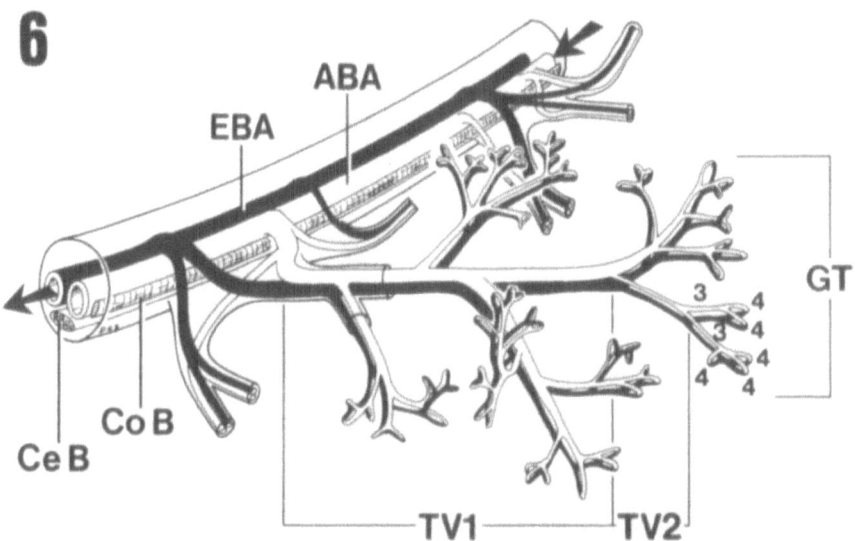

Fig. 6. Diagram of gill tuft showing pattern of connections of afferent (ABA) and efferent (EBA) branchial arteries with primary (TV1) and secondary (TV2) tuft vessels. Higher order branches numbered 3, 4.

CeB: ceratobranchial cartilage;
CoB: coracobranchials muscle;
GT: gill tuft.

It is worth noting in passing that blood flow in the distinctly tubular capillary loops should be well described by the Poiseuille relationship (Folkow and Neil, 1970); the sheet flow theory used to describe pressure-flow relationships in cat lungs (Sobin, Tremer and Fung, 1970) and lingcod gills (Farrell, 1980) is inappropriate here, as it is in regard to cane toad (*Bufo marinus*) lungs (Smith and Campbell, 1976).

Gill filters. – In discussing tadpole gills an interesting complication arises from the presence of an extensive filter apparatus in many species. In *L. ewingi*, the filter epithelium is thrown into a complex honeycomb of folds which is interrupted by a series of filter tracts (Figs. 11, 12). The blood supply is from the ventral aorta, i.e. blood of equivalent composition perfuses both gill tufts and filters. Although this implies a potential for gas exchange, any such potential is offset by the thickness of the filter epithelium and its covering with a substantial coat of mucus.

Ambystoma mexicanum

The description of Figge (1936) for *A. tigrinum* applies equally well to *A. mexicanum*; only detail revealed by vascular casting is highlighted here.

61

The six gill arches receive their afferent blood supply from six large ABAs which run the length of each arch to supply the 50 or so lamellae on each side (Figs. 4, 13). Blood drains from the lamellae via a reciprocal system of efferent vessels which eventually unite to form the dorsal aorta.

The exchange unit consists of a flat, plate-like structure, approximately termed a lamella, but which differs from a fish gill lamella in having two closely applied

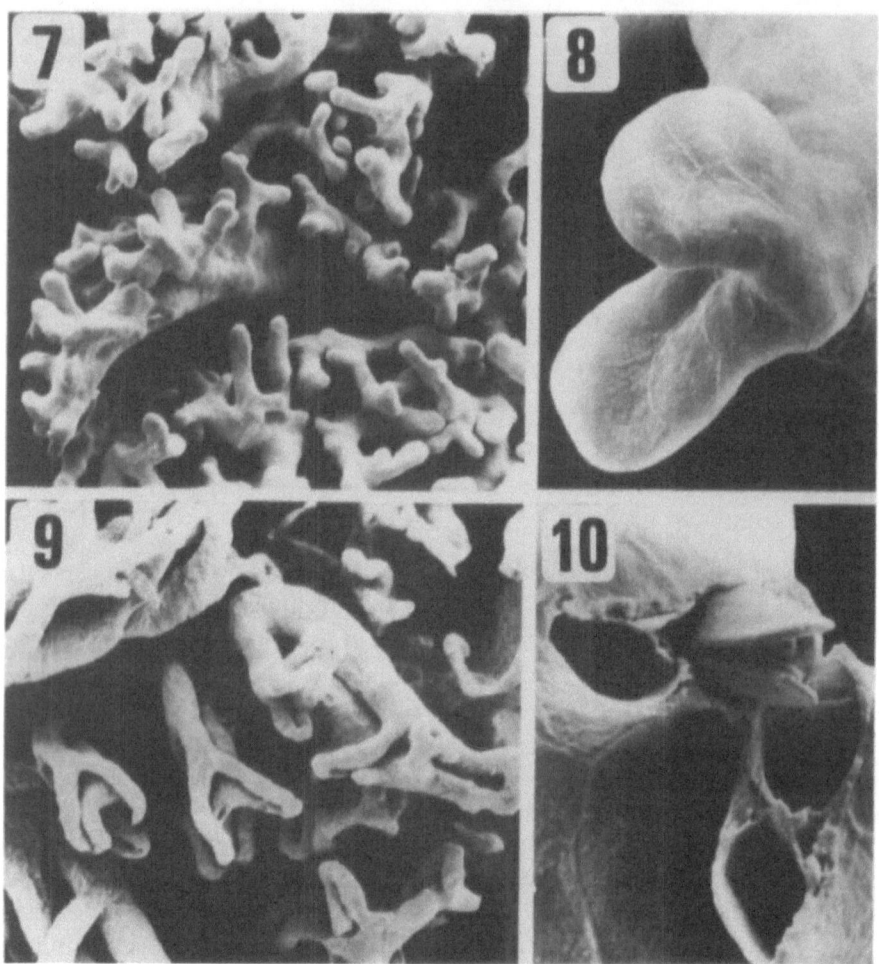

Fig. 7. Surface view of CPD gill tuft showing branching patterns of capillary loops. ×240.

Fig. 8. Detail of surface features of capillary loop. CPD preparation. ×1216

Fig. 9. Corrosion cast of gill tuft vasculature in region comparable with that shown in Fig. 7. ×384

Fig. 10. Fractured region of CPD specimen showing erythrocytes protruding from capillary loop. ×2160.

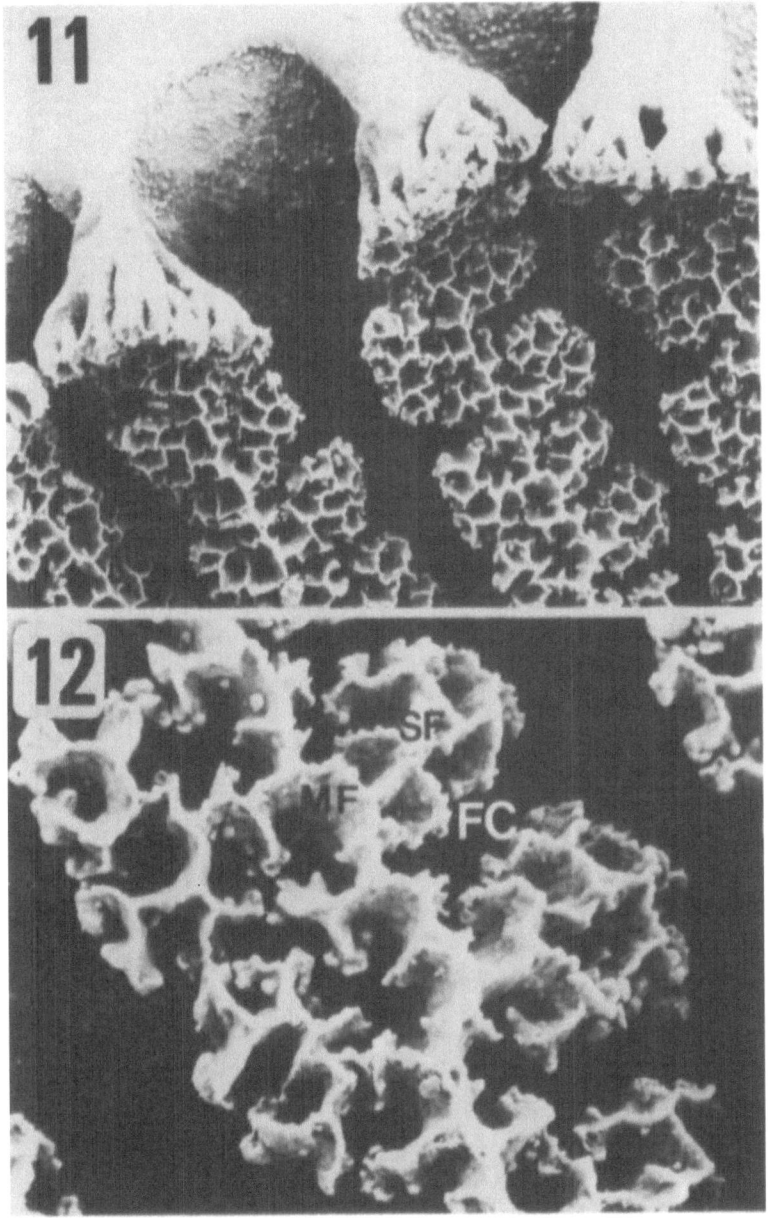

Fig. 11. Surface features of CPD filter apparatus showing ruffled epitherlium of filter fold rows separated by filter canals. ×150.

Fig. 12. SEM of CPD preparation showing filter folds at higher magnification. Note general orders of branching: main fold (MF). secondary fold (SF). tertiary fold (TF). SF and TF enclose filter niches (FN) and filter crevices (FC) separate neighboring filter folds. ×416.

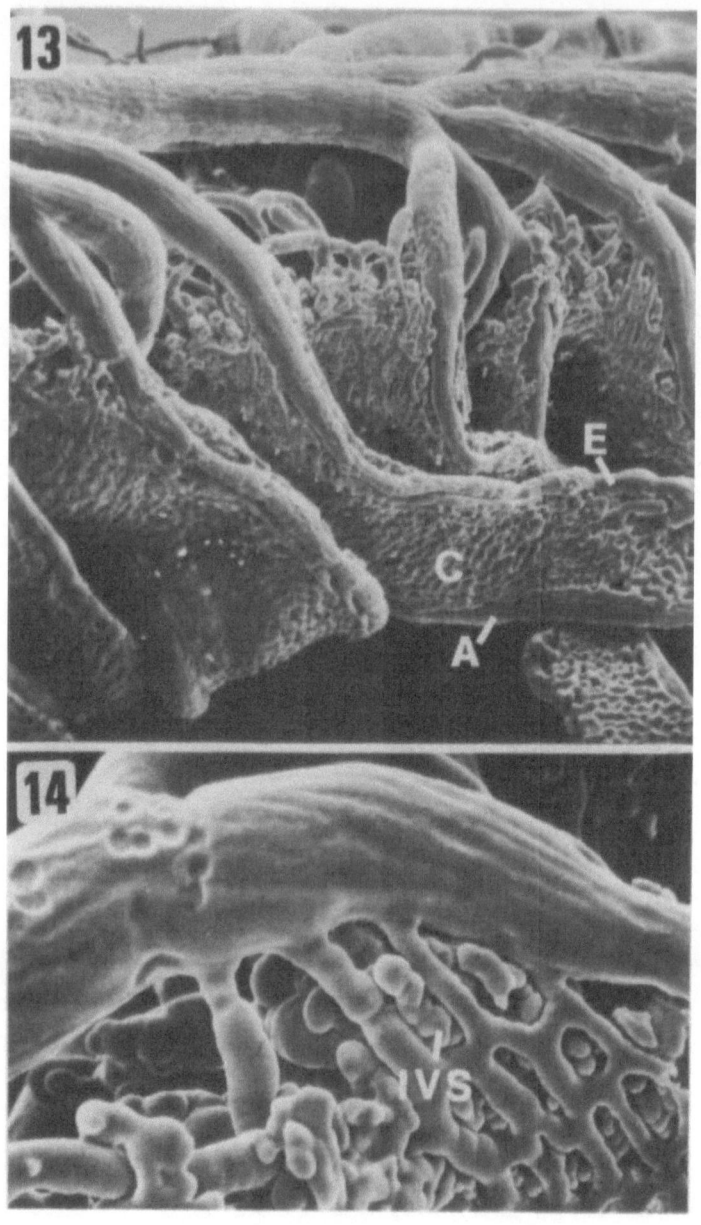

Fig. 13. Detail of cast of gill lamella in *A. mexicanum* showing afferent (A) and efferent (E) lamellar arteries, and double-layer of respiratory capillaries (C). ×715.

Fig. 14. Close-up of surface of lamella sheet showing intralamella vascular space (VS) visible between meshes of capillaries. ×4550

64

sheets of capillary-sized vessels (Figs. 13, 14). On anatomical grounds it would be surprising if the sheet flow theory referred to above did not apply in this vascular bed. However, exchange of gases is limited in the sense that only one face of each sheet is exposed to the water flow.

As noted by Figge (1936) the direction of blood flow is essentially anterior in each lamella; if the outer or exposed edge of the lamella is regarded as the 'leading' edge when the arch is abducted, then blood and water has a contraflow relationship during abduction, but flows in the same direction during adduction.

There is a conspicuous interstitial space within each lamella; this space is readily filled with methacrylate or Microfil (Canton Biomedical, Boulder Colorado) and is apparently derived from the large afferent or efferent lamellar vessels (Fig. 14).

No vessels draining this central vascular space have yet been observed. Thus, the space may be functionally equivalent to the corpus cavernosum of elasmobranch gills, in which arterial pressure is seconded to provide filamental, or in this case lamellar, rigidity.

Gill shunts

Respiratory by-passes or shunts are well demonstrated in larval amphibians and in air-breathing fishes (Johansen, 1970; Satchell, 1976). Despite the almost legendary difficulty of ascribing various physiological 'shunt-like effects' to anatomically definable pulmonary or respiratory by-pass vessels (McIlroy, 1965), shunts have been clearly demonstrated in the lungs of *Xenopus laevis* (Smith and Rapson, 1977) and in the gills of *A. tigrinum* (Figge, 1936) and *R. esculenta* (De Saint Aubain, 1981).

Our studies have extended these observations to include branchial shunts in *L. ewingi* tadpoles and in *A. mexicanum* (Figs. 4, 15). The location of these shunts is essentially similar in both cases; short, fairly substantial connections are formed between afferent and efferent branchial arteries, thereby forming low-resistance paths which, if patent, enable blood to bypass the branchial exchange units.

The carotid labyrinth of *L. ewingi* is located in such a way as to form a 'shunt' between the ABA and EBA of the first arch (Fig. 16); its functional importance as a shunt will almost certainly be slight and, in any event, will only apply to exchange units on arch I.

The importance of gills to total gas exchange

The case in favor of a respiratory role for axolotl gills is compelling. The detailed study of Figge (1936) has shown that both ventilation and perfusion of the gills are under presumably autonomic control and are sensitive to O_2 and CO_2 tensions in

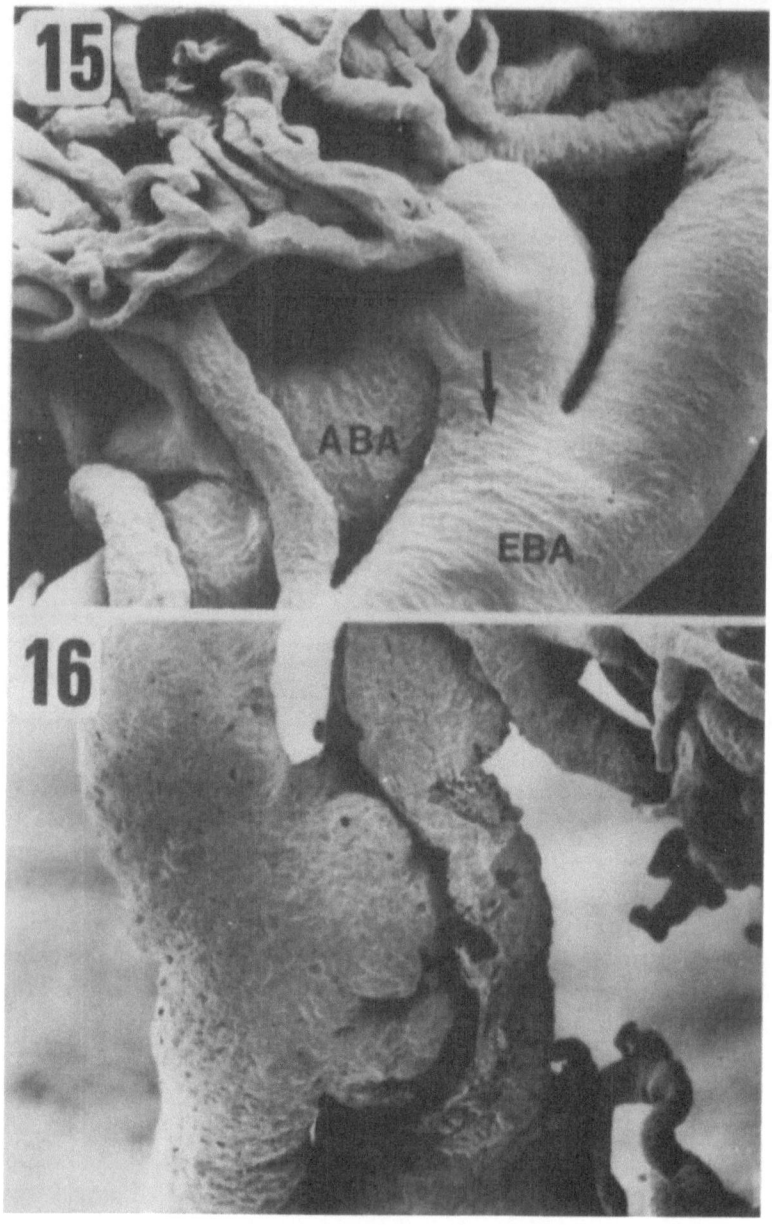

Fig. 15. Vascular cast showing direct connection (arrow) forming shunt between afferent (ABA) and efferent (EBA) branchial arteries. ×280

Fig. 16. Vascular cast of carotid labyrinth in stage 41 tadpoles. ×385

the ambient water. In passing across the lamella, blood undergoes a pronounced color shift consistent with increasing oxygenation. Further detail could be added by application of modern methods of micro-gasometry and micro-pressure recording, but it must be acknowledged that Figge capitalised on the favorable geometry of these lamellae with direct observations of flow patterns through the compound microscope.

Most attempts to determine the relative importance of different gas exchange sites in tadpoles have been confined to determinations of the capillary density of these respiratory surfaces (Strawinsky, 1956; Czopek, 1957, 1965; Gradwell, 1969). This approach is inadequate because it ignores many of the factors which govern the extent of gas exchange at any particular site. These factors include the O_2 tension difference across the exchange surface, the adequacy of ventilation, and the thickness of the blood-water barrier. Many of these inadequacies have been overcome in a recent study in which oxygen uptake across gills, skin and lung was measured directly in larval *R. catesbeiana* (Burggren and West, 1982).

We argue that the role of the gill filters in gas exchange is severely limited by the thickness of the filter epithelium. This conclusion seems reasonable for *L. ewingi* at stage 41 but it must be emphasised that such may not be the case at earlier stages of development, nor may it apply in other species. In *X. laevis*, there are no gill tufts but an extensive filter apparatus which may have respiratory importance (Millard, 1945).

An interesting controversy has recently been revitalised concerning the contribution of the skin to total gas exchange. Earlier studies, based on the microscopic analysis of capillary density, were equivocal; Strawinsky (1956), Czopek (1957) and Gradwell (1969) all favored the skin as the primary site of O_2 uptake and this finding has been supported by the physiological study of Burggren and West (1982) which showed that the skin of *Rana catesbeiana* tadpoles accounted for about 2/3 of total O_2 uptake. This value may be an over-estimate of the amount of O_2 taken up by the skin and subsequently available for use by non-cutaneous tissues; Nonnote (1981) has shown that, in ten species of teleost fish, including the eel, *Anguilla anguilla*, O_2 utilisation by the skin equals or exceeds the amount of O_2 entering the skin from the surrounding water. If this effect is shown to apply in bullfrog tadpoles, then the relative importance of their skin in total gas exchange will be correspondingly reduced.

De Saint-Aubain (1981) noted that, compared to the skins of some adult anuran or urodele larvae, the skin of *Rana temporaria* and *Bufo bufo* tadpoles was only sparsely vascularised. We have made similar, qualitative observations for *L. ewingi,* and conclude that relatively poor cutaneous ventilation, and the likely perfusion of the skin with blood already arterialised in the gills, will mitigate against the skin as an exchanger.

Many factors may interact to produce the great variability in gill and lung development in tadpoles, including the species, the temperature at which larvae are reared, the habitat and the developmental stage under investigation. As an example, Burggren and Mwalukoma (1983) have clearly demonstrated increased cutaneous vascularisation and cutaneous epithelial thinning in bullfrog tadpoles reared in hypoxic water: rearing in hyperoxic water produces the opposite effects.

Acknowledgements

This study was funded, in part, by a grant to DGS from the Australian Research Grants Scheme.

References

Burggren, W.W. and West, N.H. (1982). Changing respiratory importance of gills, lungs and skin during metamorphosis in the bullfrog *Rana catesbeiana*. *Respir. Physiol.* 47; 151–164.

Burggren, W.W. and Mwalukoma, A. (1983). Influence of chronic hypoxia and hyperoxia on gas exchange processes in larval and adult bullfrogs (*Rana catesbeiana*). I. Morphological responses of lungs, skin and gills. *J. Exp. Biol.* 105: 191–203.

Czopek, J. (1957). The vascularisation of respiratory surfaces in *Ambystoma mexicanum* (Cope) in ontogeny. *Zool. Polon.* 8: 131–149.

Czopek, J. (1965). Quantitative studies on the morphology of the respiratory surfaces in amphibians. *Acta. Anat.* 62: 296–323.

De Saint-Aubain, M.L. (1981). Shunts in the gill filament in tadpoles of *Rana temporaria* and *Bufo bufo* (Amphibia, Anura). *J. Exp. Zool.* 217: 143–145.

Farrell, A.P. (1980). Gill morphometrics, vessel dimensions, and vascular resistance in ling cod, *Ophiodon elongatus. Can. J. Zool.* 58: 807–818.

Figge, F.H.J. (1936). The differential reaction of the blood vessels of a branchial arch of *Ambystoma tigrinum* (Colorado Axolotl). I. The reaction to adrenaline, oxygen and CO_2. *Physiol. Zool.* 9: 79–101.

Folkow, B. and Neil, E. (1970). Circulation. London: Oxford U.P.

Gosner, K.L. (1960). A simplified table for staging anuran embryos and larvae with notes on identification. *Herpetologica* 16: 183–190.

Gradwell, N. (1969). The respiratory importance of the tadpole operculum in *Rana catesbeiana. Can. J. Zool.* 47: 1239–1243.

Gradwell, N. (1972a). Gill irrigation in *Rana catesbeiana*. Part I. On the anatomical basis. *Can. J. Zool.* 50: 481–499.

Gradwell, N. (1972b). Gill irrigation in *Rana catesbeiana*. Part II. On the musculoskeletal mechanism. *Can. J. Zool.* 50: 481–499.

Gradwell, N. (1975). Experiments on oral suction and gill breathing in five species of Australian tadpoles (Anura: Hylidae and Leptodactylidae). *J. Zool.* 177: 81–98.

Hutchison, V.H., Whitford, W.G. and Kohl, M. (1968). Relation of body size and surface area to gas exchange in anurans. *Physiol. Zool.* 41: 65–85.

Johansen K. (1970). Air breathing in fishes. In: Fish Physiology. W.S. Hoar and D.J. Randall, eds., Academic Press, N.Y., London, pp. 361–411.

Kenny, J.S. (1969). Feeding mechanisms in anuran larvae. *J. Zool.* 157: 225–246.

Marshall, A.M. (1893). The development of the frog. In: Vertebrate Embryology. Smith, Elder and Co., London, pp. 160–185.

Marshall, A.M. (1932). The frog: an introduction to anatomy, histology and embryology. Macmillan and Co., Ltd., London, pp. 131–140.

McIlroy, M.B. (1965). Pulmonary shunts. In: Handbook of Physiology, Section 3, Respiration, Volume II. W.O. Fenn and H. Rahn, eds., Amer. Physiol. Soc., Washington D.C., pp. 1519–1524.

Millard, N. (1945). The development of the arterial system of *Xenopus laevis*, including experiments on the destruction of the larval arches. *Trans. R. Soc. S. Afr.* 30: 217–234.

Nonnotte, G. (1981). Cutaneous respiration in six freshwater teleosts. *Comp. Biochem. Physiol.* 70A: 541–544.

Satchell, G.H. (1976). The circulatory system of air breathing fish. In: Respiration in Amphibious Vertebrates. G.M. Hughes, ed., Academic Press, London, pp. 105–123.

Schulze, F.E. (1889). Uber die inneren Kiemen der Batrachierlarven. I. Mittheilung. Uber das epithel der Lippen, der Mundrachen und Kiemenhohle erwachsener Larven von *Pelobates fuscus*. *Abh. Akad. Wiss. Berl.* 1: 58.

Smith, D.G. and Campbell, G.D. (1976). Anatomy of the pulmonary vascular bed in the toad *Bufo marinus. Cell Tiss. Res.* 165: 199–213.

Smith, D.G. and Rapson, L. (1977). Differences in pulmonary microvascular anatomy between *Bufo marinus* and *Xenopus laevis. Cell Tissue Res.* 178: 1–15.

Sobin, S.S., Tremer, H.M. and Fung, Y.C. (1970). Morphometric basis of the sheet flow concept of the pulmonary alveolar microcirculation in the cat. *Circ. Res.* 26: 397–414.

Strawinski, S. (1956). Vascularisation of respiratory surfaces in ontogeny of the edible frog, *Rana catesbeiana. Zool. Pol.* 7: 327–365.

Wassersug, R. (1980). Internal oral features of larvae from eight Anuran families: functional, systematic, evolutionary and ecological considerations. Univ. of Kansas Museum of Nat. Hist., Misc. Publ. No. 68.

Wassersug, R.J., Paul, R.D. and Feder, M.E. (1981). Cardiorespiratory synchrony in anuran larvae (*Xenopus laevis, Pachymedusa dacnicolor* and *Rana berlandieri*). *Comp. Biochem. Physiol.* 70A: 329–334.

5. Consequences of aerial respiration for amphibian larvae

Martin E. Feder

Department of Anatomy and Committee on Evolutionary Biology, The University of Chicago, 1025 East 57th Street, Chicago, Illinois 60637, U.S.A.

Abstract

Aerial respiration is a common feature of many amphibian larvae, especially anuran larvae. The widespread occurrence of air breathing in amphibian larvae raises the question of what selective advantages should have led to its evolution in so many diverse forms. Laboratory measurements suggest that aerial respiration is an important response to aquatic hypoxia. Aquatic respiration is primarily cutaneous, larvae lose O_2 to hypoxic water, and air breathing replaces lost O_2. Moreover, air breathing larvae undergo little anaerobiosis in aquatic hypoxia and tolerate low P_{O_2} well. However, resistance to hypoxia is not a sufficient explanation of aerial respiration in larvae for several reasons: larvae breathe air in normoxic water even though aquatic gas exchange is adequate to meet O_2 needs and some lungless larvae tolerate hypoxia well. Facilitation of locomotor activity is also not a sufficient explanation. Although air breathing improves locomotor stamina in some species, its effects on buoyancy interfere with swimming in others. Also, larvae may seldom sustain locomotion for long enough for air breathing to be of benefit. Aerial respiration may allow high ingestion rates by compensating for decreases in buccopharyngeal respiration necessitated by suspension feeding. Alternatively, aerial respiration in larvae may simply reflect precocious development of the lungs.

The diversity of amphibian larvae

Amphibians of many species begin life in an aquatic habitat and eventually assume a terrestrial or semi-terrestrial existence. This transition in habitat and habits necessitates a major reorganization of the respiratory physiology and morphology of amphibians (Burggren, present volume). A surprising aspect of metamorphosis is that the respiratory changes substantially precede the shift in habitat and habits. Larvae of many species of amphibians, for example, begin to ventilate their lungs at the earliest free-swimming stages (see below), while major alterations in the locomotor apparatus, feeding structures, external morphology,

Seymour, R.S., (ed.) Respiration and metabolism of embryonic vertebrates.
© *1984, Dr W. Junk Publishers, Dordrecht/Boston/London. ISBN-13:978-94-009-6538-6*

and habitat utilization lag far behind. However, the precedence of the respiratory transition is seldom recognized, even by many workers in the field and by most textbooks of biology, herpetology, and comparative respiratory physiology (e.g. Goins and Goins, 1971; Porter, 1972; Dejours, 1981; Feder, 1981). Here I survey the ontogeny of aerial respiration in amphibians, describe the partitioning of O_2 uptake between air and water in anuran larvae, and consider what selective factors may account for the occurrence of aerial respiration early in the ontogeny of so many amphibians.

Characterization of the respiratory patterns of amphibian larvae is problematic because the larvae are so diverse. Amphibian larvae that are most familiar to biologists superficially resemble fishes; they are aquatic and have large tails with high fins. Most such larvae begin life as aquatic eggs and initially undergo aquatic respiration alone, first by diffusion through the jelly coat and later via the gills and skin (Burggren, present volume). The aquatic respiratory structures are diverse (Medvedev, 1937; Strawinski, 1956; Gradwell, 1969; Saint-Aubain, 1982; Burggren and Mwalukoma, 1983; McIndoe and Smith, present volume). Ventilation of the buccopharynx (hereafter termed 'gill ventilation') simultaneously ventilates the buccopharyngeal walls, operculum, gill filters, gill filaments, and skin in the path of the effluent, each of which may be specialized as a gas exchanger. However, larval respiration does not remain exclusively aquatic; larvae of most species develop lungs (Noble, 1931). Although the larval lungs are less elaborate than in the adult (Atkinson and Just, 1975; Burggren and Mwulakoma, 1983), larvae clearly ventilate them in every species that has been studied (see Fig. 1 and below). Ontogenetic observations of aerial respiration are relatively few for anuran larvae, but document the initiation of air breathing well before metamorphic climax. The species and earliest reported occurrences are (Gosner (1960) developmental stages in Arabic numerals and Taylor and Kollros (1946) stages in Roman numerals): *Hymenochirus boettgeri* (25, I); *Pseudacris triseriata* (37, XII), *Rana berlandieri* (25, I), *Rana catesbeiana* (25–39, I–XIV), *Rana cyanophylictis* (30, V), *Rana pipiens* (29, IV), *Scaphiopus bombifrons* (35, X), and *Xenopus laevis* (25, I) (Wassersug and Seibert, 1975; Marian et al., 1980; Burggren et al., 1983; Feder, 1983b; Feder and Wassersug, 1984; Feder, unpublished data). For reference, the earliest free-swimming stage (with no limb buds) is 25 (I), the forelimb emerges at stage 42 (XX), and metamorphosis is complete at stage 46 (XXX). Data for *Pseudacris, Rana pipiens, R. cyanophylictis*, and *Scaphiopus* represent the earliest stages examined rather than the initiation of aerial respiration in the middle of development. Thus, anurans may undergo bimodal gas exchange for a substantial portion of the larval period, if not during all free-swimming stages. Aerial respiration in salamanders also begins during the larval period (Whitford and Sherman, 1968; Branch and Taylor, 1977).

The larvae described above, although familiar to biologists, are by no means typical of amphibians (Noble, 1931). Departures occur from this patterns towards both decreased and increased reliance upon aerial gas exchange. Ranid tadpoles cannot breathe air for months when overwintering beneath ice (Bradford, 1983). Some anuran larvae develop lungs only at the completion of metamorphosis, and thus always rely solely upon aquatic respiration while larvae. These include bufonids and some inhabitants of torrential streams (e.g. *Ascaphus*) (Wassersug and Seibert, 1975). Aquatic larvae of some salamanders (e.g. some plethodontids, *Rhyacotriton*) also lack lungs. Several levels of increased reliance upon aerial gas exchange are evident. Many larvae simultaneously use gills and lungs (see below). Some pipid and hylid tadpoles, however, have lost portions of the branchial apparatus (Noble, 1931, Feder and Wassersug, 1983). Other tadpoles actually live out of water, but in damp microhabitats immediately adjacent to streams (Wassersug and Heyer, 1983). Some tadpoles occur out of water in foam nests or on the backs of terrestrial parents (Noble, 1931). In the extreme case, some terrestrial salamanders (Plethodontidae) and anurans (e.g., *Eleuthrodactylus*) lay terrestrial eggs that develop to the adult form without free-living larvae (Noble, 1931).

This variety of life histories and associated respiratory characteristics has several important implications. The partitioning of gas exchange in amphibious vertebrates is known mainly from experimentation with adult fishes, amphibians and reptiles (Hughes, 1976; Randall et al., 1981). Larval amphibians are a test case of the generality of conclusions based upon adults (Feder, 1983b). Second, amphibians are not necessarily restricted to aquatic environments for reproduction or respiration. Third, larval amphibians present a unique opportunity for respiratory experimentation; one can choose species with almost any desired set of respiratory organs. Last, experimental protocols in which larvae are prevented from breathing either water or air may yield misleading results (Feder, 1981).

Respiratory responses of anuran larvae in normoxic and hypoxic water

Work in my laboratory has considered how anuran larvae partition gas exchange between water and air, and how their patterns of gas exchange differ from those of other aquatic vertebrates and adult amphibians. We have used the diversity of gas exchangers in tadpoles to examine the functional significance of each kind of gas exchange organ: How does a species with lungs respond to hypoxia? What are the consequences of lack of lungs or gills for a tadpole? To answer such questions, we examined three diverse types of larvae. Larvae of ranids (*Rana berlandieri, R. catesbeiana, R. pipiens*) are relatively large (up to 50 g wet mass), have both lungs and gills, are not buoyant, and both filter suspended matter and graze on vegetation. Larvae of *Xenopus laevis*, the clawed frog, are smaller (<10 g wet mass), have lungs and vascularized gill filters but lack gill filaments, are buoyant, generally hover midwater continually, and are obligate suspension feeders. Toad larvae (*Bufo americanus, Bufo woodhousei*) are small (<3 g wet mass), lack lungs until the completion of metamorphosis, can remain midwater only by vigorous swimming, and resemble ranid larvae in their feeding habits. The lunged larvae can be made functionally lungless by interposing a screen between them and the water's surface, thereby preventing air breathing.

In tadpoles with functional lungs, aerial respiration accounts for a minority of O_2 uptake, and can be supplanted entirely or in part by aquatic O_2 uptake. Aerial respiration typically accounts for 10–20% of total O_2 consumption in normoxic tadpoles (Burggren and West, 1982; West and Burggren, 1983; Burggren et al., 1983; Feder, 1983a, b; Feder and Wassersug, 1984). However, if use of the lungs is prevented, normoxic *Rana* and *Xenopus* larvae can meet routine respiratory requirements through aquatic O_2 uptake and avoid extensive anaerobiosis; lungless tadpoles (e.g. *Bufo*) do so continually (Feder, 1983a, 1983b; Feder and Wassersug, 1984). In fact, some larvae will voluntarily breathe air less frequently if the opercular spout is cannulated or when in deep water, where swimming to the surface involves considerable effort (West and Burggren, 1982; Burggren et al., 1983; Feder and Moran, unpublished). Alternatively, some larvae increase rates of air breathing when active or disturbed (Feder, 1981; Wassersug and Feder, 1984). Aquatic gas exchange predominates in normoxic water. The skin accounts for most aquatic O_2 uptake, and is the most important gas exchanger in normoxia (Burggren and West, 1982; Burggren et al., 1983). Nonetheless, the gills and buccopharynx are ventilated continuously, sometimes at quite high rates (Burggren and West, 1982; West and Burggren, 1982; Bradford, 1983; Feder, 1983b; Feder and Wassersug, 1984). This is surprising given the branchial contribution to overall gas exchange and the failure of experimental increases in the ventilatory flow to increase branchial O_2 uptake (Burggren and West, 1982; Burggren et al., 1983). These paradoxical aspects of gill ventilation may be due to the role of the gills in food particle entrapment and osmoregulation; use of the gills in these functions may preclude a close coupling of gill ventilation with respiratory demands (Feder et al., 1984). For example, anuran larvae first increase and then decrease gill ventilation rates in increasing concentrations of suspended food particles (Seale et al., 1982). In dense suspensions, branchial ventilation nearly ceases.

Exposure to hypoxia initiates several responses. Tadpoles of *Rana berlandieri* become more active (which may lead them from hypoxic habitats), and tadpoles of *Rana temporaria* specifically seek out oxygenated water (Savage, 1935; Costa, 1967; Feder, 1983b). However, tadpoles may forgo these responses and sometimes occur in hypoxic microhabitats even when less hypoxic water is nearby (Noland and Ultsch, 1981). Larvae of many species increase pulmonary ventilation in proportion to the degree of hypoxia and breathe air several times per minute in severe hypoxia (Fig. 1; Wassersug and Seibert, 1975; West and Burggren, 1982; Feder, 1983b; Feder and Wassersug, 1984). Gill ventilation, buccal stroke volume, and heart rate may increase in moderate hypoxia, but all of these responses may not occur in a given species (Wassersug et al., 1981; West and Burggren, 1982; Feder, 1983a,b; Feder and Wassersug, 1984). Lung and gill

74

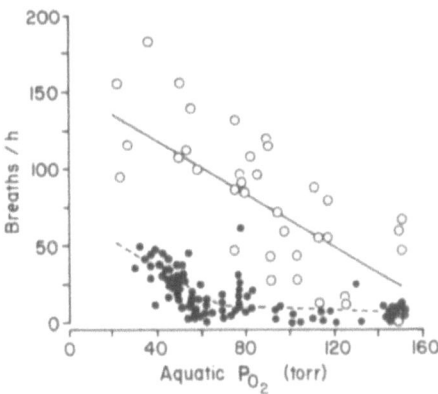

Fig. 1. Effect of aquatic P_{O_2} on lung ventilation rate in anuran larvae. *Rana* larvae increase rates of air breathing in hypoxic water, both when active (solid line and open circles) and when resting (broken line and closed circles) (Feder, 1983b; Wassersug and Feder, 1983). Active larvae were swimming at 9.2 cm · sec^{-1} for more than 15 min. Note that most larvae breathe air even in normoxic water.

ventilation may reflexly interact (West and Burggren, 1982, 1983). Anaerobic metabolism (i.e., lactate accumulation) is apparently a response of last resort. Tadpoles of *Rana berlandieri* and *Xenopus laevis* do not accumulate lactate in hypoxic water (P_{O_2}<20 torr) unless they are prevented from breathing air or otherwise stressed (Feder, 1983b; Feder and Wassersug, 1984). Other anaerobic pathways have not been examined in tadpoles, but are seldom utilized in adult amphibians (Bennett, 1978). Chronic exposure to hypoxia increases respiratory surface area and decreases respiratory diffusion barriers in *Rana* larvae (Burggren and Mwalukoma, 1983).

Many of these responses to hypoxia (e.g. air breathing, movement to oxygenated microhabitats) surely are effective. However, others (e.g. increased gill ventilation) may be either inefficient, ineffective, or actually deleterious (see below). As the aquatic P_{O_2} decreases, the P_{O_2} gradient across the aquatic gas exchangers tends to reduce aquatic O_2 uptake or promote loss of O_2 to the water; increased gill ventilation may consequently increase O_2 loss (West and Burggren, 1982; Feder, 1983b; Feder and Wassersug, 1984). Air breathing *Rana* tadpoles minimize these problems by curtailing gill ventilation at low P_{O_2} (West and Burggren, 1982; Feder, 1983b). However, air breathing *Xenopus* tadpoles maintain similar gill ventilatory rates regardless of P_{O_2}, and *Bufo* tadpoles increase gill ventilation as the aquatic P_{O_2} declines (Feder, 1983a; Feder and Wassersug, 1984).

Decreases in gill ventilation such as in *Rana* larvae cannot totally prevent O_2 loss to hypoxic water because the skin, a major gas exchanger, is always exposed. Potential means of reducing cutaneous O_2 loss to hypoxic water include the shunting of blood from the skin and increasing hemoglobin O_2 affinity. Adult amphibians may shunt blood to and from the skin (Moalli et al., 1980). However,

75

documentation of shunting in anuran larvae is anecdotal (Burggren et al., 1983). Tadpole hemoglobins have a high affinity for O_2 (McCutcheon, 1936; Pinder and Burggren, 1983). Accordingly, the O_2 loss to the water may represent a slight unloading of O_2 bound to hemoglobin in the lungs, with the vast majority of O_2 transported to other tissues (Feder and Wassersug, 1984).

Metabolic measurements suggest that lungs are crucial to tadpoles in aquatic hypoxia (Burggren and West, 1982; West and Burggren, 1982; Feder, 1983b; Feder and Wassersug, 1983). At high P_{O_2}, aquatic O_2 consumption accounts for 60–80% of total O_2 consumption in *Rana* and *Xenopus* larvae. As the P_{O_2} declines, aquatic O_2 uptake drops steadily and larvae lose O_2 to the water at low P_{O_2}. By contrast, aerial O_2 uptake increases steadily, and comes to account for >100% of total O_2 consumption at low P_{O_2}.

What, then, are the consequences of the absence of pulmonary function, either in larvae that are prevented from using lungs (here termed 'exclusively aquatic' larvae) or in naturally lungless larvae such as *Bufo*? In *Rana* and *Xenopus* larvae, elimination of pulmonary function has few immediate consequences in normoxic water; total rates of O_2 consumption may decrease slightly or not at all (Feder, 1983b; Feder and Wassersug, 1983). Exclusively aquatic *Xenopus* survive several weeks in normoxic water. In hypoxia, total O_2 consumption declines more in exclusively aquatic larvae than in bimodal larvae, but exclusively aquatic larvae lose relatively little O_2 to the water (Feder, 1983b; Feder and Wassersug, 1983).

Responses to hypoxia change when larvae are exclusively aquatic (Feder, 1983b; Feder and Wassersug, 1983). Relative to bimodal larvae, exclusively aquatic larvae of *Rana* decrease gill ventilatory rates at high P_{O_2} and increase gill ventilation at low P_{O_2}. Bimodal *Xenopus* larvae show no relationship between gill ventilatory frequency and aquatic P_{O_2}, but exclusively aquatic tadpoles increase gill ventilation in moderate hypoxia. *Xenopus* larvae also undergo extensive anaerobiosis when exclusively aquatic, but not when bimodel (Feder and Wassersug, 1983). Exclusively aquatic *Rana* larvae do not accumulate lactate except at very low P_{O_2} (Feder, 1983b).

Bufo larvae, which lack lungs, are clearly superior to exclusively aquatic larvae of *Rana* and *Xenopus* in extraction of O_2 from the water, especially in hypoxia (Feder, 1983a). However, bimodal larvae are better able than *Bufo* to regulate O_2 consumption in hypoxia (Fig. 2). These different abilities do not translate directly into tolerance of hypoxia. All three species are tolerant of hypoxia when air is available. When exclusively aquatic, *Xenopus* are least resistant to hypoxia, but *Rana* larvae, which normally breathe air, tolerate hypoxia as well as *Bufo* larvae (Feder, 1983a). Of course, tolerance is an inadequate index of hypoxia's long term effects. The data on O_2 consumption suggest that hypoxia might well affect growth rates differently in larvae of the three species.

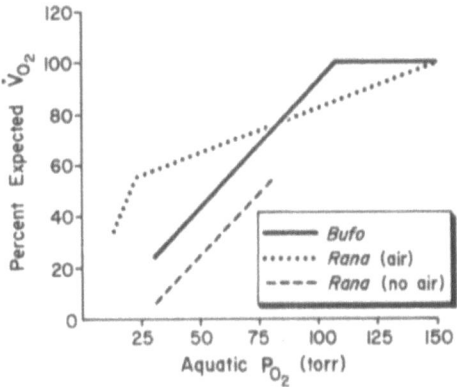

Fig. 2. Effect of aquatic hypoxia on metabolic rates of lunged and lungless anuran larvae. *Rana* larvae breathing air can maintain greater rates of O_2 consumption in hypoxic water than can lungless *Bufo* larvae. When prevented from breathing air, the O_2 consumption of *Rana* larvae decreases precipitously in hypoxia (Feder, 1983a, 1983b). Rates of O_2 consumption are plotted as a percentage of rates expected for resting larvae of similar size in normoxic water.

Functional consequences of air breathing in aquatic anuran larvae

Aerial respiration is commonly considered an adaptation in those aquatic vertebrates in which it occurs. A major but often unstated assumption underlies this consideration: that aerial respiration improves organismal performance and thereby improves Darwinian fitness. Seldom is this assumption tested explicitly. In part, this stems from the impracticality of measuring fitness (but see Arnold, 1983). However, organismal performance can be measured and one can falsify the proposition that aerial respiration improves organismal performance in a given situation. For example, recent applications of this approach to fishes and salamanders have shown that aerial respiration may have neutral or negative consequences for organismal performance (Kramer, 1983). Here I use measurements of organismal performance and metabolism to evaluate six general explanations of the adaptive significance of aerial respiration for aquatic anuran larvae.

(1) *The physical properties of water necessitate air breathing in larval anurans.* Water has a lower capacitance for O_2 than air, and O_2 diffuses less rapidly through water than air. Moreover, water is more dense and more viscous than air (Dejours, 1981). These differing properties of water and air suggest that water is intrinsically less suitable than air for gas exchange, which may have led to the evolution of aerial respiration (reviewed by Kramer, 1983). One observation in support of this scenario is that some aquatic vertebrates capable of bimodal respiration breathe primarily water at low temperatures (at which gas exchange requirements are relatively small) but breathe primarily air at high temperatures (at which gas exchange requirements are large and water contains less dissolved

O_2) (Burggren et al., 1983). We tested the generality of this argument by partitioning O_2 uptake among lungs, gills, and skin in larvae of two ranid species at different temperatures (Burggren et al., 1983). Surprisingly, these larvae showed the same proportions of pulmonary, branchial, and cutaneous O_2 consumption at all temperatures, even though the total O_2 consumption increased considerably. Many fishes and adult amphibians also do not conform to the expected pattern (Burggren et al., 1983).

If the physical properties of water necessitate air breathing, larvae denied air should have greatly reduced rates of O_2 consumption. Contrary to this prediction, O_2 consumption declines slightly in *Xenopus* larvae and either is unchanged or increases in *Rana* larvae when air breathing is prevented (Feder, 1983b; Feder and Wassersug, 1984). When active without air, both species can increase O_2 consumption. Lungless *Bufo* larvae have rates of O_2 consumption similar to those of other tadpoles (Feder, 1982). Thus, anuran larvae can consume O_2 at high rates without breathing air, but breathe air regularly nonetheless. These results argue against a 'physical necessity' for air breathing.

(2) *The energetic cost of breathing water is prohibitive; air is a cheap alternative.* Because of the physical properties of water, extracting O_2 from water may require considerable energy for ventilating the aquatic gas exchangers (Holeton, 1980). Thus, even though larvae are physically capable of extracting O_2 from the water, the cost of doing so may impel them to breathe air instead. Kramer (1983) has recently criticized this argument as it applies to fish: if energetic cost is paramount in dictating the partitioning of gas exchange between air and water, air breathing should be favored at all aquatic P_{O_2}. Instead, air breathing fish (and tadpoles) breathe air frequently only at low aquatic P_{O_2}. Air breathing may reduce the cost of ventilation, but imposes other costs: the energetic cost of swimming to the surface to breathe; time spent swimming to breathe, which cannot be allocated to other activities (feeding, reproduction); and increased risk of both aerial and aquatic predation. Kramer and his colleagues have substantiated each of these costs for fish. All are significant in tadpoles as well. *Rana pipiens* larvae in a deep container (where the transit cost of air breathing should be large) breathe air less frequently and grow less rapidly than conspecifics in a shallow portion of the same container at identical P_{O_2} (Feder and Moran, unpublished data). The 'deep' larvae prefer to be near the surface rather than to feed on the bottom. *Rana berlandieri* larvae are attacked by aquatic predators (turtles) more frequently when moving than when still (Feder, 1983c). Air breathing entails movement, and may increase attacks by predators (Feder, 1983c).

An 'energetic costs' argument for air breathing may be less applicable to anuran larvae than other forms of aquatic vertebrates. Because of the respiratory importance of the skin, which is not actively ventilated, the cost of pumping water may be unrelated to the cost of respiration. Most anuran larvae feed by continuously pumping water through the buccopharynx (Wassersug and Hoff, 1979). Thus, minimizing the cost of branchial ventilation by breathing air instead of water is not a viable option for tadpoles because it would eliminate feeding.

(3) *Aquatic hypoxia necessitates air breathing in anuran larvae.* The clear importance of air breathing in response to aquatic hypoxia strongly supports this explanation. Findings relevant to this conclusion are discussed at length above. In summary, lung ventilation is most frequent in hypoxia, where it compensates for reduced aquatic O_2 uptake, forestalls lactate accumulation, and offsets the loss of O_2 to the water.

This adaptive explanation has several problems, some of which stem from lack of data rather than outright contradictions. The first problem is whether exposure to aquatic hypoxia was sufficiently frequent to account for the evolution of aerial respiration in aquatic larvae. Surveys of O_2 levels in the microhabitats of larvae are few, and most seem intent on documenting environmental hypoxia rather than characterizing the average or range of microhabitats that tadpoles normally encounter (Savage, 1935; Crump, 1981; Noland and Ultsch, 1981; Bradford, 1983). Although tadpoles often live in shallow vegetation-choked bodies of standing water, which seem likely to be hypoxic (Noland and Ultsch, 1981; West and Burggren, 1982; Kramer, 1983), wind-driven mixing and photosynthesis may oxygenate such habitats. Documentation that many populations of tadpoles sometimes encounter aquatic hypoxia would greatly strengthen the case for hypoxia as a general explanation for air breathing. Microspatial analyses of aquatic P_{O_2} would be especially welcome; the bottom-dwelling habits of many tadpoles may expose them to hypoxic bottom sediments even in normoxic water.

A second problem concerns bufonid tadpoles. If aquatic hypoxia necessitates air breathing, bufonid tadpoles should be restricted to normoxic habitats. However, the only relevant study of bufonid microdistributions relative to aquatic P_{O_2} shows that some *Bufo* larvae occur in moderately to severely hypoxic water (Noland and Ultsch, 1981). Bufonid larvae typically occur in better oxygenated water than do ranid larvae, but obviously are not obligated to do so.

A third problem is evident in the many larvae that breathe air in normoxic water. Air breathing is clearly not an immediate response to hypoxia in this situation. However, normoxic larvae might well breathe air and thereby maintain an O_2 store if encounters with hypoxia are frequent (Randall et al., 1981).

(4) *Locomotor activity necessitates air breathing in anuran larvae.* Air breathing may promote sustained activity if it augments the capacity for aerobic metabolism, especially in hypoxia, and if it minimizes dependence upon anaerobiosis (Bennett, 1980). We examined the possible facilitation of swimming activity through aerial respiration by placing larvae (*Bufo*, *Rana*, and *Xenopus*) in an aquatic flow chamber with and without access to air (Wassersug and Feder, 1983). The aquatic P_{O_2} was varied to examine the combined effects of hypoxia and activity on respiratory behaviors. Our measurements yielded mixed results.

Air breathing increases in active *Rana* larvae, and clearly improves their stamina (Figs. 1, 3). This increase is proportional to the frequency of lung ventilation, the body size of the larvae, the aquatic P_{O_2}, and the interaction of

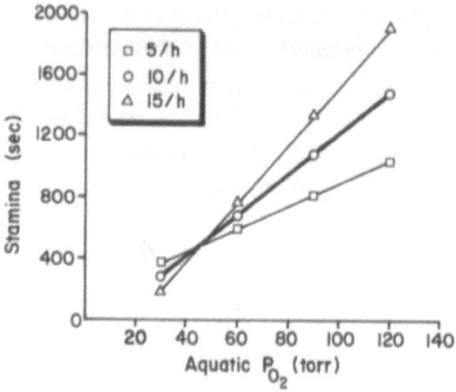

Fig. 3. Relation of lung ventilation rate to stamina in *Rana* larvae swimming at 11.6 cm · sec⁻¹. The results of a multiple regression are plotted for an 18 mg (dry mass) larva (Wassersug and Feder, 1983).

these factors. Individual *Rana* larvae vary in lung ventilation rates; those with high rates can swim for long periods, while those with low rates cannot. For *Bufo* larvae or exclusively water breathing *Rana* larvae, stamina is proportional to the P_{O_2}. Gill hyperventilation may be self-defeating. In *Xenopus* larvae, opening the mouth to ventilate the buccopharynx significantly increases the drag upon a swimming tadpole. Each time swimming *Xenopus* open their mouths, the current visibly displaces them downstream. Swimming larvae respond to this increased drag by curtailing gill ventilation, and soon exhaust (Wassersug and Feder, 1983).

Attributing air breathing in larvae to its facilitation of locomotor activity is problematic in several respects. In at least one species, air breathing directly hinders swimming ability. When tadpoles in a current surface to breathe air, they are broadside to the current instead of parallel to it. Such tadpoles typically cannot maintain trim and are displaced downstream. After inspiring air, tadpoles typically release a gas bubble as they descend. Tadpoles in a current often are either unable or unwilling to release this bubble, are buoyant, and are displaced by the current. These effects are especially pernicious in *Xenopus* larvae, which are buoyant under most conditions. The swimming ability of normoxic *Xenopus* larvae actually improves when they are prevented from breathing air. In hypoxic water, however, air breathing increases the stamina of *Xenopus* larvae (Wassersug and Feder, 1983).

It is difficult to conceive of how the effects of air breathing on locomotor ability could improve fitness. Air breathing should not affect burst activity, which is supported primarily by anaerobic metabolism in amphibians (Bennett, 1978). By contrast, air breathing should foster sustained activity and recovery from fatigue, which are supported in part by aerobic metabolism (Bennett, 1978; Feder and Olsen, 1978). For air breathing to benefit larvae through its effects on locomotor performance, sustained activity or rapid recovery from fatigue must also benefit anuran larvae.

Demonstrating the advantages of sustained activity is problematic. Largely from anecdotal data or incidental observations, most natural historians concur that tadpoles seldom undergo sustained swimming at high speeds (R.J. Wassersug, pers. comm.). A possible exception concerns encounters with predators. Do these encounters involve sustained persuits in which increased stamina due to air breathing might be essential for a tadpole's survival? Few field data exist. As a first attempt at an answer, we staged predatory encounters between tadpoles (*Rana berlandieri*) and turtles (*Chrysemys picta*) in a large enclosure (Feder, 1983c). The results do not support any selective advantage of sustained activity or air breathing. In 129 encounters between tadpoles and turtles, the mean pursuit lasted 11.4 sec, only one exceeded 30 sec, and fewer than 10% of the pursuits exceeded 20 sec. Stamina also had little bearing on the probability of tadpoles escaping turtles. Most escapes were due to sudden turns or bursts of speed. Furthermore, air breathing actually promoted attacks upon tadpoles by increasing their conspicuousness to turtles.

Air breathing might be important in tadpole habitats rich in predators, (e.g. Caldwell et al., 1980). Even though individual predatory encounters may be brief, frequent predatory encounters may necessitate repeated burst activity and rapid recovery. Substantiating this possibility would require simple field observations of tadpole activity, which are currently unavailable.

(5) *The regulation of feeding and buoyancy necessitates air breathing in anuran larvae.* The branchial apparatus of tadpoles functions in both feeding and gas exchange. These functions may conflict; a pattern of buccopharyngeal function that maximizes feeding may inhibit gas exchange. For example, food particles and mucus that accumulate on the gill filters during feeding may decrease their diffusive conductance, and the slowing of buccal pumping in dense suspensions of food decreases gill ventilation (Feder et al., 1984; Seale et al., 1982).

One way for a tadpole to resolve a functional conflict between feeding and branchial respiration is to allocate the respiratory function of the buccopharynx and gills to another organ, such as the lungs, thereby maximizing the trophic function of the gills. *Xenopus* larvae do exactly that (Feder et al., 1984). In dense suspensions of food particles, *Xenopus* larvae decrease gill ventilatory frequency and simultaneously increase the lung ventilatory frequency. The increased lung ventilation is not due to an increased metabolism associated with feeding, for the rate of O_2 consumption is not elevated in dense food suspensions. Thus, lungs may be important in allowing high rates of feeding in anuran larvae. We tested this conclusion by measuring feeding rates in larvae prevented from breathing air (Fig. 4). Bimodal and exclusively aquatic *Xenopus* larvae have similar ingestion rates at normoxic P_{O_2}. At hypoxic P_{O_2}, ingestion rates of exclusively aquatic larvae (but not bimodal larvae) decline dramatically, even though gill ventilation increases. Apparently, *Xenopus* larvae can use their gill filters for either feeding or respiration, depending on whether the lungs are available for gas exchange.

Although *Xenopus* is the only tadpole observed with regard to the interaction

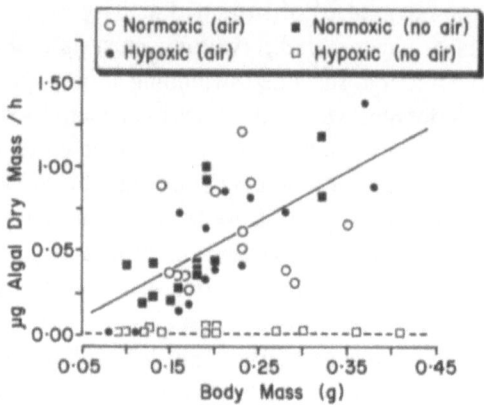

Fig. 4. Effects of hypoxia and lung ventilation on ingestion rates of *Xenopus* larvae (see Feder et al., 1984). *Xenopus* larvae ingest algae at high rates in most circumstances, and increase air breathing rates when feeding. If larvae in hypoxic water ($P_{O_2} = 45$–50 torr) are prevented from breathing air, little ingestion occurs even though gill ventilation rates increase.

of feeding and respiration, this species may not be unique in facing this functional conflict. Many larvae live in turbid, particle-laden water, and frequent the bottom sediments. Dwelling on the bottom may result in clogging of the gill filters; indeed, a common aspect of preparing specimens for study is freeing the gills of detritus (R.J. Wassersug, pers. comm.). Larvae of some species (e.g. *Xenopus* and *Pachymedusa*) hover midwater at all times, which may prevent clogging of the gill filters (Feder and Wassersug, 1984). In *Xenopus* larvae, the buoyancy afforded by the lungs is essential for this hovering. *Xenopus* larvae normally are positively buoyant, and scull continuously with the tail filament to achieve neutral buoyancy. If the sculling is prevented, larvae float to the surface. Larvae in which air breathing is prevented soon sink to the bottom, and can remain suspended only by vigorous swimming. As soon as these larvae regain access to air, they resume their normal hovering midwater.

To be sure, hovering does not occur in most tadpoles. The generality of this explanation for air breathing is unclear at present, and deserves further study.

(6) *The occurrence of air breathing is related to post-metamorphic use of lungs.* Many structures used primarily by post-metamorphic stages (e.g. limbs) begin their development well before metamorphosis; other structures are used by larvae, modified during metamorphosis, and used by post-metamorphic stages (e.g. the gut, eyes, and buccopharynx). Thus, if lungs had no function before metamorphosis and no deleterious effects (e.g. excessive buoyancy), it would not be surprising if their development preceded metamorphosis, as it does in many species. If lungs are present, they can be used according to the requirements of a given species or simply used because they are there. The evolution of pulmonary function in larvae as a developmental happenstance might well be inconsistent

with a single 'adaptive' explanation.

Inasmuch as the preceding explanations for air breathing in larvae suffer from lack of consistency, generality, or supportive environmental data, this explanation is not as far-fetched as it first may seem. An obvious objection is that air breathing is costly and should not occur before metamorphosis unless its benefits outweigh its costs. However, other structures besides lungs (e.g. limb buds) appear well before metamorphic climax despite their cost, and growing lungs early in development may cost little more than later in development. The costs of maintaining the respiratory structures themselves and the risk of predation due to air breathing might be negligible to a tadpole if its food is abundant, air breathing is relatively inconspicuous, and if predation is slight. There could be little natural selection to eliminate air breathing that is due to historical or developmental factors (Feder, 1983c). While physiologists seldom endorse this hypothesis, it is useful as a null hypothesis and point of departure for future work.

Conclusion

Several aspects of methodology deserve emphasis here. Ecologically relevant measures of organismal performance are used where possible to verify lower level explanations. For example, in flow chamber measurements aerial respiration enhances stamina, yet in trials with live predators aerial respiration appears irrelevant or deleterious (Feder, 1983c; Wassersug and Feder, 1983). The analysis considers multiple adaptive explanations simultaneously, examines phylogenetically diverse species, and focuses on falsification of these explanations rather than on description of adaptations in various species. Finally, the analysis assumes that the consequences of aerial respiration are robust; i.e. the consequences will manifest themselves similarly in diverse species despite differing evolutionary histories. The analysis should be sensitive to general consequences of aerial respiration but insensitive to special cases and ad hoc explanations.

No single explanation of the adaptive significance of aerial respiration in anuran larvae is entirely satisfactory. Resolution of some difficulties could be achieved through relatively simple field observations (e.g. characterization of P_{O_2} variation in typical tadpole microhabitats, frequency of predation on tadpoles, tadpole antipredator behavior) and more complex laboratory experimentation (e.g., the distribution and control of the cutaneous circulation). However, some explanations (unsuitability of water as a gas exchange medium, costliness of aquatic respiration, enhancement of locomotor performance) are clearly less satisfactory than others.

Acknowledgements

I gratefully acknowledge the assistance and advice of Warren Burggren, Allen Gibbs, Katie Moran, Timothy Holtsford, Juan Markin, Rashel Paul, Nancy Ronczy, Dianne Seale, and Richard Wassersug. Research was supported by NSF Grant DEB 78-23896 and The Louis Block Fund, The University of Chicago.

References

Arnold, S.J. (1983. Morphology, performance and fitness. *Am. Zool.* 23: 347–361.

Atkinson, B.G. and Just, J.J. (1975). Biochemical and histological changes in the respiratory system of *Rana catesbeiana* larvae during normal and induced metamorphosis. *Develop. Biol.* 45: 151–165.

Bennett, A.F. (1978). Activity metabolism of the lower vertebrates. *Ann. Rev. Physiol.* 40: 447–469.

Bennett, A.F. (1980). The metabolic foundations of vertebrate behavior. *BioScience* 30: 452–456.

Bradford, D.F. (1983). Winterkill, oxygen relations, and energy metabolism of a submerged dormant amphibian, *Rana muscosa. Ecology* 64: 1171–1183.

Branch, L.C. and Taylor, D.H. (1977). Physiological and behavioral responses of larval spotted salamanders (*Ambystoma maculatum*) to various concentrations of oxygen. *Comp. Biochem. Physiol.* 58A: 269–274.

Burggren, W.W., Feder, M.E. and Pinder, A. (1983). Temperature and the balance between aerial and aquatic respiration in larvae of *Rana berlandieri* and *Rana catesbeiana. Physiol. Zool.* 56: 263–273.

Burggren, W. and Mwalukoma, A. (1983). Influence of chronic hypoxia and hyperoxia on gas exchange processes in larval and adult bullfrogs (*Rana catesbeiana*). I. Morphological responses of lungs, gills, and skin. *J. Exp. Biol.* 105: 191–203.

Burggren, W.W. and West, N.H. (1982). Changing respiratory importance of gills, lungs and skin during metamorphosis in the bullfrog, *Rana catesbeiana. Respir. Physiol.* 47: 151–164.

Caldwell, J.P., Thorp, J.H. and Jervey, T.O. (1980). Predator-prey relationships among larval dragonflies, salamanders, and frogs. *Oecologia* 46: 285–289.

Costa, H.H. (1967). Avoidance of anoxic water by tadpoles of *Rana temporaria. Hydrobiologia* 30: 374–384.

Crump, M.L. (1981). Variation in propagule size as a function of environmental uncertainty for tree frogs. *Am. Nat.* 117: 724–737.

Dejours, P. (1981). Principles of Comparative Respiratory Physiology, 2nd ed. North-Holland, Amsterdam.

Feder, M.E. (1981). Effect of body size, trophic state, time of day, and experimental stress on oxygen consumption of anuran larvae: an experimental assessment and evaluation of the literature. *Comp. Biochem. Physiol.* 70A: 497–508.

Feder, M.E. (1982). Effect of body size and developmental stage on oxygen consumption of anuran larvae: a reappraisal. *J. Exp. Zool.* 220: 33–42.

Feder, M.E. (1983a). Effect of hypoxia and body size on the energy metabolism of lungless tadpoles, *Bufo woodhousei*, and air-breathing anuran larvae. *J. Exp. Zool.* 228: 11–19.

Feder, M.E. (1983b). Responses to acute aquatic hypoxia in larvae of the frog *Rana berlandieri. J. Exp. Biol.* 104: 79–95.

Feder, M.E. (1983c). The relation of air-breathing and locomotion to predation on tadpoles (*Rana berlandieri*) by turtles. *Physiol. Zool.* 56: 522–531.

Feder, M.E. and Olsen, L.E. (1978). Behavioral and physiological correlates of recovery from exhaustion in the lungless salamander *Batrachoseps attenuatus* (Amphibia: Plethodontidae). *J. Comp. Physiol.* 128: 101–107.

Feder, M.E., Seale, D.B., Boraas, M.E., Wassersug, R.J. and Gibbs, A.G. (1984). Functional conflicts between feeding and gas exchange in suspension-feeding tadpoles, *Xenopus laevis. J. Exp. Biol.* (in press).

Feder, M.E. and Wassersug, R.J. (1984). Aerial versus aquatic oxygen consumption in larvae of the clawed frog, *Xenopus laevis. J. Exp. Biol.* 108: 231–245.

Goin, C.J. and Goin, O.B. (1971). Introduction to Herpetology, 2nd ed. W.H. Freeman, San Francisco.

Gosner, K.L. (1960). A simplified table for staging anuran embryos and larvae with notes on identification. *Herpetologica* 16: 183–190.

Gradwell, N. (1969). The respiratory importance of vascularization of the tadpole operculum in *Rana catesbeiana* Shaw. *Can. J. Zool.* 47: 1239–1243.

Holeton, G.F. (1980). Oxygen as an environmental factor of fishes. In: Environmental Physiology of Fishes. M.A. Ali, ed., Plenum, New York and London, pp. 7–32.

Hughes, G.M. (1976). Respiration of Amphibious Vertebrates. Academic Press, London.

Kramer, D.L. (1983). The evolutionary ecoloy of respiratory mode in fishes: an analysis based on the costs of breathing. *Env. Biol. Fish.* 9: 145–158.

Marian, M.P., Sampath, K., Nirmala, A.R.C. and Pandian, T.J. (1980). Behavioral response of *Rana cyanophylictis* tadpole exposed to changes in dissolved oxygen concentration. *Physiol. Behav.* 25: 35–38.

McCutcheon, F.H. (1936). Hemoglobin function during the life history of the bullfrog. *J. Cell. Comp. Physiol.* 8: 63–81.

Medvedev, L. (1937). The vessels of the caudal fin in amphibian larvae and their respiratory function. *Zool. Journal* 16:393–403.

Moalli, R., Meyers, R.S., Jackson, D.C. and Millard, R.W. (1980). Skin circulation of the frog, *Rana catesbeiana*, distribution and dynamics. *Respir. Physiol.* 40: 137–148.

Noble, G.K. (1931). The Biology of the Amphibia. Mc-Graw Hill, New York.

Noland, R. and Ultsch, G.R. (1981). The roles of temperature and dissolved oxygen in microhabitat selection by the tadpoles of a frog (*Rana pipiens*) and a toad (*Bufo terrestris*). *Copeia* 1981: 645–652.

Pinder, A. and Burggren, W. (1983). Respiration during chronic hypoxia and hyperoxia in larval and adult bullfrogs. II. Changes in respiratory properties of whole blood. *J. Exp. Biol.* 105: 205–213.

Porter, K.R. (1972). Herpetology. W.B. Saunders, Philadelphia.

Randall, D.J., Burggren, W.W., Farrell, A.P. and Haswell, M.S. (1981). The Evolution of Air Breathing in Vertebrates. Cambridge University Press, Cambridge.

Saint-Aubain, M.L. (1982). The morphology of amphibian skin before and after metamorphosis. *Zoomorphology* 100: 55–63.

Savage, R.M. (1935). Ecology of young tadpoles, with special reference to the habit of mass spawning in *Rana temporaria temporaria* Linn. *Proc. Zool. Soc. London* 3: 605–610.

Seale, D.B., Hoff, K. and Wassersug, R. (1982). *Xenopus laevis* larvae (Amphibia, Anura) as model suspension feeders. *Hydrobiologia* 87: 161–169.

Strawinski, S. (1956). Vascularization of respiratory surfaces in ontogeny of the edible frog, *Rana esulenta* [sic] L. *Zoologica Poloniae* 7: 327–365.

Taylor, A.C. and Kollros, J. (1946). Stages in the normal development of *Rana pipiens* larvae. *Anat. Rec.* 94: 7–24.

Wassersug, R.J. and Feder, M.E. (1983). The effects of aquatic oxygen concentration, body size, and respiratory behaviours on the stamina of obligate aquatic (*Bufo americanus*) and facultative air-breathing (*Xenopus laevis* and *Rana berlandieri*) anuran larvae. *J. Exp. Biol.* 105: 173–190.

Wassersug, R.J. and Heyer, W.R. (1983). Morphological correlates of subaerial existence in leptodactylid tadpoles associated with flowing water. *Can. J. Zool.* 61: 761–769.

Wassersug, R.J. and Hoff, K. (1979). A comparative study of the buccal pumping mechanism of tadpoles. *Biol. J. Linn. Soc.* 12: 225–259.

Wassersug, R.J., Paul, R.D. and Feder, M.E. (1981). Cardio-respiratory synchrony in anuran larvae

(*Xenopus laevis, Pachymedusa dacnicolor,* and *Rana berlandieri*). *Comp. Biochem. Physiol.* 70A: 329–334.

Wassersug, R.J. and Seibert, E.A. (1975). Behavioral responses of amphibian larvae to variation in dissolved oxygen. *Copeia* 1975: 86–103.

West, N.H. and Burggren. W.W. (1982). Gill and lung ventilatory responses to steady-state aquatic hypoxia and hyperoxia in the bullfrog tadpole (*Rana catesbeiana*). *Respir. Physiol.* 47: 165–176.

West, N.H. and Burggren, W.W. (1983). Reflex interactions between aerial and aquatic gas exchange organs in larval bullfrogs. *Am. J. Physiol.* 244: R770–R777.

Whitford, W.G. and Sherman, R.E. (1968). Aerial and aquatic respiration in axolotl and transformed *Ambystoma tigrinum. Herpetologica* 24: 233–237.

6. Physiological features of embryonic development in terrestrially-breeding plethodontid salamanders

David F. Bradford

Department of Biology, University of California, Los Angeles, California 90024, U.S.A.

Abstract

Terrestrial breeding in plethodontid salamanders, in comparison to aquatic breeding in amphibians in general, is associated with long incubation time, large ovum and hatchling size, and high degree of development at hatching. These features are interrelated with each other and two presently described features. (1) The rate of embryonic development in terrestrially-breeding plethodontids is slow. The time to develop to the end of neurulation at 14–15 °C is 18.6, 26.6, 30.4, and 40.2 d in *Desmognathus ochrophaeus, Aneides ferreus, A. flavipunctatus,* and *Ensatina eschscholtzii,* respectively. (2) The time an embryo spends developing beyond neurulation, relative to the time spent previously (= 'relative incubation time', RIT), is large. RIT in *D. ochrophaeus* and *E. eschscholtzii* averages 28% longer than that of non-plethodontid salamanders. Data for oxygen consumption during embryonic development of *E. eschscholtzii* indicates that the large ovum size is probably due mostly to selection for large hatchling size, rather than the need for energy during the protracted development. Large ovum size is associated with a slow rate of embryonic development which, together with the long RIT, accounts for the long incubation time.

Rate of development increases with temperature within the temperature tolerance range of about 10–25 °C for *D. ochrophaeus, A. ferreus,* and *A. flavipunctatus.* Oxygen consumption of *E. eschscholtzii* at 14 °C increases in an approximately exponential manner during embryonic development, and does not change significantly immediately after hatching. The yolk reserves at hatching are large; hatchling *D. ochrophaeus* are able to subsist on yolk reserves for a period nearly equal to the incubation time.

Introduction

The embryo in most terrestrially-breeding salamanders in the family Plethodontidae undergoes direct development, in which it bypasses the ecological larval stage, and hatches in a form similar to the adult (Salthe and Mecham, 1974). This

Seymour, R.S., (ed.) Respiration and metabolism of embryonic vertebrates.

feature, which allows complete independence of standing water, has evolved independently in several amphibian families, and has been a key element in the dramatic radiation of the plethodontids and other groups (Wake and Lynch, 1976). Despite the importance of this feature, however, little attention has been given to the physiological consequences and limitations of direct development in amphibians. This is due largely to difficulty in inducing oviposition in the laboratory, high mortality of embryos in the laboratory, and small clutch size.

The present study elucidates interrelationships among several features of terrestrial breeding in plethodontids. First, incubation time is long, relative to amphibians in general, ranging from about 2 to 9 months (Hanken, 1979). This long period represents a costly energetic drain in brooding females (L.D. Houck, pers. comm.). Second, ovum size and hatchling size are large relative to adult body size. Ovum volume in plethodontids with direct development is several times that for aquatically-breeding salamanders of the same body size (Salthe, 1969). Third, the large ovum size may also be associated with a slow rate of embryonic development. Ovum size and rate of development are roughly negatively correlated in non-plethodontid amphibians (Salthe and Mecham, 1974). Fourth, hatchlings emerge in an advanced state of development. As a consequence, the developmental period may be protracted.

Four questions are specifically addressed concerning the embryonic dvelopment of terrestrially-breeding plethodontids, relative to aquatically-breeding amphibians. (1) Is rate of development inversely related to ovum size? (2) Is the long incubation time associated with a slow rate of development? (3) Is the long incubation time associated with a protraction in the duration of development beyond a stage through which all amphibians pass? (4) If embryonic development is protracted, does the associated increased energy demand account for the large ovum size?

The plethodontids studied represent two types of terrestrial breeding: (1) direct development – *Aneides ferreus, A. flavipunctatus*, and *Ensatina eschscholtzii*, and (2) nearly direct development – *Desmognathus ochrophaeus*. Individuals of the latter species undergo embryonic development on land, and then enter the water after hatching for a relatively brief aquatic larval period (Tilley, 1968, 1973). The embryonic stages beyond gastrulation have been documented in *D. ochrophaeus* and *D. fuscus* (Tilley, 1972; Montague, 1977), whereas only a few stages have been described in the other species (*E. eschscholtzii*; Stebbins, 1954).

Material and methods

Gravid females were collected as follows: *A. ferreus* and *A. flavipunctatus*, (Mendocino Co., California, March, 1978); *E. eschscholtzii* (Santa Monica Mountains, Los Angeles Co., California, February, 1977); and *D. ochrophaeus*, (near Highlands, Macon Co., and Brevard, Transylvania Co., North Carolina,

June, 1978). Individuals were kept at 14–15 °C under controlled light cycle (14L:10D) in the laboratory in plastic shoeboxes with moist paper towelling and pottery fragments. Oviposition was spontaneous in one *E. eschscholtzii* (May), one *A. flavipunctatus* (May), and three *D. ochrophaeus* (June). Oviposition was induced in one *A. flavipunctatus* (June), two *A. ferreus* (June), and two *D. ochrophaes* (July) by subcutaneous injection of 2–6 IU · g[-1] of human chorionic gonadotropin (Pregnyl) in 0.7% saline, and/or subcutaneous implantation of 1–3 frozen pituitary glands from breeding female *Taricha* spp. These techniques produced fertile eggs in about 20% of the gravid females.

Eggs were treated in the laboratory in one of three ways. Eggs in Treatment I were oviposited in the laboratory, and subsequently separated from the clutch, staged (see below) and placed on moist paper towelling (sphagnum moss for *E. eschscholtzii*) at constant temperatures of 5, 10, 15, 20, 25, and 28.6 °C (14 °C only in *E. eschscholtzii*). Eggs from the same clutch in *Aneides* spp. and *D. ochrophaeus* were placed at 2–3 selected temperatures. Eggs were rinsed in spring water and rotated every 1–2 d. Ovum diameter of several eggs was measured in water with a ruler and dissecting microscope.

Eggs in Treatments II and III (*D. ochrophaeus* only) were collected in the field as 19 whole clutches. Clutches were kept on moist filter paper in Petri dishes, half with attending females, at 3–8 °C for up to 7 d. Embryos were staged and placed at constant temperatures of 10, 15, 20, and 25 °C. During the following 13 d, embryo mortality and parental desertion were high. Eggs in clutches that were no longer brooded were separated and placed in spring water, 3 cm deep (Treatment II). Three clutches at 15 °C and two at 20 °C were left with attending females on filter paper (Treatment III).

Embryos were staged according to Harrison for *Ambystoma maculatum* (Rugh, 1962). Rate of development was defined as the inverse of the time between stage 1 (oviposition) and stage 21 (completion of neurulation), a stage common to all amphibians (Bachmann, 1969). Similarly the rates to hatching and yolk depletion in unfed larvae were defined as the inverse of time between stage 1 and hatching (i.e. incubation time) and stage 1 and yolk depletion, respectively. In cases where the initial stage observed was not stage 1, rate values were corrected based on data for complete staging at 15 °C. At the time of placement at the experimental temperature, embryos in Treatment I were in stages 1–8 (median = 1) for *D. ochrophaeus*, stages 1–3 (median = 1) for *A. ferreus*, stages 1–8 (median = 4.5) for *A. flavipunctatus*, and stage 1 for *E. eschscholtzii*. Initial stages of *D. ochrophaeus* in Treatment II were 9–14 (median = 9.5), and in Treatment III, 10–17 (median = 15).

Oxygen consumption ($\dot{V}_{O_2}$) was measured for embryos of *E. eschscholtzii* at 14 °C in the manner of Vleck et al. (1979). Two to eight embryos were placed on moist paper towelling in an acrylic syringe fitted with a nylon stopcock. The volume of air in the syringe was adjusted to between 85 and 150 ml, and the stopcock closed. After 30–120 h undisturbed in the dark, the fractional concentrations of O_2 in the sample syringe and a blank syringe were measured with a Beckman E2 paramagnetic oxygen analyzer. Fractional O_2 concentration in the sample chamber declined less than 2%. It was assumed that the initial air was saturated with water. The $\dot{V}_{O_2}$ of the outer jelly layer of two egg capsules, removed from healthy eggs, was measured four times between 60 and 140 d of incubation. $\dot{V}_{O_2}$ of the jelly capsule appeared to be constant at $21\,\mu l\ O_2(STP)\cdot d^{-1}$, and was substracted from the $\dot{V}_{O_2}$ of intact egg capsules.

Results

Embryonic survival and temperature tolerance

Temperature significantly affected survival through gastrulation (stage 13) and through neurulation (stage 21) in *D. ochrophaeus* in Treatment I (χ^2_{3df}, P<0.05; Fig. 1). Temperature also significantly affected survival through neurulation in the two *Aneides* species combined (χ^2_{3df}, P<0.05). Survival through neurulation in the three species, and survival to hatching in *D. ochrophaeus*, were greatest at 15 and 20 °C. For *E. eschscholtzii* at 14 °C, survival through gastrulation, neurulation, and hatching was similar to that for *D. ochrophaeus* at 15 °C.

Mortality in Treatment I at 15 and 20 °C occurred mostly between stage 21 and hatching. This mortality was often associated with the formation of adhesions

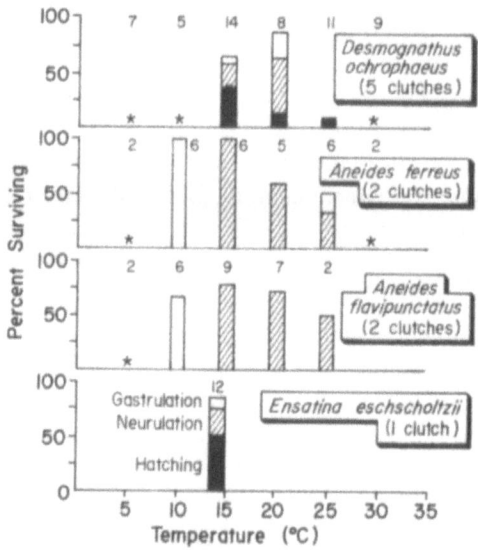

Fig. 1. Survivorship through gastrulation (stage 13; top of open box), neurulation (stage 21; top of hatched box), and hatching (top of shaded box), in fertile eggs of four species of plethodontids in Treatment I at 5–28.6 °C. Asterisks designate zero survivorship; numbers of individuals are given at top.

between the embryo and egg capsule. None of the embryos of *Aneides* species hatched, although development in some individuals proceeded for many weeks to beyond the stage of hind limb buds. At 10 °C survival beyond stage 21 did not occur in *D. ochrophaeus* and *Aneides* spp. in Treatment I, and at 25 °C it was low (Fig. 1). At these temperatures, however, a substantial proportion of individuals survived beyond gastrulation. At 5 and 28.6 °C, no individuals developed through gastrulation.

Embryos of *D. ochrophaeus* in Treatment II survived to hatching in a greater proportion than those subjected to Treatment I at 10–25 °C (Figs. 1 and 2, χ^2_{1df}, P<0.05). Survival to hatching for individuals subjected to Treatment II was not significantly related to temperature within the range of experimental temperatures (10–25 °C; χ^2_{3df}, P≫0.05). Much of the mortality in this group occurred prior to placement of eggs in water. Adhesions did not commonly occur in embryos in Treatment II after placement in water.

The mean survivorship to hatching of brooded clutches of *D. ochrophaeus* (Treatment III) was 84% at 15 °C (N = 3 clutches), and 47% at 20 °C (N = 2 clutches).

Rate of embryonic development

The rate of embryonic development between stages 1 and 21 increased with temperature for *D. ochrophaeus*, *A. ferreus*, and *A. flavipunctatus* (Fig. 3). The Q_{10} between 15 and 20 °C was 3.2, 2.2, and 2.5 in the above species, respectively. At both 15 and 20 °C, the order in rate of development was *D. ochrophaeus* > *A. ferreus* > *A. flavipunctatus* (P<0.05 in all *a posteriori* comparisons except for the two *Aneides* species at 20 °C: Least Significant Range Test, ANOVA, Sokal and Rohlf, 1969). *E. eschscholtzii* at 14 °C developed more slowly than the extrapolated rates for the other three species at this temperature. The 'adaptive temperature' ($T_0 + 10$), where T_0 = the intersection of the rate-temperature curve and the temperature axis (e.g. Fig. 3) (Bachmann, 1969), lies within the 15–20 °C range in

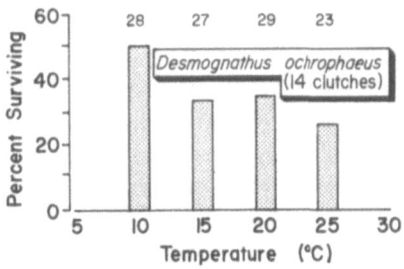

Fig. 2. Survivorship to hatching for embryos of *D. ochrophaeus* subjected to Treatment II. Numbers of individuals are given at top.

91

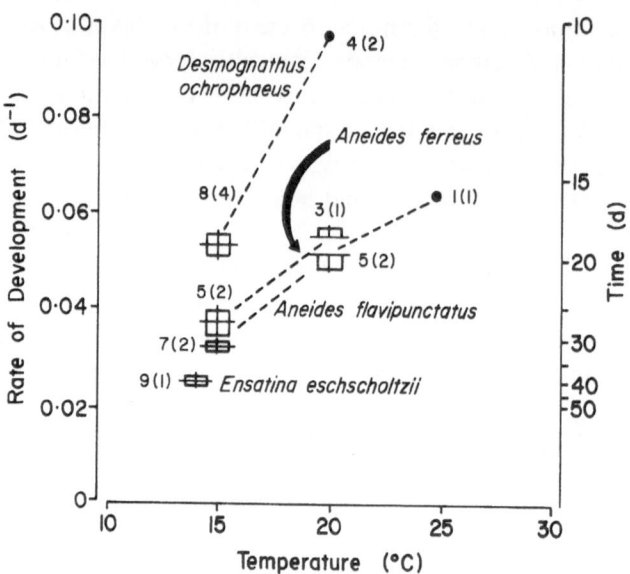

Fig. 3. Rate of embryonic development (d⁻¹) between stage 1 (oviposition) and stage 21 (end of neurulation) for individuals from eggs oviposited in the laboratory (Treatment I). Boxes designate $\bar{x} \pm$ s.d.; vertical lines represent range. The unshown minimum value for *A. ferreus* at 20 °C was 53.7, and the unshown maximum value for *A. flavipunctatus* at 20 °C was 54.8. Numbers of individuals are given above each symbol; numbers of clutches are given in parentheses.

three species: 18.3 °C for *D. ochrophaeus*, 14.4 °C for *A. ferreus*, and 16.4 °C for *A. flavipunctatus*.

Ovum diameter averaged 3.2, 4.0, and 5.1 mm in freshly deposited eggs of *D. ochrophaeus*, *A. ferreus*, and *A. flavipunctatus*, respectively. Ovum diameter in *E. eschscholtzii* averaged 6.6 mm on day 6, which corresponds to an estimated 6.4 mm diameter at oviposition (Kaplan, 1979). Rate of development in these four species was negatively related to ovum size (Spearman Rank Correlation, one-tailed, P<0.05; Siegel, 1956).

Rate to hatching and yolk depletion

In *D. ochrophaeus*, embryos in the three treatment groups did not differ in rate to hatching, or rate to yolk depletion (based on clutch means, ANOVA, P≫0.05). Therefore, data for the three treatment groups are combined in Fig. 4. The rates to hatching and yolk depletion were both directly related to temperature. Hatching of *D. ochrophaeus* at 10, 15, 20 and 25 °C occurred at 137, 69.4, 40.5, and 34.0 d after oviposition, respectively. The Q_{10} for rate to hatching was 3.9, 2.9, and 1.4 at 10–15, 15–20, and 20–25 °C, respectively. An abundance of yolk was visible in hatchling *D. ochrophaeus*, and the unfed hatchlings lived for an extended period

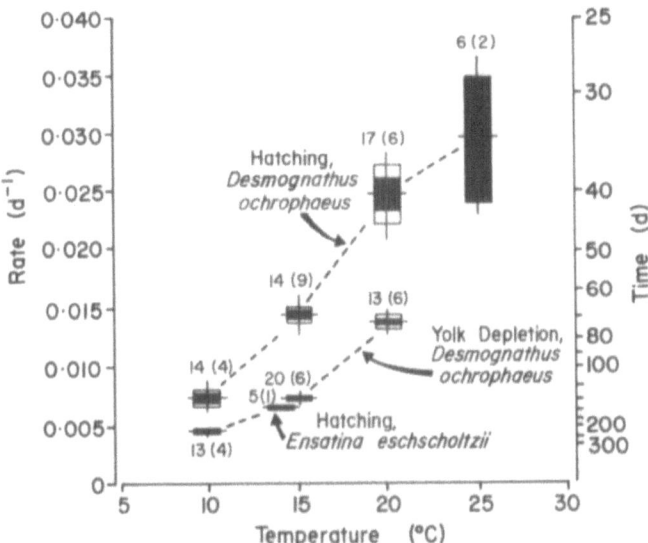

Fig. 4. Rate to hatching and yolk depletion in *D. ochrophaeus* at 10–25 °C, and rate to hatching in *E. eschscholtzii* at 14 °C. *D. ochrophaeus* embryos were subjected to Treatment I, II and III, whereas *E. eschscholtzii* embryos were subjected to Treatment I. Symbols as in Fig. 3, except the solid box represents 95% confidence limits of mean.

on energy reserves. The time between hatching and yolk depletion at 10, 15, and 20 °C was 67%, 100%, and 82%, respectively, of the time between oviposition and hatching (Fig. 4).

The rate to hatching in *E. eschscholtzii* at 14 °C was substantially lower than the interpolated rate for *D. ochrophaeus* at 14 °C. *E. eschscholtzii* hatched in a mean of 154 d after oviposition (Fig. 4). Hatchlings of *E. eschscholtzii* contained an abundance of yolk, fed readily, and survived for many months after hatching.

Oxygen consumption ($\dot{V}_{O_2}$)

The $\dot{V}_{O_2}$ of *E. eschscholtzii* embryos increased throughout development, and could be approximated by the exponential equation, $\dot{V}_{O_2} = 18.0e^{0.0183t}$ where $\dot{V}_{O_2}$ is $\mu l \cdot d^{-1}$ and t is days (Fig. 5). The $\dot{V}_{O_2}$ of hatchlings did not differ significantly from that predicted by regression analysis for the embryos.

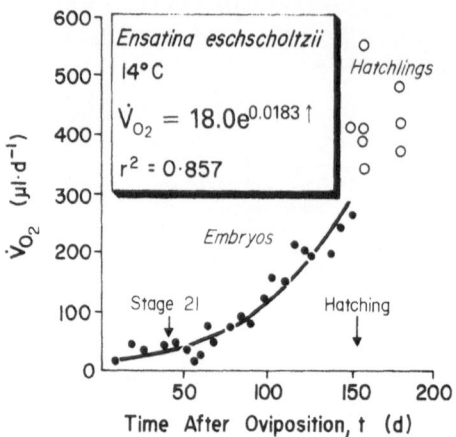

Fig. 5. Oxygen consumption ($\dot{V}_{O_2}$) during embryonic development in *E. eschscholtzii*. The number of embryos in the sample chamber was 2–8 (median = 6). $\dot{V}_{O_2}$ of hatchlings was measured individually. Hatchlings weighed 274–350 mg ($\bar{x}$ = 308 mg) on days 152–159. The exponential equation for embryonic $\dot{V}_{O_2}$ was computed from semi-log transformed data fitted by least squares regression. The 95% confidence limits of the slope were 0.0183 ± 0.0035.

Discussion

Rate of embryonic development

The rates of embryonic development in the four plethodontids studied are slow in comparison to aquatically-breeding amphibians. At 20 °C, *D. ochrophaeus, A. ferreus,* and *A. flavipunctatus* take 10.4, 18.3, and 19.5 d, respectively, to develop from stage 1 to stage 21. In *E. eschscholtzii,* this development probably takes even longer (Fig. 3). In aquatically-breeding amphibians, however, development to stage 21 at 20 °C is accomplished in less than 5 d in most species (e.g. 1.0, 2.2, 4.7, and 3.1 d in *Scaphiopus hammondii, Rana pipiens, Ambystoma opacum,* and *Triturus alpestris,* respectively (Knight, 1938; Moore, 1939; Brown, 1967).

A comparison of the 'relative developmental time', as defined by Bachmann (1969), corroborates the finding that terrestrially-breeding plethodontids have slow rates of development. The relative developmental time is the inverse slope of the line relating rate of development and temperature. It is inversely related to developmental rate, and is a measure that is independent of developmental temperature. Relative developmental times for *D. ochrophaeus, A. ferreus,* and *A. flavipunctatus* in the present study are 2908, 6474, and 6022 h · C, respectively. These values are much larger than those reported for nearly all other amphibians (Bachmann, 1969). However, the Q_{10} for rate of development between 15 and 20 °C in the plethodontids is similar to that for other amphibians in the middle of the temperature range for normal development.

Much of the difference in rate of embryonic development between the terrestrially-breeding plethodontids and other amphibians can be attributed to differences in ovum size (Question 1). The data for plethodontids extend the range in ovum size for comparing rates of development in amphibians, and corroborate the negative relationship between rate of development and ovum size (Salthe and Mecham, 1974).

Relative Incubation Time (RIT)

RIT is defined as the ratio, time between stage 1 and hatching/time between stage 1 and stage 21. RIT is thereby normalized among amphibians by rate of development, which is determined by the time to stage 21, and provides a measure of the protraction of the developmental period beyond a stage through which all amphibians pass. RIT is computed from literature for nine species of non-plethodontid salamanders and two species of terrestrially-breeding plethodontids (Table 1). Individuals of the non-plethodontid species have extensive aquatic larval periods, although eggs of *A. opacum* are usually laid on land, and subsequently flooded. RIT for the two terrestrially-breeding plethodontids ($\bar{x} = 3.83$) is significantly greater than that for the non-plethodontids ($\bar{x} = 3.00$) (P<0.01, t-test, unequal variance; Sokal and Rohlf, 1969). Consistent with this finding, approximations of RIT in two other terrestrially-breeding plethodontids, *Batrachoseps wrighti* and *Desmognathus fuscus* (Stebbins, 1949; Montague, 1977) are above 4.0.

The 28% greater RIT in the terrestrially-breeding plethodontids, in comparison to non-plethodontids, may account for part of the large difference in ovum volume between direct-developing plethodontids and aquatically-breeding salamanders of the same body mass (Question 4). A 28% increase in RIT should correspond to an even greater increase in energy metabolism, due to the continual increase in energy metabolism during embryonic development in amphibians (Salthe and Mecham, 1974). For example, in embryos of *E. eschscholtzii* (Fig. 5), as much energy is consumed after 121 d as is consumed until that time. The time of 121 d corresponds to an RIT of 3.00, the value of aquatically-breeding species. Nevertheless, this increase in energy consumption is small relative to the approximate 4–16 fold difference in ovum volume between aquatically-breeding salamanders and direct-developing plethodontids (Salthe, 1969; Kaplan and Salthe, 1979). Presumably the large ovum size in terrestrially-breeding plethodontids has resulted primarily from selection for a large body size at hatching (Salthe, 1969).

Incubation time

Terrestrially-breeding plethodontid salamanders have long incubation periods relative to amphibians in general. Eggs of one species, *Bolitoglossa compacta*,

Table 1. Relative Incubation Time (RIT) in caudate amphibians. RIT = time from stage 1 to hatching/time from stage 1 to stage 21. In most cases information was gleaned from more than one source.

Species	RIT	Reference
Non-plethodontids:		
Necturus maculosus	2.20	Eycleshymer and Wilson, 1910
Proteus anguinus	3.51	Briegleb, 1962; Vandel et al., 1966
Euproctus asper	3.01	Gasser, 1964; Gallien and Durocher, 1957; Gallien and Bidaud, 1959
Pleurodeles waltl	3.74	Gallien and Durocher, 1957
Triturus helveticus	2.66	Gallien and Bidaud, 1959
T. vulgaris	3.26	Glaesner, 1925, Glucksohn, 1931
Ambystoma maculatum	3.56	Hamburger, 1960; Harrison, 1969
A. mexicanum	2.40	Stauffer, 1945; Løvtrup and Werdenius 1957; Gallien and Durocher, 1957
A. opacum	2.64	Moore, 1939; Kaplan and Salthe, 1979; Kaplan, 1980
	$\bar{x} = 3.00 \pm 0.43$ (95% C.L.)	
Plethodontids:		
Desmognathus ochrophaeus	3.81	present study
Ensatina eschscholtzii	3.84	present study
	$\bar{x} = 3.83 \pm 0.19$ (95% C.L.)	

required 250 d to hatch in the laboratory at 13 °C, a temperature believed to be near its natural incubation temperature (Hanken, 1979). The incubation periods of the few other terrestrially-breeding plethodontids that have been observed or estimated are between 2 and 6 months (Hanken, 1979; present study). In contrast, many aquatically-breeding caudates hatch in less than 12 d at 20 °C, and many aquatically-breeding anurans hatch in less than 6 d at 20 °C.

The long incubation time of terrestrially-breeding plethodontids, relative to amphibians in general, is accounted for by two features: (1) slow rate of embryonic development (Question 2), and (2) large RIT (Question 3). For example, incubation time in *D. ochrophaeus* at 18–20 °C is 5.3 times that for *Triturus helveticus*, an aquatically-breeding caudate, and 8.0 times that for *Rana pipiens*, an aquatically-breeding anuran (Shumway, 1940; Gallien and Bidaud, 1959). Rate of development in *D. ochrophaeus* is much slower than in these two species (factors = 0.271 and 0.238, respectively), whereas RIT is larger (factors = 1.43 and 1.91, respectively). The slower rate of development in *D. ochrophaeus* accounts for most of the difference in incubation time from that of *T. helveticus* and *R. pipiens* (86% and 78%, respectively). The larger RIT in *D. ochrophaeus* accounts for the remaining 14% and 22%, respectively.

The incubation time for *D. ochrophaeus* in the present study (69.4 d at 15 °C) is similar to the times observed for this species and *D. fuscus* in other studies (69–77 d at 14–16 °C; Tilley, 1972, Montague, 1977). The 154 d incubation period for *E. eschscholtzii* at 14 °C is consistent with the estimated incubation time in the field of 4–5 months (Stebbins, 1954).

Interaction of traits

Terrestrial breeding in plethodontids, relative to aquatic breeding in amphibians in general, has apparently resulted in a large increase in ovum size, and a small increase in RIT. The increased ovum size is probably due mostly to selection for larger hatchlings, rather than the need for more energy during the extended development. Increased ovum size has resulted in a slow rate of development which, together with the lengthened RIT, has greatly increased incubation time. This slowing of rate functions represents a substantial cost in time and energy to brooding females.

Acknowledgements

Research was supported by NSF grants DEB 78–03174 and 81–03513 administered by G.A. Bartholomew, NIH Integrative and Systems Biology Training Grant GM 07191, and a UCLA Graduate Division Grant. I am grateful to D.C. Forester for providing information on collecting localities, D. Rochlin for assistance in the laboratory, and L.D. Houck and an anonymous reviewer for reading an early version of the manuscript. The Highlands Biological Station was used during collection and initial handling of *D. ochrophaeus*.

References

Bachmann, K. (1969). Temperature adaptations of amphibian embryos. *Am. Nat.* 103: 115–130.
Briegleb, W. (1962). Zur Biologie und Ökologie des Grottenolms (*Proteus anguinus* Laur. 1768). *Z. Morphol. Oekol. Tiere.* 51: 271–334.
Brown, H.A. (1967). Embryonic temperature adaptations and genetic compatibility in two allopatric populations of the spadefoot toad, *Scaphiopus hammondi*. *Evolution* 21: 742–761.
Eycleshymer, A.C. and Wilson, J.M. (1910). Normal plates of the development of *Necturus maculosus*. F. Keibel's Normentafeln zur Entwicklungsgeschichte der Wirbeltiere, H. 11.
Glaesner, L. (1925). Normentafel zur Entwicklungsgeschichte des gemeinen Wassermolches (*Molge vulgaris*). F. Keibel's Normentafeln zur Entwicklungsgeschichte der Wirbeltiere, H. 14.
Gallien, L. and Bidaud, O. (1959). Table chronologique du développement chez *Triturus helveticus* Razoumowsky. *Bull. Soc. Zool. Fr.* 84: 22–32.
Gallien, L. and Durocher, M. (1957). Table chronologique du développement chez *Pleurodeles waltlii* Michah. *Bull. Biol. Fr. Belg.* 91: 97–114.

Gasser, F. (1964). Observations sur les stades initiaux du développement de l'urodèle Pyrénéen *Euproctus asper. Bull. Soc. Zool. Fr.* 89: 423–428.

Glücksohn, S. (1931). Äussere Entwicklung der Extremitäten und Stadieneinteilung der Larvenperiode von *Triton taeniatus* Leyd. und *Triton cristatus* Laur. *Arch. Entwicklungsmech. Org.* 125: 341–405.

Hamburger, V. (1960). A Manual of Experimental Embryology. Univ. of Chicago Press, Chicago.

Hanken, J. (1979). Egg development time and clutch size in two neotropical salamanders. *Copeia* 1979: 741–744.

Harrison, R.G. (1969). Organization and Development of the Embryo. Yale Univ. Press, New Haven.

Kaplan, R.H. (1979). Ontogenetic variation in 'ovum' size in two species of *Ambystoma. Copeia* 1979: 348–350.

Kaplan, R.H. (1980). Ontogenetic energetics in *Ambystoma. Physiol. Zool.* 53: 43–56.

Kaplan, R.H. and Salthe, S.N. (1979). The allometry of reproduction: An empirical view in salamanders. *Am. Nat.* 113: 671–689.

Knight, F.C.E. (1938). Die Entwicklung von *Triton alpestris* bei verschiedenen Temperaturen, mit Normentafel. *Wilhelm Roux' Arch. Entwicklungsmech. Org.* 137: 461–473.

Løvtrup, S. and Werdinius, B. (1957). Metabolic phases during amphibian embryogenesis. *J. Exp. Zool.* 135: 203–220.

Montague, J.R. (1977). Note on the embryonic development of the dusky salamander, *Desmognathus fuscus* (Caudata: Plethodontidae). *Copeia* 1977: 375.

Moore, J.A. (1939). Temperature tolerance and rates of development in the eggs of amphibia. *Ecology* 20: 459–478.

Rugh, R. (1962). Experimental Embryology: Techniques and Procedures. Burgess, Minneapolis.

Salthe, S.N. (1969). Reproductive modes and the number and sizes of ova in the urodeles. *Am. Midl. Nat.* 81: 467–489.

Salthe, S.N. and Mecham, J.S. (1974). Reproductive and courtship patterns. In: Physiology of the Amphibia, Volume II. B. Lofts. ed., Academic Press, New York, pp. 309–521.

Shumway, W. (1940). Stages in the normal development of *Rana pipiens. Anat. Rec.* 78: 139–147.

Siegel, S. (1956). Nonparametric Statistics for the Behavioral Sciences. McGraw-Hill, New York.

Sokal, R.R. and Rohlf, F.J. (1969). Biometry. W.H. Freeman, San Francisco.

Stauffer, E. (1945). Versuche zur experimentellen Herstellung haploider Axolotl-Merogone. *Rev. Suisse de Zool.* 52: 231–327.

Stebbins, R.C. (1949). Observations on laying, development, and hatching of the eggs of *Batrachoseps wrighti. Copeia* 1949: 161–168.

Stebbins, R.C. (1954). Natural history of the salamanders of the plethodontid genus *Ensatina. Univ. Calif. Publ. Zool.* 54: 47–124.

Tilley, S.G. (1968). Size-fecundity relationships and their evolutionary implications in five desmognathine salamanders. *Evolution* 22: 806–816.

Tilley, S.G. (1972). Aspects of parental care and embryonic development in *Desmognathus ochrophaeus. Copeia* 1972: 532–540.

Tilley, S.G. (1973). Observations on the larval period and female reproductive ecology of *Desmognathus ochrophaeus* (Amphibia: Plethodontidae) in western North Carolina. *Am. Midl. Nat.* 89: 394–407.

Vandel, A., Durand, J. and Bouillon, M. (1966). Contribution a l'étude du développement de *Proteus anguinus* Laurenti. (Batraciens, Urodèles). *Ann. Speleol.* 21: 609–619.

Vleck, C.M., Hoyt, D.F. and Vleck, D. (1979). Metabolism of avian embryos: patterns in altricial and precocial birds. *Physiol. Zool.* 52: 363–377.

Wake, D.B. and Lynch, J.F. (1976). The distribution, ecology, and evolutionary history of plethodontid salamanders in tropical America. *Nat. Hist. Mus. Los Angeles C. Sci. Bull.* 25: 1–65.

7. Coupling of physiology of embryonic turtles to the hydric environment

Gary C. Packard and Mary J. Packard

Department of Zoology and Entomology, Colorado State University, Fort Collins, Colorado 80523, U.S.A.

Abstract

Embryonic turtles developing in flexible-shelled eggs consume more of their yolk and grow larger before hatching when incubated in relatively wet environments than they do when incubated in relatively dry settings. These differences in size of young result from differences in rates of embryonic metabolism and growth in some species and from differences in duration of incubation in others. However, neither the specific physical factor eliciting the responses nor the underlying physiological mechanism has been established unequivocally for any species. Because embryos in relatively wet environments have different patterns of net water-exchange than do embryos in drier settings, most attention has been directed at means by which such exchanges could influence their physiology. Water exchanges may exercise control over oxidative metabolism by altering bulk water in cytoplasm of cells of growing animals, by affecting concentrations of urea in body fluids of embryos, by influencing growth of the allantois, and, indirectly, by affecting incubation temperature; water fluxes may influence duration of incubation via effects on water potential in compartments such as the yolk. However, the evidence to support these hypotheses is fragmentary, and no single mechanism is applicable to all species studied to date.

Introduction

Eggs of many species of turtles, including all cheloniids, chelydrids, and dermochelyids and most pelomedusids and emydids (Ewert, 1979), have flexible, highly porous shells (Feder et al., 1982; G.C. Packard et al., 1979b, 1981a; Tracy, 1982). Embryos developing in eggs like these consume more of their yolk and grow larger before hatching when incubated in relatively wet environments than they do when incubated in relatively dry settings (G.C. Packard et al., 1980, 1981a, b, 1982; Tracy et al., 1978), thereby indicating that conditions of the hydric environment have important physiological effects on developing young. In the paragraphs that follow, we shall briefly review studies elucidating the organismal

Seymour, R.S., (ed.) Respiration and metabolism of embryonic vertebrates.
© *1984, Dr W. Junk Publishers. Dordrecht/Boston/London. ISBN-13:978-94-009-6538-6*

responses to incubation in different hydric conditions, and shall then consider in detail several hypotheses that might account for these responses. Only passing mention will be made of the rigid-shelled eggs produced by many other species of turtles (Ewert, 1979), because incubating rigid-shelled eggs in different hydric environments has little effect on developing embryos (G.C. Packard et al., 1979a, 1981c).

Organismal responses to the hydric environment

Embryos of only two species have been studied in detail, and all of the work on these animals has been performed in our laboratory. One of the species is the common snapping turtle (Chelydridae: *Chelydra serpentina*), and the other is the painted turtle (Emydidae: *Chrysemys picta*). Although we shall, of necessity, focus on the published record for these two forms, we recognize the possibility that responses of these species may not be representative of embryonic turtles generally.

In our experiments, eggs are incubated under conditions intended to simulate the hydric environment to which they are exposed in nature (G.C. Packard et al., 1981c). Different amounts of distilled water are added to dry vermiculite in plastic containers to produce substrates differing in water potential. Eggs then are placed on each substrate so that approximately half of the shell comes into contact with vermiculite, with the other half of the shell being exposed to air inside the container. Other eggs are placed on a wire platform in each container so that essentially the entire surface of the eggshell is exposed to air inside the box. Placing some eggs on platforms and others on substrates presumably approximates conditions encountered by eggs at the center and periphery, respectively, of natural nests (G.C. Packard et al., 1981c), and permits us to make qualitative statements concerning exchanges of water in liquid and vapor phases (G.C. Packard et al., 1981a). Containers are covered with aluminum foil to maintain high humidities in air trapped inside (Tracy et al., 1978), and then are incubated at a constant temperature of 29 °C. This protocol elicits net exchanges of water between eggs and their environment over the course of incubation that are similar to those occurring in natural nests (M.J. Packard et al., 1982).

When eggs are incubated according to this protocol, those in different experimental treatments undergo very different patterns of temporal change in mass in consequence of different patterns of net water-flux across their shells (Fig. 1). Eggs contacting substrates generally increase in mass during the first half of incubation (reflecting net absorption of water) and decrease in mass during the second half of development (reflecting net loss of water). In contrast, eggs resting on platforms generally change little in mass during the first half of incubation and decline rapidly in mass thereafter. However, the specific pattern of response depends on the water potential of the substrate to which eggs are exposed. Eggs

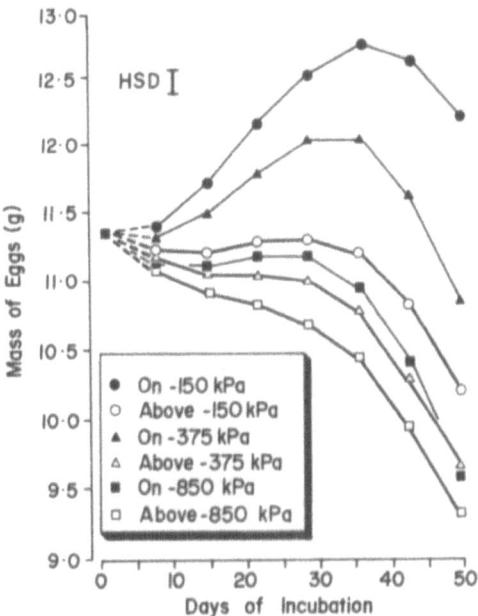

Fig. 1. Temporal changes in mass of eggs of snapping turtles incubating in contact with and above substrates differing in water potential (redrawn from Morris et al., 1983). A water potential of −150 kPa is characteristic of a very wet (but not saturated) substrate, whereas −850 kPa is characteristic of a fairly dry medium. Analysis of covariance indicated that temporal change in mass of eggs varied significantly as a complex function of water potential and position of eggs on or above substrates (Morris et al., 1983). Means for a given day that differ by the Honestly Significant Difference (HSD) are significantly different at alpha = 0.05 (Snedecor and Cochran, 1967). Data for eggs of painted turtles are identical in general outline to values summarized here (G.C. Packard et al., 1981a, 1983b).

on platforms are exposed to greater desiccation stress than are eggs resting on vermiculite in the same container, but eggs in containers with dry substrates are exposed to greater dehydration stress than are eggs in boxes with wet substrates. Thus, full appreciation of the hydric environment to which an egg is exposed depends on knowledge of both the position of the egg in the container and the water potential of the substrate.

Embryonic snapping turtles developing in eggs in relatively wet environments consume energy reserves in the yolk more rapidly (Fig. 2) and grow faster (Fig. 3) during the last one-third of incubation than do embryos in eggs exposed to relatively dry environments (Morris et al., 1983). Assuming that the rate of depletion of yolk is a reasonable index to the sustained level of oxidative metabolism (Kleiber, 1961), embryos in wet environments have higher rates of metabolism during later stages of incubation than do embryos in dry settings. The rates of growth of embryos are correlated, in turn, with metabolism, because turtles that grow the fastest are the ones that consume their yolk most rapidly (Figs. 2, 3). However, eggs in wet environments also incubate longer than do eggs in dry

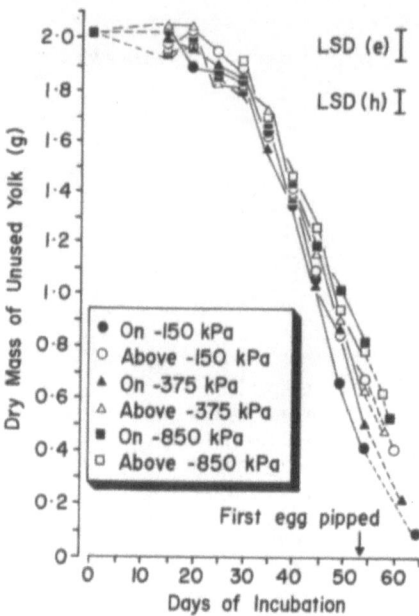

Fig. 2. Temporal variation in dry mass of unused yolk in viable eggs of snapping turtles incubating in different hydric environments (redrawn from Morris et al., 1983). Means for eggs exposed to a particular treatment are connected by solid lines; means for eggs on day 55 of incubation are connected with corresponding means for residual yolk of hatchlings by dashed lines. Values for hatchlings are plotted on the average day of emergence for animals in each group. Analysis of covariance indicated that the pattern of temporal decline in mass of yolks varied significantly in relation to water potential of substrates and to position of eggs on or above substrates (Morris et al., 1983); substrate water potential and position of eggs had significant effects also on the amount of residual yolk in hatchlings (Morris et al., 1983). The LSD (e) is the Least Significant Difference for comparing means for incubating eggs, whereas the LSD (h) is for comparing means for hatchlings. Means differing by the appropriate LSD are significantly different at alpha = 0.05 (Snedecor and Cochran, 1967).

settings, thereby giving embryos in the former more time to grow before hatching than is available to embryos in the latter (Fig. 3). The differences between embryos in wet and dry environments in rates of growth and in duration of incubation both contribute to differences in size of hatchlings (Fig. 3), and both responses are indicative of an important influence of the hydric environment on physiology of developing young.

Certain of the organismal responses of embryonic painted turtles to different hydric environments are similar to those recorded for snapping turtles, but others are not. For example, embryonic painted turtles in wet environments consume energy reserves in the yolk (Fig. 4) and increase in dry mass (Fig. 5) at rates indistinguishable from those characterizing embryos in dry conditions (G.C. Packard et al., 1983b). Considered together, these data indicate that overall

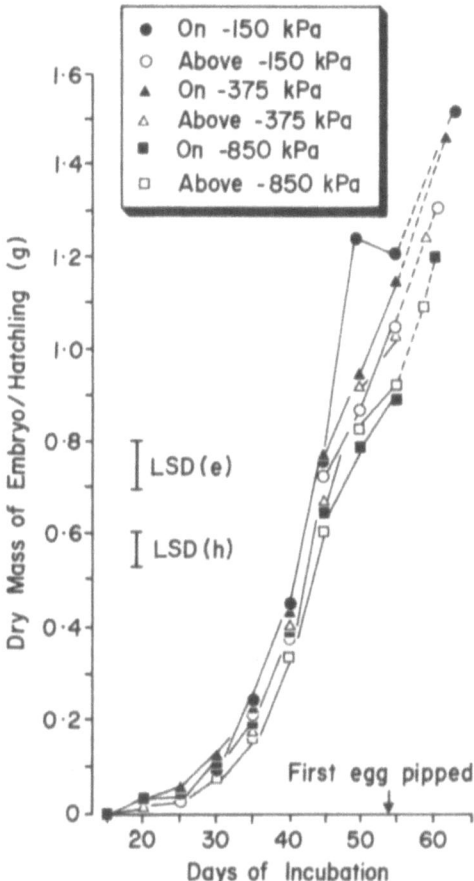

Fig. 3. Temporal variation in dry mass of carcasses (free of yolk) of embryonic snapping turtles incubating in different hydric environments (redrawn from Morris et al., 1983). Means for embryos exposed to a particular treatment are connected by solid lines; means for embryos on day 55 of incubation are connected with corresponding means for hatchlings by dashed lines. Values for hatchlings are plotted on the average day of emergence for animals in each group. Analysis of covariance indicated that the pattern of temporal increase in mass of carcasses varied significantly in relation to water potential of substrates and to position of eggs on or above substrates (Morris et al., 1983); substrate water potential and position of eggs had significant effects also on the mass of hatchlings (Morris et al., 1983). See also the caption to Fig. 2.

metabolism of embryos in this species is not influenced by the hydric environment. Nevertheless, eggs in wet settings develop longer than those in dry conditions, and hatchlings emerging from eggs in wet environments therefore are larger than those coming from eggs in dry conditions (Fig. 5). Thus, the hydric environment has no apparent influence on oxidative metabolism of embryonic painted turtles, but it does influence the duration of incubation. An effect of the hydric environment on physiology of embryos is therefore apparent.

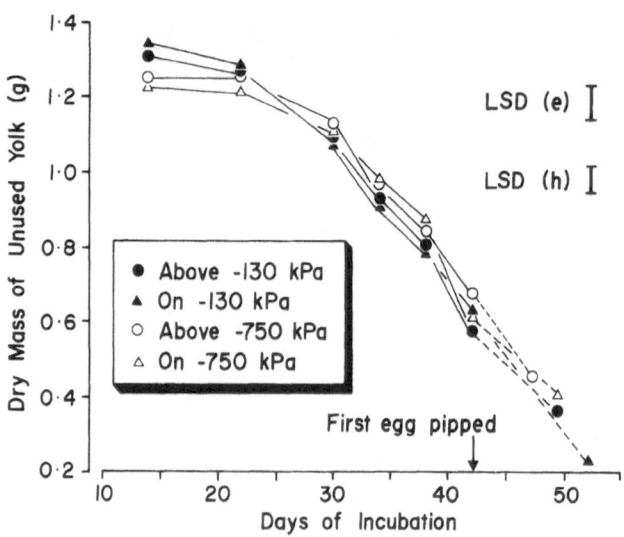

Fig. 4. Temporal variation in dry mass of unused yolk in viable eggs of painted turtles incubating in different hydric environments (redrawn from G.C. Packard et al., 1983b). Means for eggs exposed to a particular treatment are connected by solid lines; means for eggs on day 42 of incubation are connected with corresponding means for residual yolk of hatchlings by dashed lines. Values for hatchlings are plotted on the average day of emergence for animals in each group. Analysis of covariance indicated that the pattern of temporal decline in mass of yolks was not affected by water potential of substrates or by position of eggs on or above substrates (G.C. Packard et al., 1983b); however, the dry mass of residual yolk in hatchlings was influenced significantly by both of these factors (G.C. Packard et al., 1983b). See also the caption to Fig. 2.

Physiological basis for organismal responses

There are several possible explanations for the aforementioned findings for snapping turtles and painted turtles. Differences in the availability of water to developing young, differences in temperature of eggs in the various experimental groups, or differences in gas exchange by embryos in different hydric environments may underlie the observed differences in metabolism and/or duration of incubation. We shall now explore these several possibilities and evidence for and against each of them. Hypotheses for control of metabolism will be treated separately from those pertaining to control of duration of incubation because of the likelihood that these responses have different physiological bases.

Influence of the hydric environment on metabolism

Effects of temperature. – The process of sexual differentiation in developing young of most turtles is subject to a measure of genetic control via a system of

104

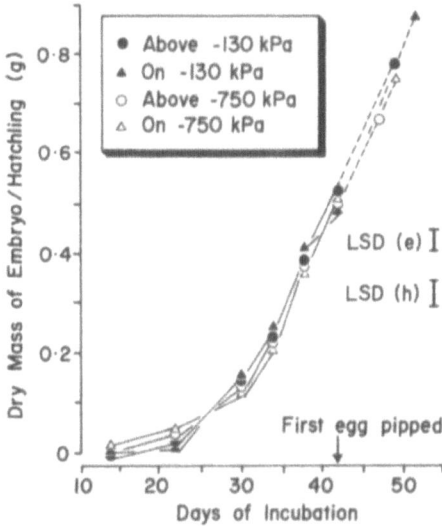

Fig. 5. Temporal variation in dry mass of carcasses (free of yolk) of embryonic painted turtles incubating in different hydric environments (redrawn from G.C. Packard et al., 1983b). Means for embryos exposed to a particular treatment are connected by solid lines; means for embryos on day 42 of incubation are connected with corresponding means for hatchlings by dashed lines. Values for hatchlings are plotted on the average day of emergence for animals in each group. Analysis of covariance indicated that the pattern of temporal increase in mass of carcasses was not affected by water potential of substrates or by position of eggs on or above substrates (G.C. Packard et al., 1983b); however, the dry mass of hatchlings was influenced significantly by both of these factors (G.C. Packard et al., 1983b). See also the caption to Fig. 2.

H-Y antigens similar to that characterizing sexual differentiation in mammals (Engel et al., 1981), but the genetic control is easily overridden by environmental factors (Zaborski et al., 1982). The factor on which most attention has focused is ambient temperature (Bull, 1980; Vogt and Bull, 1982). In general, exposing eggs to temperatures above 30 °C during the middle one-third of development elicits differentiation of phenotypic females, whereas exposing eggs to 24–28 °C leads to differentiation of phenotypic males. Both males and females arise from eggs exposed to 28–30 °C, but proportions vary from predominantly male near 28 °C to predominantly female near 30 °C.

The process of sexual differentiation is pertinent to the present discussion, because eggs of snapping turtles used in our study of embryonic metabolism and growth were incubated at 29 °C (Morris et al., 1983) – that is, at a temperature centered in the reported transitional range for sexual differentiation in this species (Bull, 1980). Small but consistent differences in temperature arising incidental to incubating eggs in wet and dry environments could have elicited differentiation of different proportions of males and females in the samples. Additionally, because male embryos have different titers of steroid hormones in

105

their blood than do females (Pieau et al., 1982), there could also have been hormonally based differences between the sexes in rates of metabolism and growth (Norris, 1980). These factors acting together could have led to variation in rates of metabolism of the type we have recorded (Fig. 2).

For example, when eggs of snapping turtles are incubated in relatively dry surroundings, they experience higher rates of net evaporation during the middle one-third of development (i.e., the period when sexual differentiation occurs) than do eggs in wetter settings (see data in Fig. 1 for eggs held on platforms). Accordingly, eggs in dry environments presumably experience more evaporative cooling than do eggs in wet environments, and the former may therefore be somewhat cooler than the latter. Unfortunately, measurements of temperature have not been taken from eggs in different hydric environments, so we are unable to say exactly how much the temperature of eggs might be reduced in dry settings. Nonetheless, if such cooling is measurable – perhaps on the order of a few tenths of a degree – more embryos might differentiate as males in dry environments than in wet ones.

We recently incubated eggs of snapping turtles at a temperature of 29 °C to assess this possibility (G.C. Packard et al., 1984). Interestingly, all of the embryos in this experiment differentiated as females, indicating that the temperature of incubation was slightly above the transitional range characterizing the population with which we work. More importantly, different rates of evaporation from eggs in the different hydric environments were not sufficient to alter temperature of eggs to the extent required to affect sexual differentiation. Nonetheless, incubation was longer and hatchlings were larger among eggs in the wet environment than among those from the dry settings (G.C. Packard et al., 1984). These findings are consistent with those reported previously (Fig. 3), and indicate that observed differences in metabolism are not secondary to an effect of treatments on sexual differentiation.

If all eggs develop at temperatures above the transitional range, however, it still is possible that a small reduction in temperature in the dry environment could affect metabolism without altering the pattern of sexual differentiation. Embryonic snapping turtles are, after all, ectotherms, and the level of metabolism they sustain presumably is correlated directly with temperature. A lower temperature among eggs incubated in a dry environment might cause a reduction in metabolism of embryos below that sustained by embryos developing in wetter conditions. A small difference in metabolism between embryos in wet and dry environments could, in turn, cause animals to follow different growth trajectories, thereby leading to substantial differences in size of young late in incubation and at hatching (Fig. 3).

We address this point indirectly by considering data for painted turtles (G.C. Packard et al., 1983b). Eggs incubated in wet and dry environments exhibit the same patterns of net water-exchange as do the eggs of snapping turtles (see Fig. 1), so temperatures of painted turtle eggs presumably differ also between wet and dry settings. Nonetheless, these differences in temperature must be minor, because there is no effect of the hydric environment on rates of consumption of energy reserves (Fig. 4) or on rates of growth in dry mass of embryos (Fig. 5). If temperature of eggs were reduced enough in dry environments to affect metabolism of snapping turtles, a similar reduction in temperature and metabolism should also occur in painted turtles. It does not.

One final point concerns this hypothesis and others that invoke differences in temperature among eggs in different hydric environments as the underlying cause of different patterns of metabolism and growth by embryos. A thermally induced increase in metabolism should be accompanied by a decrease in duration of incubation, whereas a thermally induced decrease in metabolism should be accompanied by an increase in the time required for completion of development (Bustard and Greenham, 1968; Dodge et al., 1978; Goode and Russell, 1968; Miller and Limpus, 1981; Yntema, 1978). In our experiments on embryonic snapping turtles and painted turtles, this requirement for a negative correlation between metabolic rate and duration of incubation is not satisfied (Figs. 2, 4). Thus, neither the present hypothesis nor others based on putative differences in temperature is adequate to explain the data. We therefore conclude that differences in temperature of flexible-shelled eggs in wet and dry environments are too small to elicit the observed differences in metabolism and growth of embryonic turtles.

Diffusion of respiratory gases. – By the later stages of incubation, the allantois has grown downward over the yolk sac and into the lower hemisphere of the egg (Mitsukuri, 1890, 1891), and a portion of the oxygen exchange by the embryo therefore must occur across the lower aspect of the eggshell. However, oxygen diffuses more slowly when moving through a porous substrate than it does in the free gas phase (Marshall, 1959; van Bavel, 1952), and it diffuses more slowly also through a wet substrate than through a dry one (Flegg, 1953). Consequently, transport of oxygen to exchange surfaces may be impaired (or the tension of oxygen at exchange surfaces may be reduced) among eggs on substrates compared to eggs on platforms, and the effect may be more pronounced for eggs on wet substrates than for those on dry ones. Any attendant reduction in metabolism among embryos on substrates would presumably lead to an increase in the length of incubation (Ackerman, 1981). Our observation that incubation is longer for eggs on substrates than for those on platforms is consistent with this scheme (Figs. 3, 5).

Our data for rates of metabolism in embryonic snapping turtles and painted turtles provide evidence to refute this hypothesis, however. Painted turtles developing in eggs on wet substrates consume energy reserves and grow in dry mass at the same rates as young developing in eggs above substrates (Figs. 4, 5), indicating that putative differences in gas transport have no effect on metabolism of embryos. Moreover, snapping turtles developing in eggs resting on wet substrates actually have higher rates of metabolism and growth than embryos in eggs on drier substrates or on platforms (Figs. 2, 3), which is an outcome just the opposite of what is required by the present hypothesis. Other hypotheses come to mind that invoke differences in rates of transport of oxygen to surfaces of eggshells as the underlying basis for differences in metabolism of developing embryos. However, such hypotheses require that there be an inverse relationship between the rate of metabolism by embryos and the duration of incubation (Ackerman, 1981). Because this condition is not met by data from our investigations (Figs. 2, 4), we conclude that differences in rates of diffusion of respiratory gases to surfaces of eggshells cannot account for effects of the hydric environment on metabolism of embryonic turtles.

Inhibition of chorioallantoic development. – An intriguing possibility has been raised in connection with studies of avian development concerning the possible effect of water exchanges on growth of the chorioallantois. According to Snyder and Birchard (1982), the degree of hydration of the (inner) shell membrane early in incubation may be a determinant of growth of the chorioallantois. Because the forming chorioallantois must slide along the inner surface of the shell membrane, these authors suppose that growth of the chorioallantois is promoted when the shell membrane has a high water content (i.e., affording relatively low resistance to movement of the chorioallantois) and retarded when the shell membrane has a low water content (i.e., affording relatively high resistance to movement). An egg sustaining a high rate of transpiration early in incubation has a relatively dry shell membrane, growth of the chorioallantois therefore is impeded, and gas exchange by the embryo consequently is inhibited (Tazawa, 1980). Inhibition of gas exchange presumably retards metabolism and growth of the developing bird (Tazawa, 1980).

This hypothesis is supported only in part by available data for turtles. Among embryonic snapping turtles, transpiration of water from exposed surfaces of eggs is higher in dry environments than it is in wet environments (Fig. 1). This difference in rates of transpiration could affect development of the chorioallantois, with embryos in dry environments having stunted chorioallantoic membranes and those in wet environments having normal membranes. Oxygen uptake could therefore be lower among embryos in dry settings than among those in wet settings, thereby inhibiting metabolism of embryos in dry conditions (Ackerman, 1981) and leading to the observed patterns of consumption of yolk (Fig. 2) and growth in dry mass (Fig. 3) in different hydric environments.

If this hypothesis has merit, however, duration of incubation of embryos in dry environments should be longer than that characterizing embryos in wet conditions (Ackerman, 1981). As noted previously, incubation is actually shorter among embryos in dry settings than among those in wet environments (Fig. 3). Additionally, because the hypothesis is based on physical principles that should apply also to painted turtles, we expect some influence to be manifested on metabolism and growth of embryos of this species as well. Contrary to expectation, however, embryonic painted turtles consume nutrient reserves in the yolk and grow in dry mass at the same rates in wet and dry environments (Figs. 4, 5). These observations are inconsistent with the present hypothesis, and lead us to question its applicability to eggs and embryos of any species of turtle.

Variation in amount of bulk water in cells. – Water inside cells of animals exists in two distinct, aqueous phases: vicinal water and bulk water (Clegg, 1981a, b). Vicinal water is the phase adjacent to surfaces of the endoplasmic reticulum and cytoskeleton (i.e., microtubules, microfilaments), and in which all of the 'soluble' metabolic pathways of the cytosol are localized. Bulk water is the phase of relatively high solvent capacity that is more remote from surfaces of the ER and cytoskeleton, and that provides for diffusion of metabolites among intracellular compartments.

Our focus is on intracellular water in the bulk phase, for two reasons. First, bulk water must be present for conventional metabolism to proceed inside individual cells (Clegg, 1981a). Second, variation in the amount of bulk water inside cells leads to variation also in the sustained level of metabolism (Clegg, 1981a).

We earlier suggested that the different patterns of net water-exchange experienced by flexible-shelled eggs incubating in wet and dry environments (Fig. 1) might lead to differences in amounts of bulk water in cells of developing embryos (Morris et al., 1983; G.C. Packard et al., 1983b). By this hypothesis, smaller amounts of bulk water in cells of turtles developing in dry settings would lead to a reduction in metabolism and growth relative to rates sustained by embryos developing in wetter settings.

Although this hypothesis has not been tested explicitly, it receives little support from our recent work on levels of hydration of snapping turtles (Morris et al., 1983) and painted turtles (G.C. Packard et al., 1983b). As noted already, snapping turtles developing in wet environments consume energy reserves in the yolk more rapidly (Fig. 2) and grow faster (Fig. 3) than do young developing in dry settings, as would be predicted by the preceding account. However, the percentage of body mass that is water does not vary among embryos developing in different hydric environments (Fig. 6), indicating that similar amounts of water are likely to occur in the bulk phase in cells of embryos developing in both wet and dry settings. Admittedly, it still is possible that intracellular water in the bulk phase differs among hydric environments in the manner expected from the present hypothesis. For this circumstance to be realized, however, extracellular water would have to be higher among young developing in dry settings than among those in wet settings. Such a possibility is not parsimonious and therefore seems unlikely to us.

Among painted turtles, embryos developing in wet environments have larger percentages of water in their bodies than do young incubating in dry settings (Fig. 7), an outcome that is consistent with the hypothesis concerning bulk water. However, indices of metabolism and growth in this species do not vary among eggs held in different hydric environments (Figs. 4, 5). Thus, differences in the

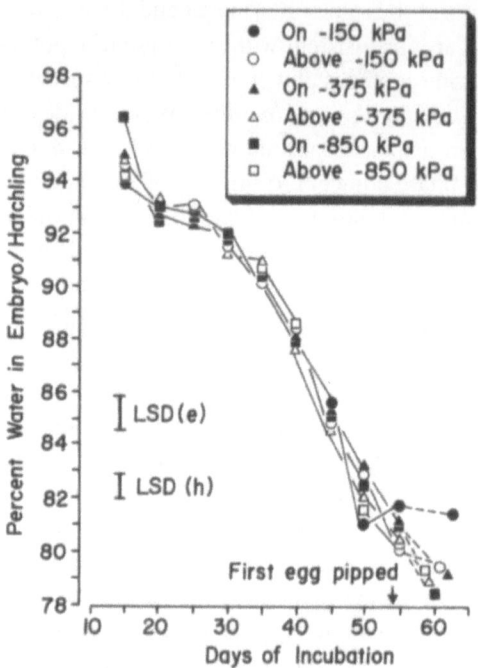

Fig. 6. Temporal variation in the percentage of body mass of embryonic snapping turtles that is water (redrawn from Morris et al., 1983). Means for embryos exposed to a particular treatment are connected by solid lines; means for embryos on day 55 of incubation are connected with corresponding means for hatchlings by dashed lines. Values for hatchlings are plotted on the average day of emergence for animals in each group. Analysis of covariance indicated that the pattern of temporal decline in percentage of water was not affected by water potential of substrates or by position of eggs on or above substrates (Morris et al., 1983); there was significant variation among hatchlings, however, caused by the high value for animals incubated on wet substrates (Morris et al., 1983). See also the caption to Fig. 2.

amount of water present in bodies of developing young are not accompanied by differences in rates of metabolism and growth. The extra water in bodies of turtles developing in wet environments probably is located in the extracellular compartment rather than the intracellular compartment, so it is possible that bulk water inside cells does not vary in different hydric environments. If so, the data for painted turtles do not refute the hypothesis, but they do not lend much support to it either.

Because this hypothesis receives no direct support in experimental data currently available, we suspect that it does not apply to embryonic turtles. Animals subjected to desiccation probably defend intracellular water at the expense of extracellular fluids (see Schmidt-Nielsen, 1964), so we anticipate that no differences will be detected in bulk water in cells of turtle embryos developing in different hydric environments.

Accumulation of urea inside eggs. – Embryonic snapping turtles detoxify most

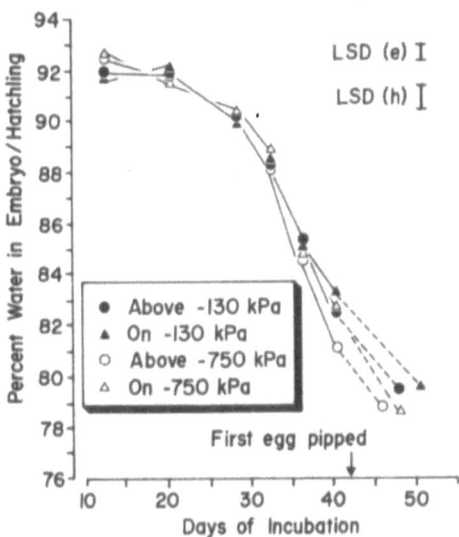

Fig. 7.Temporal variation in the percentage of body mass of embryonic painted turtles that is water (redrawn from G.C. Packard et al., 1983b). Means for embryos exposed to a particular treatment are connected by solid lines; means for embryos on day 42 of incubation are connected with corresponding means for hatchlings by dashed lines. Values for hatchlings are plotted on the average day of emergence for animals in each group. Analysis of covariance indicated that the pattern of decline in percentage of water was affected significantly by water potential of substrates and by position of eggs on or above substrates (G.C. Packard et al., 1983b); significant variation was also detected among hatchlings (G.C. Packard et al., 1983b). See also the caption to Fig. 2.

of the ammonia produced in catabolism of proteins by converting it to urea, which then accumulates within the confines of the egg (G.C. Packard et al., 1983a). Because embryos developing in wet environments have higher rates of metabolism late in incubation than do embryos in dry conditions (Fig. 2), the total amount of urea produced by embryos in the former setting is greater than that produced by those in the latter (G.C. Packard et al., 1983a).

As was noted previously, however, eggs in wet and dry environments differ appreciably in pattern of net water-exchange with the surroundings (Fig. 1), so different amounts of liquid are present inside eggs at different points in incubation to serve as solvent for the nitrogenous waste. The amount of water inside an egg can be estimated from the initial mass of the egg, which is 70% water (Cunningham and Hurwitz, 1936; G.C. and M.J. Packard, 1980; Ricklefs and Burger, 1977), and the change in mass of the egg between oviposition and the day of sampling (see Fig. 1). Assuming that urea is not sequestered in the allantois but instead is distributed proportionately among all compartments inside eggs (see Clark, 1953; Clark and Fischer, 1957; Clark et al., 1957), concentrations of this metabolite can easily be estimated (G.C. Packard et al., 1983a). Interestingly, the highest concentrations of urea are attained inside eggs in dry environments, and the lowest concentrations occur in eggs in wet conditions (Fig. 8).

Urea inhibits activity of numerous enzymes catalyzing key steps in metabolic pathways (Rajagopalan et al., 1961), even when it is present at concentrations as low as those characterizing fluid compartments in eggs of snapping turtles (Giordano et al., 1962; Hand and Somero, 1982). Thus, metabolism of embryos possibly is subject to slight inhibition in consequence of rising concentrations of urea late in incubation, and the degree of inhibition may be greater among embryos in dry environments than among those in wet settings (Fig. 8). This would account for the different patterns of metabolism and growth characterising embryos in different hydric environments (Figs. 2, 3).

On the other hand, concentrations of urea inside eggs of painted turtles also vary with the hydric environment (Fig. 9). Concentrations of this metabolite are higher after day 30 of incubation among eggs exposed to dry conditions than among those in wetter settings (Fig. 9), and inhibition of metabolism presumably should be greater in dry settings than in wet conditions. Nevertheless, metabolism of developing young seems not to be affected by the urea (see Fig. 4), and there is no difference in growth rates among embryos from wet and dry surroundings (Fig. 5).

Although the findings for painted turtles might appear to be cause for rejecting this hypothesis for control of metabolism, we believe that it is premature to do so.

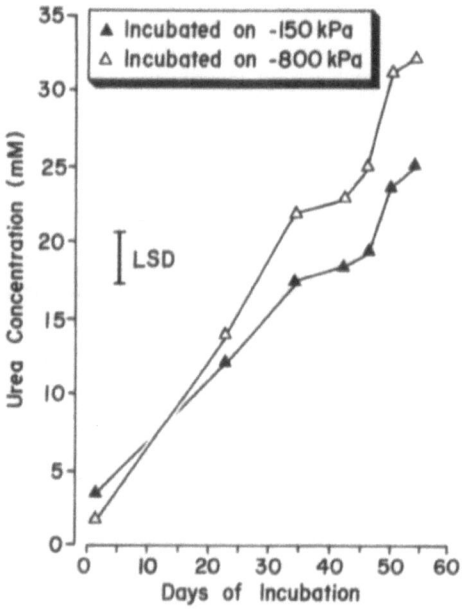

Fig. 8. Temporal variation in the concentration of urea inside eggs of snapping turtles incubating on wet and dry substrates (redrawn from G.C. Packard et al., 1983a). Analysis of covariance indicated that the pattern of increase in concentration of this metabolite varied significantly between the treatments (G.C. Packard et al., 1983a). Means differing by the Least Significant Difference (LSD) are significantly different at alpha = 0.05 (Snedecor and Cochran, 1967).

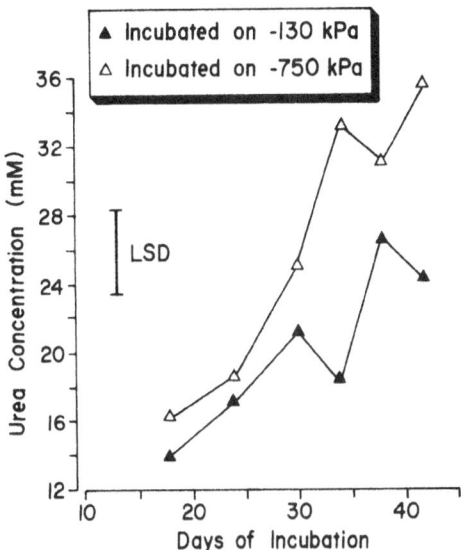

Fig. 9. Temporal variation in the concentration of urea inside eggs of painted turtles incubating on wet and dry substrates. Data were from the study by G.C. Packard et al. (1983b), and computations were as described earlier (G.C. Packard et al., 1983a). Analysis of variance indicated that the pattern of increase in concentration of urea varied significantly between the treatments [$F(5,63) = 4.47$, $P = 0.002$]. Means differing by the Least Significant Difference (LSD) are significantly different at alpha = 0.05 (Snedecor and Cochran, 1967).

The debilitating effects of urea are counteracted in numerous vertebrates by naturally occurring methylamines (Yancey et al., 1982), so embryonic painted turtles may be protected against urea-induced inhibition of metabolism by an as yet undetected compound. This possibility needs to be investigated before dismissing the hypothesis that metabolism of embryonic turtles generally is controlled by accumulation of urea.

Influence of the hydric environment on duration of incubation

Chemical potential of water in the yolk. – The proximate source of water to sustain developing embryos is liquid in the yolk sac (Morris et al., 1983; G.C. Packard et al., 1983b). We therefore studied the chemical potential of water in yolk of eggs of painted turtles held on wet and dry substrates, to determine if there are important differences in water potential that are related to the different patterns of net water-exchange characterizing eggs in different hydric conditions (G.C. Packard et al., 1983b).

Chemical potential of water in the yolk increases during the first 10 days of incubation as water flows into the vitelline sac/yolk sac from the albumen (Ewert, 1979; G.C. Packard et al., 1981a, 1983b), but declines thereafter as water is

113

extracted from the yolk by the developing embryo (Fig. 10). However, at any time during development, water potential is higher in yolk of eggs held in wet environments than it is in yolk of eggs from dry settings (Fig. 10).

The two sets of curves illustrated in Figure 10 are essentially parallel (G.C. Packard et al., 1983b). When straight lines are fit to the data for day 10 onward and then extrapolated to the expected time of pipping, values of -728 kPa ($= 320$ mOsm) and -744 kPa ($= 327$ mOsm) are obtained for residual yolk of young in eggs on wet and dry substrates, respectively. Because these values for water potential of yolk at the time of pipping are statistically indistinguishable, the possibility exists that water potential of this (or some other) compartment provides a cue initiating hatching in painted turtles (G.C. Packard et al., 1983b).

By this hypothesis, embryonic painted turtles in different hydric environments have similar rates of metabolism and growth. However, a hatching reflex is initiated in embryos when water potential of yolk – or of some other water compartment having a water potential correlated with that of yolk – reaches a threshold level. The threshold is reached sooner in eggs incubated in dry environments than in eggs held in wet conditions (Fig. 10), thereby accounting for the shorter incubation observed in the former than in the latter (Fig. 5).

This hypothesis cannot be evaluated fully owing to the absence of data for other species. The hypothesis is attractive, however, in that it can be applied with little modification to embryonic development in snapping turtles, where incubation also is longer in wet environments than it is in dry settings (Fig. 3). On the other hand, despite the fact that differences in water potential recorded in Figure 10 are

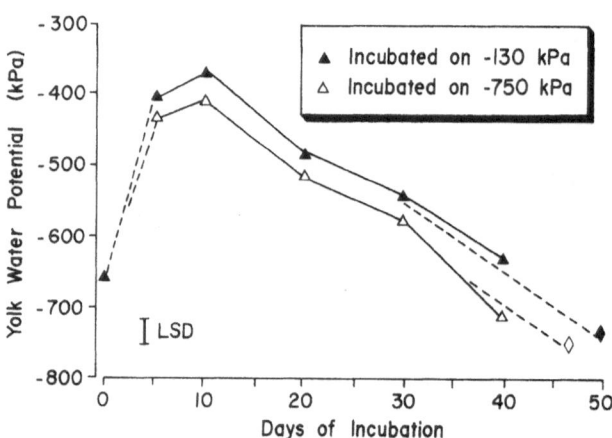

Fig. 10. Temporal variation in chemical potential of water in the yolk of eggs of painted turtles incubating on wet and dry substrates (after G.C. Packard et al., 1983b). Analysis of variance indicated that water potential in eggs on wet substrates was consistently higher than that of eggs on dry substrates (G.C. Packard et al., 1983b). Means differing by the Least Significant Difference (LSD) are significantly different at alpha = 0.05 (Snedecor and Cochran, 1967). The dashed lines at the right are extrapolations from the data to the average times of pipping (diamond symbols) for turtles in each group.

consistent, the absolute difference between eggs in different hydric environments is small (43 kPa = 19 mOsm), raising the possibility that the phenomenon has no biological significance whatsoever. Nonetheless, we believe that the hypothesis merits further study, if only to disprove it.

Summary and conclusions

The preceding discussion illustrates the many problems involved in designing and executing experiments in which one – and only one – of the important factors of the physical environment is varied while others are held constant. However, despite the fact that uncertainties remain concerning the possible influence in our experiments of subtle differences among groups in temperature and/or gas exchange, the weight of evidence indicates that neither of these factors varied sufficiently to account for the physiological responses of developing embryos. The factor that varied most owing to the design of the experiments was availability of water to sustain embryogenesis (Fig. 1), and it is this factor that apparently elicited the major physiological responses we documented. Strong support for this contention comes from the observation that size of hatchling snapping turtles and painted turtles is highly correlated with the net water-exchanges experienced by their eggs – even when the eggs have been incubated in the same hydric environment where variation among eggs in temperature and availability of oxygen is likely to be minimal (M.J. Packard et al., 1982).

One physiological effect of available water is on the duration of incubation, and another is on the rate of metabolism. However, these effects seem to be independent of one another, for both effects need not be operative in any given species. Moreover, the actual means by which water intervenes in the physiology of developing embryos is as yet unknown. As a working hypothesis, we suggest that metabolism in embryos lacking protection against urea is modulated by accumulations of this metabolic waste in body fluids, thereby coupling metabolism and growth of embryos to the hydric environment and to water exchanges experienced by their eggs. We suggest also that the duration of incubation is determined by the chemical potential of water in yolk or some other compartment, with pipping being stimulated when water potential declines to a critical, threshold level.

Embryos of turtles producing eggs with rigid shells do not exhibit different patterns of growth in different hydric environments, nor do they have different periods of incubation (G.C. Packard et al., 1979a, 1981c). However, the eggshells in these species are relatively impervious both to liquid water (G.C. Packard et al., 1979a, 1981c) and to vapor (G.C. Packard et al., 1979b; Tracy, 1982), so there are constraints on the amounts of water that these eggs can exchange with their environment. Eggs incubating in very different hydric conditions therefore do not differ appreciably in their patterns of net water-exchange with the environment

(M.J. Packard et al., 1982), and the potential for physiology to be altered in different hydric environments is thereby minimized. Consequently, we suspect that there are no fundamental differences between embryos in flexible-shelled eggs and those in rigid-shelled eggs with respect to the physiological role for water.

Current evidence indicates that at least three different components of the physical environment – temperature (Bustard and Greenham, 1968; Dodge et al., 1978; Miller and Limpus, 1981; Yntema, 1978), water potential (Morris et al., 1983), and partial pressure of respiratory gases (Ackerman, 1981) – influence metabolism and growth of turtles developing in flexible-shelled eggs. However, all of the investigations yielding information on which this generalization is based were designed to vary one component of the physical environment (i.e., the one of interest) and to hold the others constant. Whereas such studies will continue to be of value in elucidating other factors that may affect physiology of developing embryos and in sharpening focus for studies to be performed in the field, they probably will not yield major new insights concerning the effects of the different variables in nature. Eggs in natural nests are exposed simultaneously to variation in temperature (Alho and Pádua, 1982; Burger, 1976; Goode and Russell, 1968, Hendrickson, 1958; Pieau, 1982; Vogt, 1980; Wilhoft et al., 1983), in water potential (G.L. Paukstis and G.C. Packard, unpublished), and in tensions of respiratory gases (Ackerman, 1977; Thompson, 1981) over the time course of both a single day and the duration of incubation. Thus, new research on the ways physical variables interact to influence physiology of embryos is certainly needed. Such studies may hold surprises for those of us who tend mentally to compartmentalize the effects of different environmental variables (see Gutzke and Paukstis, 1983).

On the other hand, studies to assess the interactions of these several environmental variables empirically will be large and extraordinarily complex. For example, an experiment incorporating three levels of each of the three environmental variables currently of interest (i.e., the minimum number of levels required to establish a pattern of response for each variable by itself) would dictate a total of 27 experimental groups! The number of eggs required for such an experiment is likely to be prohibitive except in extraordinary circumstances, and logistics become a real challenge. Consequently, we anticipate that investigators of the physiological ecology of reptilian eggs and embryos will turn with increasing frequency to mathematical modelling of physiology (Tracy, 1982), followed by selected experiments to test specific, important predictions emerging from the models. Such an approach has been used with considerable success by Muth (1980) in studies of eggs of the lizard *Dipsosaurus dorsalis*, and holds great promise for the future.

Acknowledgements

Our thoughts on this subject were sharpened in the course of discussions with R.A. Ackerman, R.D. Gettinger, W.H.N. Gutzke, and G.L. Paukstis, to all of whom we are grateful. We also thank T.J. Boardman and J.R. zumBrunnen for assistance with statistical analyses and R.D. Gettinger, J.D. Miller, A. Muth, and M.B. Thompson for critically evaluating drafts of the manuscript. A version of the manuscript was expertly typed by Viola Watt.

References

Ackerman, R.A. (1977). The respiratory gas exchange of sea turtle nests (*Chelonia, Caretta*). *Respir. Physiol.* 31: 19–38.

Ackerman, R.A. (1981). Growth and gas exchange of embryonic sea turtles (*Chelonia, Caretta*). *Copeia* 1981: 757–765.

Alho, C.J.R. and Pádua, L.F.M. (1982). Reproductive parameters and nesting behavior of the Amazon turtle *Podocnemis expansa* (Testudinata: Pelomedusidae) in Brazil. *Can. J. Zool.* 60: 97–103.

Bull, J.J. (1980). Sex determination in reptiles. *Q. Rev. Biol.* 55: 3–21.

Burger, J. (1976). Temperature relationships in nests of the northern diamondback terrapin, *Malaclemys terrapin terrapin. Herpetologica* 32: 412–418.

Bustard, H.R. and Greenham, P. (1968). Physical and chemical factors affecting hatching in the green turtle, *Chelonia mydas* (L.). *Ecology* 49: 269–276.

Clark, H. (1953). Metabolism of the black snake embryo. I. Nitrogen excretion. *J. Exp. Biol.* 30: 492–501.

Clark, H. and Fischer, D. (1957). A reconsideration of nitrogen excretion by the chick embryo. *J. Exp. Zool.* 136: 1–15.

Clark, H., Sisken, B. and Shannon, J.E. (1957). Excretion of nitrogen by the alligator embryo. *J. Cell. Comp. Physiol.* 50: 129–134.

Clegg, J.S. (1981a). Metabolic consequences of the extent and disposition of the aqueous intracellular environment. *J. Exp. Zool.* 215: 303–313.

Clegg, J.S. (1981b). Intracellular water, metabolism, and cellular architecture. *Collect. Phenom.* 3: 289–312.

Cunningham, B. and Hurwitz, A.P. (1936). Water absorption by reptile eggs during incubation. *Am. Nat.* 70: 590–595.

Dodge, C.H., Dimond, M.T. and Wunder, C.C. (1978). Effect of temperature on the incubation time of eggs of the eastern box turtle (*Terrapene carolina carolina* Linné). *Florida Mar. Res. Publ.* 33: 8–11.

Engel, W., Klemme, B. and Schmid, M. (1981). H-Y antigen and sex-determination in turtles. *Differentiation* 20: 152–156.

Ewert, M.A. (1979). The embryo and its eggs: development and natural history. In: Turtles: Perspectives and Research. M. Harless and H. Morlock, eds., Wiley, New York, pp. 333–413.

Feder, M.E., Satel, S.L. and Gibbs, A.G. (1982). Resistance of the shell membrane and mineral layer to diffusion of oxygen and water in flexible-shelled eggs of the snapping turtle (*Chelydra serpentina*). *Respir. Physiol.* 49: 279–291.

Flegg, P.B. (1953). The effect of aggregation on diffusion of gases and vapours through soils. *J. Sci. Food Agric.* 4: 104–108.

Giordano, C., Bloom, J. and Merrill, J.P. (1962). Effects of urea on physiological systems. I. Studies on monoamine oxidase activity. *J. Lab. Clin. Med.* 59: 396–400.

117

Goode, J. and Russell, J. (1968). Incubation of eggs of three species of chelid tortoises, and notes on their embryological development. *Aust. J. Zool.* 16: 749–761.

Gutzke, W.H.N. and Paukstis, G.L. (1983). Influence of the hydric environment on sexual differentiation of turtles. *J. Exp. Zool.* 226: 467–469.

Hand, S.C. and Somero, G.N. (1982). Urea and methylamine effects on rabbit muscle phosphofructokinase. Catalytic stability and aggregation state as a function of pH and temperature. *J. Biol. Chem.* 257: 734–741.

Hendrickson, J.R. (1958). The green sea turtle. *Chelonia mydas* (Linn.) in Malaya and Sarawak. *Proc. Zool. Soc. London* 130: 455–535.

Kleiber, M. (1961). The Fire of Life/An Introduction to Animal Energetics. Wiley, New York.

Marshall, T.J. (1959). The diffusion of gases through porous media. *J. Soil Sci.* 10: 79–82.

Miller, J.D. and Limpus, C.J. (1981). Incubation period and sexual differentiation in the green turtle *Chelonia mydas* L. In: Proceedings of the Melbourne Herpetological Symposium. C.B. Banks and A.A. Martin, eds., Zoological Board of Victoria, Parkville, Australia, pp. 66–73.

Mitsukuri, K. (1890). On the foetal membranes of Chelonia. *Anat. Anz.* 5: 510–519.

Mitsukuri, K. (1891). On the foetal membranes of Chelonia. *J. Coll. Sci. Imp. Univ. Tokyo* 4: 1–53.

Morris, K.A., Packard, G.C., Boardman, T.J., Paukstis, G.L. and Packard, M.J. (1983). Effect of the hydric environment on growth of embryonic snapping turtles (*Chelydra serpentina*). *Herpetologica* 39: 272–285.

Muth, A. (1980). Physiological ecology of desert iguana (*Dipsosaurus dorsalis*) eggs: temperature and water relations. *Ecology* 61: 1335–1343.

Norris, D.O. (1980). Vertebrate Endocrinology. Lea & Febiger, Philadelphia.

Packard, G.C. and Packard, M.J. (1980). Evolution of the cleidoic egg among reptilian antecedents of birds. *Am. Zool.* 20: 351–362.

Packard, G.C., Taigen, T.L., Boardman, T.J., Packard, M.J. and Tracy, C.R. (1979a). Changes in mass of softshell turtle (*Trionyx spiniferus*) eggs incubated on substrates differing in water potential. *Herpetologica* 35: 78–86.

Packard, G.C., Taigen, T.L., Packard, M.J. and Shuman, R.D. (1979b). Water-vapor conductance of testudinian and crocodilian eggs (class Reptilia). *Respir. Physiol.* 38: 1–10.

Packard, G.C., Taigen, T.L., Packard, M.J. and Boardman, T.J. (1980). Water relations of pliable-shelled eggs of common snapping turtles (*Chelydra serpentina*). *Can. J. Zool.* 58: 1404–1411.

Packard, G.C., Packard, M.J. and Boardman, T.J. (1981a). Patterns and possible significance of water exchange by flexible-shelled eggs of painted turtles (*Chrysemys picta*). *Physiol. Zool.* 54: 165–178.

Packard, G.C., Packard, M.J., Boardman, T.J. and Ashen, M.D. (1981b). Possible adaptive value of water exchanges in flexible-shelled eggs of turtles. *Science* 213: 471–473.

Packard, G.C., Taigen, T.L., Packard, M.J. and Boardman, T.J. (1981c). Changes in mass of eggs of softshell turtles (*Trionyx spiniferus*) incubated under hydric conditions simulating those of natural nests. *J. Zool.* 193: 81–90.

Packard, G.C., Packard, M.J. and Boardman, T.J. (1982). An experimental analysis of the water relations of eggs of Blanding's turtles (*Emydoidea blandingii*). *Zool. J. Linn. Soc.* 75: 23–34.

Packard, G.C., Packard, M.J. and Boardman, T.J. (1983a). Influence of hydration of the environment on the pattern of nitrogen excretion by embryonic snapping turtles (*Chelydra serpentina*). *J. Exp. Biol.* (in press).

Packard, G.C., Packard, M.J., Boardman, T.J., Morris, K.A. and Shuman, R.D. (1983b). Influence of water exchanges by flexible-shelled eggs of painted turtles *Chrysemys picta* on metabolism and growth of embryos. *Physiol. Zool.* 56: 217–230.

Packard, G.C., Packard, M.J. and Boardman, T.J. (1984). Effects of the hydric environment on metabolism of embryonic snapping turtles do not result from altered patterns of sexual differentiation. *Copeia* (in press).

Packard, M.J., Packard, G.C. and Boardman, T.J. (1982). Structure of eggshells and water relations of reptilian eggs. *Herpetologica* 38: 136–155.

Pieau, C. (1982). Modalities of the action of temperature on sexual differentiation in field-developing embryos of the European pond turtle *Emys orbicularis* (Emydidae). *J. Exp. Zool.* 220: 353–360.

Pieau, C., Mignot, T., Dorizzi, M. and Guichard, A. (1982). Gonadal steroid levels in the turtle *Emys orbicularis* L.: a preliminary study in embryos, hatchlings, and young as a function of the incubation temperature of eggs. *Gen. Comp. Endocrinol.* 47: 392–398.

Rajagopalan, K.V., Fridovich, I. and Handler, P. (1961). Competitive inhibition of enzyme activity by urea. *J. Biol. Chem.* 236: 1059–1065.

Ricklefs, R.E. and Burger, J. (1977). Composition of eggs of the diamondback terrapin. *Am. Midl. Nat.* 97: 232–235.

Schmidt-Nielsen, K. (1964). Desert Animals. Oxford Univ. Press, London.

Snedecor, G.W. and Cochran, W.G. (1967). Statistical Methods. Sixth edition. Iowa State University Press, Ames.

Snyder, G.K. and Birchard, G.F. (1982). Water loss and survival in embryos of the domestic chicken. *J. Exp. Zool.* 219: 115–117.

Tazawa, H. (1980). Adverse effect of failure to turn the avian egg on the embryo oxygen exchange. *Respir. Physiol.* 41: 137–142.

Thompson, M. (1981). Gas tensions in natural nests and eggs of the tortoise *Emydura macquarii*. In: Proceedings of the Melbourne Herpetological Symposium. C.B. Banks and A.A. Martin, eds., Zoological Board of Victoria, Parkville, Australia, pp. 74–77.

Tracy, C.R. (1982). Biophysical modeling in reptilian physiology and ecology. In: Biology of the Reptilia, Vol. 12. Physiology C. Physiological Ecology. C. Gans and F.H. Pough, eds., Academic Press, London, pp. 275–321.

Tracy, C.R., Packard, G.C. and Packard, M.J. (1978). Water relations of chelonian eggs. *Physiol. Zool.* 51: 378–387.

van Bavel, C.H.M. (1952). Gaseous diffusion and porosity in porous media. *Soil Sci.* 73: 91–104.

Vogt, R.C. (1980). Natural history of the map turtles *Graptemys pseudogeographica* and *G. ouachitensis* in Wisconsin. *Tulane Stud. Zool. Bot.* 22: 17–48.

Vogt, R.C. and Bull, J.J. (1982). Temperature controlled sex-determination in turtles: ecological and behavioral aspects. *Herpetologica* 38: 156–164.

Wilhoft, D.C., Hotaling, E. and Franks, P. (1983). Effects of temperature on sex determination in embryos of the snapping turtle. *Chelydra serpentina. J. Herpetol.* 17: 38–42.

Yancey, P.H., Clark, M.E., Hand, S.C., Bowlus, R.D. and Somero, G.N. (1982). Living with water stress: evolution of osmolyte systems. *Science* 217: 1214–1222.

Yntema, C.L. (1978). Incubation times for eggs of the turtle *Chelydra serpentina* (Testudines: Chelydridae) at various temperatures. *Herpetologica* 34: 274–277.

Zaborski, P., Dorizzi, M. and Pieau, C. (1982). H-Y antigen expression in temperature sex-reversed turtles (*Emys orbicularis*). *Differentiation* 22: 73–78.

8. Hydrostatic pressure, shell compliance and permeability to water vapor in flexible-shelled eggs of the colubrid snake *Elaphe obsoleta*

Harvey B. Lillywhite[1] and Ralph A. Ackerman[2]

Abstract

Eggshells of the colubrid snake *Elaphe obsoleta* absorb water vapor and expand reversibly in response to increasing levels of ambient water vapor density. Compliance of the eggshell increases several fold as a function of hydration. Conversely, dehydration of the eggshell causes shrinkage of the shell matrix and reduces its compliance. Pressures inside eggs that are incubated in moist sand average 11.3 kPa just before hatching and increase greatly (>57 kPa) if eggs are exposed to drier conditions. The higher pressures may rupture eggshells at foci of stress where the outer calcareous layer has exfoliated from underlying membranes.

Permeability of the eggshell to water vapor also varies with the hygric environment, increasing greatly in humid atmospheres. Ultrastructural observations indicate that conformational changes within the eggshell attend changes in its state of hydration. However, the relationship between egg permeability and ambient water vapor is apparently unrelated to structural changes within the eggshell. The gross shape of the curve relating permeability to ambient water vapor is similar to that of a wet agar egg of similar size and changes little or not at all during development when size of the egg increases. Permeabilities of snake eggs are at least two orders of magnitude greater than those of avian eggs, irrespective of the ambient water vapor, and this fact has profound ecological importance.

Introduction

Eggs of reptiles differ strikingly from those of contemporary birds, some of the more obvious differences being the mechanical properties of the eggshell. In

[1] *Department of Zoology, University of Florida, Gainesville, Florida 32611, U.S.A.*
[2] *Department of Zoology, Iowa State University, Ames, Iowa 50011, U.S.A.*

Seymour, R.S., (ed.) Respiration and metabolism of embryonic vertebrates.
© *1984, Dr W. Junk Publishers, Dordrecht/Boston/London. ISBN-13: 978-94-009-6538-6*

many lizards and snakes, as well as some turtles, the shell lacks a heavily mineralized structure and is termed 'flexible' or 'parchment-like' (Packard et al., 1977; Packard and Packard, 1980) in contrast to the predominantly mineral shell of some turtles, crocodilians and birds, which is 'rigid'. Whereas rigid eggshells retain their shape throughout incubation, flexible shells may shrink or expand depending on the magnitude and direction of net water exchange with the nest environment. Consequently, eggs with a compliant shell may swell enormously due to uptake of water during incubation (e.g., Badham, 1971; Tracy, 1980; Andrews and Sexton, 1981; Packard et al., 1982). Although numerous studies have examined the water relations of flexible-shelled eggs, little information is yet available concerning eggshell compliance as it relates to 'turgor' of the egg and physiology of the developing embryo.

Here we report observations concerning the compliance of reptilian eggshells and the mass transfer characteristics of the eggs. Specific objectives are to quantify in vivo compliance and to relate changes in this variable to hydration and permeability of the eggshell, and hydrostatic pressures inside eggs.

Material and methods

Eggs from three captive snakes (*Elaphe obsoleta*) were obtained immediately after oviposition. Seventeen eggs were oviposited by two newly captured females on 4 and 5 July, 1982 (clutches 1 and 2); a third clutch of twelve eggs was oviposited by a third captive female on 2 February, 1983. All of the eggs were incubated completely buried in fine, moist sand and spaced more than 2 cm apart. The sand filled about half of the volume of a plastic container $30 \times 16.5 \times 9$ cm $(L \times W \times H)$ which was covered with a glass plate and maintained inside a darkened cabinet. Small volumes of distilled water were added to the sand periodically so that the uppermost layer remained moist without causing significant condensation of water droplets on the glass plate. This procedure maintained the moisture level in sand at ca. 2.5% by mass. The water potential of sand containing 2.5% water is typically between -100 and 0 kPa (unpublished data) but cannot be measured accurately with thermocouple techniques (Hillel, 1980). Eggs were removed from the sand periodically and weighed to the nearest 0.01 g. Two clutches were incubated at a mean temperature of 24.0 °C (range 22–28 °C); the third clutch was incubated at a mean temperature of 22.8 °C (range 21.4–23.4 °C)

Hydrostatic pressures of eggs

Measurements of pressures within extraembryonic compartments of eggs were obtained in two ways. Single or instantaneous measurements were made by impaling the upper eggshell of incubating eggs with a 23 gauge hypodermic needle fitted to PE-50 polyethylene tubing connected to a Gould-Statham P23-ID pressure transducer. A small area of shell ($<$1 cm^2) was temporarily exposed and then recovered with moist sand after the needle was in place. The volume of the measuring system from transducer to needle tip was filled with 0.9% saline. Repetitive or continuous recordings of pressure were obtained from indwelling catheters (PE-50) implanted into eggs by insertion through a needle puncture that was subsequently sealed with cyanoacrylate cement. Pressures were recorded on chart paper with a Grass model RPS 7C polygraph. It was impractical to determine the precise location of catheter tips, but in all cases the catheters contacted relatively clear, viscous fluid, probably within the allantoic or amniotic sacs.

Pressures were measured in 16 eggs buried in sand and in five eggs while they were removed from sand and subjected temporarily to a range of ambient moisture conditions. Water vapor pressure external to eggs was controlled by placing eggs inside plastic containers ($30 \times 16.5 \times 9$ cm) having variable quantities of anhydrous calcium sulfate (Drierite) or water placed in small plastic weigh trays. The container was covered with a plastic lid, but was not sealed completely (air tight) from outside room air. Thus, water vapor leaked slowly into or out of the container. Different amounts of water or Drierite varied the internal water vapor and established a range of steady state conditions within the container. Pressures were also measured separately in three eggs that were alternately painted with water and allowed to dry in room air. Temperature within one of the wetted eggs was measured with a 40 gauge copper-constantan thermocouple, implanted similarly as a catheter, and coupled to a Honeywell potentiometric recorder.

Eggshell compliance

Compliance of eggshells was determined in vivo in nine eggs that were buried in sand, two weeks or less prior to hatching, and in four eggs implanted with catheters and subjected to a range of ambient vapor pressures as described above. Because egg pressure varied with humidity (see below), each egg was 'adjusted' to a common reference pressure (13.33 kPa) by adding or subtracting fluid via the catheter prior to measurement at any ambient condition. Compliance was calculated from the change in pressure observed following the sequential injection of three 50 μl volumes of saline into the egg at 1 min intervals.

Eggshell hydration

The amount of water vapor absorbed at different ambient vapor pressures was determined for empty eggshells dried to constant mass in a desiccator (0 kPa) at

23 °C. Shells of snake eggs were thoroughly cleaned of internal contents after hatching and dried initially by flowing room air through the egg. Air under pressure entered the egg through a piece of Tygon tubing at one end and exited through a small slit made by the eggtooth of the hatchling snake. The air pressure minimized wrinkling and distortion of the eggshell during drying. For purposes of comparison, mineralized eggshells were obtained by 'blowing out' eggs of domestic chickens (*Gallus domesticus*). All openings or holes in both types of eggshells were sealed with silastic following complete drying. Hydration of closed and empty eggshells with water vapor was quantified by repeated measurements of mass (to 0.1 mg) at equilibrium during exposure to increments of ambient water vapor pressure established within the glass draft shield ($18 \times 15 \times 18$ cm) of a Mettler AC 100 electronic balance. Humidity levels were established by adjusting quantities (and surface areas) of anhydrous calcium sulfate or water held in plastic weigh trays within the draft shield, as described for plastic containers. Relative humidity was maintained within 0.5% and measured with a PCRC-11T humidity sensor positioned within the draft shield and coupled to a Humeter (Phys-Chemical Research Corp., New York). Air temperature was maintained within 1 °C and measured with a 40 gauge thermocouple and a Honeywell potentiometric recorder.

Equilibrium water content of eggshells was expressed as percentage of dry mass of eggshell and as mass per unit volume of eggshell. For the latter measurements, eggshell area was estimated by comparing the total mass of an individual shell with that of a representative piece of shell cut to a known area. It was assumed that the thickness of any individual shell was approximately uniform. Eggshell thickness was measured from small fragments using a dial caliper (precision 0.01 mm).

Permeability to water vapor

Permeability of eggshells to water vapor was determined from five previously undisturbed eggs, each positioned diagonally across the upper edges of a plastic weigh tray within the draft shield of the Mettler AC 100 balance. Changes of mass (to 0.1 mg) were determined at 5 or 10 min intervals following at least 60 min equilibration and the establishment of a constant rate of water loss. Such measurements were conducted at ambient temperatures ranging from 24.8 to 28.4 °C ($\bar{x} = 27.0 \pm 0.88$ s.d.) for different eggs. Steady state vapor pressures within the draft shield were varied using drying agent or water as described above. Although the draft shield was somewhat 'leaky' with respect to water vapor, we could measure no convection with a hot wire anemometer (Alnor) and regard the setup as a free convection system. Humidity and temperatures were measured as described for the hydration experiments. Eggshell temperatures, in addition to air temperatures, were measured with a 40 gauge thermocouple that was adpressed lengthwise against the eggshell to maximize contact. The conductance of eggs to water vapor was calculated by dividing the mass loss by the difference of water vapor pressure between the egg (assumed saturated at eggshell temperature) and ambient atmosphere. These results were then expressed per unit of surface area (permeability) to facilitate comparisons of different aged eggs. Surface areas of eggs were estimated using the formula for a prolate spheroid and measurements of the principal dimensions of eggs.

Similar measurements were made using a 'wet egg' model cast in 2% agar from a hollow snake egg (see Tracy and Sotherland, 1979).

Surface morphology of eggshells

Pieces of shell from three hatched eggs were examined with a Philips 501 scanning electron microscope. Specimens were dried in room air (critical point drying rendered no improvement) and coated with gold-palladium in a Technics Hummer II Sputter Coater. Pieces of shell (each ca. 4 mm^2) were cut in pairs from the same area and each examined in one of two conditions. One of each pair was mounted 'dry'; the other was first hydrated between layers of wet paper, and then fixed in 2% buffered glutaraldehyde before mounting. Some 'dry' specimens were also fixed in glutaraldehyde to reveal any effects of the fixative. Four specimens were exposed briefly to sulfuric acid to remove the mineralized (outer) shell prior to further treatment.

Results

General observations, development and hatching

Eggs in clutches 1 and 2 weighed 14.8 ± 1.22 g (s.d.) and those in clutch 3 weighed 20.03 ± 1.04 g at the beginning of incubation. All eggs increased in mass during development due to uptake of water from the incubation medium (Packard et al., 1977). Rates of water uptake were initially rapid and remained relatively constant until the latter third of the incubation period when the mass of eggs stabilized (Fig. 1). Hatchling snakes averaged $90.7 \pm 0.03\%$ (s.d.) of the initial egg mass in clutches 1 and 2 and $49.3 \pm 9.72\%$ of initial egg mass in clutch 3. Development of snakes was curtailed or abnormal in the latter clutch, and only four of these eggs (25%) hatched. All of the eggs in clutches 1 and 2 hatched, including those that were impaled with catheters and several that were either ruptured or opened prematurely for examination. The cause of developmental problems in clutch 3 is uncertain.

Uptake of water resulted in swelling and considerable enlargement of eggs, sometimes causing the outer shell layer to fracture and separate during the latter

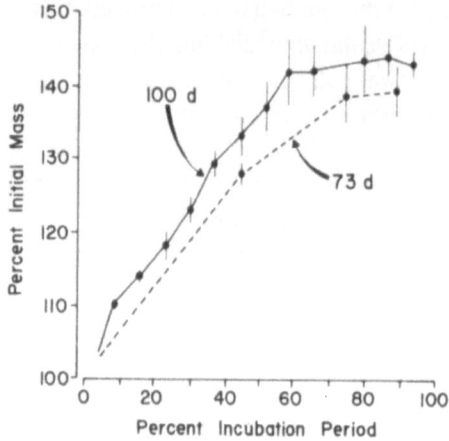

Fig. 1. The percent of original mass of eggs plotted as a function of the percent of incubation completed. Data are means and two standard errors on either side of the mean for clutch 3 (solid line, N = 6) and clutches 1 + 2 (dashed line, N = 11). Numbers adjacent to the lines indicate the average length of incubation for the two data sets (hatching of different eggs within a clutch occurred over four days). Differences of incubation times and mass changes are probably the result of different incubation temperatures (see text). Data are omitted for eggs in which the mass was altered by experimental procedures. Three eggs in clutch 3 were opened artificially; two were dead in an advanced state of development.

half of incubation (Fig. 2). The underlying shell membranes protruded abnormally in a few eggs where exfoliation was extensive. After hatching, eggshells were noted to shrink considerably as they dried.

Hydrostatic pressures, hydration and compliance of eggshells

Pressures in buried eggs ranged from 8.5 to 17.1 kPa ($\bar{x} = 11.3 \pm 2.77$ s.d.) during the latter two weeks of incubation and varied inversely with eggshell hydration when eggs were alternately wetted and then allowed to dry (Figs. 3, 4). Exposure of eggs to <1.2 kPa water vapor pressure increased pressures to levels exceeding 57 kPa and sometimes ruptured eggs where the outer eggshell had exfoliated. Pressure changes induced by hydration of eggshells were reversible and not due to cooling of the eggshell surface (Fig. 4). Eggs were highly sensitive to external water vapor; small and transient changes of ambient water vapor pressure or even partial exposure of buried eggs produced abrupt changes of egg pressure.

Empty shells absorbed substantial quantities of water vapor, even at low ambient humidity. In comparison with mineralized eggshells of domestic chickens, those of *E. obsoleta* absorbed greater than ten-fold more water vapor per unit mass or volume (Fig. 5) and did so more rapidly (Fig. 6). Compliance of eggshells determined in vivo averaged 50.65 μl $\cdot$ kPa^{-1} ($\bar{x} \pm 14.73$ s.d., N = 9) in

Fig. 2. Photograph of egg that has completed 80% of incubation. Note fracturing of the outer shell layer. Bar equals 1 cm.

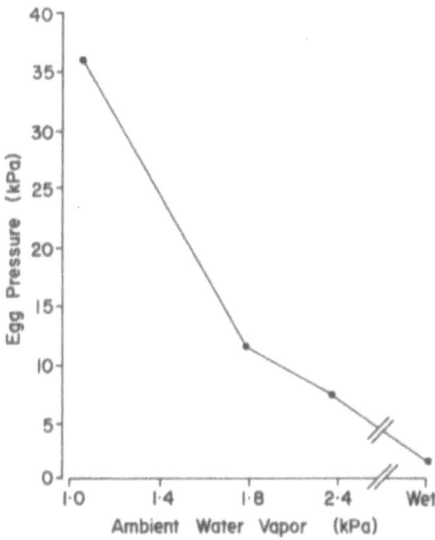

Fig. 3. Data for a single egg showing the relationship between internal pressure and ambient water vapor pressure. Air temperature was 22–23 °C. The egg was exposed to each condition for ca. 1 h during which internal pressure became relatively stable. Because of continuous water loss from the egg, however, pressure measurements are not true equilibria. The last data point to the right on the abscissa depicts pressure measured in the egg after it had been painted with water (at 22.6 °C). Egg had completed 97% of its incubation.

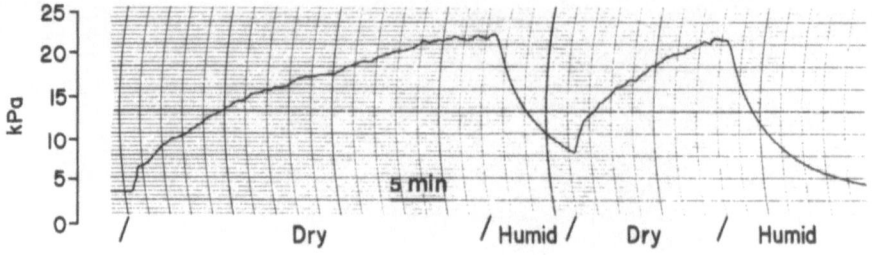

Fig. 4. Record of internal pressure measured in an egg that was alternately exposed to 'dry' (3% RH) and 'humid' (90% RH) air. Air temperature was 22 °C. Egg had completed 75% of its incubation. Similar patterns of pressure resulted when eggs were wetted with water and then allowed to dry. Varying the water temperature (22–40 °C) did not change the pattern of pressure change.

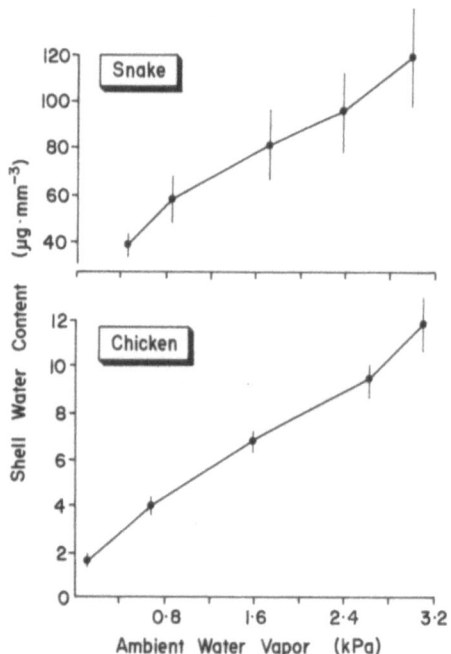

Fig. 5. Water content of empty eggshells equilibrated at selected ambient water vapor pressures. Mean air temperature was 26.4 (±2.5 s.d.) °C. Data points are means and two standard errors on either side of the mean, plotted for five snake eggs (*Elaphe obsoleta*) and five domestic chicken eggs (*Gallus domesticus*).

128

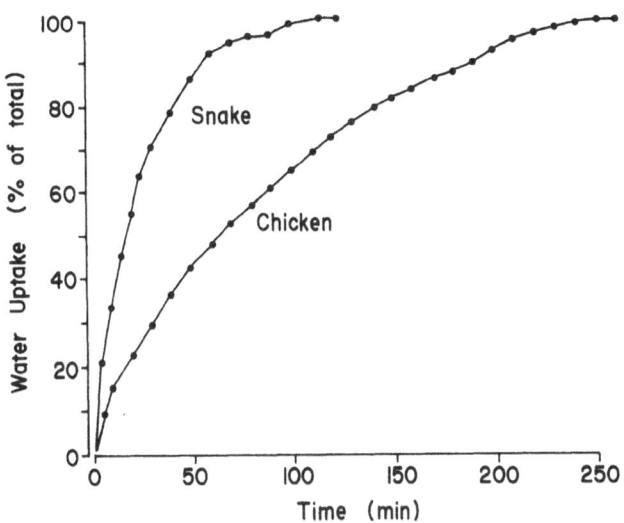

Fig. 6. Time course of water uptake measured in single eggs of *Elaphe obsoleta* and *Gallus domesticus*. Eggs were initially dry and exposed to an ambient water vapor pressure of 0.7 kPa (25% RH, air temperature 23 °C).

advanced eggs incubated in moist sand and diminished markedly when eggs were exposed to drier environments (Fig. 7). Compliance increased three-fold (to $140 \, \mu l \cdot kPa^{-1}$) in a single egg exposed to a saturated atmosphere, then wetted with liquid water.

Permeability to water vapor

Permeability of eggs to water vapor was nearly constant in dry air (<2 kPa ambient water vapor pressure) and increased exponentially at more saturated conditions (Fig. 8). Permeability did not appear to change during development of eggs, although the highest values measured were in advanced eggs at high ambient humidity. Differences between early and late stage eggs were not affirmed statistically because of the small number of eggs measured and the fact that ambient conditions differed uniquely for each egg. Permeability values for an agar egg were roughly 30% higher than those for eggs of *E. obsoleta*. Except for a more pronounced upturn at low humidity, they followed a similar pattern with respect to ambient conditions (Fig. 8).

Morphology of the eggshell

Eggshells of *E. obsoleta* consist of two parts. The outer layer is relatively rigid and

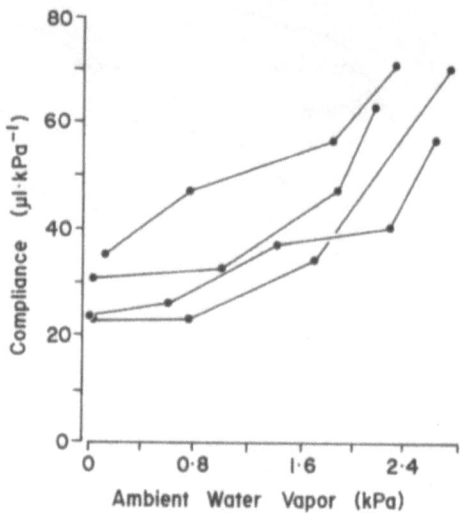

Fig. 7. Compliance of whole eggs determined in vivo at selected conditions of ambient water vapor pressure. All measurements were made within two weeks of hatching.

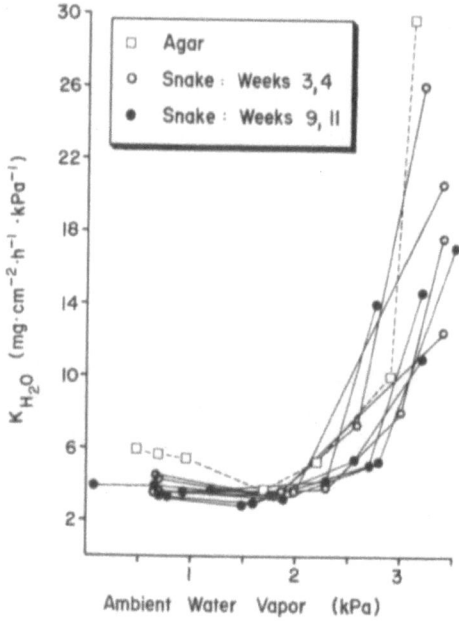

Fig. 8. Permeability to water vapor of eggs exposed to selected conditions of ambient water vapor pressure. Lines connect data points for individual eggs, and the different symbols indicate the times of measurement in relation to the total incubation period of 14 weeks. Three eggs were tested twice, once during early development and again during late development. Two additional eggs (total n = 5) were tested once, during early and late development respectively. Eggs increased in volume ca. 20% between the two periods of measurement. Boxes depict data for an artificial egg cast in agar. The agar egg presumably evaporates as a free water surface and has no internal resistance equivalent to the eggshell (Tracy and Sotherland, 1979).

130

tends to split or fragment during swelling of the egg. It consists of fiber-supported networks of spheroid particles and effervesces when placed in acid, suggesting that the layer is a calcareous matrix. The underlying or inner shell layer is fibrous (Fig. 9E). Both layers possess pores (Fig. 9A, B) which appear to be sparsely distributed and without regular pattern.

Material from shell fragments that are hydrated prior to fixation tends to separate more than that of fragments that are initially dry (Fig. 9B, C, D, F). These differences are not seen consistently in all specimens, however.

Discussion

The eggshell of *Elaphe obsoleta* has limited development of mineralized components and thus resembles eggshells of other squamate reptiles (Packard et al., 1977, 1982). In both reptilian and avian eggs the degree of mineralization determines rigidity of the shell and limits water exchange during incubation (Packard et al., 1982). Within the Reptilia, for example, strictly fibrous eggs may double or triple in mass (e.g., Andrews and Sexton, 1981) whereas calcareous eggs experience little or no increase in mass during incubation (Lynn and von Brand, 1945). The fibrous layers of the shell are proteinaceous and elastic, whereas mineralized components are sometimes virtually inflexible. In eggshells of *E. obsoleta* and certain other squamates, the calcareous layer is interspersed with fibers and therefore variably elastic. However, splitting of this layer and protrusion of the underlying membranes during development indicate that the former is less compliant.

Uptake of water by eggs expands volume against the partially mineralized shell and thus imparts pressure to the internal fluid system which lacks an air cell (Packard et al., 1977; our unpublished observations). Inasmuch as freshly oviposited eggs are relatively flaccid, we infer that pressure increases throughout incubation (see also Packard et al., 1977) and reaches maxima of ca. 11 kPa in late stage eggs of *E. obsoleta* incubated in sand. These pressures are conceivably similar to those in natural nests where eggs are buried in moist materials (Fitch, 1963). Pressures measured in snake eggs are roughly one order of magnitude greater than published estimates of pressures in turtle eggs based on impression tonometry (Quay, 1976).

Reductions of ambient humidity increase pressure in eggs that are turgid (Figs. 3, 4) and potentially rupture the eggshell. Therefore, eggs are susceptible to rather large changes of pressure if there is temporal or spatial heterogeneity in the hygric environment. Exposure of eggs to drier than 'normal' conditions can be caused by drought, disturbance of eggs by predators or other animals (H. Fitch, personal communication), and possibly 'poor' choices of nesting sites by ovipositing females. Whether or not hydrostatic pressures influence the developmental physiology of embryos remains to be investigated.

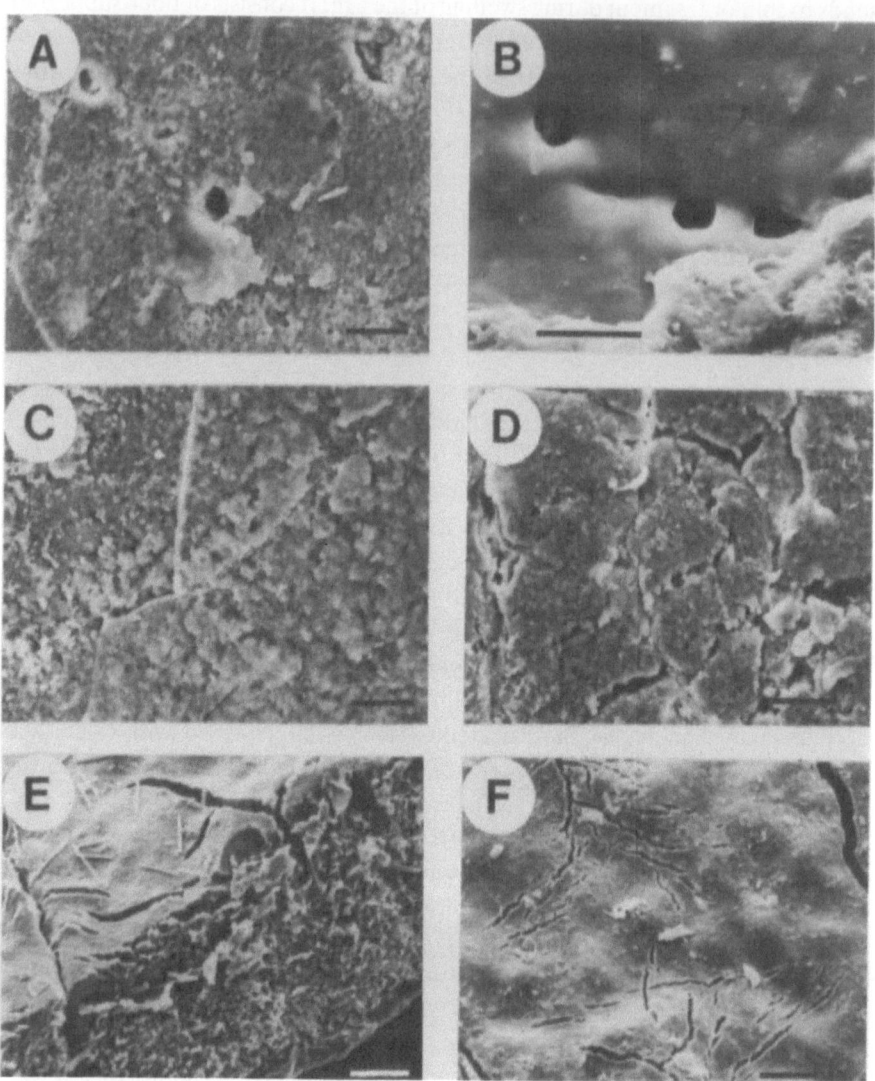

Fig. 9. Scanning electron micrographs of surfaces of eggshells of *Elaphe obsoleta* following hatching. A, outer surface of a dry shell showing calcareous layer and pores. B, inner or fibrous layer and pores of dry eggshell exposed by etching the calcareous (outer) layer with acid. C, outer surface of dry shell showing fractures which apparently open when the shell is hydrated. D, outer surface of the same shell illustrated in C, but in this photograph the sample was hydrated prior to fixation. Note that fractures appear to spread apart, and the shell has a more 'open' appearance than in C. E, radial view of a shell showing the fibrous layer and what appears to be a terminal inner membrane. The outermost calcareous matrix has been removed with acid. F. surface view of the inner or fibrous layer of a shell fragment that was etched with acid and subsequently hydrated prior to fixation. Compare features of fracturing with the appearance of the same layer in B. Scale bars; A, C through F = 50 μm; B = 10 μm.

The inverse relationship of egg pressure to ambient water vapor pressure reflects tendency of the eggshell to expand or contract about a fixed volume of fluid. Quay (1976) measured the tendency for eggshells of five species of turtles to resist deformation by impression tonometry and found that deformation was greatest (least tension) when the eggshell was moist. Compliance of flexible eggshells clearly is a function of eggshell hydration (Fig. 7) and appears to be mediated by changes in the tensile properties of the fibrous layers. Absorption of water (aqueous or vapor) by the shell disrupts or expands the fibrous matrix (Fig. 9) presumably due to swelling of protein; intramolecular configurations and intermolecular packing of shell proteins are probably altered by hydrogen bonding and the accomodation of bound water (Scheuplein and Morgan, 1967). Comparisons of water uptake in flexible and rigid eggshells (Fig. 5, 6) indicate that calcareous materials impede both the rate and absolute magnitude of water absorption by shells. Thus, hydration as well as elastic properties of eggshells appear to be adjusted by the ratio of mineral to proteinaceous components.

The foregoing suggests that structural changes caused by hydration or stretching of shells might alter the permeability of eggs (see also Feder et al., 1982). We note, however, that permeability of snake eggs to water vapor is at least two orders of magnitude greater than that of avian eggs, and rates of water loss approach those from free water surfaces. These comparisons hold over a range of ambient conditions (H.L., unpublished observations), and patterns of permeability vs atmospheric water vapor (hence eggshell hydration) are nearly identical for snake eggs and similarly shaped agar models (Fig. 8). Exceptions are that permeabilities of real eggs tend to diverge downward from those of the agar model at extreme ends of the humidity scale. Stretching of the eggshell due to water uptake (by eggs) during incubation does not have measurable effects on permeability, except possibly under conditions of high humidity where permeability appears to be reduced in late vs early stage eggs (Fig. 8). Whether the latter effect is statistically significant cannot be established at present; we can think of no compelling reason to expect such a circumstance. Thus we conclude that the gross shape of the curve relating egg permeability to ambient water vapor (Fig. 8) is not attributable to changes in the physical structure of the eggshell. Moreover, it is evident that film or boundary layer conditions must contribute significantly to the resistance of flexible-shelled eggs to losses of water vapor under conditions of free convection (Ackerman et al., in press). The increase in permeability at very high humidities is likely to be due to an effective decrease in the thickness of the layer (shell plus external film) across which water vapor is diffusing as saturation equilibrium is approached. If it is true that agar eggs have no internal resistance to water movement (Tracy and Sotherland, 1979), then the effect is likely to be attributable to reduction or disappearance of the external film.

The high water vapor conductance of flexible-shelled reptilian eggs has tremendous ecological importance. The permeability of these eggs is so great that exposure to vapor pressure environments even slightly less (<99%) than saturated will result in substantial losses of egg water. Conversely, if the environmental vapor pressure exceeds that inside the egg, the egg will take up water as vapor condenses in the shell. Thus, stability of the environment and the potential selection of nesting sites by females will influence the survival of eggs. It has been suggested that the uptake of water by eggs has survival value for the developing embryo (Packard et al., 1982, 1983). Some of this water, stored internally in diverticula of the gut, may be usable by the hatchlings (Ackerman, Ar and Dmi'el, unpublished observations).

Hydration of eggshells with liquid water reduces conductance to O_2 and potentially inhibits the development of embryos (Feder et al., 1982; Packard et al., 1977). Thus, liquid water connections between the substrate and interior of the egg which would result in flooding of the shell (saturated hydraulic conductivity) seem unlikely. The presence of morphologically distinct 'pores' in the eggshell of *E. obsoleta* (Fig. 9A, B) is especially interesting because it has been reported that lepidosaurian eggshells possess no pores (Packard et al., 1977) and the functional significance of such structures is puzzling in view of the extremely high permeability of fibrous and flexible type eggshells. It does appear possible that these pores possess hygrophobic features which provide 'dry' conductive pathways for O_2 under all hygric conditions short of complete flooding.

A survey of reptiles indicates that heavily calcified eggs have evolved in the Crocodilia, Chelonia, gekkonid and dibamid lizards, while viviparity (hence reduction of eggshell) has arisen independently at least 95 times in lizards, snakes and amphisbaenians (Shine, present volume). Because eggshell properties have obviously undergone frequent evolutionary modification and because the physiological controls of shelling appear to be relatively labile (Cuellar, 1979; Guillette, 1982), eggshell morphology is predicted to be especially responsive to selective forces that are prevalent during embryonic development. Clearly, heavily calcified and relatively non-compliant shells confer advantages of reduced water loss and structural stability to eggs that undergo development in open or potentially fluctuating environments. Thus reptilian eggs incubated in a variety of settings are characterized by a corresponding variation in permeability and shell structure (Ackerman, Ar and Dmi'el, unpublished observations).

Acknowledgements

We are grateful to Roger Seymour, Martin Feder and two anonymous referees for contributing comments that led to improvements in the manuscript. This research was supported in part by the U.S. Public Health Service, NIH grant # HL24640.

References

Ackerman, R.A., Dmi'el, R and Ar, A. (in press). Energy and water vapor exchange by parchment-shelled reptile eggs. *Physiol. Zool.*

Andrews, R.M. and Sexton, O.J. (1981). Water relations of the eggs of *Anolis limifrons*. *Ecology* 62: 556–562.

Badham, J.A. (1971). Albumen formation in eggs of the agamid *Amphibolurus barbatus barbatus*. *Copeia* 1971: 543–545.

Cuellar, H. (1979). Disruption of gestation and egg shelling in deluteinized oviparous whiptail lizards *Cnemidophorus uniparens*. *Gen. Comp. Endocrinol.* 39: 150–157.

Feder, M.E., Satel, S.L. and Gibbs, A.G. (1982). Resistance of the shell membrane and mineral layer to diffusion of oxygen and water in flexible-shelled eggs of the snapping turtle (*Chelydra serpentina*). *Respir. Physiol.* 49: 279–291.

Fitch, H.S. (1963). Natural history of the black rat snake (*Elaphe obsoleta obsoleta*) in Kansas. *Copeia* 1963: 649–658.

Guillette, L.J. Jr. (1982). The evolution of viviparity and placentation in the high elevation, Mexican lizard *Sceloporus aeneus*. *Herpetologica* 38: 94–103.

Hillel, D. (1980). *Fundamentals of Soil Physics*. Academic Press, New York.

Lynn, W.G. and von Brand, T. (1945). Studies on the oxygen consumption and water metabolism of turtle embryos. *Biol. Bull.* 88: 112–125.

Packard, G.C. and Packard, M.J. (1980). Evolution of the cleidoic egg among reptilian antecedents of birds. *Amer. Zool.* 20: 351–362.

Packard, G.C., Packard, M.J., Boardman, T.J., Morris, K.A. and Shuman, R.D. (1983). Influence of water exchanges by flexible-shelled eggs of painted turtles *Chrysemys picta* on metabolism and growth of embryos. *Physiol. Zool.* 56: 217–230.

Packard, G.C., Tracy, C.R. and Roth, J.J. (1977). The physiological ecology of reptilian eggs and embryos, and the evolution of viviparity within the class Reptilia. *Biol. Rev.* 52: 71–105.

Packard, M.J., Packard, G.C. and Boardman, T.J. (1982). Structure of eggshells and water relations of reptilian eggs. *Herpetologica* 38: 136–155.

Quay, W.B. (1976). Impression tonometry of Chelonian eggs (Reptilia, Testudines). A guide to internal pressure and membrane phenomena. *J. Herpetol.* 10: 55–62.

Scheuplein, R.J and Morgan, L.J. (1967). 'Bound water' in keratin membranes measured by a microbalance technique. *Nature, Lond.* 214: 456–457.

Tracy, C.R. (1980). Water relations of parchment-shelled lizard (*Sceloporus undulatus*) eggs. *Copeia* 1980: 478–482.

Tracy, C.R. and Sotherland, P.R. (1979). Boundary layers of bird eggs: do they ever constitute a significant barrier to water loss? *Physiol. Zool.* 52: 63–66.

9. Influence of incubation water content on oxygen uptake in embryos of the Burmese python (*Python molurus bioittatus*)

Craig Patrick Black[1], Geoffrey F. Birchard[2], Gordon W. Schuett[3] and Virginia D. Black[4]

Abstract

Flexible-shelled reptilian eggs exchange water via two routes; 1) diffusive water vapor loss through shell interstices and 2) liquid water imbibition from incubation substrate in response to a water potential gradient. Uptake of liquid water may result in the filling of some of the shell interstices which would in turn increase shell O_2 diffusion resistance. We measured whole-egg mass changes, water vapor conductance (G_{H_2O}), O_2 consumption, trans-shell P_{O_2} gradient, incubation period, and hatchling weights in four groups of Burmese python (*Python molurus bivittatus*) eggs incubated in substrates with estimated water potentials of -360, -220, -130 and -80 kPa respectively. O_2 consumption among the four groups did not differ significantly throughout incubation. Trans-shell P_{O_2} gradients measured during the periods of maximum O_2 consumption varied from 34 ± 6 (s.e.) torr in eggs incubated in the driest substrate to 64 ± 9 torr in eggs incubated in the wettest substrate, and O_2 conductance values calculated from trans-shell P_{O_2} and O_2 consumption data were only about one-tenth of that predicted from G_{H_2O}. Thus, the incubating Burmese python egg has a functional water layer in the shell, and the embryos are subject to hypoxia comparable to that observed in developing chicken embryos. This does not appear to compromise tissue O_2 delivery, however, because hatchling weights and O_2 consumption are not adversely affected by increased incubation substrate water content.

[1] *Department of Biology, University of Toledo, Toledo, Ohio 43606, U.S.A.*
[2] *Department of Physiology, Dartmouth Medical School, Hanover, New Hampshire 03755, U.S.A.*
[3] *Department of Biology, Central Michigan University, Mt Pleasant, Michigan 48859, U.S.A.*
[4] *Department of Paediatrics, Wayne State University School of Medicine, Detroit, Michigan 48201, U.S.A.*

Seymour, R.S., (ed.) Respiration and metabolism of embryonic vertebrates.
© *1984, Dr W. Junk Publishers, Dordrecht/Boston/London. ISBN-13: 978-94-009-6538-6*

Introduction

During incubation, water relations in the flexible-shell reptilian egg are quite complex. Unlike avian eggs which are normally incubated on a dry substrate and exhibit continuous diffusive water vapor loss, reptilian eggs frequently show simultaneous, bidirectional water flux which occurs through two distinct processes: 1) diffusive loss of water vapor to the surrounding atmosphere dependent upon temperature and ambient vapor pressure, and 2) imbibition of liquid water from the substrate dependent upon water potential difference between substrate and egg contents (Packard et al., 1977; Packard and Packard, 1980).

Not only is water uptake during incubation a common phenomenon in flexible-shell reptilian eggs, but it may be essential for normal development, since at least some reptilian eggs are laid without sufficient water to allow the embryo to totally mobilize yolk resources during development (Packard and Packard, 1980; Packard et al., 1977, 1981). In addition, both wet and dry hatchling weights increase with increasing water uptake during incubation in an least one species, the painted turtle, *Chrysemys picta*, apparently because embryos developing in more moist substrates can develope further prior to hatching (Packard et al., 1981, 1983). The primary determinant of liquid water uptake by the egg is substrate water potential, which is determined by substrate water content (Tracy et al., 1978).

The movement of liquid water through the shell has consequences beyond egg water relations, however. The flexible shell is penetrated by numerous interstices which provide a route through the shell for both diffusive respiratory gas exchange and the entrance of liquid water. Because the egg is constantly exchanging liquid water, many of these pores are likely filled either partially or completely with liquid water, particularly when the egg is being incubated in a medium saturated with water. Since the diffusion coefficient of O_2 through atmosphere is approximately 10^4 times greater than that through water, a layer of liquid water in the shell interstices would result in increased resistance to the diffusive movement of O_2 across the shell, perhaps sufficient to adversely affect metabolism and growth.

Two alternative responses by the embryo to this situation may be hypothesized: 1) the embryo could decrease O_2 consumption and thereby maintain a P_{O_2} under the shell adequate for tissue oxygenation, or 2) it could maintain O_2 consumption at normal levels which would produce an increase in the P_{O_2} gradient across the shell proportional to the decrease in shell O_2 conductance. Since water uptake by flexible-shelled eggs is dependent upon incubation substrate water content, the first hypothesis would predict that O_2 consumption would decrease with increasing substrate water content while the second hypothesis would predict that the P_{O_2} gradient across the shell would increase with increasing substrate water content.

In this study, groups of eggs from a single clutch of Burmese python (*Python molurus bivittatus*) eggs were incubated during approximately the last half of the incubation period in substrates at four different levels of water content to test the preceeding two predictions.

Materials and methods

Twenty-four days following oviposition, 28 eggs were selected from a single clutch of 58 Burmese python eggs being artificially incubated at the Toledo Zoo. Eggs were separated into 4 groups, each of which was placed in a covered plastic box (40 × 24 × 15 cm with ventilation holes) containing hydrated vermiculite, and were incubated in a still-air, chest incubator at 31 °C. Prior to placement of eggs in the boxes, 500 g of dry vermiculite were added to each box, and the vermiculite was hydrated with distilled water equivalent to either 60, 80, 100, or 120% of the dry weight of the vermiculite. Instrumentation was not available to measure directly the water potential in each of these incubation substrates, but utilizing the equations relating vermiculite water content to water potential (Tracy et al., 1978), we estimated water potentials of approximately −360, −220, −130, and −80 kPa respectively in these four mixtures. Eggs were placed so that about 50% of the shell was buried in the substrate. Twice weekly, eggs were removed from the boxes, the boxes were weighed, and water was added as needed to maintain the original water content.

Hatching occurred in all groups between 62 and 70 d following initiation of incubation. One to two days prior to emergence, the snake makes several slits in the upper surface of the shell, and then using the shell as a retreat progressively begins to venture forth from the hole until the shell is finally abandoned. Hatchlings were weighed within 12 h of the time they abandoned the shell.

At intervals of 4–6 d, eggs were weighed individually to assess water flux. On day 34 of incubation, two eggs from each box were selected randomly and placed individually in desiccators containing $CaSO_4$. These desiccators were then maintained at 25 °C in a constant temperature incubator. Following the procedure of Packard et al. (1981), the eggs were allowed to equilibrate for about 1 h and then were weighed six times each over the next 7 h. The rate of decline in mass was determined using a linear regression procedure (Packard et al., 1981), and egg water conductance was calculated by the equation (Ar et al., 1974):

$$G_{H_2O} = \dot{M}_{H_2O} / \Delta P_{H_2O} \tag{1}$$

where G_{H_2O} is conductance (mg $H_2O \cdot d^{-1} \cdot kPa^{-1}$), $\dot{M}_{H_2O}$ is the rate of water loss in the desiccator (mg $H_2O \cdot d^{-1}$), and P_{H_2O} is the trans-shell vapor pressure difference (kPa). All values were adjusted to standard pressure to enable comparison with those from eggs of other vertebrate species.

For O_2 consumption measurements, three eggs were chosen randomly from each of the four boxes on day 28 of incubation, and O_2 consumption of these eggs was measured manometrically at 31 °C at weekly intervals until the last week of incubation when two measurements were made on each egg. Eggs were allowed to stabilize in the measurement chambers (500 cc glass vessels with soda lime added as CO_2 absorbant) for at least 1 h prior to recording of measurements. During both the equilibration and measurement periods, chamber P_{O_2} was kept atmospheric by addition of pure O_2 through an injection port. Measurements usually required about 3 h from the time the egg was placed in the chamber.

We utilized two different methods to determine P_{O_2} beneath the shell. First, syringes were attached to plastic hypodermic needle hubs which had been cut short and attached to the shell by epoxy (Tazawa et al., 1980). Because the flexible-shell egg does not have an air cell, the albumen layer lies immediately below the hole. This does not seem to present a problem as far as equilibration of gases dissolved in egg contents with gas in the syringe is concerned, but in eggs which are placed in wet incubation substrate the shell is stretched tightly, and the contents are under some pressure. Once a hole is made in the shell, albumen flows through it into the hub and syringe. This usually hardens, plugging the syringe and preventing equilibration between gases under the shell and syringe gas. This makes the 'syringe' method usable only on eggs with a relatively 'relaxed' shell. The contents of syringes were analyzed using a Scholander Micro-gas Analyzer (Scholander. 1947).

Under-shell P_{O_2} was also measured by inserting a 23 ga. hypodermic needle attached to a 1 cc syringe (dead space filled with mercury) immediately under the shell and removing about 0.5 cc of fluid from the albumen compartment of the egg. The P_{O_2} of this fluid was then measured in a blood-gas analyzer (Radiometer BMS 3 Mk 2 Blood Micro System). This procedure appears to have no adverse affects on the egg even when done repeatedly, since holes made in the shell by the needle are quickly sealed with hardened albumen. The disadvantage of this method lies in not knowing for certain whether the fluid sample comes from above or below the chorioallantois. Comparisons between P_{O_2} data gathered by the two methods on six eggs at day 49 of incubation showed no statistical differences among the measurements (Student's t-test), indicating that P_{O_2} values derived from albumen fluid were indicative of that between the shell and chorio-allantois.

Shell O_2 conductance was calculated with an equation analogous to Eq. 1:

$$G_{O_2} = \dot{V}_{O_2}/\Delta P_{O_2} \tag{2}$$

where G_{O_2} is oxygen consumption (ml $O_2 \cdot d^{-1} \cdot torr^{-1}$), $\dot{V}_{O_2}$ is O_2 consumption (ml $O_2 \cdot d^{-1}$), and ΔP_{O_2} is the trans-shell O_2 partial pressure gradient (torr.)

Most of the data collected in this study were compared utilizing a one-way analysis of variance (Sokal and Rohlf, 1969). Under-shell P_{O_2} data were analyzed via a two-way analysis of variance since both time of incubation and substrate water content appeared to be affecting under-shell P_{O_2}.

Results

Mortality among embryos was low and appears to show no relationship to incubation substrate water content (Table 1). Mean incubation period values do not show a statistically significant change with increasing incubation substrate water content (one-way analysis of variance). However, the increase in neonate live mass with increasing substrate water is significant.

Shell water conductance values show no relationship to substrate water content (Table 1), and by a one-way analysis of variance there is no difference among these means. Assuming that a 'typical' Burmese python egg weighs about 200 g, water conductance is about 36 times greater than that predicted for a bird egg of equal mass (Ar and Rahn, 1978).

The pattern of change in the mass of the eggs throughout the last half of incubation is shown in Fig. 1; these mass changes presumably reflect net water flux for the egg (Packard et al., 1977). Eggs incubated in the two wettest substrates gained mass until nearly the end of incubation when they lost a small amount of mass. Eggs incubated in the two drier substrates lost mass continuously, particularly those in the -360 kPa substrate. By the end of incubation, eggs in this substrate had a collapsed appearance and were very dimpled. Prior to day 27 of incubation all eggs had been incubated in vermiculite of a single water content (not determined). Analysis of variance of egg mass on day 27 shows no significant difference among eggs in the four groups. By day 42 and throughout the rest of the incubation period, mean mass for the four groups differed significantly.

Oxygen consumption values for the three eggs selected from each of the four incubation substrate groups are shown in Fig. 2. The basic pattern is similar to that observed in most other oviparous vertebrate embryos, i.e., metabolism during the first half of incubation is very low and increases quite slowly; during the last half of incubation the increase in O_2 consumption is approximately exponential until about the last 10% of incubation where the rate of increase either slows or ceases altogether (Lynn and Von Brand, 1945; Ackerman, 1980; Vleck et al., 1980). O_2 consumption measurements of eggs in the four groups were made over two-day periods at about weekly intervals. Analysis of variance showed no significant differences among the means in each group of measurements.

Table 1. Hatching success, incubation length, neonatal weights and shell water for eggs of *Python molurus* incubated under four levels of incubation substrate water content.

Group #	Medium Water Potential (kPa)	Hatching Success (# Hatched/ # Incubated)	Incubation Time days ($\bar{x} \pm$ s.e.)	Neonate mass g ($\bar{x} \pm$ s.e.)	G_{H_2O} (m mol $H_2O \cdot$ d$^{-1} \cdot$ kPa^{-1}) ($\bar{x} \pm$ s.e.)
I	-360	5/7	64 ± 2.2	94 ± 3.9	476 ± 34
II	-220	6/7	66 ± 1.7	104 ± 5.8	453 ± 25
III	-130	5/7	67 ± 3.0	111.8 ± 2.0	445 ± 35
IV	-80	6/7	68 ± 2.7	116 ± 5.3	474 ± 27

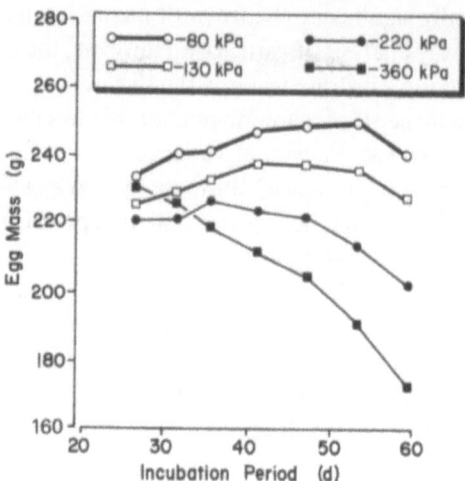

Fig. 1. Mean mass values during the last half of incubation for four groups of Burmese python eggs incubated in substrates at four different water potentials.

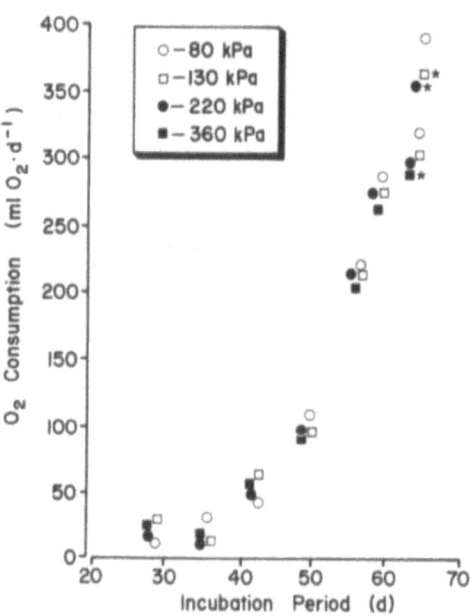

Fig. 2. Mean O_2 consumption values during the last half of incubation for four groups of Burmese python eggs incubated in substrates at four different water potentials. Symbols are as in Fig. 1; symbols with stars indicate measurements on eggs which had been slit by the hatching embryo.

142

Fig. 3a shows mean under-shell P_{O_2} values for each of the four groups of eggs during the last quarter of incubation. Under-shell P_{O_2} decreases both with increasing O_2 consumption by the embryo (a direct reflection of incubation date) and with substrate water content. A two-way analysis of variance examining under-shell P_{O_2} against incubation date and substrate water content shows that both of these factors are related to a highly significant variation in under-shell P_{O_2}. The interaction term in the ANOVA is significant showing that as incubation proceeds, the effect of differences in substrate water content increases. This is of course due to the exponential increase in O_2 consumption by the embryo during the portion of the incubation period when under-shell P_{O_2} measurements were recorded. Shell O_2 conductance values remain approximately constant for eggs incubated at each of the substrate water levels, but the values are smaller at higher substrate water content (Fig. 3b).

Discussion

Our results support the idea that water exchange between the reptilian egg and its surrounding substrate and atmosphere during incubation influences embryonic growth (Packard et al.,1977, 1980). Although increased substrate water content does not affect O_2 consumption in Burmese pythons (Fig. 2), it is significantly correlated with increased hatchling size. Packard et al. (1981) suggest that embryos in wetter substrate attain larger size by extending the incubation period; although our data are suggestive of this, the trend is not statistically significant.

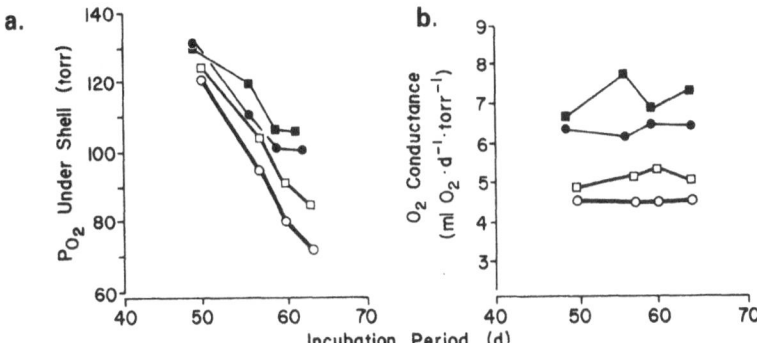

Fig. 3a. Mean values for under-shell P_{O_2} during the last third of incubation for four groups of Burmese python eggs incubated in substrates at four different water potentials. Symbols are as in Fig. 1.

*Fig. 3b.*Calculated mean values for shell O_2 conductance (Eq. 2) during last third of incubation for four groups of Burmese python eggs incubated in substrates at four different water potentials. Symbols are as in Fig. 1.

One reason may lie in the difficulty of determining precisely when hatch occurs in the Burmese python since the hatchling may remain in the egg for up to several days after the first slits appear in the shell.

Studies on water exchange in a number of reptilian species have yielded a tremendous range of G_{H_2O} values. Among reptilian eggs, those with calcareous shells seem to have much lower conductances than those of flexible-shell eggs (Ackerman, 1980; Packard et al., 1979). This difference is probably related to differences in shell structure both in terms of greater thickness of the calcareous shell and a decreased diffusional surface area. Flexible-shell eggs are composed of an amorphous matrix of fibers, while calcareous shells have true pores (Packard et al., 1977). Differences in G_{H_2O} among flexible-shell eggs are less easily explained, however. Indeed, they may have no adaptive significance at all, or as Packard et al. (1981) suggest, they may be related to inadequate measurement techniques. However, just as in birds, shell conductance may be under the evolutionary influence of environmental factors. Shells of painted turtles and snapping turtles (*Chelydra serpentina*) have conductance of $70\times$ and $55\times$ that of comparably-sized bird eggs (Packard et al., 1981, 1979), while the value for Burmese python eggs is $36\times$. Both turtle species bury their eggs in subsurface chambers which contain a saturated or nearly saturated atmosphere. Thus, water loss via diffusion is minimal, and shell conductance for O_2 and CO_2 also may be increased (Ackerman, 1980). Burmese Python eggs are deposited in a pile on the surface of the ground and then incubation heat is supplied via shivering thermogenesis from the female which coils herself completely around and over the pile of eggs (Hutchinson et al., 1966; Van Mierop and Barnard, 1976). Thus, python eggs are exposed to an atmosphere which although humid is probably not fully saturated and are faced with a somewhat larger trans-shell P_{H_2O} gradient than eggs incubated in cavities.

In addition to affecting embryonic growth, water exchange also influences O_2 transport. In the avian egg there is a close correspondence between measured G_{O_2} and G_{O_2} predicted from the ratio of the binary diffusion coefficients for O_2 and H_2O vapor in N_2 (Paganelli et al., 1978). This 'coupling' of G_{O_2} and G_{H_2O} occurs because both gases are diffusing through shell pores which contain only atmospheric gas and are therefore free of liquid H_2O. G_{H_2O} measurements for all Burmese python eggs average $8.3\,g\ H_2O \cdot d^{-1} \cdot kPa^{-1}$. Assuming the ratio of the binary diffusion coefficients of O_2 to H_2O vapor is 0.85, G_{O_2} is predicted to be $65\,ml\ O_2 \cdot d^{-1} \cdot torr^{-1}$, or about an order of magnitude greater than the measured values (Fig. 3b). A functional layer of liquid water is therefore present in the shell of the Burmese python egg, and although this water layer does decrease O_2 conductance, it does not seem to adversely affect growth and metabolism.

144

Acknowledgements

Funds in support of this research were provided by the Faculty Research and Fellowship Fund of the University of Toledo Graduate School awarded to CPB. Burmese python eggs were provided by the Toledo Zoological Society. We gratefully acknowledge the support of P. Tolson and W. Dennler.

References

Ackerman, R.A. (1980). Physiological and ecological aspects of gas exchange by sea turtle eggs. *Am. Zool.* 20: 575–584.

Ar, A., Paganelli, C.V., Reeves, R.B., Green, D.G. and Rahn, H. (1974). The avian egg: water vapor conductance, shell thickness, and functional pore area. *Condor* 76: 153–158.

Ar, A., and Rahn, H (1978). Interdependence of gas conductance, incubation length, and weight of the avian egg. In: Respiratory Function in Birds, Adult and Embryonic. J. Piiper, ed., Springer-Verlag, New York, pp. 227–236.

Hutchison, V.H., Dowling, H.G. and A. Vinegar (1966). Thermoregulation in a brooding female Indian Python, *Python molurus* bivittatus. *Science* 151: 694–696.

Lynn, W.G. and Von Brand, T. (1945). Studies on the oxygen consumption and water metabolism of turtle embryos. *Biol Bull.* 88: 112–125.

Packard, G.C. and Packard, M.J. (1980). Evolution of the cleidoic egg among reptilian antecedents of birds. *Am. Zool.* 20: 351–362.

Packard, G.C., Packard, M.J. and Boardman, T.J. (1981). Pattern and possible significance of water exchange by flexible-shelled eggs of painted turtles (*Chrysemys picta*). *Physiol. Zool.* 54: 165–178.

Packard, G., Packard, M., Boardman, T., Morris, K. and Shuman, R. (1983). Influence of water exchanges by flexible-shelled eggs of painted turtles (*Chrysemys picta*). *Physiol. Zool.* 56: 217–230.

Packard, G.C., Taigen, T.L., Packard, M.J. and Boardman, T.J. (1980). Water relations of pliable-shelled eggs of common snapping turtles (*Chelydra serpentina*). *Can J. Zool.* 58: 1404–1411.

Packard, G.C., Taigen, T.L., Packard, M.J. and Shuman, R.D. (1979). Water conductance of testudinian and crocodilian eggs (Class Reptilia). *Respir. Physiol.* 38: 1–10.

Packard, G.C., Tracy, C.R. and Roth, J.J. (1977). The physiological ecology of reptilian eggs and embryos, and the evolution of viviparity within the class Reptilia. *Biol. Rev.* 52: 71–105.

Paganelli, C.V., Ackerman, R.A. and Rahn, H. (1978). The avian egg: in vivo conductances to oxygen, carbon dioxide, and water vapor in late development. In: Respiratory Function in Birds, Adult and Embryonic. J. Piiper, ed., Springer-Verlag, New York, pp. 212–218.

Scholander, P.F. (1947). Analyzer for accurate estimation of respiratory gases in one-half cubic centimeter samples. *J. Biol. Chem.* 167: 235–250.

Sokal, R.R. and Rohlf, F.J. (1969). Biometry. W.H. Freeman and Co., San Francisco.

Tazawa, H., Ar, A., Rahn, H. and Piiper, J. (1980). Repetitive and simultaneous sampling from the air cell and blood vessels in the chick embryo. *Respir. Physiol.* 39: 265–272.

Tracy, C.R., Packard, G.C. and Packard, M.J. (1978). Water relations of chelonian eggs. *Physiol. Zool.* 51: 378–387.

Van Mierop, L.H.S. and Barnard, S.M. (1976). Observations on the reproduction of *Python molurus* bivittatus (Reptilia, Serpentes, Boidae). *J. Herpetol.* 10: 333–340.

Vleck, C.M., Vleck, D. and Hoyt, D.F. (1980). Patterns of metabolism and growth in avian embryos. *Am. Zool.* 20: 405–416.

10. Physiological and ecological questions on the evolution of reptilian viviparity

Richard Shine

Department of Zoology, University of Sydney, New South Wales, 2006, Australia

Abstract

Viviparity has evolved independently at least 95 times in squamates, and in some cases more than once within a single genus. This high frequency of evolutionary origins means that there may be a wide diversity of physiological adaptations to live-bearing in different (even congeneric) viviparous species. Also, hypotheses requiring comparison of oviparous and viviparous forms may best be carried out on closely-related forms. Viviparity apparently has evolved most often in species occupying cold climatic areas. Possible physiological protoadaptations to viviparity in cold-climate reptiles, and the apparent evolutionary conservatism of embryonic development rates and thermal tolerances, deserve investigation.

All turtles and crocodilians are oviparous, and retain eggs in utero only briefly. Even when turtles retain eggs in utero due to stress of captivity, embryonic development apparently is arrested at the late gastrula stage. In contrast, most oviparous squamates retain eggs for almost half of the total duration of embryonic development. The consistency between diverse squamates in this regard suggests that some general physiological limitation may be involved.

Introduction

The dichotomy between oviparity (egg-laying) and viviparity (live-bearing, including both 'ovoviviparity' and 'euviviparity') is a central feature of reptilian reproductive biology. There is general consensus that oviparity is the primitive condition and viviparity the derived one, so that viviparity has evolved from oviparity rather than vice versa. The evolution of reptilian viviparity has been the subject of considerable scientific speculation, dating from seminal contributions of Mell (1929), Weekes (1933) and Sergeev (1940). Attention has been focussed both on the selective pressures hypothesized to favour the evolution of viviparity, and on the morphological and physiological adaptations and protoadaptations to this shift in reproductive mode (e.g. Greer, 1968; Greene, 1970; Fitch, 1970; Packard, 1966; Packard et al., 1977; Tinkle and Gibbons, 1977; Shine and Bull,

Seymour, R.S., (ed.) Respiration and metabolism of embryonic vertebrates.
© *1984, Dr W. Junk Publishers, Dordrecht/Boston/London. ISBN-13:978-94-009-6538-6*

1979; Thompson, 1981; Yaron, 1984; Grigg and Harlow, 1981). Recent evolutionary-ecological work has raised several problems, as yet unsolved, which may benefit from studies of embryonic physiology.

My work on reptilian viviparity has been aimed at documenting the various independent evolutionary origins of the trait in different reptilian taxa, in formulating and testing hypotheses on the selective forces responsible for the evolution of the trait, and in examining the role of intermediate stages of prolonged oviducal retention of eggs in oviparous squamates. In the present paper, I summarize some of my findings and speculations on these topics, with particular reference to their implications for studies on embryonic respiration and metabolism. In particular, I focus on three main questions:

(i)   How many independent evolutionary origins of viviparity have occurred in reptiles, and in what taxa have these origins occurred?

(ii) What environmental or species characteristics have been important selective forces in the evolution of reptilian viviparity?

(iii) To what degree do oviparous reptiles retain developing eggs in utero prior to oviposition?

Evolutionary origins of viviparity

All living species of turtles and crocodilians are oviparous, but approximately one-fifth of squamates (lizards, snakes and amphisbaenians) are viviparous. Viviparity is not restricted to only a few families or genera, but instead is scattered in occurrence through most families. In many cases, both viviparity and oviparity occur within a single family, genus, or even species. These situations show that viviparity has evolved independently in many separate reptilian lineages.

Although there is a general appreciation of the fact that viviparity has evolved several times in reptiles, most authors probably have underestimated the number of independent evolutionary origins. A recent analysis of literature on phylogenetic relationships and reproductive modes of squamate reptiles reveals at least 95 independent origins of viviparity (Shine, 1984). These cases include 65 examples in which both viviparity and oviparity occur in a single genus; in ten of these taxa, both reproductive modes are exhibited by a single species. These species include the scincid lizards *Ablepharus bivittatus* (Terentev and Chernov, 1965), *Lerista bougainvillii* (A. Greer, pers. comm.), *Saiphos equalis* (Greer, 1983) and *Sphenomorphus nigricaudus* (A. Greer, pers. comm.); the iguanid lizard, *Sceloporus aeneus* (Guillette, 1982); the colubrid snakes *Helicops angulatus* (D.A. Rossman, pers. comm.), *Opheodrys vernalis* (Blanchard, 1933), *Psammophylax variabilis* (Broadley, 1977); and the viperid snakes *Echis carinatus* (Kramer and Schnurrenberger, 1963) and *Vipera xanthina* (Kratzer, 1968).

The 95 identified origins of viviparity encompass all three squamate suborders, and are distributed among 20 families (Shine, 1984). However, analysis reveals some taxonomic biases: viviparity has evolved more frequently in some taxa than in others (e.g. anguid versus teiid lizards), and multiple origins of viviparity often have occurred within the same genus (e.g. five origins within the scincid genus *Sphenomorphus*) (Shine, 1984).

The extraordinarily high number of independent evolutionary origins of squamate viviparity has important implications for the study of embryonic physiology. Firstly, there is no a priori reason to expect that taxa which have evolved viviparity independently should have done so by way of the same physiological and morphological adaptations. Hence, apparently inconsistent results between studies on different viviparous species, may well reflect real biological differences rather than methodological ones. This may be true even with studies of congeneric live-bearers, because they may represent independent acquisitions of the viviparous trait.

Secondly, hypotheses on embryonic or maternal protoadaptations or adaptations to viviparity, may be tested most validly on closely-related oviparous and viviparous forms. Guillette's (1982) work on oviparous and viviparous subspecies of *Sceloporus aeneus* demonstrates the value of such a closely controlled comparison.

The adaptiveness of viviparity

Life-history theory predicts that viviparity may evolve from oviparity where (i) offspring survivorship is thereby increased , and (ii) the prolonged retention of offspring in utero does not greatly decrease either food intake or vulnerability of the reproducing female to predators. Hence, reptilian viviparity may evolve in response to factors which kill eggs in the nest but not in utero, and is most likely to be favored in species where females are not greatly burdened when gravid.

Many authors have speculated on the selective forces likely to have been important in favoring the evolution of reptilian viviparity. Most attention has been focussed on factors likely to kill eggs in the nest but not in utero. Such factors include extremes of temperature, desiccation, flooding, fungal attack and predation (Mell, 1929; Weekes, 1933; Sergeev, 1940; Neill, 1964; Greer, 1968; Greene, 1970; Packard et al., 1977; Tinkle and Gibbons, 1977; Shine and Bull, 1979; Blackburn, 1982; Shine, 1983a, 1984). Despite the wide range of potential mortality sources, there is general (but not universal) agreement that low temperatures are of particular importance in the evolution of reptilian viviparity.

The 'cold-climate hypothesis' suggests that in cold areas, soil temperatures may be too low to permit complete embryonic development over a single summer. Hence, some eggs fail to hatch and are then killed by low winter temperatures. Eggs retained in utero, however, develop at higher temperatures (due to the behavioral thermoregulation of the female reptile) and thus are able to complete their development in the available time. This hypothesis seems to have been developed independently by Mell (1929) in German, Weekes (1933) in English, and Sergeev (1940) in Russian.

Available data provide good support for the 'cold-climate hypothesis':

(i)   A study on montane Australian skinks confirms the validity of several major assumptions of the hypothesis. For example, female body temperatures are higher than nest temperatures, eggs develop much more rapidly at the higher temperature, and eggs which are delayed in hatching are killed by low temperatures (Shine, 1983a). Also, gravid females of many reptile species are physically burdened to the point of increased vulnerability to predation, and decreased food intake (Shine, 1980).

(ii) A comparison of the climatic conditions occupied by closely-related oviparous and viviparous taxa shows a strong trend for the viviparous form to occupy a colder area than does its closest oviparous relative (Shine, 1984).

(iii) Present-day distributions of viviparous squamates are highly correlated with environmental temperatures (e.g. Sergeev, 1940; Greer, 1968; Greene, 1970; Shine and Berry, 1978).

Other biases are evident in the evolution of viviparity; for example, the trait appears to have evolved more often in snakes than lizards, and especially in venomous snakes. Also, viviparity tends to evolve in taxa showing parental care of eggs (Shine and Bull, 1979; Shine, 1984). However, the bias to cold climates is by far the most clear-cut. Although this trend has been recognized and interpreted by ecologists, its possible physiological implications have rarely been addressed. For example, do cold-climate reptile species show any characteristic physiological adaptations which might preadapt them for viviparity? One suggestion in this line is that high elevations (a common correlate of cold climate) are associated with low oxygen availability, and that this condition may have favored the evolution of elaborate placental mechanisms for maternal-foetal oxygen exchange (Guillette et al., 1980). The force of this hypothesis may be reduced by the fact that oxygen tensions remain fairly high even at elevations above those inhabited by most reptiles, but little is known about the extent to which oxygen tension may be limiting in reptile eggs at any elevation.

Another interesting aspect of the 'cold-climate hypothesis' is its reliance upon the assumption that embryonic developmental rates and thermal tolerances are evolutionarily inflexible (Shine, 1983b). If this were not true, then uterine retention of eggs would not be the only way in which development time could be reduced in cold climates. Instead, natural selection could favor more rapid embryogenesis at any given temperature, so that development could occur within a single summer even at low soil temperatures. Analogous adaptations are seen in anuran tadpoles which develop in temporary pools (Zweifel, 1968). Alternatively, the evolution of greater embryonic tolerances to low temperatures might obviate the need for uterine retention. The only evidence to support the idea that embryonic developmental rates and thermal tolerances are evolutionarily inflexible, is the similarity in these factors between related species (Shine, 1983a). Further data on this topic would be valuable.

Intermediate stages leading to viviparity

Presumably, natural selection in an oviparous species does not give rise to viviparity via a single macromutation. Instead, viviparity arises because selection favors progressively longer retention of eggs in utero, with development proceeding prior to oviposition. It has long been known that some oviparous squamates retain eggs for so long that they are on the verge of viviparity: for example, *Opheodrys vernalis* (Blanchard, 1933), *Saiphos equalis* (Bustard, 1964), and *Typhlops bibronii* (Erasmus and Branch, 1983). However, there have been few comparative data to document the usual stage of embryonic development at oviposition in squamates. The conventional assumption has been that most oviparous reptiles retain eggs in utero only briefly before laying, and that embryos at this stage are very early in development (e.g. Tinkle and Gibbons, 1977; Shine and Bull, 1979).

A recent review challenges this assumption (Shine, 1983b). Data on turtles and crocodilians confirm that eggs are retained only briefly, and embryos are at or near the late gastrula stage at oviposition (Ewert, 1979). However, oviparous squamates typically retain eggs in utero for weeks or months, so that more than 40% of embryonic development occurs in utero (Shine, 1983b). Embryos at oviposition typically have reached approximately stage 30 on the Dufaure and Hubert (1961) scale. The squamates which oviposit about halfway though embryonic development include agamid, anguid, gekkonid, iguanid, lacertid, pygopodid, and scincid lizards, and boid, colubrid, elapid and typhlopid snakes (Shine, 1983b). Also, they come from a wide variety of habitats, including deserts, rainforests and high mountains. Data on one skink species failed to show an elevational cline in the duration of uterine retention, although this phenomenon has previously been reported in anoline lizards (Huey, 1977). At least one squamate has been recorded to show only brief oviducal retention of eggs (*Chamaeleo lateralis*: Blanc, 1974).

The finding that oviparous squamates generally retain eggs in utero for almost half of development, is a very surprising one. Speculations on the evolution of viviparity generally have assumed that prolonged retention of eggs should be common only in severely cold climates. Why then is the duration of uterine retention so long, and why is it apparently not correlated with climatic conditions occupied by the species? The consistency in the duration of retention in so many squamates suggests that a single factor may be responsible. Possible explanations include: (i) at about stage 30, embryonic growth rate accelerates markedly (e.g. Thompson, 1981; Guillette, 1982). This onset of 'exponential' growth (Guillette, 1982) results in a rapid increase in egg weight due to the influx of water to the developing embryo. Even in species which are truly 'ovoviviparous' (i.e. lack placental transfer of nutrients), egg weight increases up to three-fold during development (e.g. Thompson, 1981). This weight and volume increase is most obvious at about stage 30, and may result in a significant physical burdening of the ovigerous female. Data on scincid and lacertid lizards show that an increased clutch weight slows the female down, and may decrease her ability to evade predators (Shine, 1980; Bauwens and Thoen, 1981). Hence, selection may favor uterine retention of eggs only until it becomes too 'expensive' to the female: (ii) the increased growth rate of the embryo is correlated with a sudden increase in oxygen demand (Guillette, 1982). An inability of the oviducal tissue to meet this demand may determine the time of oviposition.

The common trend for prolonged oviducal retention of eggs in ecologically dissimilar squamates, and the apparent absence of such retention in turtles and crocodilians, suggest that explanations for these phenomena may best be looked for in terms of basic physiological properties of each group. For example, turtles seem to be able to retain eggs in utero with embryogenesis arrested at the gastrula stage; in contrast, squamate eggs retained in utero continue to develop normally (brief review in Shine, 1983b). The physiological mechanisms underlying such differences should be a fertile field of study.

Physiological constraints also may have been important in preventing the evolution of viviparity in several major taxa. The absence of viviparous chelonians and crocodilians has been explained in this way (Packard et al., 1977). These authors argue that the reliance of embryonic chelonians and crocodilians on calcium from the egg-shell, has prevented any evolutionary pathways involving loss of the shell. Presumably, this precludes viviparity. In contrast, the yolk serves as the major calcium store during embryogenesis in squamate reptiles, so that adaptive thinning or loss of the shell is feasible in these animals. This hypothesis also is consistent with the evolution of viviparity within a squamate family, the Gekkonidae. Live-bearing has evolved twice within the diplodactylines (which usually lay parchment-shelled eggs), but not at all within the other gekkonid subfamilies (which produce eggs with highly calcified shells). However, available data are equivocal as to whether gekkonids with calcified egg-shells rely upon the shell for calcium during embryonic development (Jenkins and Simkiss, 1968; Blackburn, 1982; Shine, 1984).

The hypothesis that embryonic calcium metabolism constrains the evolution of viviparity, has been vigorously attacked by Tinkle and Gibbons (1977). These authors maintain that such constraints are intrinsically implausible: natural selection works in an opportunistic fashion, and there should be alternative pathways by which the 'optimal' reproductive strategy could evolve. Nonetheless, the evolutionary-ecological work on reptilian viviparity has been unable to explain the lack of viviparity in many groups – particularly among the squamates - in which it might realistically be expected to occur. One example of such a group is the Pythoninae, which rely upon shivering thermogenesis of brooding females to maintain high egg temperatures (Vinegar et al., 1970). This surely is more metabolically expensive than behavioural thermoregulation: why have not the pythons, like the boas, evolved viviparity? The answers to such questions may well lie in the realm of embryonic physiology.

Acknowledgements

I thank Roger Seymour for inviting me to participate in the conference, Gordon Grigg for comments on the manuscript, Terri and James Shine for typing, and the Australian Research Grants Scheme for funding the work.

References

Bauwens, D. and Thoen, C.. (1981). Escape tactics and vulnerability to predation associated with reproduction in the lizard *Lacerta vivipara*. *J. Anim. Ecol.* 50: 733–743.

Blackburn, D.G. (1982). Evolutionary origins of viviparity in the Reptilia. I. Sauria. *Amphibia-Reptilia* 3: 185–205.

Blanc, F. (1974). Table de developpement de *Chamaeleo lateralis* Gray, 1831. *Ann. Embryol. Morphol.* 7: 99–115.

Blanchard, F.N. (1933). Eggs and young of the smooth green snake, *Liopeltis vernalis* (Harlan). *Pap. Michigan Acad. Sci. Arts Lett.* 17: 493–508.

Broadley, D.G. (1977). A revision of the African snakes of the genus *Psammophylax* Fitzinger (Colubridae). *Occ. Pap. nat. Mus. Monum. Rhodesia*, B 6: 1–43.

Bustard, H.R. (1964). Reproduction in the Australian rain forest skinks, *Saiphos equalis* and *Sphenomorphus tryoni*. *Copeia* 1964: 715–716.

Dufaure, J.P. and Hubert, J. (1961). Table de developpement du lezard vivipare: *Lacerta (Zootoca) vivipara* Jacquin. *Arch. Anat. Microscop. Morphol. Exp.* 50: 309–328.

Erasmus, H. and Branch, W.R. (1983). Egg retention in the South African blind snake *Typhlops bibroni*. *J. Herpetol.* 17: 97–99.

Ewert, M.A. (1979). The embryo and its egg: development and natural history. In: Turtles, Perspectives and Research. M. Harless and H. Morlock, eds.. John Wiley and Sons, New York, pp. 333–413.

Fitch, H.S. (1970). Reproductive cycles in lizards and snakes. Univ. Kansas Mus. Nat. Hist., Misc. Publ. 52: 1–247.

Greene, H.W. (1970). Modes of reproduction in lizards and snakes of the Gomex Farias region, Tamaulipas, Mexico. *Copeia* 1970: 565–568.

Greer, A.E. (1968). Mode of reproduction in the squamate faunas of three altitudinally correlated life zones in east Africa. *Herpetologica* 24: 229–232.

Greer, A.E. (1983). The Australian scincid lizard genus *Calyptotis* De Vis: resurrection of the name, description of four new species, and discussion of relationships. *Rec. Aust. Mus.* 35: 29–59.

Grigg, G.C. and Harlow, P. (1981). A fetal-maternal shift of blood oxygen affinity in an Australian viviparous lizard, *Sphenomorphus quoyii* (Reptilia, Scincidae). *J. Comp. Physiol.* 142: 495–499.

Guillette, L.J. Jr., Jones, R.E.. Fitzgerald, K.T. and Smith, H.M. (1980). Evolution of viviparity in the lizard genus *Sceloporus*. *Herpetologica* 36: 201–215.

Guillette, L.J. Jr. (1982). The evolution of viviparity and placentation in the high elevation Mexican lizard genus *Sceloporus*. *Herpetologica* 36: 201–215.

Huey, R.B. (1977). Egg retention in some high-altitude *Anolis* lizards. *Copeia* 1977: 373–375.

Jenkins, N.K. and Simkiss, K. (1968). The calcium and phosphate metabolism of reproducing reptiles with particular reference to the adder (*Vipera berus*). *Comp. Biochem. Physiol.* 26: 865–876.

Kramer, E. and Schnurrenberger. H. (1963). Systematik, verbreitung und okologie der Libyschen schlangen. *Rev. Suisse Zool.* 70: 453–568.

Kratzer, H. (1968). Zur fortpflanzung von *Vipera lebetina mauritanica*. *Aqua Terra-Z* 21: 380–382.

Mell, R. (1929). Beitrage zur fauna sinica IV. Grundzuge einer okologie der Chinesischen reptilien und einer herpetologischen Tiergeographe Chinas. Walter de Gruyter Co., Berlin.

153

Neill, W.T. (1964). Viviparity in snakes: some ecological and zoogeographical considerations. *Am. Nat.* 48: 35–55.

Packard, G.C. (1966). The influence of ambient temperatures and aridity on modes of reproduction and excretion of amniote vertebrates. *Am. Nat.* 100: 667–682.

Packard, G.C., Tracy, C.R. and Roth, J.J. (1977). The physiological ecology of reptilian eggs and embryos, and the evolution of viviparity within the class Reptilia. *Biol. Rev.* 52: 71–105.

Sergeev, A.M. (1940). Researches on the viviparity of reptiles, *Moscow Soc. Nat. (Jubilee Issue)*: 1–34.

Shine, R. (1980). 'Costs' of reproduction in reptiles. *Oecologia* 46: 92–100.

Shine, R. (1983a). Reptilian viviparity in cold climates: testing the assumptions of an evolutionary hypothesis. *Oecologia* 57: 397–405.

Shine, R. (1983b). Reptilian reproductive modes: the oviparity-viviparity continuum. *Herpetologica* 39: 1–8.

Shine, R. (1984). The evolution of viviparity in reptiles: an ecological analysis. In: Biology of the Reptilia, Vol. 15. C. Gans and F. Billett, eds., Wiley-Interscience, New York.

Shine, R. and Berry, J.F. (1978). Climatic correlates of live-bearing in squamate reptiles. *Oecologia* 33: 261–268.

Shine, R. and Bull, J.J. (1979). The evolution of live-bearing in lizards and snakes. *Am. Nat.* 113: 905–923.

Terentev, P.V. and Chernov, S.A. (1965). Key to Amphibians and Reptiles. Translation from Israel Program for Scientific Translations, Jerusalem.

Thompson, J. (1981). A study of the sources of nutrients for embryonic development in a viviparous lizard, *Sphenomorphus quovii. Comp. Biochem. Physiol.* 70A: 509–518.

Tinkle, D.W. and Gibbons, J.W. (1977). The distribution and evolution of viviparity in reptiles. *Misc. Publ. Mus. Zool., Univ. Michigan* 154: 1–55.

Vinegar, A., Hutchinson, V.H. and Dowling, H.G. (1970). Metabolism, energetics and thermoregulation during brooding of snakes of the genus *Python* (Reptilia, Boidae). *Zoologica* 55: 19–48.

Weekes, H.C. (1933). On the distribution, habitat and reproductive habits of certain European and Australian snakes and lizards, with particular regard to their adoption of viviparity. *Proc. Linnaean Soc. N.S.W.* 58: 270–274.

Yaron, Z. (1984). Reptilian placentation and gestation: structure, function and endocrine control. In: Biology of the Reptilia, Vol. 15. C. Gans and F. Billett, eds., Wiley-Interscience, New York.

Zweifel, R.G. (1960). Reproductive biology of anurans of the arid southwest, with emphasis on adaptation of embryos to temperature. *Bull. Amer. Mus. Nat. Hist.* 140: 1–64.

11. Comparative aspects of calcium metabolism in embryonic reptiles and birds

Mary J. Packard and Gary C. Packard

Department of Zoology and Entomology, Colorado State University, Fort Collins, Colorado 80523, U.S.A.

Abstract

Eggs of oviparous, amniotic vertebrates must be endowed at oviposition with all of the organic and most of the inorganic components required for embryonic growth. A major inorganic constituent of these eggs is calcium used for ossification of the skeleton. The two main sources of this element are the yolk and the eggshell, but the proportion of calcium supplied by these compartments varies among species. Some of the calcium absorbed from the eggshell by embryos of domestic fowl is stored in the yolk, causing the calcium content of this compartment to increase appreciably during incubation. In contrast, the calcium content of yolk declines throughout incubation in eggs of reptiles, and the yolk that is withdrawn into the abdominal cavity just prior to hatching contains only small quantities of this element. Thus, major differences in the pattern of calcium metabolism characterize avian and reptilian embryos, and studies of embryos of domestic fowl may not provide a broadly-based model with which to characterize calcium metabolism in embryos of other species. Control of calcium transport across the cellular epithelia (the yolk sac and chorioallantois) that separate embryos from their sources of calcium (yolk and eggshell) represents one aspect of control of calcium metabolism during embryogenesis, but this process has been examined only in eggs of domestic fowl and only in the chorioallantois. Calcium transport across the chorioallantois of embryonic chicks is influenced by a vitamin K-dependent calcium-binding protein, carbonic anhydrase, the level of calcium to which the chorioallantois is exposed, and vitamin D. However, a complete story concerning control of calcium transport across the chorioallantois and its relationship to calcium regulation in embryos of domestic fowl is not yet possible.

Introduction

Embryos of birds and oviparous reptiles develop in nutritional isolation from the female parent and exchange only oxygen, carbon dioxide, and water (as liquid and/or vapor) with their environment. The young of many species emerge from

Seymour, R.S., (ed.) Respiration and metabolism of embryonic vertebrates.
© *1984, Dr W. Junk Publishers, Dordrecht/Boston/London. ISBN-13:978-94-009-6538-6*

incubation as miniature copies of adults and must be capable of immediate, independent existence. In consequence, eggs must be endowed with all of the organic and most of the inorganic constituents required for formation of the organ systems characteristic of fully developed individuals.

Two inorganic components required in large quantities during development are phosphorous and calcium, used for ossification of the skeleton. Most research has concerned the requirement for calcium, thus reflecting the influence of a century-long debate, beginning in 1822, concerning whether or not avian embryos obtain calcium from their eggshells (Needham, 1963). This emphasis on calcium has continued in recent times because of the potentially disruptive effects of calcium on metabolic function and the requirement that calcium metabolism be carefully controlled in consequence (Matthews et al., 1981).

Unfortunately, calcium metabolism during embryogenesis has been examined primarily in eggs of domestic fowl. As a result, a fully comparative approach to this subject is not yet possible. In the sections to follow, our emphasis will be on several interrelated components of calcium metabolism in embryos of oviparous, amniotic vertebrates. These include the pattern of mobilization of calcium during development, the relative importance of the yolk and eggshell as sources of calcium, and the transport of calcium to embryos from these extraembryonic compartments. We will discuss both birds and reptiles regarding the first two aspects of calcium metabolism and primarily birds regarding the third. We anticipate that many of the generalizations made and hypotheses posed will require modification as more is learned about calcium metabolism in avian and reptilian embryos.

Sources of calcium and patterns of its mobilization

Embryos of oviparous, amniotic vertebrates generally have two main sources of calcium available to them: one source is the yolk and the other is the eggshell (Needham, 1963; Romanoff, 1967; Simkiss, 1967). However, the relative importance of these two compartments as sources of calcium during embryogenesis varies among classes and perhaps among species as well (Simkiss, 1962, 1967; Jenkins and Simkiss, 1968; Bustard et al., 1969; Jenkins, 1975; M.J. Packard et al., 1984a, b).

Avian eggs and embryos

Our comments concerning avian eggs and embryos apply only to embryos of domestic fowl, for this is the only species studied in detail. This situation is unfortunate because we do not know to what extent selection for egg production, shell strength, and other characteristics may have caused departures from the native state.

156

A hen's egg (exclusive of the eggshell) contains 20–35 mg of calcium, with 3–4 mg being in the albumen and the remainder in the yolk (Needham, 1963; Romanoff, 1967; Simkiss, 1967; Crooks and Simkiss, 1974). A much larger reservoir of calcium resides in the thick, calcareous matrix of the eggshell, which contains 2–2.5 g of calcium (Needham, 1963; Simkiss, 1967). Thus, at the beginning of embryogenesis, and embryo of domestic fowl has a relatively small amount of calcium available to it in the yolk and albumen, but an almost unlimited supply is available from the shell.

Our focus for discussion is on movements of calcium within eggs containing 20–24 mg of calcium in the yolk at oviposition or 23–27 mg of calcium in yolk and albumen (Fig. 1; Johnston and Comar, 1955; Crooks and Simkiss, 1974; Dunn and Boone, 1977). During the first 9–10 days of incubation, the calcium content of yolk declines, and there are only minor changes in calcium content of embryos during the same interval (Johnston and Comar, 1955; Romanoff, 1967; Crooks and Simkiss, 1974; Dunn and Boone, 1977). Between days 10 and 12, the calcium content of yolk begins a slow but steady increase that is maintained until about day 16 (Fig. 1). From that time on, there is little or no change in calcium content of

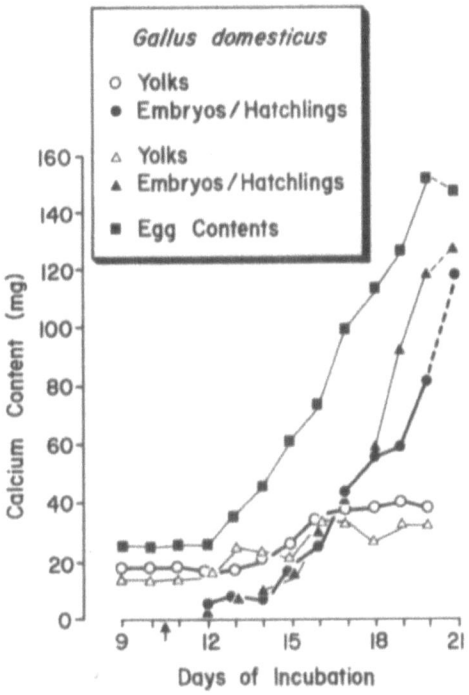

Fig. 1. Calcium content of yolks, embryos/hatchlings, and combined egg contents of eggs of *Gallus domesticus* during the latter part of incubation. Calcium content of hatchlings (day 21) includes the calcium contained within the residual yolk. Circles represent data from Crooks and Simkiss (1974); triangles data from Johnston and Comar (1955); and rectangles data from Dunn and Boone (1977).

this compartment, and on day 20 of incubation, one day prior to hatching, the calcium content of yolk is about 30–32 mg, which represents an increase of about 44% over that present at oviposition (Fig. 1).

Between days 11 and 12 of incubation, calcium content of embryos increase from less than 1 mg to about 2.5 mg (Fig. 1). For the remainder of incubation, the increase in calcium content of embryos is more-or-less linear (Fig. 1). When young emerge from their eggs on day 21, they have a calcium content of about 125 mg (Fig. 1), some of which is found in residual yolk withdrawn into the abdominal cavity prior to hatching (Romanoff, 1967).

Overall, the calcium content of eggs (i.e., calcium in yolk, albumen, and embryo) increases by about 100 mg between oviposition and hatching, as calcium is absorbed from the shell and incorporated into both yolks and embryos. Calcium content of eggs increases about 80% during incubation and yolk-free hatchlings contain approximately 75% more calcium than was present in yolks alone at oviposition (Romanoff, 1967). The increase in calcium content of both yolks and embryos indicates that embryos absorb calcium in excess of the immediate demands of embryogenesis and that this excess is stored in the yolk. The calcium contained within the residual yolk presumably aids in growth of chicks in the first few days after hatching (Romanoff, 1967).

Reptilian eggs and embryos

Most studies of reptilian eggs have been limited to an examination of eggs only at the beginning and the end of incubation, and relatively few investigators have reported on changes in calcium content of yolks and embryos throughout development (Simkiss, 1962, 1967; Jenkins and Simkiss, 1968; Bustard et al., 1969; Jenkins, 1975; M.J. Packard et al., 1984a, b). In general, eggs of crocodilians and chelonians are similar to those of birds in that yolk represents the primary store of calcium within eggs and the eggshell represents a much larger reservoir (Table 1). Eggs of only one oviparous squamate have been examined, and in that case the situation is reversed: the eggshell contains relatively little calcium compared to that in egg contents (Table 1). Nonetheless, embryos of each of the species examined thus far clearly obtain a portion of their calcium from the eggshell, for in each case the calcium in egg contents (hatchling plus residual yolk) at the end of incubation exceeds that present in egg contents (yolk plus albumen plus embryo) at oviposition (Table 1).

The proportion of calcium obtained from eggshells by embryos is highly variable, ranging from a high of 80% in a sea turtle to a low of 20% in a colubrid snake (Table 1). These differences in the degree to which reptilian embryos rely on the eggshell for a portion of the calcium required during embryogenesis may reflect differences in amounts of calcium available in eggshells (Table 1). Eggs laid by most squamate reptiles have poorly calcified shells compared to eggs of

Table 1. Calcium content of various compartments of eggs of several species of oviparous reptiles. Where possible, values are given for calcium of both yolk and albumen (Alb) at oviposition. Egg contents refers to total calcium within eggs (exclusive of the eggshell) at oviposition (Ovip) and at hatching (Hat). The last two columns give the quantity of calcium and the percent of total calcium obtained by the embryo from the shell.

| | Calcium content (mg) | | | | | | |
| | Egg contents | | | | | | |
	Yolk	Alb	Ovip	Hat	Shell	mg from shell	% from shell
Caretta caretta[a]	14	0.4	14	71	–	57	80
Dermochelys coriacea[b]	31	3	34	139	–	105	76
Chelonia mydas[c]	46	1	47	122	412	75	62
Caretta caretta[c]	33	2	35	88	282	53	60
Chelydra serpentina[d]	12	0.2	12	27	338	15	56
Crocodylus novaeguinae[c]	–	–	124 (164)	280	2253	156 (116)	56 (41)
Coluber constrictor[f]	–	–	30	38	10	8	21

References: [a] Simkiss (1967), species = *Thalassochelys corticata;* [b] Simkiss (1962); [c] Bustard et al. (1969); [d] M.J. Packard et al. (1984b); [e] Jenkins (1975), author presented two sets of values; [f] M.J. Packard et al. (1984a).

crocodilians and chelonians (Ferguson, 1982; M.J. Packard et al., 1982), and squamate embryos obtain relatively little calcium from this source (Table 1).

The pattern of change in calcium content of yolks and embryos of reptilian eggs is known for the turtles *Caretta caretta* and *Chelydra serpentina* and the snake *Coluber constrictor* (Simkiss, 1967; M.J. Packard et al., 1984a,b). However, in each species the pattern of mobilization and storage of calcium is similar (Figs. 2, 3; Simkiss, 1967; M.J. Packard et al., 1984a, b). Therefore, we shall limit our discussion to eggs of *Chelydra* and *Coluber*, the two species for which the most complete data are available.

In both species, the calcium content of yolk changes relatively little for the first half of incubation, when embryos are small and (presumably) have not yet begun osteogenesis (Figs. 2, 3). During the second half of incubation, however, there is a precipitous decline in calcium content of yolk (Figs. 2, 3). At the end of incubation, residual yolk removed from the abdominal cavity of hatchlings contains considerably less calcium than was present in this compartment at oviposition (Figs. 2, 3). In *Chelydra*, calcium content of yolk is reduced by about 97% and in *Coluber* it is reduced by about 90% between oviposition and hatching.

Calcium content of embryos of both species increases more-or-less linearly during the second half of incubation (Figs. 2, 3). However, embryos of *Chelydra*

159

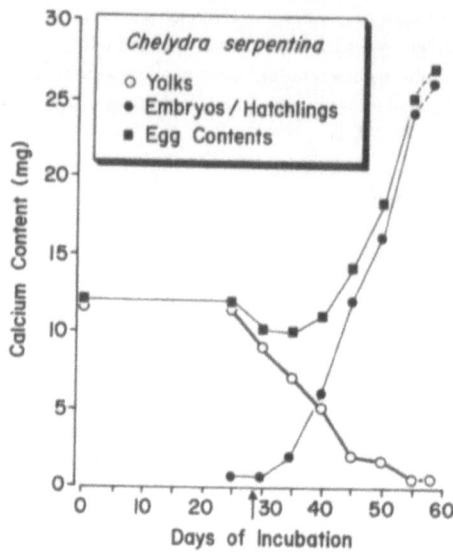

Fig. 2. Calcium content of yolks, embryos/hatchlings, and combined egg contents of eggs of *Chelydra serpentina* during incubation. Calcium in egg contents is the sum of calcium in yolk and albumen at oviposition and yolk and embryos/hatchlings at all other points. The apparent decline in calcium present in egg contents between days 25 and 40 is not statistically significant. The arrow denotes the mid-point of the 58-d incubation period. See M.J. Packard et al. (1984b) for details.

accumulate calcium at a somewhat slower rate (1.1 mg/day) overall than do embryos of *Coluber* (1.6. mg/day; M.J.Packard et al., 1984a,b). At the end of incubation, eggs of both species contain appreciably more calcium than was present at oviposition, with all of the increase being attributable to increases in calcium content of embryos (Figs. 2, 3).

The similarities between crocodilian, chelonian, squamate, and avian embryos concerning sources of calcium for embryonic development indicate a need to re-evaluate the assumption that squamate embryos obtain calcium exclusively from the yolk, with embryos of other oviparous amniotes obtaining calcium from both yolk and eggshell (Simkiss, 1967; Jenkins and Simkiss, 1968). Squamates invest a relatively large quantity of calcium in egg contents at oviposition, and this situation has made it seem unnecessary to postulate additional sources of calcium for use in embryonic development (Table 1; Simkiss, 1967; Jenkins and Simkiss, 1968). In addition, the poorly calcified eggs laid by most squamates have made it seem unlikely that the shell would be used as a source of calcium by embryos. However, it is apparent that embryos of oviparous squamates do have the capacity to absorb calcium from eggshells (Fig. 3; M.J. Packard et al., 1984a).

If a dichotomy concerning calcium metabolism does exist among embryos of oviparous, amniotic vertebrates, avian embryos may constitute one group with reptilian embryos comprising the other. Only in eggs of domestic fowl does the

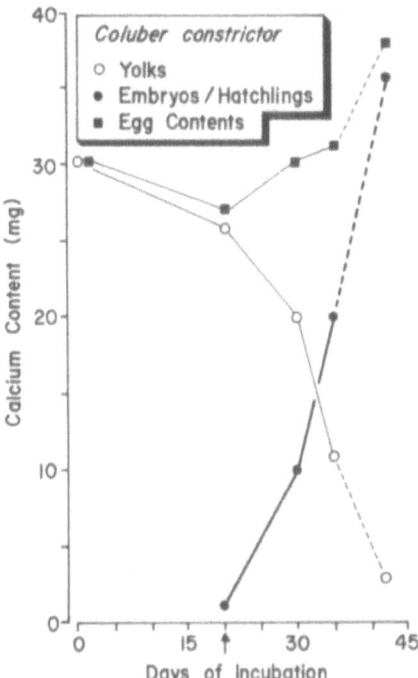

Fig. 3. Calcium content of yolks, embryos/hatchlings, and egg contents of eggs of *Coluber constrictor* during incubation. Small embryos were present in eggs at oviposition, but were not analyzed separately. Eggs contained no albumen. Thus, calcium in egg contents is the sum of calcium in yolks and embryos/hatchlings at each point. The apparent decline in calcium in egg contents is not statistically significant. The arrow denotes the mid-point of the 41-d incubation period. See M.J. Packard et al. (1984a) for details.

calcium content of both the embryo and the yolk increase during incubation (Fig. 1). In snake and turtle embryos, there is no increase in yolk calcium during incubation (Figs. 2, 3; Simkiss, 1967). Apparently, reptilian embryos do not absorb calcium from eggshells in excess of the demands of embryogenesis, and it is unlikely that yolk withdrawn into the body cavity of a reptilian embryo just prior to hatching would support substantial growth of the skeleton after hatching.

Sources of calcium and the evolution of viviparity

Eggs of *Coluber constrictor* contain more calcium at any given point in incubation than do the similarly-sized eggs of *Chelydra serpentina* (Figs. 2, 3; Table 2). Moreover, the quantity of calcium in the *Coluber* eggs at oviposition is within the range of values reported for the much larger eggs of domestic fowl. In contrast, calcium content of shells of *Coluber* eggs is considerably lower than that of shells of other oviparous, amniotic vertebrates, and the *Coluber* embryos obtain only a

Table 2. Dry mass, calcium content, and calcium concentration of yolks and carcasses of embryos/ hatchlings of *Chelydra serpentina* and *Coluber constrictor* at comparable points in incubation. The average length of incubation of the *Chelydra* eggs was 58 d and that of the *Coluber* eggs was 41 d. See M.J. Packard et al. (1984a, b) for details of experimental procedures.

| | Percent of incubation[a] | | | | | |
| | *Chelydra* | | | *Coluber* | | |
	0	50	100	0	50	100
Yolk						
dry mass (g)	2.2	1.8	0.8	1.4	1.2	0.2
calcium content (mg)	12.0	9.0	0.4	30.0	26.0	3.0
calcium concentration						
(mg · g^{-1} dry mass)	5.0	5.0	0.5	21.0	22.0	15.0
Embryos-hatchlings						
dry mass (g)	–	0.1	1.2	–	0.1	1.0
calcium content (mg)	–	0.6	26.0	–	1.0	36.0
calcium concentration						
(mg/g dry mass)	–	6.0	22.0	–	10.0	36.0

[a] 0 = Day of oviposition. 100 = day of hatching.

small proportion of their calcium from this source (Table 1; Fig. 3).

The observations outlined above provide support for an earlier suggestion that the absence of viviparity among crocodilians and chelonians may stem in part from a reliance of embryos of these groups on the eggshell for a major portion of the calcium needed for embryogenesis (G.C. Packard et al., 1977). It is unlikely that viviparity could evolve in reptiles in which eggshells supply large quantities of calcium to developing young (G.C. Packard et al., 1977; Thompson, 1982). Thus, we propose that the large mass of yolk characterizing the relatively naked eggs of the earliest reptiles contained all of the calcium necessary for embryonic growth and development. The evolution of more highly derived eggshells with increasing quantities of calcium (G.C. and M.J. Packard, 1980) presumably enabled embryos to obtain calcium from both yolk and eggshell and ultimately led to a diminution of the importance of the yolk as a source of calcium among chelonians and crocodilians. Viviparity, a mode of reproduction derived from oviparity, could arise among squamates because, with their calcium-rich eggs and poorly calcified shells, they have departed relatively little from the ancestral condition and have therefore retained the versatility necessary for evolution of viviparous reproduction.

Transport of calcium to embryos

Embryos of all oviparous, amniotic vertebrates clearly must mobilize relatively large quantities of calcium for use in skeletal formation. In each group, the calcium to support development is stored in extraembryonic compartments (the yolk and the eggshell) and must cross cellular barriers (the yolk sac and the chorionic portion of the chorioallantois) to gain access to embryos. Thus, the cellular epithelia across which calcium is transported may be major sites for control of calcium metabolism during embryogenesis (Clark and Simkiss, 1980). Certainly the sensitivity of cells to changes in ionic calcium dictates that calcium transport across epithelia be carefully controlled irrespective of the relationship between control of transport and control of calcium metabolism.

Differences between avian and reptilian embryos concerning patterns of mobilization of calcium and relative importance of the yolk and eggshell as sources of calcium may indicate differences in the relative importance of the yolk sac and chorioallantois as sites for control of calcium metabolism. Control of calcium transport across the chorionic portion of the chorioallantois may prove to be a major factor in regulating calcium metabolism in embryos that rely on the eggshell as the primary source of calcium. In contrast, calcium transport across the yolk sac membrane may be more critical for control of calcium metabolism in embryos that obtain most of their calcium from the yolk.

Yolk sac

There presently is only indirect evidence that calcium transport in the yolk sac is under physiological control. In embryonic turtles and snakes, the concentration of calcium in the yolk sac declines with incubation, as does the total quantity of this element in the yolk, indicating that calcium is being removed selectively (M.J. Packard et al., 1984a, b). This observation strongly indicates that embryonic snakes and turtles exercise some control over removal of calcium from the yolk, but the mechanisms involved in this process are unknown.

Indirect evidence for the involvement of the yolk sac in mediation of calcium metabolism in embryonic birds comes from studies of embryos maintained in shell-less culture where the yolk is the only source of calcium during development. Embryos developing under these conditions remove virtually all of the calcium from the yolk during embryogenesis, and calcium content of this compartment does not increase as it does during normal incubation (Dunn and Boone, 1977). Thus, embryos maintained in shell-less culture may switch from mechanisms favoring deposition of calcium in the yolk to mechanisms favoring absorption of calcium from this source (Clark and Simkiss, 1980).

It is not known if differences between normal embryos and embryos in shell-less culture represent an all-or-none switch from deposition of calcium in the yolk in one case to absorption of calcium in the other. Nonetheless, calcium must cross the cellular epithelium of the yolk sac whether it is being deposited in this compartment or withdrawn from it (Lambson, 1970; Juurlink and Gibson, 1973; Mobbs and McMillan, 1979). Presumably, this process must be regulated, but the mechanisms involved have not been identified.

Chorioallantois

Transport of calcium to embryos has been examined only in eggs of domestic fowl and only in the chorionic portion of the chorioallantois. A complete story concerning control of calcium transport across the chorion and its relationship to calcium metabolism is not yet possible. Nonetheless, a number of factors clearly affect transport of calcium across the chorionic epithelium, and each of these has the potential to affect control of calcium metabolism as a result. In the paragraphs to follow, we will emphasize the structure of the chorionic portion of the chorio-allantois and its function as a calcium-transporting epithelium.

Formation and differentiation of the chorioallantois. – The chorioallantois arises from a progressive fusion of two extraembryonic membranes, the chorion, which consists of ectodermal and mesodermal derivatives, and the allantois, which consists of endodermal and mesodermal derivatives (Romanoff, 1960; Balinsky, 1975). The chorioallantois completely lines the inner surface of the inner shell membrane by approximately day 10 of incubation, but full cellular differentiation is not complete until days 13–14 of development (Romanoff, 1960; Coleman and Terepka, 1972a; Rieder et al., 1980).

Between days 10 and 14 of incubation, the chorioallantois thickens consider-ably and distinctive cell types appear in the allantoic and chorionic portions of the membrane (Fig. 4; Stewart and Terepka, 1969; Coleman and Terepka, 1972a; Rieder et al., 1980). In its mature form the chorioallantois is approximately 110–140 μm thick (calculated from Coleman and Terepka, 1972a and Rieder et al., 1980), with its outer portion firmly attached to the inner surface of the inner shell membrane and its inner portion forming the lining of the allantoic sac (Fig. 4; Romanoff, 1960).

The allantois serves as a repository for nitrogenous wastes, various ions, and water, and the endodermal portion of the allantois is morphologically and func-tionally analagous to a toad bladder (Fig. 4; Choi, 1963; Romanoff, 1967; Free-man and Vince, 1974; Hoyt, 1979; Leaf, 1982). Prior to the end of incubation, most of the water and solutes are absorbed across the allantoic epithelium and transported to the embryo (Stewart and Terepka, 1969; Hoyt, 1979; Simkiss, 1980; Murphy et al., 1982).

The mesodermal derivative of the chorioallantois forms from a fusion of chorionic and allantoic mesoderm and separates the chorionic and allantoic derivatives (Fig. 4). This portion of the chorioallantoic membrane undergoes

164

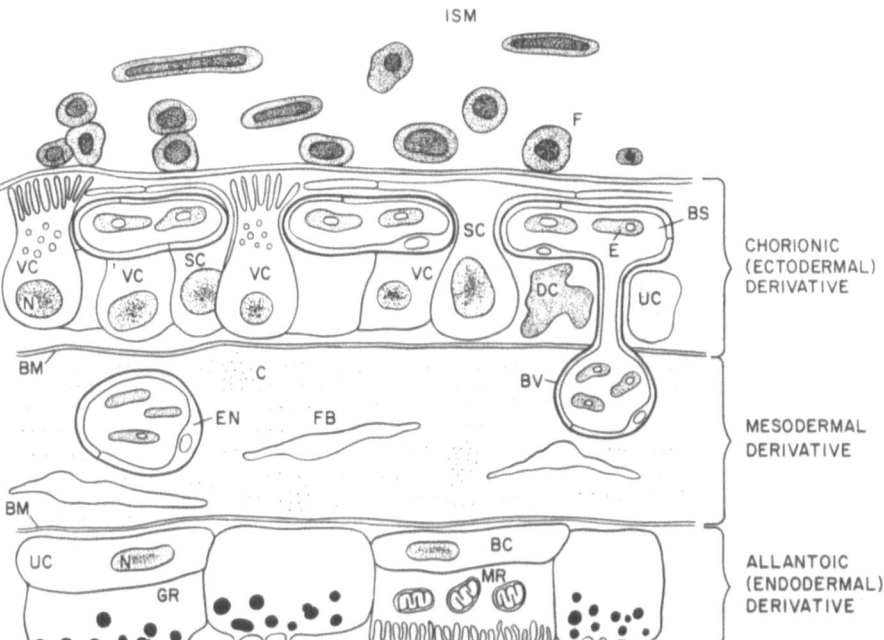

ISM

CHORIONIC
(ECTODERMAL)
DERIVATIVE

MESODERMAL
DERIVATIVE

ALLANTOIC
(ENDODERMAL)
DERIVATIVE

ALLANTOIC SAC

Fig. 4. Schematic representation of structure of the mature chorioallantoic membrane, sectioned perpendicular to its surface. In such preparations, the vascular supply of the chorion appears to be composed of capillaries when in reality it consists of a single large blood sinus (see Narbaitz, 1977). Drawing is not to scale, and size relationships have been exaggerated to emphasize details. The mesodermal derivative is much thicker in relation to the chorionic and allantoic derivatives than shown here (see Wangensteen and Weibel, 1982). ISM, inner shell membrane; F, fibers of inner shell membrane; VC, villus cavity cell; N, nucleus; SC, sinus covering cell; DC, dark cell; E, erythrocyte; BS, lumen of blood sinus; UC, undifferentiated cell; BM, basement membrane; EN, endothelial cell; C, collagen fibrils; FB, fibroblast; BV, blood vessel; GR, granule-rich cell; BC, basal cell; MR, mitochondria-rich cell.

little differentiation (but increases in thickness) and consists primarily of connective tissue, fibroblasts, and blood vessels throughout incubation (Leeson and Leeson, 1963; Stewart and Terepka, 1969; Coleman and Terepka, 1972a; Rieder et al., 1980; Wangensteen and Weibel, 1982). The mesoderm lacks capillaries, and blood vessels from this layer open directly into the chorionic blood sinus (Fig. 4; Rieder et al., 1980). Presumably, the absence of capillaries in this region dictates that water and solutes absorbed from the allantoic sac enter the chorionic vasculature for transport to embryos to occur.

The major cell types of the chorionic portion of the membrane are sinus covering cells (formerly called capillary covering cells) and villus cavity cells (Leeson and Leeson, 1963; Owczarzak, 1971; Coleman and Terepka, 1972a; Narbaitz, 1972, 1977; Narbaitz and Tellier, 1974; Saleuddin et al., 1976; Rieder et

165

al., 1980; Wangensteen and Weibel, 1982). Characteristic features of the sinus covering cell are the long cellular processes interposed between the vascular channels of the chorion and the inner surface of the inner shell membrane (Fig. 4; Coleman and Terepka, 1972a; Narbaitz, 1972, 1977; Rieder et al., 1980; Wangensteen and Weibel, 1982). These cellular processes are firmly attached to the inner surface of the inner shell membrane, and removal of the shell membrane may strip away the plasmalemma of sinus covering cells (Coleman and Terepka, 1972a; Anderson et al., 1981). A basal lamina separates sinus covering cells from the endothelial cells of the vascular sinus (Coleman and Terepka, 1972a; Narbaitz, 1977; Rieder et al., 1980; Wangensteen and Weibel, 1982).

The villus cavity cells of the chorion are characterized by numerous microvilli projecting from the apical surface of the cell to the inner surface of the shell membrane (Fig. 4; Coleman and Terepka, 1972a; Narbaitz, 1972, 1977; Rieder et al., 1980). The microvilli are attached to the inner surface of the inner shell membrane and removal of the shell membrane also disrupts these cells (Anderson et al., 1981). Villus cavity cells are more numerous than sinus covering cells in most areas of the chorioallantois, but sinus covering cells are predominant in the region beneath the air cell (Narbaitz, 1972, 1977). Early work identified the sinus covering cell as the predominant cell of the chorion because observations were limited to the chorioallantois from beneath the air cell (Coleman and Terepka, 1972a).

Other cell types in the chorion include relatively undifferentiated cells and darkly staining cells that may be in the process of degenerating (Fig. 4; Coleman and Terepka, 1972a; Rieder et al., 1980). Desmosomes occur at regions of contact between all cells in the chorionic portion of the membrane, and desmosomes and tight junctions occur between sinus covering and villus cavity cells at points of contact near the shell membrane (Owczarzak, 1971; Coleman and Terepka, 1972a; Rieder et al., 1980). An amorphous layer covers the entire outer surface of the chorion, and this layer is continuous with the fuzzy coat surrounding the fibers of the inner shell membrane (Wangensteen and Weibel, 1982).

The vascular system of the chorionic portion of the chorioallantois consists of a single, large blood sinus separated from the inner surface of the inner shell membrane by the vascular endothelium and extensions of sinus covering cells (Narbaitz, 1977). Earlier interpretations that a capillary network exists in the chorion were based on examination of only transverse sections of the chorioallantoic membrane (Wangensteen et al., 1970/71; Coleman and Terepka, 1972a; Tazawa and Ono, 1974; Narbaitz, 1977; Tazawa, 1978; Wangensteen and Weibel, 1982). When the chorioallantois is sectioned parallel to its surface rather than perpendicular to it, it is apparent that the vascular supply is a blood sinus interrupted by columns of cells with no subdividing septum (Narbaitz, 1977).

Calcium transport across the chorion

Characteristics of the transport system. – Transport of calcium across the chorionic ectoderm into the vascular supply involves several processes: entry of calcium into the cells of the chorion, transport through the cells, and extrusion from the cells. These interrelated processes generally have not been examined independently of one another as they have in the intestinal epithelium, the best known of the calcium-transporting epithelia (Bronner, 1982). Thus, it sometimes is difficult to determine which of these processes – uptake, transport, and/or extrusion – is perturbed by a given experimental treatment.

Transport phenomena in the chorioallantois have been studied using both in vivo and in vitro techniques. The former involves mounting a small chamber on the inner shell membrane, usually in the region of the air cell. Small quantities of liquid containing the materials of interest can then be applied to the surface of the shell membrane, and appearance of the sample in the peripheral blood supply, in the embryo, in the yolk, and/or in the chorioallantoic membrane itself can be monitored (Crooks and Simkiss, 1975; Crooks et al., 1976; Kyriakides and Simkiss, 1975; Saleuddin et al., 1976; Tuan and Zrike, 1978; Tuan, 1983). Calcium in the sample of membrane analyzed in such studies presumably reflects uptake, transport, and extrusion, and is present bound to cells, within cells, and in the vascular supply.

Some work done in vitro involves mounting pieces of the chorioallantois in a Ussing chamber, bathing both sides of the membrane with appropriate solutions, and monitoring transport of specific ions from chorionic side to allantoic side or vice versa (Moriarty and Terepka, 1969; Terepka et al., 1969; Moriarty and Hogben, 1970; Garrison and Terepka, 1972a, b; Dunn et al., 1981). Such in vitro procedures examine transport as it occurs across the chorionic ectoderm through the mesodermal portion, across the allantoic epithelium, and into the allantoic sac, a route much different from that followed by materials in intact eggs. During incubation, calcium normally is transported across the chorion, into the vascular supply of the blood sinus, and to the embryo via the allantoic vein. Nonetheless, experiments performed in vitro have contributed importantly to our understanding of transport phenomena in the chorioallantois, and most work is in good agreement with studies performed in vivo (Dunn et al., 1981).

In some in vitro experiments, the transport systems of the chorioallantois have been analyzed indirectly by monitoring the effects of specific substances on oxygen consumption of the chorion and allantois independently or of the entire chorioallantoic membrane (Garrison and Terepka, 1972a, b). A significant fraction of total oxygen consumption by the membrane is attributable to stimulation of the calcium transport system in the chorion (Garrison and Terepka, 1972a).

The mature chorioallantois exhibits the classic properties of active transport processes: maintenance of a potential difference across the membrane, dependence of transport on oxygen, inhibition of transport by metabolic inhibitors, and transport against large electrical or chemical gradients (Terepka et al., 1969; Garrison and Terepka, 1972a, b; Moriarty and Terepka, 1969; Moriarty and Hogben, 1970; Kyriakides and Simkiss, 1975). The transport properties of the membrane are age-dependent and are not manifested until full cytological differentiation of the membrane has occurred (Garrison and Terepka, 1972a, b; Coleman and Terepka, 1972a; Tuan and Zrike, 1978). In addition, normal function of the membrane depends on the structural integrity of the chorion. Stripping away the overlying shell membrane, a procedure that disrupts the plasma membranes of both major cell types in the chorionic ectoderm (Anderson et al., 1981), abolishes the transport of calcium (Moriarty and Terepka, 1969; Terepka et al., 1969; Moriarty and Hogben, 1970; Garrison and Terepka, 1972a, b).

Factors influencing calcium transport/general considerations. – Sodium/potassium and calcium ATPase activity can be demonstrated in the chorioallantois (Saleuddin et al., 1976). Activity of the former is required for calcium transport to occur (see below), but the involvement of the latter in this process has not been demonstrated. There is little correlation between calcium transport in the chorion and the activity of the calcium-stimulated ATPase, and the activity of this enzyme does not increase with age as do the other transport properties of the membrane (Saleuddin et al., 1976; Tuan and Scott, 1977).

The chorionic side of the chorioallantois actively transports calcium, sodium, and chloride ions in vitro, and the transport of calcium is inhibited by ouabain and by the absence of either sodium or chloride from the medium bathing the chorioallantois membrane (Moriarty and Terepka, 1969; Terepka et al., 1969; Moriarty and Hogben, 1970; Garrison and Terepka, 1972a, b; Kyriakides and Simkiss, 1975). Thus, activity of the sodium/potassium ATPase is required for calcium transport to occur normally, and the transport of calcium is intimately tied to the presence, if not the transport, of both sodium and chloride. The exact relationship between calcium transport and transport of other ions during normal function is not clear, however.

The first step in calcium transport in the chorion is binding of calcium to the cells of the chorion (Terepka et al., 1976, also see below). Binding is a passive process, but much of the initial entry of calcium into the cells of the chorion requires energy and is linked to the presence of normal sodium gradients (Terepka et al., 1976). Thus, the inhibition of transport by ouabain or by the absence of sodium may stem from a perturbation of the entry of calcium ions into the chorion. We do not know whether a sodium/calcium interaction is required for extrusion of calcium from the cells of the chorion, and we cannot speak to which aspects of the overall transport mechanism may be linked to chloride ions. Transport of calcium, sodium, chloride, hydrogen, and/or bicarbonate ions may be linked in complex ion-exchange at the interface of the chorion with the inner shell membrane as well as at the interface between the cells of the chorion and the vascular supply, but resolution of these points awaits further study (see below).

168

The calcium-transport mechanism is saturated at 1 mM calcium both in vivo and in vitro, and higher concentrations of calcium lead to no additional increments in calcium transport (Terepka et al., 1969; Garrison and Terepka, 1972a; Crooks et al., 1976). The concentration of calcium to which the chorion is exposed at its interface with the inner shell membrane is not known, but is generally believed to be low (Terepka et al., 1971).

Calcium-binding protein. – A calcium-binding protein has recently been isolated from the chorioallantois and is assumed to function in the calcium-transport mechanism of the membrane (Tuan and Scott, 1977; Tuan et al., 1978a, b, c). This calcium-binding protein is unique to the chorion and shares few similarities with the vitamin D-dependent calcium-binding proteins from the shell gland, kidney, or intestine of adult chickens (Tuan and Scott, 1977). These differences between the chorionic calcium-binding protein and the calcium-binding proteins of adult animals may stem in part from the fact that the chorion is an extraembryonic derivative that shares no embryologic origin with the tissues elaborating the calcium-binding proteins of the adult (Balinsky, 1975). There are no differences in binding activity between preparations of the chorioallantois taken from the equator or air cell, and it is unlikely that the protein is localized specifically in one cell type (Tuan et al., 1978a).

The chorionic calcium-binding protein is highly specific for calcium ions and seems to be associated with the outer surface of the chorion (Tuan et al., 1978a, b, c; Tuan, 1979). Approximately 95% of the calcium-binding activity of the chorion is attributable to this protein, and the calcium-binding activity of the membrane shows the same age-dependent developmental expression as do the transport functions of the membrane (Tuan and Scott, 1977). Moreover, calcium-binding activity in the chorion is inhibited by sulfhydryl-blocking agents as is calcium transport (Garrison and Terepka, 1972a; Tuan et al., 1978a).

Given that much of the work on transport of calcium across the chorion is consistent with the idea that binding of calcium to the cells of the chorion is the first step of the transport process, the chorionic calcium-binding protein provides a logical mechanism by which this step is accomplished (Garrison and Terepka, 1972a, b; Terepka et al., 1976; Tuan et al., 1978b). However, the involvement of the protein in calcium transport or the requirement that it be present in active form for transport, as opposed to binding, to occur has been shown. It would be interesting to know what effect treatment of chorion with anticalcium-binding protein antisera has on calcium transport and/or calcium metabolism in vivo since this treatment abolishes much of the calcium-binding activity in extracts of the chorion (Tuan, 1980).

Carbonic anhydrase. – Studies concerning the effect of carbonic anhydrase on calcium transport must be interpreted with caution, because they were performed for the most part on preparations of the chorioallantois from beneath the air cell of an egg and this region of the membrane transports calcium at a reduced rate compared to the membrane from equatorial regions (Dunn et al., 1981). Nonetheless, the chorioallantois from the air cell region does transport calcium, and it is reasonable to assume that calcium transport in both regions of the chorion is under similar control.

The quantity of carbonic anhydrase in the chorioallantois increases in an age-dependent manner as do a number of the transport properties of the membrane (Tuan and Zrike, 1978; Gay, 1980). Most of the enzyme is non-vascular in origin, an observation indicating that its primary function is unrelated to respiratory exchange of carbon dioxide (Tuan and Zrike, 1978; Gay, 1980). Activity of the enzyme is highest in preparations of the membrane taken from the equatorial region of the egg (Narbaitz et al., 1981), and immunocytochemical and auto-radiographic techniques show that the enzyme is localized primarily in villus cavity cells, the most common cell type in the equatorial region of the chorion (Narbaitz, 1972; Rieder et al., 1980; Anderson et al., 1981; Narbaitz et al., 1981).

Inhibition of carbonic anhydrase activity interferes with maintenance of the potential difference across the chorioallantois, and this effect is exacerbated if the chloride concentration of the bathing medium is reduced (Kyriakides and Sim-kiss, 1975). Inhibition of the enzyme also reduces uptake of calcium by the chorion to about 50% of normal, but has no effect on the ability of the chorio-allantois to bind calcium (Tuan and Zrike, 1978). These effects of carbonic anhydrase inhibition are consistent with an effect on an active component of calcium transport, perhaps one involving ion-exchange processes (see below).

It has long been assumed that hydration of carbon dioxide by carbonic an-hydrase aids in dissolution of the eggshell during embryogenesis by providing a source of hydrogen ions used to dissolve the calcium carbonate matrix of the eggshell (Coleman and Terepka, 1972a; Dawes, 1975; Gay, 1980; Rieder et al., 1980; Anderson et al., 1981; Narbaitz et al., 1981). According to this scheme, calcium diffuses through the shell membrane to the chorion and ultimately is transported across the chorion into the embryonic blood supply (Dawes, 1975). There is little direct experimental evidence to support this hypothesis, and the perturbation of calcium uptake after inhibition of carbonic anhydrase activity occurs rather quickly, an observation making it unlikely that effects on the calcium-transport system stem from an inhibition of shell dissolution (Tuan and Zrike, 1978).

An active role for carbonic anhydrase in dissolution of the eggshell may be difficult to detect. The hydration of carbon dioxide and formation of hydrogen and bicarbonate ions in consequence occur at a significant rate even in the absence of catalysis by carbonic anhydrase (Maren, 1967). Thus, some dissolution of the eggshell may occur simply in consequence of the elevated levels of carbon dioxide characteristic of hen's eggs late in incubation and the humidity generated by the constant loss of water vapor. This scenario requires that the fibers of the shell membranes be associated, at the very least, with a molecular layer of water in which the reaction can proceed and through which calcium can diffuse from the site of dissolution to the chorion. If inhibition of carbonic anhydrase does perturb dissolution of the eggshell and hence provision of calcium ions for transport, the effect may be too small or too protracted to be detectable given the variability inherent in studies of calcium transport across the chorion.

One function of carbonic anhydrase in a number of tissues is elaboration of hydrogen and bicarbonate ions used in a variety of ion-exchange processes (Maren, 1967; Carter, 1972). The absence of these ions could diminish calcium transport across the chorion if either or both of the ions is required for an ion exchange with calcium. Under these circumstances, dissolution of the eggshell could be an indirect consequence of a more general ion exchange that requires carbonic anhydrase for proper function. These suggestions are consistent with experimental evidence and represent a shift in emphasis rather than a full departure from previous interpretations (Tuan and Zrike, 1978; Gay, 1980; Rieder et al., 1980; Anderson et al., 1981).

Interactions between the chorion and calcium. – Full development of a number of properties of the chorion related to its transport capabilities seems to depend on an interaction between the developing membrane and the eggshell, or perhaps between the chorion and calcium contained within the shell membranes (Dunn et al., 1981; Tuan, 1983). The chorioallantoic membrane normally develops without contact with the eggshell beneath the air cell of an intact egg. In this region of the egg, the chorioallantois is overlain by only the inner shell membrane whereas in all other areas both the inner and outer shell membranes as well as the calcareous layer overlie the chorioallantois. Thus, there is probably little or no transport of calcium across the chorion beneath the air cell in vivo, and the chorioallantois from this region transports calcium at about half the rate of the chorioallantois from the equatorial region of an egg in vitro (Dunn et al., 1981; Tuan, 1983).

The chorioallantois also develops without contact with the eggshell when eggs are maintained in shell-less culture or when an artificial air space is created between the chorioallantois and the inner shell membrane during normal incubation of eggs (Narbaitz and Tellier, 1974; Dunn and Boone, 1976; Narbaitz and Jande, 1978; Tuan and Zrike, 1978; Dunn and Fitzharris, 1979; Tuan, 1980). Under these conditions, the membrane undergoes normal cytodifferentiation, but several biochemical and physiological properties of the membrane do not develop normally (Narbaitz and Tellier, 1974; Narbaitz and Jande, 1978; Tuan and Zrike, 1978; Dunn and Fitzharris, 1979; Tuan, 1980, 1983; Dunn et al., 1981).

Preparations of chorioallantoic membrane which have had no contact with the eggshell transport significantly less calcium than preparations from undisturbed eggs (Dunn et al., 1981; Tuan, 1983). Transport can be enhanced if the membrane of the eggs in shell-less culture is allowed contact with pieces of intact eggshell (calcareous layer plus shell membranes), with pieces of combined inner and outer shell membrane, or with pieces of calcareous layer alone, but calcium transport does not attain the levels characteristic of membranes from eggs incubated normally (Dunn et al., 1981; Tuan, 1983). In addition, uptake of calcium by membranes from eggs maintained in shell-less culture does not increase in the age-dependent fashion shown by the chorioallantois from undisturbed eggs (Tuan, 1980). The level of calcium-binding protein present in the chorion from eggs in shell-less culture is higher than in the chorion from control eggs, but calcium-binding activity is less (Tuan, 1980). It appears that embryos in shell-less culture synthesize more calcium-binding protein than normal embryos, but that the protein is inactive (Tuan, 1980).

These biochemical and physiological differences between the chorioallantois that develops without contact with the eggshell and the normal membrane indicate the existence of a possible feedback mechanism between levels of calcium and the function of the chorionic transport system. The calcium content of shell membranes of eggs in shell-less culture or of eggs with an artificial air space cannot be replenished by calcium mobilized from the eggshell, and the chorioallantois presumably is exposed to lower than normal concentrations of calcium (Dunn et al., 1981; Tuan, 1983). How, or if, this situation influences the normal development of mechanisms involved in calcium transport remains to be determined.

Embryos developing in shell-less culture are smaller than embryos developing normally and may exhibit a number of developmental anomalies compared to normal embryos (Dunn and Boone, 1976, 1977; Narbaitz and Jande, 1978; Burke et al., 1979; De Gennaro et al., 1980; Slavkin et al., 1980). The most striking of these is the extreme calcium deficiency of the cultured embryos (Dunn and Boone, 1977; Narbaitz and Jande, 1978; Tuan, 1980). Perhaps calcium deficiency in these embryos acts as a signal to induce synthesis of additional calcium-binding protein, but activation of the protein is prevented or inhibited by the lower than normal concentrations of calcium to which the chorionic epithelium is exposed.

Vitamin K. – During normal incubation, formation of active calcium-binding protein in the chorion may depend on the presence of vitamin K (Tuan et al., 1978c). Injection of eggs with warfarin, an inhibitor of vitamin K activity, results in a reduction in calcium-binding activity in the chorioallantois with no change in the amount of calcium-binding protein (Tuan, 1979). Moreover, incubation of explants of the chorioallantoic membrane with vitamin K enhances calcium-binding activity and leads to an increase in the total quantity of protein present (Tuan et al., 1978c). Taken in concert these results indicate that vitamin K may play a role in induction of new protein as well as in activation of existing protein (Tuan et al., 1978c; Tuan, 1979). Thus, formation or accumulation of an inactive form of calcium-binding protein during shell-less culture may stem from improper mobilization of vitamin K from yolk and/or inadequate activity of this vitamin.

It has not been shown, however, that vitamin K is essential to normal development of the calcium-transport mechanism nor that vitamin K affects calcium transport. Vitamin K given to embryos in vivo leads to no enhancement of calcium-binding activity in the chorion (Tuan, 1979), and it has not been demonstrated that treatment of the chorion with inhibitors of vitamin K activity reduces the ability of the membrane to transport calcium. Thus, we cannot speak to potential mechanisms leading to accumulation of inactive calcium-binding protein in the chorion of embryos in shell-less culture nor to the role of vitamin K during normal incubation.

Vitamin D. – The active metabolite of vitamin D is now considered to be a steroid hormone in the classic sense of the term (DeLuca, 1981). This hormone is required for normal transport of calcium in the intestinal epithelium where it induces synthesis of calcium-binding proteins and influences permeability of cell membranes to calcium through effects on membrane phospholipids (Schachter and Kowarski, 1981, 1982; Bronner et al., 1982; DeLuca et al., 1982; Norman et al., 1982; Rasmussen et al., 1982). Vitamin D may also affect calcium transport in the chorion although its mechanism of action remains to be elucidated.

The chorioallantois contains a binding protein and putative target cells for vitamin D and can convert 25-hydroxyvitamin D_3, a generally less active form of the hormone (DeLuca, 1981), to 1,25-dihydroxyvitamin D_3, the active metabolite (Narbaitz et al., 1980; Puzas et al., 1980; Coty et al., 1981). Moreover injecting eggs with the active metabolite of vitamin D leads to an elevation of carbonic anhydrase activity in the chorioallantois (Narbaitz et al., 1981). Increased activity of an enzyme does not necessarily imply that the enzyme is present in higher quantity, but this observation is in keeping with a potential effect of the hormone on synthesis or activity of carbonic anhydrase, an enzyme whose inhibition does influence uptake of calcium by the chorion (Tuan and Zrike, 1978).

The involvement of vitamin D in calcium transport or in overall calcium metabolism of embryos clearly requires additional research. The active form of this hormone apparently is not transferred from females to their eggs (Sunde et al., 1978), for females receiving the active metabolite as their only source of vitamin D lay eggs that give rise to abnormal embryos and exhibit much reduced hatchability (Ameenuddin et al., 1982). In contrast, hens receiving 25-hydroxyvitamin D_3 lay eggs that have high hatching success and give rise to normal chicks (Henry and Norman, 1978; Ameenuddin et al., 1982). Indeed, several analogs of vitamin D given to hens as their only source of this hormone can support normal embryonic development (Ameenuddin et al., 1982).

Thus, eggs must be endowed with some form of vitamin D for embryogenesis to proceed normally. Although analogs other than the active form of vitamin D are biologically active (DeLuca, 1981), 1,25-dihydroxyvitamin D_3 may be the active metabolite during embryogenesis for titers of this form of vitamin D gradually increase during incubation (Seino et al., 1982). If this assumption is correct, it is possible that the conversion of 25-hydroxyvitamin D_3 to 1,25-dihydroxyvitamin D_3 occurs in the chorioallantois and that a portion of the hormone is used in situ to affect aspects of calcium transport by the chorion.

Localization of calcium transport to specific cells. – Early work identified sinus covering cells as the cells responsible for transport of calcium across the chorion (Coleman et al., 1970; Coleman and Terepka, 1972b, c). However, this work was done using preparations of the chorioallantois from beneath the air cell, a region that consists predominantly of sinus covering cells and that transports calcium at a lower rate than does the chorioallantois from the equator (Coleman and Terepka, 1972b, c; Narbaitz, 1972, 1977; Dunn et al., 1981). These observations, coupled with the observation that villus cavity cells are the most common cell type in equatorial preparations of the membrane, indicate that villus cavity cells may be responsible for calcium transport across the chorion (Narbaitz, 1972; Dunn et al., 1981).

Differences in levels of calcium-binding protein cannot be invoked to explain differences in calcium transport in the two regions of the chorion, because levels of calcium-binding activity are similar in preparations of the chorioallantois from both the equator and the air cell (Tuan et al., 1978a). However, the higher levels of carbonic anhydrase in equatorial preparations of the chorioallantois, coupled with its localization primarily in villus cavity cells and its effects on calcium uptake, may underlie some of these differences in calcium transport between the equatorial and air cell regions of the chorion (Rieder et al., 1980; Anderson et al., 1981; Narbaitz et al., 1981).

Alternatively, both villus cavity and sinus covering cells may transport calcium in vivo, but sinus covering cells may simply transport calcium less efficiently than villus cavity cells. Indeed, calcium has been localized in both cell types (Coleman and Terepka, 1972b, c; Narbaitz, 1972), and such data obviate a definitive statement concerning localization of calcium transport to a single kind of cell.

Summary remarks

It is apparent from the preceding comments that a large number of interrelated factors operate to influence calcium metabolism in eggs of oviparous, amniotic vertebrates, that most research in this area concerns eggs of domestic fowl, and that there are few comparative data on calcium metabolism during embryogenesis. It is not clear how applicable the research done with eggs of domestic fowl will prove to be to eggs of other oviparous amniotes, because differences in calcium metabolism between avian and reptilian embryos already are apparent.

The large size and ready availability of eggs of domestic fowl make them attractive for study of mechanisms of calcium transport and its control, but these advantages should not blind us to the potential advantages of a more comparative and broadly based approach to problems of calcium metabolism during embryonic development.

For example, have problems associated with mobilizing and transporting large quantities of calcium during embryogenesis placed constraints on the evolution of the morphology and physiology of the calcium-transporting epithelia? Are there morphological and physiological similarities between the yolk sac epithelium of species that rely on the yolk as the major source of calcium and the chorionic epithelium of species that obtain most of their calcium from the eggshell? Are there similarities between the chorioallantois and yolk sac of oviparous reptiles and the chorioallantoic and yolk sac placentae of viviparous reptiles? Do factors influencing calcium transport in eggs of domestic fowl also influence mobilization of calcium in reptilian eggs? The answers to these and other questions concerning calcium metabolism during embryogeneis should provide exciting insights into target organs involved in control of calcium metabolism, mechanisms of control of this process, and the evolution of these mechanisms.

Acknowledgements

We thank R. Narbaitz and R.S. Tuan for allowing us to examine preprints of manuscripts. N.B. Clark, B.E. Dunn, T.A. Gorell, and K. Simkiss reviewed drafts of the manuscript. Our efforts were supported in part by the Department of Zoology and Entomology and by grants from the Webster-Barnes Foundation of the Department of Physiology and Biophysics, Colorado State University (to MJP) and the National Science Foundation (DEB79–11546 to GCP).

References

Ameenuddin, S., Sunde, M. DeLuca, H.F., Ikekawa, N. and Kobayashi, Y. (1982). 24-hydroxylation of 25-hydroxyvitamin D_3: is it required for embryonic development in chicks? Science 217: 451–452

Anderson, R.E., Gay, C.V. and Schraer, H. (1981). Ultrastructural localization of carbonic anhydrase in the chorioallantoic membrane by immunocytochemistry. J. Histochem. Cytochem. 29: 1121–1127.

Balinsky, B.I. (1975). An introduction to Embryology. 4th edition. W.B. Saunders Co., Philadelphia.

Bronner, F. (1982). Intestinal calcium absorption and transport. In: Membrane Transport of Calcium. E. Carafoli, ed., Academic Press, London, pp. 237–262.

Bronner, F., Lipton, J., Pansu, D., Buckley, M., Singh, R., and Miller, A., III (1982). Molecular and transport effects of 1,25-dihydroxyvitamin D_3 in rat duodenum. Fed. Proc., Fed. Am. Soc. Exp. Biol. 41: 61–65.

Burke, B., Narbaitz, R. and Tolnai, S. (1979). Abnormal characteristics of the blood from chick embryos maintained in 'shell-less' culture. Rev. Can. Biol. 38: 63–66.

Bustard, H.R., Simkiss, K. and Jenkins, N.K. (1969). Some analyses of artificially incubated eggs and hatchlings of green and loggerhead sea turtles. J. Zool. 158: 311–315.

Carter, M.J. (1972). Carbonic anydrase: Isoenzymes, properties, distribution, and functional significance Biol. Rev Cambridge Philos. Soc. 47: 465–513.

Choi, J.K. (1963). The fine structure of the urinary bladder of the toad, Bufo marinus. J. Cell Biol. 16: 53–71.

Clark, N.B. and Simkiss, K. (1980). Time, targets and triggers: a study of calcium regulation in the bird. In: Avian Endocrinology. A. Epple, ed., Academic Press, London, pp. 191–208.

Coleman, J.R., DeWitt, S.M., Batt, P. And Terepka, A.R. (1970). Electron probe analysis of calcium distribution during active transport in chick chorioallantoic membrane. Exp. Cell Res. 63: 216–220.

Coleman, J.R. and Terepka, A.R. (1972a). Fine structural changes associated with the onset of calcium, sodium and water transport by the chick chorioallantoic membrane. J. Membr. Biol. 7: 111–127.

Coleman, J.R. and Terepka, A.R. (1972b). Electron probe analysis of the calcium distribution in cells of the embryonic chick chorioallantoic membrane/ I. A critical evaluation of techniques. *J. Histochem. Cytochem.* 20: 401–413.

Coleman, J.R. and Terepka, A.R. (1972c). Electron probe analysis of the calcium distribution in cells of the embryonic chick chorioallantoic membrane/II. Demonstration of intracellular location during active transcellular transport. *J. Histochem. Cytochem.* 20: 414–424.

Coty, W.A., McConkey, C.L., Jr. and Brown, T.A. (1981). A specific binding protein for $1\alpha,25$-dihydroxyvitamin D in the chick embryo chorioallantoic membrane. *J. Biol. Chem.* 256: 5545–5549.

Crooks, R.J. and Simkiss, K. (1974). Respiratory acidosis and eggshell resorption by the chick embryo. *J. Exp. Biol.* 61: 197–202.

Crooks, R.J. and Simkiss, K. (1975). Calcium transport by the chick chorioallantois *in vivo. Quart. J. Exp. Physiol.* 60: 55–63.

Crooks, R.J., Kyriakides, C.P.M. and Simkiss, K. (1976). Routes of calcium movement across the chick chorioallantois. *Quart. J. Exp. Physiol.* 61: 265–274.

Dawes, C.M. (1975). Acid-base relationships within the avian egg. *Biol. Rev. Cambridge Philos. Soc.* 50: 351–371.

De Gennaro, L.D., Packard, D.S., Jr., Stach, R.W. and Wagner, B.J. (1980). Growth and differentiation of chicken embryos in simplified shell-less cultures under ordinary conditions of incubation. *Growth* 44: 343–354.

DeLuca, H.F. (1981). The transformation of a vitamin into a hormone: the vitamin D story. *The Harvey Lectures, Series 75,* pp. 333–379.

DeLuca, H.F., Franceschi, R.T., Halloran, B.P. and Massaro, E.R. (1982). Molecular events involved in 1,25-dihydroxyvitamin D_3 stimulation of intestinal calcium transport. *Fed. Proc., Fed Am. Soc. Exp. Biol.* 41: 66–71.

Dunn, B.E. and Boone, M.A. (1976). Growth of the chick embryo *in vitro. Poult. Sci.* 55: 1067–1071.

Dunn, B.E. and Boone, M.A. (1977). Growth and mineral content of cultured chick embryos. *Poult. Sci.* 56: 662–672.

Dunn, B.E. and Fitzharris, T.P. (1979). Differentiation of the chorionic epithelium of chick embryos maintained in shell-less culture. *Dev. Biol.* 71: 216–227.

Dunn, B.E., Graves, J.S. and Fitzharris, T.P. (1981). Active calcium transport in the chick chorioallantoic membrane requires interaction with the shell membrane and/or shell calcium. *Dev. Biol.* 88: 259–268.

Ferguson, M.W.J. (1982). The structure and composition of the eggshell and embryonic membranes of *Alligator mississippiensis. Trans. Zool. Soc. London* 36: 99–152.

Freeman, B.M. and Vince, M.A. (1974). Development of the Avian Embryo/ A Behavioural and Physiological Study. Chapman and Hall, London.

Garrison, J.C. and Terepka, A.R. (1972a). Calcium-stimulated respiration and active calcium transport in the isolated chick chorioallantoic membrane. *J. Membr. Biol.* 7: 128–145.

Garrison, J.C. and Terepka, A.R. (1972b). The interrelationships between sodium ion, calcium transport and oxygen utilization in the isolated chick chorioallantoic membrane. *J. Membr. Biol.* 7: 146–163.

Gay, C.V. (1980). Recent developments in carbonic anhydrase localization: occurrence of carbonic anhydrase in calcium-mobilizing tissues. In: Biopshysics and Physiology of Carbon Dioxide. C. Bauer, G. Gros and H. Bartels, eds., Springer-Verlag, Berlin, pp. 399–405.

Henry, H.L. and Norman, A.W. (1978). Vitamin D: two dihydroxylated metabolites are required for normal chicken egg hatchability. *Science* 201: 835–837.

Hoyt, D.W. (1979). Osmoregulation by avian embryos: the allantois functions like a toad's bladder. *Physiol. Zool.* 52: 354–362.

Jenkins, N.K. (1975). Chemical composition of the eggs of the crocodile (*Crocodylus novaeguineae*). *Comp. Biochem. Physiol.* 51A: 891–895.

Jenkins, N.K. and Simkiss, K. (1968). The calcium and phosphate metabolism of reproducing reptiles with particular reference to the adder (*Vipera berus*). *Comp. Biochem. Physiol.* 26: 865–876.

Johnston, P.M. and Comar. C.L. (1955). Distribution and contribution of calcium from the albumen, yolk and shell to the developing chick embryo. *Am. J. Physiol.* 183: 365–370.

Juurlink, B.H.J. and Gibson. M.A. (1973). Histogenesis of the yolk sac in the chick. *Can. J. Zool.* 51: 509–519.

Kyriakides, C.P.M. and Simkiss, K. (1975). Transmembrane potential differences across the chorio-allantois. *Comp. Biochem. Physiol.* 51A: 875–879.

Lambson, R.O. (1970). An electron microscopic study of the entodermal cells of the yolk sac of the chick during incubation and after hatching. *Am. J. Anat.* 129: 1–20.

Leaf, A. (1982). From toad bladder to kidney. *Am. J. Physiol.* 242: F103–F111.

Leeson, T.S. and Leeson. C.R. (1963). The chorio-allantois of the chick/Light and electron micro-scopic observations at various times of incubation. *J. Anat.* 97: 585–595.

Maren, T.H. (1967). Carbonic anhydrase: chemistry, physiology, and inhibition. *Physiol. Rev.* 47: 595–781.

Matthews, J.L., VanderWeil. C.J. and Talmage. R.V. (1981). Intracellular calcium regulation and transport leading to calcium control of physiological processes. In: Advanced Cell Biology. L.M. Schwartz and M.M. Azar. eds., Van Nostrand Reinhold Co., New York, pp. 417–432.

Mobbs, I.G. and McMillan. D.B. (1979). Structure of the endodermal epithelium of the chick yolk sac during early stages of development. *Am. J. Anat.* 155: 287–309.

Moriarty, C.M. and Hogben. C.A.M. (1970). Active Na⁺ and Cl⁻ transport by the isolated chick chorioallantoic membrane. *Biochem. Biophys. Acta.* 219: 463–470.

Moriarty, C.M. and Terepka. A.R. (1969). Calcium transport by the isolated chick chorio-allantoic membrane. *Arch. Biochem. Biophys.* 135: 160–165.

Murphy, M.J., Brown. S.C. and Brown. P.S. (1982). Hydromineral balance in the chick embryo: effects of hypophysectomy. *J. Exp. Zool.* 220: 321–330.

Narbaitz, R. (1972). Cytological and cytochemical study of the chick chorionic epithelium. *Rev. Can. Biol.* 31: 259–267.

Narbaitz, R. (1977). Structure of the intra-chorionic blood sinus in the chick embryo. *J. Anat.* 124: 347–354.

Narbaitz, R., Belanger, L.F. and Hunt. B.F. (1973). Calcium regulation by the chick embryo/An experimental approach. *Calcif. Tissue Res.* 11: 238–241.

Narbaitz, R. and Jande. S.S. (1978). Ultrastructural observations on the chorionic epithelium, parathyroid glands and bones from chick embryos developed in shell-less culture. *J. Embryol. Exp. Morphol.* 45: 1–12.

Narbaitz, R., Kacew, S. and Sitwell. L. (1981). Carbonic anhydrase activity in the chick embryo chorioallantois: regional distribution and vitamin D regulation. *J. Embryol. Exp. Morphol.* 65: 127–137.

Narbaitz, R., Stumpf, W., Sar, M., DeLuca, H.F. and Tanaka, Y. (1980). Autoradiographic demon-stration of target cells for 1,25-dihydroxycholecalciferol in the chick embryo chorioallantoic mem-brane, duodenum, and parathyroid glands. *Gen. Comp. Endocrinol.* 42: 283–289.

Narbaitz, R. and Tellier. P.P. (1974). The differentiation of the chick chorionic epithelium: an experimental study. *J. Embryol. Exp. Morphol.* 32: 365–374.

Needham, J. (1963). Chemical Embryology. Vol. III. Hafner Publishing Co., New York.

Norman, A.W., Putkey, J.A. and Nemere. I. (1982). Intestinal calcium transport: pleiotropic effects mediated by vitamin D. *Fed. Proc., Fed. Am. Soc. Exp. Biol.* 41: 78–83.

Owczarzak, A. (1971). Calcium-absorbing cell of the chick chorioallantoic membrane/I. Morphology, distribution and cellular interactions. *Exp. Cell Res.* 68: 113–129.

Packard, G.C. and Packard. M.J. (1980). Evolution of the cleidoic egg among reptilian antecedents of birds. *Am. Zool.* 20: 351–362.

Packard, G.C., Tracy, C.R. and Roth. J.J. (1977). The physiological ecology of reptilian eggs and embryos, and the evolution of viviparity within the class Reptilia. *Biol. Rev. Cambridge Philos. Soc.* 52: 71–105.

Packard, M.J., Packard, G.C. and Boardman, T.J. (1982). Structure of eggshells and water relations of reptilian eggs. *Herpetologica* 38: 136–155.

Packard, M.J., Packard, G.C. and Gutzke, W.H.N. (1984a). Calcium metabolism in embryos of the oviparous snake *Coluber contrictor. J. Exp. Biol.* in press.

Packard, M.J., Short, T.M., Packard, G.C. and Gorell, T.A. (1984b). Sources of calcium for embryonic development in eggs of the common snapping turtle *Chelydra serpentina. J. Exp. Zool.* in press.

Puzas, J.E., Turner, R.T., Forte, M.D., Kenny, A.D. and Baylink, D.J. (1980). Metabolism of $25(OH)D_3$ to $1,25(OH)_2D_3$ and $24,25(OH)_2D_3$ by chick chorioallantoic cells in culture. *Gen. Comp. Endocrinol.* 42: 116–122.

Rasmussen, H., Matsumoto, T., Fontaine, O. and Goodman, D.B.P. (1982). Role of changes in membrane lipid stucture in the action of 1,25-dihydroxyvitamin D_3. *Fed. Proc., Fed. Am. Soc. Exp. Biol.* 41: 72–77.

Rieder, E., Gay, C.V. and Schraer, H. (1980). Autoradiographic localization of carbonic anhydrase in the developing chorioallantoic membrane. *Anat. Embryol.* 159: 17–31.

Romanoff, A. (1960). The Avian Embryo. The Macmillan Co., New York.

Romanoff, A. (1967). Biochemistry of the Avian Embryo. John Wiley and Sons, New York.

Saleuddin, A.S.M., Kyriakides, C.P.M., Peacock, A. and Simkiss, K. (1976). Physiological and ultrastructural aspects of ion movements across the chorioallantois. *Comp. Biochem. Physiol.* 54A: 7–12.

Schachter, D. and Kowarski, S. (1981). IMCal: vitamin D-dependent intestinal membrane calcium-binding protein. *Ann. N.Y. Acad. Sci.* 372: 530–538.

Schachter, D. and Kowarski, S. (1982). Isolation of the protein IMCal, a vitamin D-dependent membrane component of the intestinal transport mechanism for calcium. *Fed. Proc., Fed. Am. Soc. Exp. Biol.* 41: 84–87.

Seino, Y., Yamaoka, K., Ishida, M., Yabuuchi, H., Ichikawa, M., Ishige, H., Yoshino, H. and Avioli, L.V. (1982). Biochemical characterization of $1,25(OH)_2D_3$ receptors in chick embryonal duodenal cytosol. *Calcif. Tissue Int.* 34: 265–269.

Simkiss, K. (1962). The sources of calcium for the ossification of the embryo of the giant leathery turtle. *Comp. Biochem. Physiol.* 7: 71–99.

Simkiss, K. (1967). Calcium in Reproductive Physiology. Reinhold Publishing Corp., New York.

Simkiss, K. (1980). Water and ionic fluxes inside the egg. *Am. Zool.* 20: 385–393.

Slavkin, H.C., Slavkin, M.D. and Bringas, P., Jr.(1980). Mineralization during long-term cultivation of chick embryos *in vitro. Proc. Soc. Exp. Biol. Med.* 163: 249–257.

Stewart, M.E. and Terepka, A.R. (1969). Transport functions of the chick chorio-allantoic membrane/I. Normal histology and evidence for active electrolyte transport from the allantoic fluid, *in vivo. Exp. Cell Res.* 58: 93–106.

Sunde, M.L., Turk, C.M. and DeLuca, H.F. (1978). The essentiality of vitamin D metabolites for embryonic chick development. *Science.* 200: 1067–1069.

Tazawa, H. (1978). Gas transfer in the chorioallantois. In: Respiratory Function in Birds, Adult and Embryonic. J. Piiper, ed., Springer-Verlag, Berlin, pp. 274–291.

Tazawa, H. and Ono, T. (1974). Microscopic observation of the chorioallantoic capillary bed of chicken embryos. *Respir. Physiol.* 20: 81–89.

Terepka, A.R., Coleman, J.R., Armbrecht, H.J. and Gunter, T.E. (1976). Transcellular transport of calcium. In: Calcium in Biological Systems. *Symp. Soc. Exp. Biol.* 30: 117–140.

Terepka, A.R., Coleman, J.R., Garrison, J.C. and Spataro, R.F. (1971). Active transcellular transport of calcium by embryonic chick chorioallantoic membrane. In: Cellular Mechanisms for Calcium Transfer and Homeostasis. G. Nichols, Jr. and R.H. Wasserman, eds., Academic Press, New York, pp. 371–389.

Terepka, A.R., Stewart, M.E. and Merkel, N. (1969). Transport functions of the chick chorioallan-toic membrane/II. Active calcium transport, *in vitro*. *Exp. Cell Res.* 58: 107–117.

Thompson, J. (1982). Uptake of inorganic ions from the maternal circulation during development of the embryos of a viviparous lizard, *Sphenomorphus quoyii*. *Comp. Biochem. Physiol.* 71A: 107–112.

Tuan, R.S. (1979). Vitamin K-dependent γ-glutamyl carboxylase activity in the chick embryonic chorioallantoic membrane. *J. Biol. Chem.* 254: 1356–1364.

Tuan, R.S. (1980). Calcium transport and related functions in the chorioallantoic membrane of cultured shell-less chick embryos. *Dev. Biol.* 74: 196–204.

Tuan, R.S. (1983). Supplemented eggshell restores calcium transport in chorioallantoic membrane of cultured shell-less chick embryos. *J. Embryol. Exp. Morphol.* 78: in press.

Tuan, R.S. and Scott, W.A. (1977). Calcium-binding protein of chorioallantoic membrane: identifica-tion and developmental expression. *Proc. Natl. Acad. Sci. U.S.A.* 74: 1946–1949.

Tuan, R.S., Scott, W.A. and Cohn, Z.A. (1978a). Purification and characterization of calcium-binding protein from chick chorioallantoic membrane. *J. Biol. Chem.* 253: 1011–1016.

Tuan, R.S., Scott, W.A. and Cohn, Z.A. (1978b). Calcium-binding protein of the chick chorioallan-toic membrane/I. Immunohistochemical localization. *J. Cell Biol.* 77: 743–751.

Tuan, R.S., Scott, W.A. and Cohn, Z.A. (1978c). Calcium-binding protein of the chick chorioallan-toic membrane/II. Vitamin K-dependent expression. *J. Cell Biol.* 77: 752–761.

Tuan, R.S. and Zrike, J. (1978). Functional involvement of carbonic anhydrase in calcium transport of the chick chorioallantoic membrane. *Biochem. J.* 176:67–74.

Wangensteen, D. and Weibel, E.R. (1982). Morphometric evaluation of chorioallantoic oxygen transport in the chick embryo. *Respir. Physiol.* 47: 1–20.

Wangensteen, O.D., Wilson, D. and Rahn, H. (1970/71). Diffusion of gases across the shell of the hen's egg. *Respir. Physiol.* 11: 16–30.

12. Respiratory gas transport system: similarities between avian embryos and lungless salamanders

Johannes Piiper[1] and Peter Scheid[2]

Abstract

The respiratory gas transport systems of avian embryos and lungless salamanders show common features, but substantial differences from that of adult mammals or birds.

The role of ventilation in limiting gas exchange in adult mammals and birds is functionally replaced by the diffusive resistance of the egg shell pores and of the skin epithelium, respectively. However, due to differing diffusion media, air in egg shell pores and tissue in salamander skin, the CO_2-O_2 relationships in arterialized blood differ considerably.

By sizeable admixture of venous blood, due to the arrangement of the large blood vessels, the 'arterial' blood (i.e. blood perfusing the tissues) is markedly different from the 'arterialized' blood leaving the gas exchange organ. Gas exchange organ blood flow is different from (lower than) the systemic blood flow. With respect to blood circulation, the gas exchange organ and the body tissues are arranged essentially in parallel (not in series as in adult mammals and birds). The 'cardiac output', which essentially is equal to the sum of total systemic and gas exchange organ flows, is functionally comparable to double the 'cardiac output' in the sense used in mammalian and human physiology.

Introduction

The respiratory gas transport, i.e. transport of O_2 from the environment to the cells and that of CO_2 from the cells to the environment, is conventionally analyzed in physiology using a schematic diagram as depicted in Fig. 1. The steps of O_2 and CO_2 transport are represented by the drop of O_2 partial pressure, and increase of CO_2 partial pressure, from inspired via alveolar, arterial, (mixed) venous to tissue

[1] Abteilung Physiologie, Max-Planck-Institut für experimentelle Medizin, D-3400 Göttingen, FRG

[2] Physiologisches Institut der Ruhr-Universität, D-4630 Bochum, FRG

Seymour, R.S., (ed.) Respiration and metabolism of embryonic vertebrates.
© 1984, Dr W. Junk Publishers. Dordrecht/Boston/London. ISBN-13:978-94-009-6538-6

values. The values shown in Fig. 1 approximately correspond to values measured in a mammal in resting conditions.

In vertebrates other than adult mammals and birds a number of quantitative and qualitative modifications of this diagram are required. The following factors appear to be of particular importance (cf. Piiper and Scheid, 1982): (1) respiratoty medium: water vs. air; (2) chemical combination of O_2 and CO_2 in blood; (3) functional type of the gas exchange organ; (4) limiting process in gas transfer; and (5) connection of gas exchange organ to the circulatory system.

In this paper, the factors 4 and 5 will be analyzed comparatively in the chicken embryo and in the lungless salamander *Desmognathus fuscus*, using the mammalian gas exchange system as reference, because it is by these factors that the characteristic parallelisms and differences between the respiratory gas exchange and transport systems of these animals are produced.

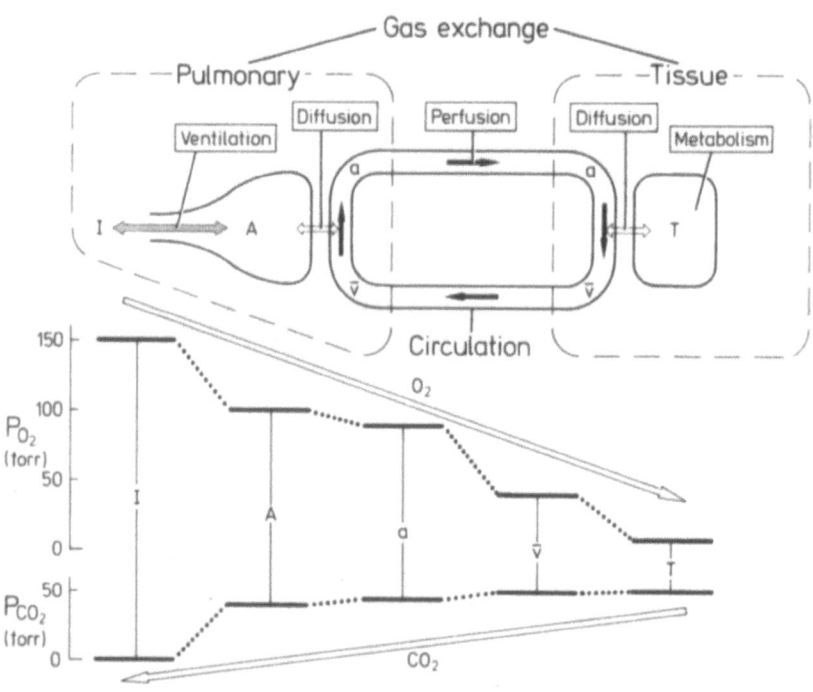

Fig. 1. Gas transport between environment and tissues. Schema valid for lung-breathing vertebrates with completely divided heart (mammals and birds). The indices I, A, a, v̄ and T refer to inspired, alveolar, arterial, mixed venous, and tissue, respectively.

Gas exchange limited by ventilation and/or diffusion

General

A basic functional schema of a vertebrate gas exchange organ (gills, lungs or skin) is shown in Fig. 2. There are generally three component processes involved: (1) convection by medium flow (ventilation): (2) diffusion between medium and blood; and (3) convection by blood flow (perfusion).

1. *Convection by medium flow.* This can either be ventilatory flow along the conducting airways of lungs (e.g. in mammals) or motion of the environmental medium at the surface of the avian egg or the salamander skin.

2. *Diffusion between medium and blood.* In general, diffusion occurs, in sequential order, across three media with different physico-chemical properties.

(2a) *Diffusion within the respiratory medium.* Since the medium is in motion (ventilation), various degrees of interaction between convection and diffusion are expected to occur. In lungs, the role of convective mixing is expected to gradually decrease and give way to simple diffusion in the most distal airways.

(2b) *Diffusion through the medium-blood barrier.* The barrier is composed of several layers, viz. an epithelium, an interstitium, an endothelium. In its

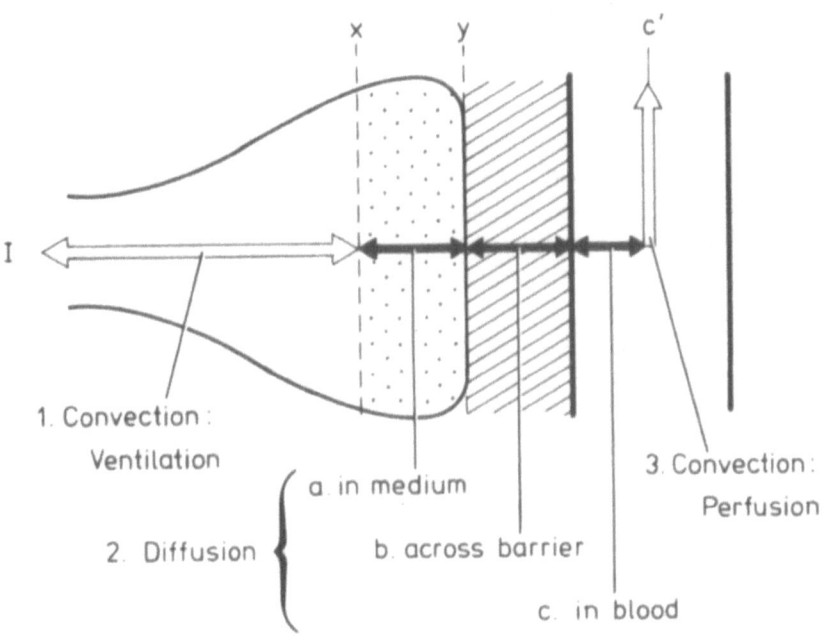

Fig. 2. Schema of gas transport mechanism in a gas exchange organ. The boundaries x and y delimit transport by ventilation, by diffusion in medium and by diffusion across the medium/blood barrier. I, inspired medium; c', end-capillary blood.

physico-chemical properties the tissue barrier may be viewed as a water (saline) layer with somewhat reduced diffusivity for gases.

(2c) *Diffusion in blood.* Respiratory gases must diffuse through plasma, red cell membrane and red cell interior. Also, chemical reactions may exert a limiting effect, thus giving rise to a reaction resistance analogous to diffusion resistance. For O_2, the chemical reaction is with hemoglobin. The physico-chemical processes that may limit CO_2 elimination include hydration/dehydration reactions, accelerated by carbonic anhydrase, and transfer of ions through the red cell membrane.

3. *Convection by blood flow.* The processes listed in (2c) take place in blood flowing through the gas exchange organ capillaries and, therefore, are combined with convective transport by blood as determined by the blood flow rate and by the blood CO_2 and O_2 dissociation curves.

In this study we should like to concentrate the analysis on the processes (1) ventilation, (2a) diffusion within medium and (2b) diffusion through the medium-blood barrier, since these show major differences between the gas exchange organs to be compared. The three gas exchangers to be considered are schematically represented in Fig. 3.

Mammalian lungs (Fig. 3A)

Gas transport along the conducting airways is by ventilation, but in distal airways it is gradually taken over by diffusion (diffusion in gas phase). The barrier between alveolar gas and pulmonary blood ('alveolar membrane') constitutes a resistance to diffusion (diffusion in tissue). However, both functional and morphometric studies show that the conditions for diffusion are such that there is, for normal resting gas exchange at least, little diffusion limitation compared to ventilation limitation. Thus the gas transport between inspired gas (I) and end-capillary blood (c') my be considered in first approximation to be limited by ventilation alone.

It is useful to define the total gas transfer conductance, G, as gas transfer rate ($\dot{V}$; positive for O_2 uptake, negative for CO_2 output) divided by the total partial pressure difference between environment (I) and end-capillary blood, i.e. blood leaving the gas exchange area, (c'):

$$G = \dot{V}/(P_1 - P_{c'}) \tag{1}$$

With exclusive ventilation limitation, G can be set equal to the ventilatory conductance, $G_{vent(g)}$, which is given by the effective ventilation, $\dot{V}_{eff}$, and the 'capacitance coefficient' of the ventilation medium, gas, β_g (Piiper et al., 1971):

$$G_{vent(g)} = \dot{V}_{eff} \cdot \beta_g \tag{2}$$

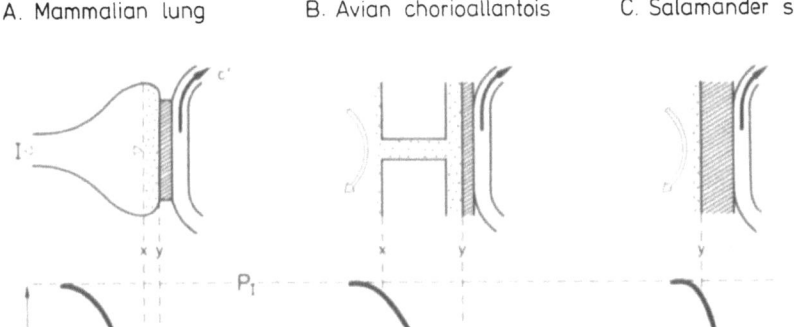

A. Mammalian lung   B. Avian chorioallantois   C. Salamander skin

Fig. 3. Models for the three gas exchange organs (above) and schematic profiles of O_2 partial pressure (below). For symbols, see Fig. 2.

Avian embryo (Fig. 3B)

The main resistance to diffusion resides in the pores of the mineral egg shell and, to a much lesser extent, in the outer and inner shell membranes which also are air-filled (diffusion in gas phase). In still air within the nest there may occur additional gradients of O_2 and CO_2 outside the egg, but in most conditions these are believed to be unimportant (dramatic exception: eggs of mound-breeding birds, see Seymour and Rahn, (1978).

Furthermore, the chorioallantoic endothelium and a liquid film between this and inner shell membrane constitute a barrier to diffusion, particularly for O_2 (diffusion in liquid phase). However, in first approximation the gas exchange may be considered as limited by diffusion through the mineral shell pores with the diffusive conductance $G_{diff(g)}$ (Wangensteen et al., 1970/71):

$$G_{diff(g)} = D_g \cdot \beta_g \cdot A_p/L_p \tag{3}$$

where D_g is the diffusion coefficient of the diffusing gas (e.g. O_2 or CO_2) in air; A_p is the total cross-sectional area of the pores, and L_p is the pore length.

Salamander skin (Fig. 3C)

The main resistance to diffusion resides in the tissue barrier of the skin, between the cutaneous capillaries and the environment. This barrier is mainly made up by the multilayered epithelium (diffusion in tissue). In addition, some stratification may develop in skin areas that are in contact with the ground or in still water. In

185

air, however, the main resistance to gas is provided by the resistance to diffusion of the tissue layer, corresponding to the diffusive conductance $G_{diff(t)}$ (Piiper et al., 1976).

$$G_{diff(t)} = D_t \cdot \alpha_t \cdot A/L \tag{4}$$

where D_t, diffusion coefficient, α_t, solubility coefficient, A, effective area, and L, effective diffusion path length, refer to the tissue barrier.

CO_2-O_2 relationships

Of particular interest is the O_2/CO_2 conductance ratio, G_{O_2}/G_{CO_2}, because it determines, along with the respiratory quotient, $R = \dot{V}_{CO_2}/\dot{V}_{O_2}$ the relationship between P_c for CO_2 and for O_2. This relationship may be expressed as the slope, s, of a 'medium R line' in the CO_2-O_2 diagram (Fig. 4):

$$s \equiv \frac{(P_c - P_I)_{CO_2}}{(P_I - P_c)_{O_2}} = R \cdot \frac{G_{O_2}}{G_{CO_2}} \tag{5}$$

(1) The line for ventilation with gas is the well-known 'gas R line'. Its slope $s_{vent(g)}$, is approximately equal to R, because β_g is equal for all (ideal) gases:

$$s_{vent(g)} \simeq R \tag{6}$$

Strictly, the inspired values P_I in Eq. (5) should be multiplied by the 'N_2 correction factor', $(P_c/P_I)_{N_2}$, to account for the effects of the volume changes of a gas mixture produced by CO_2 and O_2 exchange at $R \neq 1$ in N_2 equilibrium.

(2) The slope of the line for 'medium diffusion', $s_{diff(g)}$, is given by the ratio of the diffusivities for O_2 and CO_2 in the gas-filled pores of the mineral shell (Wangensteen and Rahn, 1970/71; Rahn et al, 1971).

$$s_{diff(g)} = \frac{D_{O_2}}{D_{CO_2}} \cdot \beta_g \cdot R \tag{7}$$

Since $(D_{O_2}/D_{CO_2})_g$ is about 1.39 (Worth and Piiper, 1978), the 'diffusion R line' is considerably steeper than that of the 'ventilation R line' for the same R.

The difference between the s values for the limiting situations of ventilation and medium diffusion limitation has the interesting consequence, experimentally confirmed by Visschedijk and Rahn (1983), that 'ventilation' of the chorioallantois by room air via the air cell in submerged eggs does not allow to adjust P_c for CO_2 and O_2 simultaneously to the normal values; either P_{O_2} for P_{CO_2} in the air cell can be made the same as under normal conditions, but not both at the same time.

(3) The slope of the 'tissue diffusion R line' is

$$s_{diff(t)} = \frac{(D \cdot \alpha)_{O_2(t)}}{(D \cdot \alpha)_{CO_2(t)}} \cdot R \tag{8}$$

186

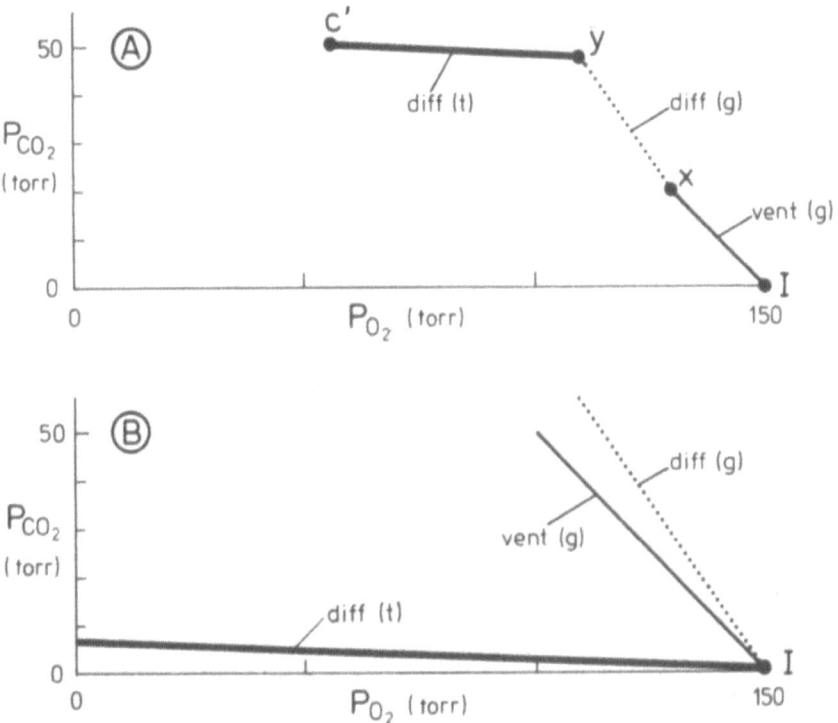

Fig. 4. CO_2-O_2 relationships visualized the in CO_2-O_2 diagram. Thin continuous line, ventilation, vent(g); thin broken line, diffusion in medium, diff(g); thick line, diffusion across tissue barrier, diff(g). A. The 'medium R line' for combined limiting effects. B. The 'medium R lines' for single limiting mechanisms. All lines drawn for R = 1.

The ratio $(D \cdot \alpha)_{O_2}/(D \cdot \alpha)_{CO_2}$ for tissue is of the same order as for water, and has been determined for skeletal muscle tissue to be about 21 (Kawashiro et al., 1975). The resulting small s value means that the maximum value of P_{CO_2} in arterialized blood reached at complete depletion of O_2 ($P_{c'}(O_2) = 0$) is about 6 torr (= 0.85 · 150 torr/21) which is in agreement with experimental data: mean P_{CO_2} in mixed blood of *Desmognathus fuscus*, 6.2 torr (Piiper et al., 1976).

For the general model of gas exchange organ, in which all the three steps (ventilation, diffusion in medium, diffusion across the medium-tissue barrier) exert rate-limiting effects, the 'medium R line' contains components of different slopes (Fig. 4A).

Gas exchange organ and circulation

Venous admixture (shunt)

In mammalian lungs, a functional shunt of a few per cent of the pulmonary blood flow (= cardiac output) is usually found on the basis of gas exchange efficiency (Fig. 5A). It is mainly attributed to perfusion of hypoventilated or non-ventilated, collapsed lung regions, or to bronchial and myocardial venous outflow into the right heart, rather than to perfusion of true shunt vessels (arterio-venous anastomoses). This shunt may be termed 'intrinsic' venous admixture, because it is in part intrinsic to the gas exchange organ.

In the chorioallantois of the chick embryo a similar intrinsic venous admixture of about 10 to 15% was estimated (Piiper et al., 1980). Its mechanism is unknown and may reflect perfusion of parts of the chorioallantois with obstructed shell membrane.

But in the bird embryo, as in the mammalian fetus, there is a considerable 'extrinsic' venous admixture due to the arrangement of the circulatory system (Fig. 5B). Due to this shunt, O_2 saturation and P_{O_2} in arterial blood, i.e. in blood perfusing body tissues, are much lower than in the arterialized blood leaving the gas exchange organ. Tazawa (1978) has estimated that in a chick embryo of 16 days the O_2 saturation reached in chorioallantoic veins, 92%, falls to 27% and 17% for arterial blood perfusing the cranial and the caudal part of the body, respectively.

Also in the lungless salamander, the venous admixture is large, estimated at more than 50% (Piiper et al., 1976). Its value results fom the proportion of blood returning from the skin and from metabolizing tissues (see below). The O_2

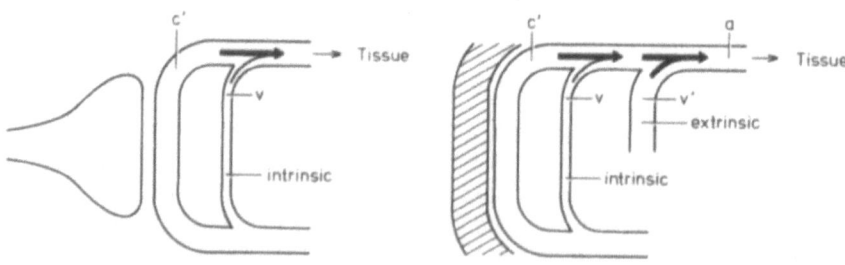

Fig. 5. Intrinsic and extrinsic venous admixture (shunt) in adult mammals (birds), A; and in embryonic birds (mammals) and in lungless salamanders, B. The intrinsic venous admixture is due to imperfect gas exchange in parts of the gas exchange organ, the extrinsic venous admixture to arrangement of the circulatory system. The possible differences in the degree of venous admixture to blood supplying the cranial and the caudal body regions is not considered.

saturation of arterialized blood in cutaneous veins was estimated at 82–89%, whereas in the mixed arterial blood it averaged 70% (Piiper et al., 1976).

Arrangement of the circulatory system

The serial arrangement of the lungs and the tissue in adult mammals (and birds) is well known. The arterial blood perfusing tissues is essentially identical with the arterialized blood leaving the lungs.

The arrangement in avian embryos (and mammalian fetuses) is fundamentally different (Fig. 6). The chorioallantois in bird embryos receives blood from the arterial system and the arterialized blood flows into the systemic veins which also carry venous tissue blood to the heart. Probably there is no complete mixing of chorioallantoic and systemic venous blood in the heart, the anterior part receiving more highly arterialized blood than the caudal parts of the body, as known for mammals (Tazawa 1978). In any case, the basic arrangement of the chorioallantois and other tissues is in parallel. The reason for this seemingly unreasonable parallel arrangement should in part be sought in the relatively easily achieved switching from chorioallantoic to pulmonary gas exchange at hatching (see Fig. 6).

In the lungless plethodontid salamanders this basically parallel arrangement of arterializing and metabolizing organs is present in its simplest form. It is highly probable that there occurs complete mixing of venous blood, returning from tissues, and of arterialized blood, returning from cutaneous veins, in the venous system and in the heart (with single atrium and ventricle). The product of mixing may be considered 'mixed venous' blood because it enters the gas exchange organ (cutaneous capillaries) but at the same time it is functionally 'arterial' blood since it perfuses body tissues.

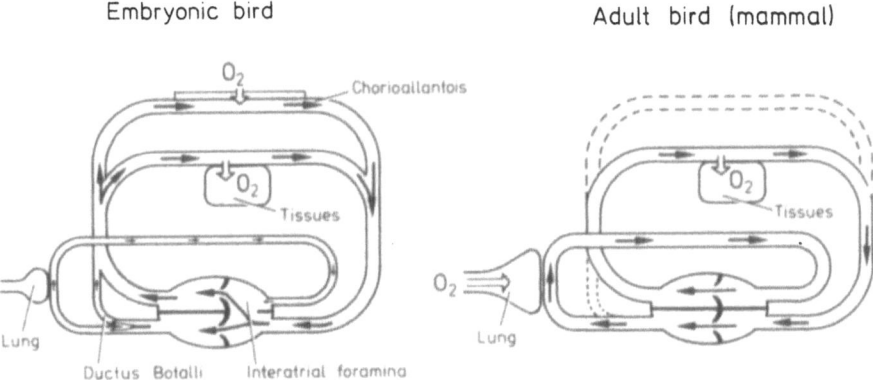

Fig. 6. Schematic simplified models for the circulatory gas transport in the avian embryo (left) and in the bird after hatching (right), to show the changes occurring at hatching: increased perfusion of lungs, closure of Ductus Botalli and of interatrial foramina, cessation of chorioallantoic flow.

The fundamentally different arrgangements of the circulatory systems create semantic problems regarding the meaning of the term 'cardiac output'.

In (normal) mammalian and avian adults the cardiac output is evidently equal to the systemic output and to the pulmonary output (not considering the minor difference due to bronchial and myocardial venous outflow into the right heart) (Fig. 7). But in the chick embryo (and in the mammalian embryo) and in the lungless salamander the cardiac output is the sum of the flow to gas exchange organ (chorioallantoic or cutaneous output) and the flow to other tissues (system output). Thus, the mammalian cardiac output corresponds to one half the cardiac output of the chick embryo and of the lungless salamander.

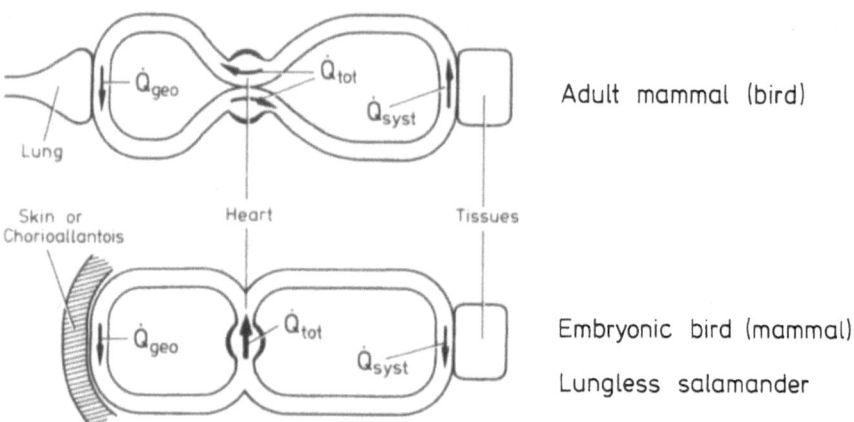

Fig. 7. Simplified schema showing the serial (left) and the parallel arrangement (right) of the gas exchange organ and the body tissues in the animals considered. $\dot{Q}_{tot}$, total cardiac output (= total blood flow from the heart cavities); $\dot{Q}_{geo}$, gas exchange organ blood flow; $\dot{Q}_{syst}$, systemic output (= total blood flow to body tissues).

References

Kawashiro, T., Nüsse, W. and Scheid, P. (1975). Determination of diffusivity of oxygen and carbon dioxide in respiring tissue: results in rat skeletal muscle. *Pflügers Arch.* 359: 231–251.

Piiper, J., Dejours, P., Haab, P. and Rahn, H. (1971). Concepts and basic quantities in gas exchange physiology. *Respir. Physiol.* 13: 292–304.

Piiper, J., Gatz, R.N. and Crawford, E.C. Jr. (1976). Gas transport characteristics in an exclusively skin-breathing salamander, *Desmognathus fuscus* (Plethodontidae). In: Respiration in Amphibious Vertebrates. G.M. Hughes, ed., Academic Press, Cambridge, pp. 339–356.

Piiper, J. and Scheid, P. (1982). Models for a comparative functional analysis of gas exchange organs in vertebrates. *J. Appl. Physiol.: Respirat. Environ. Exercise Physiol.* 53: 1321–1329.

Piiper, J., Tazawa, H., Ar, A. and Rahn, H. (1980). Analysis of chorioallantoic gas exchange in the chick embryo. *Respir. Physiol.* 39: 273–284.

Rahn, H., Wangensteen, O.D. and Farhi, L.E. (1971). Convection and diffusion gas exchange in air or water. *Respir. Physiol.* 12: 1–6.

Seymour, R.S. and Rahn, H. (1978). Gas conductance in the eggshell of the mound-building brush turkey. In: Respiratory Function in Birds, Adult and Embryonic. J. Piiper, ed. Springer-Verlag, Berlin, Heidelberg, New York, pp. 243–246.

Tazawa, H. (1978). Gas transfer in the chorioallantois. In: Respiratory Function in Birds, Adult and Embryonic. J. Piiper, ed., Springer-Verlag, Berlin, Heidelberg, New York, pp. 274–291.

Visschedijk, A.H.J. and Rahn, H. (1983). Replacement of diffusive by convective gas transport in the developing hen's egg. *Respir. Physiol.* 52: 137–147.

Wangensteen, O.D. and Rahn, H. (1970/71). Respiratory gas exchange by the avian embryo. *Respir. Physiol.* 11: 31–45.

Wangensteen, O.D., Wilson, D. and Rahn, H. (1970/71). Diffusion of gases across the egg shell of hen's egg. *Resir. Physiol.* 11: 16–30.

Worth, H. and Piiper, J. (1978). Model experiments on diffusional equilibration of oxygen and carbon dioxide between inspired and alveolar gas. *Respir. Physiol.* 35: 1–7.

13. Adult and embryonic metabolism in birds and the role of shell conductance

C.V. Paganelli and H. Rahn

*Department of Physiology, State University of New York at Buffalo
Buffalo, N.Y. 14214, U.S.A.*

Abstract

The mass-specific metabolic rate (SMR) of the non-passerine avian embryo at the pre-internal pipping (pre-IP) stage is about half that predicted for an adult bird of the same body mass. It has been suggested that the typical embryo at this stage of development has an SMR which is similar to that of the incubating adult even though its body mass is ca. 15–20 times smaller (Hoyt and Rahn, 1980). The low embryonic metabolic rate at pre-IP may be set by the limit imposed on gas transport by shell conductance. Increased metabolic demand beyond pre-IP for the final hatching act cannot be met by gas diffusion through the shell without profound hypoxia; it must be satisfied by the initiation of pulmonary gas exchange. However, the low embryonic metabolic rate at pre-IP appears to be precisely matched in each species to shell conductance so that similar air cell O_2 and CO_2 tensions are found across a wide spectrum of embryonic size and incubation period. These values are ca. 100 and 40 torr for P_{O_2} and P_{CO_2}, respectively, at the pre-IP stage of development.

Introduction

The relation between body mass and basal metabolic rate in homeotherms is a much-studied phenomenon in physiology, summarized by Kleiber (1965) among others. The relation may be expressed by the equation

$$B = 70W^{3/4} \tag{1}$$

where B is basal metabolic rate in $kJ \cdot d^{-1}$ and W is body mass in kg. A corollary to this law is obtained by expressing metabolic rate per unit of body mass:

$$B/W = 70W^{-1/4} \tag{2}$$

B/W is the mass-specific metabolic rate (SMR) in $kJ \cdot kg^{-1} \cdot d^{-1}$ or similar units. According to Eq. 2, SMR increases with decreasing body size. For example, an animal whose body mass is 0.1 kg should have a metabolic rate per unit body mass

Seymour, R.S., (ed.) Respiration and metabolism of embryonic vertebrates.
© *1984, Dr W. Junk Publishers, Dordrecht/Boston/London. ISBN-13: 978-94-009-6538-6*

which is about 3-fold that of a 10 kg animal. However, the measured SMR of a fetus is very much lower than predicted from Eq. 2. Christian Bohr, whose portrait appears in Fig. 1c and who is perhaps best known for his work on the influence of CO_2 on the hemoglobin-O_2 dissociation curve, was one of the first to call attention to this fact from his experiments on fetal and maternal metabolism in guinea pigs (Bohr, 1900). In these fetal animals, as well as in goat, sheep, rat and human embryos, estimates of SMR yield values which are similar to those of the parents (Barcroft et al., 1934; Dawes and Mott, 1959; Kleiber et al., 1943; Dawes, 1968). As Kleiber (1965) expressed it: 'The metabolic rate of the fetus *in utero* is essentially controlled by the maternal regulatory systems, although fetal hormones may also play a minor role . . . At birth, the connections to the maternal regulators . . . become suddenly severed.'

The situation with an avian embryo is obviously different from the mammal: there can be no influence of maternal regulatory systems on embryonic metabolism. Nevertheless, the same restraint on avian embryonic O_2 consumption seems to apply. SMR of the precocial avian embryo during the last stages of development is approximately equal to that of the parent bird, in spite of a 15-fold difference in mass (Hoyt and Rahn, 1980). Hatchlings, on the other hand, even after only 1 or 2 days, achieve a mass-specific O_2 consumption which is about 80% that of an adult bird of the hatchling's mass (Dawson et al., 1976). Interestingly, A.K. Hasselbalch, who like his mentor, Bohr, is famous for his contributions in mammalian respiratory physiology, first reported this phenomenon in the chicken more than 80 years ago (Hasselbalch, 1900). For their historical interest we have reproduced both his portrait (Fig. 1b) and the first data ever obtained on the metabolic rate of the chicken egg throughout incubation (Fig. 1a).

We here present data on metabolic rates of embryos and hatchlings, which taken together with air cell gas tension values lead us to conclude that shell conductance by itself is insufficient to supply the metabolic cost of the embryo beyond the pre-internal pipping (pre-IP) stage. Pulmonary ventilation must then commence to provide the O_2 necessary for the act of hatching and life outside the egg.

Adult and embryonic metabolic rates

For purposes of this presentation we have used the relation between basal metabolic rate and adult body mass for non-passerine birds (Lasiewski and Dawson, 1967) and converted it to SMR:

$$\dot{V}_{O_2} = 113 \, A^{-0.28} \qquad (3)$$

where $\dot{V}_{O_2} = ml \, O_2 \cdot g^{-1} \cdot d^{-1}$ and A = adult body mass, g.

A similar equation relating metabolic rate of non-passeriform embryos at the pre-IP stage to initial egg mass was developed by Hoyt and Rahn (1980). If we

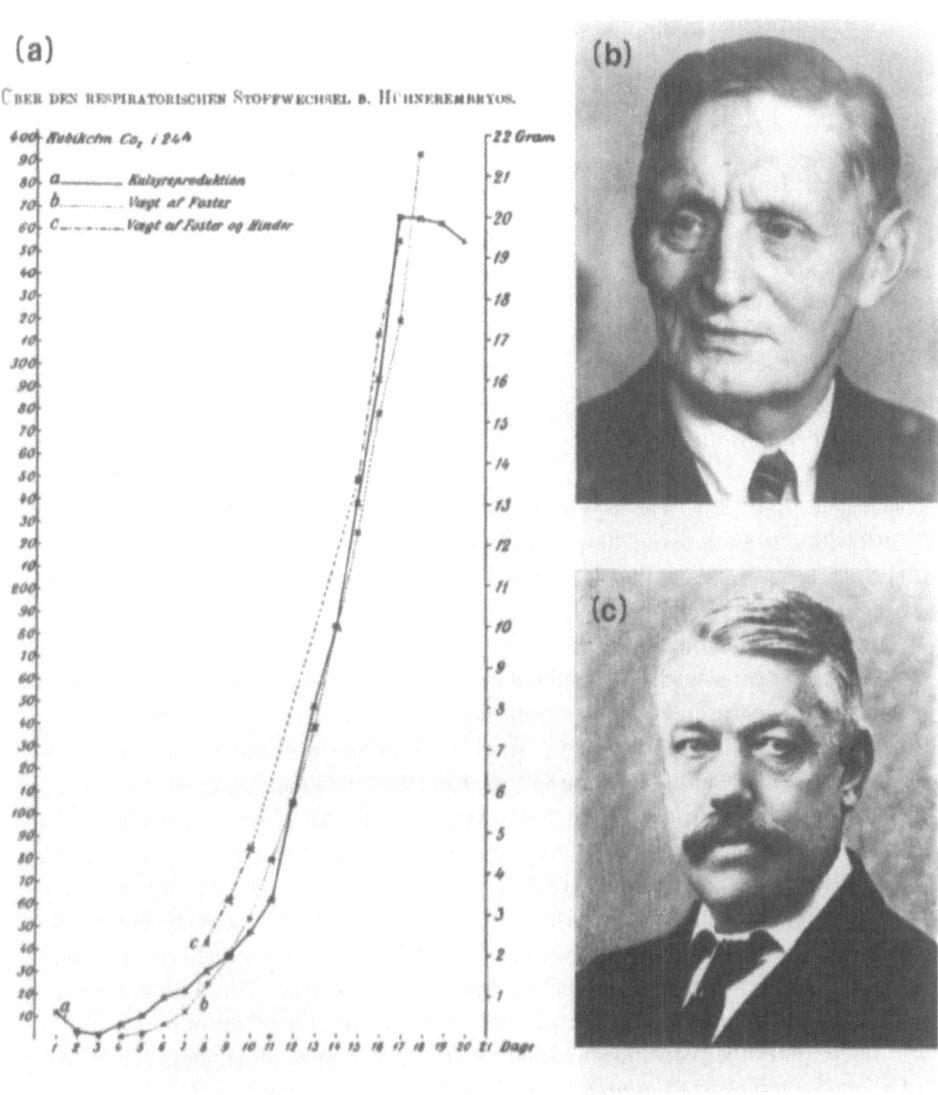

Fig. 1a. CO_2 production (curve a), weight of embryo (curve b), and weight of embryo and membranes (curve c) as functions of embryonic age. Reproduced from Hasselbalch (1900).

Fig. 1b. Karl Albert Hasselbalch (1874–1963).

Fig. 1c. Christian Bohr (1855–1911).

195

assume that at this stage the yolk-free embryo weighs 57% of the initial egg mass (Vleck et al., 1979), SMR as a function of embryo mass has the following form:

$$\dot{V}_{O_2} = 51\,E^{-0.29} \tag{4}$$

where E = yolk free embryonic mass, g.

After internal pipping, the embryo absorbs the remaining yolk sac. The mass of embryo plus yolk sac at the pre-IP stage can be estimated as 67% of initial egg mass from Heinroth's careful measurements of egg and hatchling mass in 63 species representing 14 orders (Heinroth, 1922). With this value, Eq. 4 can be rewritten with SMR expressed as a function of the mass of the pre-IP embryo plus yolk sac:

$$\dot{V}_{O_2} = 43\,E_1^{-0.29} \tag{5}$$

where E_1 = mass of embryo and yolk sac, g.

Discussion

Reduced embryonic SMR

The general concept of reduced embryonic SMR is illustrated in Fig. 2 for an adult bird with a body mass of 100 g (Circle A) which incubates a typical egg of 10 g (Rahn et al., 1975) and thus has a pre-IP embryo mass of 5.7 g (circle E). In both adult and embryo SMR is 30 ml $O_2 \cdot g^{-1} \cdot d^{-1}$ in spite of a (100/5.7) or 18-fold difference in body mass.

One may now ask when does the hatchling attain the SMR appropriate for the adult bird. Measurement of metabolic rates in Ring-billed Gull hatchlings (*Larus delawarensis*) shows them to be very close to those calculated from Eq. 3 (Dawson et al., 1976). These authors also cite hatchling metabolic rates for other species from the literature which also are in reasonable agreement with Eq. 3. These metabolic rates, to which we have added several other hatchling rates more recently reported, are listed in Table 1. In this table metabolic rates are recorded first as SMR in column 2 and then as % of the adult values predicted from Eq. 3 in column 3. The average predicted value for the 23 listed species is 82% of the adult figure (Fig. 2.). (We have assumed negligible embryo mass gain during the E-H interval. This interval is on average ca. 8% of the total incubation period.)

There is another way of estimating the increase in metabolism between E and H. Eq. 5 enables us to predict the metabolic rate of the embryo plus yolk sac as shown in column 5 of Table 1. Eq. 5 is preferable to Eq. 4 in this context since both embryo and hatchling have a yolk sac. The last column in Table 1 shows the ratio of hatchling to embryonic metabolic rate to be 2.2 on average. Thus both approaches indicate an approximate doubling in SMR from the time of internal pipping to the emerged hatchling and a hatchling O_2 consumption which is close to the one predicted by the adult metabolic equation.

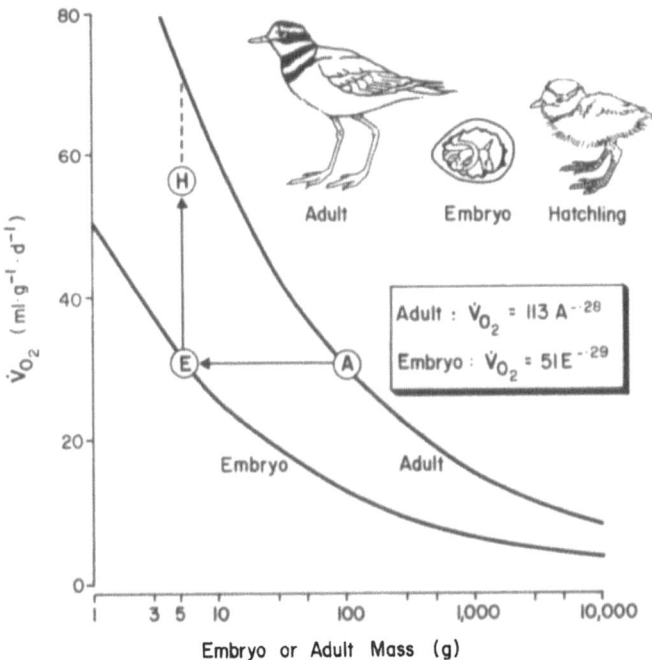

Fig. 2. Mass-specific metabolic rate (here expressed as $\dot{V}_{O_2}$ in ml (STPD) $g^{-1} \cdot d^{-1}$) as a function of adult mass in g (A) or yolk-free embryo mass in g (E). Adult curve (Eq. 3 in text) derived from Lasiewski and Dawson (1967); pre-IP embryonic curve (Eq. 4 in text) from Hoyt and Rahn (1980). As an example, point A represents the SMR of a 100 g adult bird incubating a typical 10 g egg (Rahn et al., 1975) whose yolk-free embryo of 5.7 g has the same SMR (point E). Point H represents the SMR of a typical hatchling which is about 80% of the adult value. See text for details.

Eggshell conductance

It is instructive to inquire what role eggshell conductance plays in the transition from the internal pipping stage to the end of the hatching act. During this period pulmonary gas exchange is rapidly activated and assumes the major role in meeting O_2 demand, while the function of the chorioallantois rapidly decreases (Visschedijk, 1968). One can calculate that eggshell conductance is not large enough to allow a doubling of O_2 uptake without inducing hypoxia. The normal ΔP_{O_2} between ambient atmosphere and air cell is approximately 40 torr at the pre-IP stage (Rahn and Ar, 1980). Since shell conductance is fixed by its pore structure, doubling the O_2 uptake would increase ΔP_{O_2} to 80 torr, resulting in an air cell P_{O_2} of about (146–80) or 66 torr. Since the normal ΔP_{O_2} about between air cell gas and arterialized blood is about 45 torr on day 17 of incubation, an air cell P_{O_2} of 66, other factors remaining the same, would reduce the P_{O_2} of arterialized blood returning from the chorioallantois to a value close to that normally found in *venous* blood near the end of incubation (Tazawa et al., 1980; Tazawa et al., 1983) and gives an indication of the degree of hypoxia which the embryo would experience if pulmonary ventilation did not intervene.

Table 1. Comparison of SMR of hatchlings and embryos at the pre-IP stage in 23 species. See text for details.

	Mass g	Hatchling SMR ml $O_2 \cdot g^{-1} \cdot d^{-1}$	% of adult	Ref.	Embryo SMR ml $O_2 \cdot g^{-1} \cdot d^{-1}$	SMR hatchling embryo
	(1)	(2)	(3)	(4)	(5)	(6)
Procellariiformes						
Pterodroma hypoleuca	31.8	23.8	56	(1)	15.8	1.5
Puffinus pacificus	39.1	21.5	53	(2)	14.9	1.4
Diomedea nigripes	215	17.1	68	(3)	9.1	1.9
Diomedea immutabilis	208	18.4	72	(3)	9.2	2.0
Anseriformes						
Anas crecca	16.8	41.0	81	(4)	19.0	2.2
Anas penelope	26.4	38.2	85	(4)	16.6	2.3
Anas platyrhynchos	28.8	28.3	65	(4)	16.2	1.8
Aythya ferina	40.1	31.9	81	(4)	14.7	2.2
Aythya fuligula	34.1	31.9	77	(4)	15.5	2.1
Bucephala clangula	32.4	37.9	90	(4)	15.7	2.4
Melanitta fusca	54.7	31.9	88	(4)	13.5	2.4
Mergus merganser	46.2	29.8	78	(4)	14.2	2.1
Mergus serrator	44.2	29.5	76	(4)	14.3	2.1
Somateria molissima	61.4	30.0	85	(4)	13.0	2.3
Galliformes						
Gallus domesticus	33.0	27.6	66	(5)	15.6	1.8
Tetrao urogallus	31.9	47.8	114	(6)	15.8	3.0
Laridae						
Larus argentatus	57.4	33.1	92	(7)	13.3	2.5
Larus atricilla	28.4	47.8	109	(8)	16.3	2.9
Larus delawarensis	34.6	38.9	94	(9)	15.4	2.5
Larus glaucescens	60.3	29.3	82	(7)	13.1	2.2
Larus ridibundus	26.8	54.5	123	(10)	16.6	3.3
Gygis alba	16.1	29.2	56	(11)	19.2	1.5
Alcidae						
Cepphus columba	37.5	35.5	88	(7)	15.0	2.4
Mean			82			2.2
s.e.			4			0.1

References: (1) Pettit et al. (1982a), (2) Ackerman et al. (1980), (3) Pettit et al. (1982b), (4) Koskimies and Lahti (1964), (5) Koskimies (1962), (6) Hissa et al. (1983), (7) Drent (1967), (8) Dawson et al. (1972), (9) Dawson et al. (1976), (10) Palokangas and Hissa (1971), (11) Pettit et al. (1981).

Metabolism and O_2 supply

The sensitivity of the chick embryo to change in ambient O_2 tension is well documented. A reduction to 70% of normal metabolic rate and heat production occurs when incubator O_2 tension is lowered acutely to 11% and a small but significant rise (ca. 8%) occurs when O_2 is increased to 39% (Visschedijk, 1980; Visschedijk et al., 1980). Incubation of eggs chronically in 40% and 60% O_2 in general accelerates growth in the chick embryo as a whole and most of its tissues; the degree of acceleration by day 18 is proportional to ambient O_2 concentration. However, 70% O_2 at first enhances tissue growth, but subsequently depresses it, so that embryo mass on day 18 does not differ from controls incubated in air (Stock et al., 1983). On the other hand, hypoxia, induced by covering part of the shell to render it impermeable to O_2, slows embryonic growth in air, with selectively greater effects on some organs than others (Temple and Metcalfe, 1970; Tazawa et al., 1971; McCutcheon et al., 1982). Availability of O_2 by diffusion through the shell clearly is an important determinant of embryonic growth and metabolism in this species.

The studies of growth and metabolism cited above indicate that the chick embryo is relatively hypoxic during the last stages of incubation. This conclusion is strengthened by the recent work of Ingermann et al. (1983) who measured embryonic erythrocyte organic phosphate concentrations and O_2 binding affinities as a function of ambient O_2 partial pressure. Their data indicate that erythrocyte O_2 affinity can be made to decrease reversibly with increasing ambient O_2 concentration toward the end of incubation. The inverse relation between O_2 tension and hemoglobin affinity may be viewed as a mechanism which assists in uptake of O_2 by chorioallantoic blood, and is consistent with the notion of embryonic hypoxia in late development.

In contrast, the response of wild bird eggs to reduced P_{O_2}, typically experienced at altitude, is completely different. Red-winged Blackbird embryos (*Agelaius phoeniceus*) incubated at 1600, 2400, and 2900 m showed no change in metabolic rate, incubation period, hatchling mass, or hatchling success, all indirect indicators of adequate tissue oxygenation (Carey et al., 1982). These authors speculate that reduction of the effective resistance to O_2 transfer posed by the barrier between air cell gas and chorioallantoic blood ('inner barrier' of Piiper et al., 1980) may be responsible for the wild bird embryo's ability to compensate for hypoxic air cell gas tensions.

Air cell partial pressures at pre-IP

During the early stages of development, metabolism gradually increases, but the shell conductance is fixed after the first few days of incubation (Wangensteen et al., 1970/71; Kutchai and Steen, 1971; Lomholt, 1976). Thus, O_2 tension gradually falls and CO_2 tension rises in the air cell. ΔP_{O_2} and ΔP_{CO_2} between ambient atmosphere and air cell are determined at any time by the ratio of metabolic rate and shell conductance. Table 2 lists 23 species in which air cell gas was sampled and analyzed at the pre-IP stage. On the whole, the O_2 and CO_2 values are

Table 2. Air cell gas tensions sampled and analyzed at the pre-internal pipping stage in 23 species. W = egg mass, I = incubation time, P_{O_2} = air cell O_2 tension, P_{CO_2} = air cell CO_2 tension.

	n	W g	I d	P_{O_2} torr	P_{CO_2} torr	Ref.
Struthioniformes						
Struthio camelus	12	1500	45	103	46	(1)
Procellariiformes						
Pterodroma hypoleuca	10	39	49	101	46	(2)
Puffinus pacificus	16	60	52	101	42	(3)
Diomedea immutabilis	12	285	65	106	40	(4)
Diomedea nigripes	12	305	66	106	40	(4)
Ciconiiformes						
Bubulcus ibis	8	28	23	108	42	(5)
Egretta alba	4	49	27	107	40	(5)
Eudocimus albus	3	51	22	105	33	(5)
Anseriformes						
Anas clypeata	3	40	25	92	53	(6)
Anas boscas	5	82	28	98	42	(7)
Dendrocygna autumnalis	2	51	30	122	26	(6)
Aythya fuligula	4	51	24	108	41	(6)
Anser canagicus	2	136	25	109	42	(6)
Anser domesticus	3	170	28	114	32	(7)
Galliformes						
Coturnix coturnix	5	10	17	108	38	(7)
Phasianus colchicus	20	34	24	100	45	(7)
Gallus domesticus	31	52	21	105	40	(7)
Meleagris gallopavo	7	88	28	101	45	(7)
Charadriiformes						
Gygis alba	10	23	36	100	49	(8)
Sterna hirundo	1	21	22	109	37	(7)
Sterna maxima	3	68	39	94	46	(5)
Larus argentatus	3	88	28	100	46	(7)
Columbiformes						
Columba livia	5	21	18	104	42	(7)
Mean				104	41	
s.e.				1.3	1.2	

References: (1) Meir and Ar (1982), (2) Pettit et al. (1982a), (3) Ackerman et al. (1980), (4) Pettit et al. (1982b), (5) Vleck, D.. and C.M. Vleck (unpubl.), (6) Hoyt et al. (1979), (7) Rahn et al. (1974), (8) Pettit et al. (1981).

remarkably similar and average 104 and 41 torr, respectively. This similarity in air cell gas tensions exists in spite of a range of egg mass which encompasses the 10 g quail egg and the 1500 g ostrich egg, and incubation periods from 17 to 66 d. Additionally, the maturity of hatchlings listed in Table 2 varies from the altricial pigeon and semi-altricial heron to the most precocial representatives, the waterfowl. Similarity of air cell gas tensions implies that for each species metabolic rate at the pre-IP stage is matched with shell conductance. From the average air cell partial pressures the average ΔP_{O_2} across the shell is (146–104) or 42 torr, and average ΔP_{CO_2} is 41 torr. An identical average ΔP_{CO_2} was predicted on the basis of independent measurements of $\dot{V}_{O_2}$ and water vapor conductance, D_{H_2O}, both as functions of W/I (egg mass in g)/(incubation period in d), by Rahn and Ar (1980). The data on which their predictions are based are shown in Fig. 3, where $\dot{V}_{O_2}$ (O_2 consumption in ml STPD $\cdot$ d^{-1}) at the pre-IP stage and G_{O_2} (ml STPD $\cdot$ d$^{-1}\cdot$ torr^{-1}) are both plotted on a log scale against W/I, a measure of embryonic growth rate. The G_{O_2} data were derived from independently measured values of G_{H_2O} multiplied by 1.08 (Paganelli et al., 1978). The slopes of the two regression lines are both 1; the equations whose terms are defined above are:

$$\dot{V}_{O_2} = 237 \text{ W/I} \tag{6}$$

$$\text{and } G_{O_2} = 5.6 \text{ W/I} \tag{7}$$

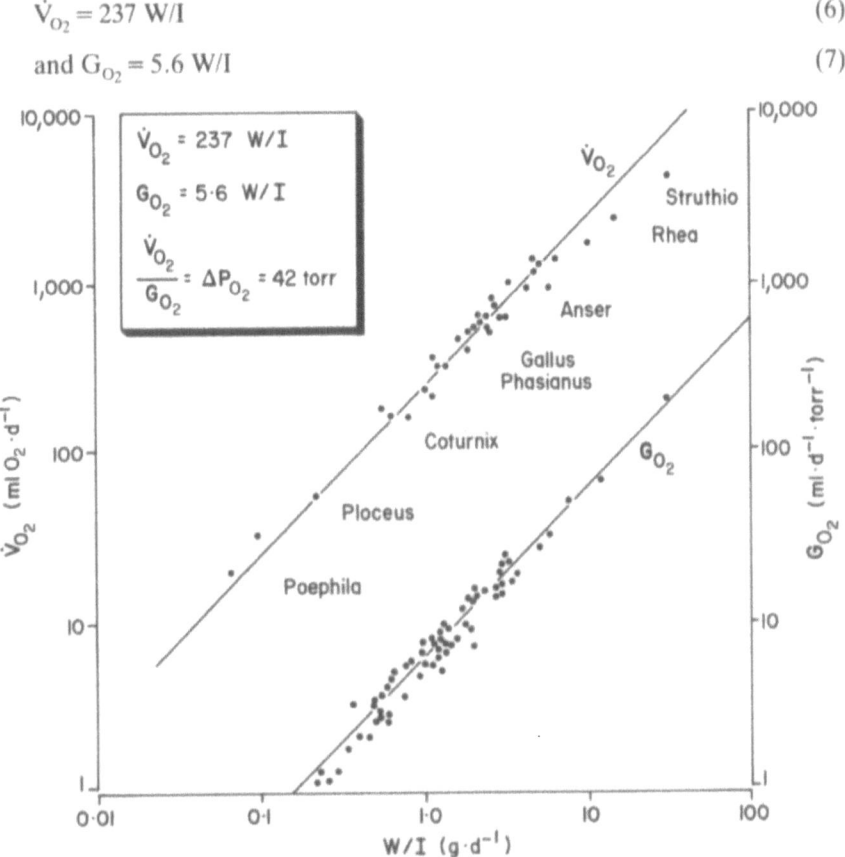

Fig. 3. Pre-IP O_2 consumption ($\dot{V}_{O_2}$) and O_2 conductance (G_{O_2}) plotted on a log scale against the ratio of egg mass (W, g) and incubation time (I, d). W/I may be considered as the average rate of development. Both lines have a slope of 1. The values of G_{O_2} are calculated from water vapor conductances presented in Rahn and Ar (1980). The ratio $\dot{V}_{O_2}/G_{O_2}$ is the average ΔP_{O_2} between external environment and air cell at the pre-IP stage. See text for details. Modified from Rahn and Ar (1980).

Thus, the quotient $\dot{V}_{O_2}/G_{O_2}$ which is the average ΔP_{O_2} across the shell, is the result of two independent sets of measurements on different groups of eggs: one for metabolic rate and one for water vapor conductance. Nevertheless, the value of ΔP_{O_2} predicted from these equations is $237/5.6 = 42$, in excellent agreement with the average ΔP_{O_2} shown in Table 2 from direct measurements of air cell gas tensions at the pre-IP stage.

It has been proposed elsewhere that the eggshell conductance assures a given water loss for optimal hatching success (Rahn, this volume). Whether regulation of O_2, CO_2, or water vapor flux is more important in selection of eggshell properties will long be debated, since each of the thousands of pores serves all gases. Although a good case can be made for control of water loss as a priority, one can also argue that it is just as important to bring about a P_{CO_2} of 40 torr prior to the final hatching act. Not only is this value similar to that of the adult bird, but it also assures normal plasma HCO_3^- and Cl^- concentrations. The absolute O_2 tension of 100 torr is probably not as significant as the absolute P_{CO_2} value, since compensations for inadequate O_2 tension can be more easily achieved by a change in hemoglobin species (Reeves, this volume), hematocrit, and circulation rate.

Acknowledgement

The authors gratefully acknowledge support of USPHS grants HL-14414 and HL-28542 for part of the data which originated in Buffalo.

References

Ackerman, R.A., Whittow, G.C., Paganelli, C.V. and Pettit, T.N. (1980). Oxygen consumption, gas exchange, and growth of embryonic Wedge-tailed Shearwaters (*Puffinus pacificus chlororhynchus*). *Physiol. Zool.* 53: 210–221.

Barcroft, J., Flexner, L.B. and McClurkin, T. (1934). The output of the foetal heart in the goat. *J. Physiol.* (London) 82: 498–508.

Bohr, Chr. (1900). Der respiratorische Stoffwechsel des Säugethierembryo. *Skand. Arch. Physiol.* 10: 413–424.

Carey, C., Thompson, E.L., Vleck, C.M. and James, F.C. (1982). Avian reproduction over an altitudinal gradient: incubation period, hatchling mass, and embryonic oxygen consumption. *Auk* 99: 710–718.

Dawes, G.S. (1968). Foetal and Neonatal Physiology. Year Book, Chicago.

Dawes, G.S. and Mott, J.C. (1959). The increase in oxygen consumption of the lamb after birth. *J. Physiol.* (London) 146: 295–315.

Dawson, W.R., Hudson, J.W. and Hill, R.W. (1972). Temperature regulation in the newly hatched Laughing Gull (*Larus Atricilla*). *Condor* 74: 177–184.

Dawson, W..., Bennet A.F. and Hudson, J.W. (1976). Metabolism and thermoregulation in hatchling Ring-billed Gulls. *Condor* 78: 49–60.

Drenth, R.H. (1967). Functional Aspects of Incubation in the Herring Gull (*Larus argentatus*). E.J. Brill, Leiden.

Hasselbalch, A.K. (1900). Ueber die respiratorischen Stoffwechsel des Hühnerembryos. *Skand. Arch. Physiol.* 10: 353–402.

Heinroth, O. (1922). Die Beziehungen zwischen Vogelgewicht, Eigewicht, Gelegegewicht und Brutdauer. *J. Ornithol.* 70: 172–285.

Hissa, R., Saarela, S., Rintamäki, H., Linden, H. and Hohtala, H. (1983). Energetics and development of temperature regulation in Capercaillie *Tetrao urogallus*. *Physiol. Zool.* 56: 142–151.

Hoyt, D.F., Board, R.G., Rahn, H. and Paganelli, C.V. (1979). The eggs of Anatidae: conductance, pore structure and metabolism. *Physiol. Zool.* 52: 438–450.

Hoyt, D.F. and Rahn, H. (1980). Respiration of avian embryos – a comparative analysis. *Respir. Physiol.* 39: 255–264.

Ingermann, R.L., Stock, M.K., Metcalfe, J. and Shih, T.-B. (1983). Effect of ambient oxygen on organic phosphate concentrations in erythrocytes of the chick embryo. *Respir. Physiol.* 51: 141–152.

Kleiber, M. (1965). Respiratory exchange and metabolic rate. In: Handbook of Physiology, Respiration, Vol. II, Ch. 35. W.O. Fenn and H. Rahn, eds. The American Physiological Society, Washington, D.C. pp. 927–938.

Kleiber, M., Cole, H.H. and Smith, A.H. (1943). Metabolic rate of rat fetuses *in vitro*. *J. Cell Comp Physiol.* 22: 167–176.

Koskimies, J. (1962). Ontogeny of thermoregulation and energy metabolism in some gallinaceous birds. *Ric. Zool. Appl. Caccia Suppl.* 4: 149–160.

Koskimies, J. and Lahti, L. (1964). Cold-hardiness of the newly hatched young in relation to ecology and distribution in ten species of European ducks. *Auk* 81: 281–307.

Kutchai, H. and Steen, J.B. (1971). Permeability of the shell and shell membranes of hens' eggs during development. Respir. Physiol. 11: 265–278.

Lasiewski, R.C. and Dawson, W.R. (1967). A reexamination of the relation between standard metabolic rate and body weight in birds. *Condor* 69: 13–23.

Lomholt, J.P. (1976). The development of the oxygen permeability of the avian egg shell and its membranes during incubation. *J. Exp. Zool.* 198: 177–184.

McCutcheon, I.E., Metcalfe, J., Metzenberg, A.B. and Ettinger, T. (1982). Organ growth in hyperoxic and hypoxic chick embryos. *Respir. Physiol.* 50: 153–163.

Meir, M. and Ar, A. (1982). Ostrich egg pre-pipping partial gas pressures. *Isr. J. Zool.* (in press).

Paganelli, C.V., Ackerman, R. and Rahn, H. (1978). The avian egg: *in vivo* conductances of O_2, CO_2 and water vapor in late development. In: Respiratory Function in Birds, Adult and Embryonic. J. Piiper, ed., Springer-Verlag, New York, pp. 212–218.

Palokangas, R. and Hissa, R. (1971). Thermoregulation in young Black-headed Gull (*Larus ridibundus L.*) (*Comp. Biochem. Physiol.*) 38A: 743–750.

Piiper, J., Tazawa, H., Ar, A. and Rahn, H. (1980). Analysis of chorioallantoic gas exchange in the chick embryo. *Respir. Physiol.* 39: 273–284.

Pettit, T.N., Grant, G.S., Whittow, G.C., Rahn, H. and Paganelli, C.V. (1981). Respiratory gas exchange and growth of White Tern embryos. *Condor* 83: 355–361.

Pettit, T.N., Grant, G.S., Whittow, G.C., Rahn, H. and Paganelli, C.V. (1982a). Respiratory gas exchange and growth of Bonin Petrel embryos. *Physiol. Zool.* 55: 162–170.

Pettit, T.N., Grant, G.S., Whittow, G.C., Rahn, H. and Paganelli, C.V. (1982b). Embryonic oxygen consumption and growth of Laysan and Black-footed Albatross. *Am. J. Physiol.* 242 (*Regulatory Integrative Comp. Physiol.* 11): R121–R128.

Rahn, H., Paganelli, C.V. and Ar, A. (1974). The avian egg: air cell gas tension , metabolism and incubation time. *Respir. Physiol.* 22: 297–309.

Rahn, H., Paganelli, C.V. and Ar, A. (1975). Relation of avian egg weight to body . *Auk* 92: 750–765.

Rahn, H. and Ar, A. (1980). Gas exchange of the avian egg: time, structure, and function. *Am. Zool.* 20: 477–484.

Stock, M.K., Francisco, D.L. and Metcalfe, J. (1983). Organ growth in chick embryos incubated in 40% or 70% oxygen. *Respir. Physiol.* 52: 1–11.

203

Tazawa, H., Ar, A., Rahn, H. and Piiper, J. (1980). Repetitive and simultaneous sampling from the air cell and blood vessels in the chick embryo. *Respir. Physiol.* 39: 265–272.

Tazawa, H., Mikami, T., and Yoshimoto, C. (1971). Effect of reducing the shell area on the respiratory properties of chicken embryonic blood. *Respir. Physiol.* 13: 352–360.

Tazawa, H., Visschedijk, A.H.J., Wittmann, J. and Piiper, J. (1983). Gas exchange, blood gases and acid-base status in the chick before, during and after hatching. *Respir. Physiol.* 53: 173–185.

Temple, G.F. and Metcalfe, J. (1970). The effects of increased incubator oxygen tension on capillary development in the chick chorioallantois. *Respir. Physiol.* 9: 216–233.

Vleck, D., Vleck, C.M. and Hoyt, D.H. (1979). Metabolism of avian embryos: patterns in altricial and precocial birds. *Physiol. Zool.* 52: 363–377.

Visschedijk, A.H.J. (1968). The air space and embryonic respiration. 1. The pattern of gaseous exchange in the fertile egg during the closing stages of incubation. *Br. Poult. Sci.* 9: 173–184.

Visschedijk, A.H.J. (1980). Effects of barometric pressure and abnormal gas mixtures on gaseous exchange of the avian embryo. *Am. Zool.* 20: 469–476.

Visschedijk, A.H.J., Ar, A., Rahn, H. and Piiper, J. (1980). The independent effects of atmospheric pressure and oxygen partial pressure on gas exchange of the chicken embryo. *Respir. Physiol.* 39: 33–44.

Wangensteen, O.D., Wilson, D. and Rahn, H. (1970/71). Diffusion of gases across the shell of the hen's egg. *Respir. Physiol.* 11: 16–30.

14. The effect of oxygen on growth and development of the chick embryo

James Metcalfe, Michael K. Stock and Rolf L. Ingermann

Heart Research Laboratory, Department of Medicine, and Department of Obstetrics and Gynecology, Oregon Health Sciences University, Portland, Oregon 97201, U.S.A.

Abstract

When eggs are incubated in 60% O_2, the weight of the 18-day embryo is greater than that of the control incubated in 21% O_2. Conversely, when O_2 availability is reduced by covering part of the eggshell, embryonic growth is retarded. The weight of the embryo at day 18 correlates with the surface area of its egg, as does the diffusing capacity for O_2 and the calculated chorioallantoic capillary surface area. Furthermore, during the last few days of chorioallantoic respiration in 21% O_2, the embryo becomes hypoxemic and its rate of growth slows. We propose that the O_2 tension in blood leaving the chorioallantoic capillaries limits the rate of growth of the chick embryo until the onset of pulmonary respiration.

Oxygen may exert its effect on growth indirectly by altering the activity of a particular group of cells that controls embryonic growth rate. Alternatively, the availability of O_2 may directly determine the rate of energy production in individual embryonic tissues; an increase in the available energy would translate into accelerated growth. The concentration of adenosine triphosphate (ATP) in the nucleated embryonic erythrocytes is responsive to changes in O_2 availability. Under normoxic conditions erythrocytic [ATP] declines between the fourteenth and eighteenth days of incubation. This is consistent with the observation that the chick embryo becomes relatively hypoxemic during the last third of its development until air breathing begins. Incubation in 70% O_2 markedly attenuates the fall in [ATP]. If the erythrocyte is characteristic of other embryonic tissues with regard to the basic mechanisms of ATP production, the growth-accelerative effect of hyperoxia may be mediated by an increased rate of ATP generation.

Our work with chick embryos began when we used them to study capillary growth and its control. Work in other laboratories had shown that when hen's eggs are incubated in environments with a high partial pressure of O_2 (P_{O_2}), development of the large chorioallantoic vessels is retarded, and the vascularity of the yolk sac and the embryonic tissues is diminished compared to eggs incubated in room air (Remotti, 1933; Flemister and Cunningham, 1940; Allen, 1963). These earlier

Seymour, R.S., (ed.) Respiration and metabolism of embryonic vertebrates.
© *1984, Dr W. Junk Publishers, Dordrecht/Boston/London. ISBN-13: 978-94-009-6538-6*

studies used morphological techniques. We measured the carbon monoxide diffusing capacity (DCO) of eggs incubated in room air and compared it with that of eggs incubated in 60% O_2 (Temple and Metcalfe, 1970).

Eggs incubated in room air have a measurable DCO by the eighth day of incubation and the diffusing capacity increases approximately four-fold between the eighth and eighteenth days in incubation (Fig. 1). Eggs incubated in 60% O_2 show a lower DCO than controls, beginning on the eighth day of incubation. The values for DCO diverge progressively as incubation continues to the eighteenth day.

We consider that the lower carbon monoxide diffusing capacity of eggs incubated in 60% O_2 compared to room-air eggs is due either to a reduction in the surface area of the capillaries of the chorioallantoic membrane or to a greater thickness of the diffusion barrier, or to a combination of these two factors (Temple and Metcalfe, 1970). A relative anemia of the hyperoxic embryos plays a small role in reducing their DCO (Bissonnette and Metcalfe, 1978).

Values for the O_2 consumption in air by eggs incubated in 21% O_2 are available from Needham (1963). The mean P_{O_2} gradient between environmental air and the

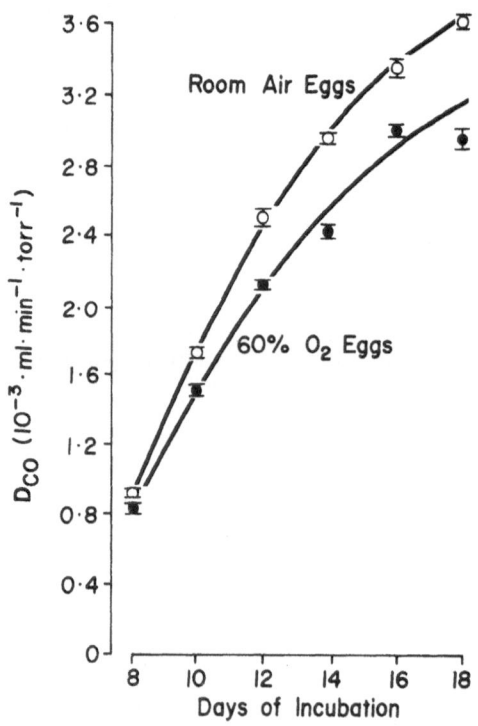

Fig. 1. Carbon monoxide diffusing capacity (D_{CO}) versus age of chicken eggs incubated in room air or 60% O_2. Bars indicate plus or minus one standard deviation. Reproduced from Temple and Metcalfe (1970) with permission.

blood in the chorioallantoic capillary network may be calculated from the diffusing capacity for O_2 (which is calculated from DCO) and O_2 consumption (Metcalfe, 1967). For eggs incubated in room air, the calculated P_{O_2} gradient would reach 75 torr by the eighteenth day of incubation (Fig. 2).

These calculations of the mean P_{O_2} gradient supported data obtained by direct measurements of gas tension in the air space that were made in 1938 by Romijn and Roos. Measurements of the P_{O_2} in single *blood* samples from the embryo were subsequently reported by Freeman and Misson (1970) and by Tazawa (1971). More recently, blood was sampled through inlying catheters; arterialized blood from the allantoic vein has a P_{O_2} of between 50 and 60 torr at the sixteenth day of incubation (Tazawa et al., 1980). These data all indicate that O_2 transfer from incubator air to chorioallantoic blood is diffusion-limited and suggest that relative hypoxemia limits the availability of O_2 to the developing chick embryo near the end of incubation before pulmonary respiration begins. The oxygenated blood of the chick embryo has a lower P_{O_2} during an important period of its development than that of the healthy adult. This embryonic hypoxemia is also characteristic of all mammals that have been studied.

Embryos from eggs incubated chronically in 60% O_2 are significantly heavier than controls from eggs incubated in room air (Temple and Metcalfe, 1970; McCutcheon et al., 1982). Conversely, chick embryo growth is retarded by incubation at 3100 meters altitude and even more retarded at 3800 meters above

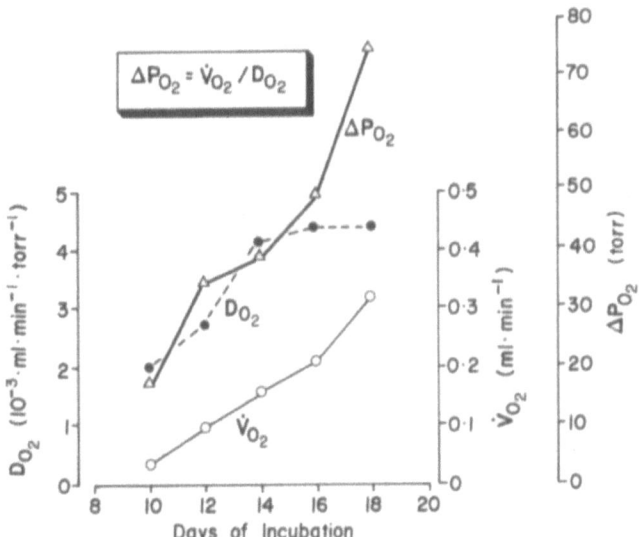

Fig. 2. The diffusing capacity for O_2 (D_{O_2}) and the O_2 consumption per minute ($\dot{V}_{O_2}$) related to age of chicken eggs incubated in 21% O_2. Data for $\dot{V}_{O_2}$ were taken from Needham (1963). Small triangles represent the calculated O_2 pressure gradient between incubator air and chorioallantoic capillary blood required to maintain $\dot{V}_{O_2}$ with the observed D_{O_2} according to the equation at the top of the figure. Taken from Metcalfe (1967) with permission.

sea level (Fig. 3; Smith et al., 1969). Similarly, chick embryo growth can be slowed by coating the part of the shell with a substance impermeable to the respiratory gases (Tazawa et al., 1971; McCutcheon et al., 1982). However, there is a fundamental difference between the observation that reduced O_2 availability *limits*, and increased O_2 availability *stimulates*, embryonic growth. Many elements are essential for normal growth and limitation of any, including O_2, will interfere with that complex process. In contrast, the demonstration that growth is stimulated by increased ambient O_2 during incubation implies that the availability of O_2 *normally* limits embryonic growth. As Robertson pointed out in 1923, 'The speed of the complex process of growth may be determined by the speed of its slowest component'. The fact that embryo growth can be accelerated by incubation in 60% O_2 suggested that a lack of O_2 normally limits the speed of the 'slowest component' and, thereby, of the chick embryo's growth.

Several additional pieces of evidence can be used to support the hypothesis that O_2 availability normally limits the growth of the chick embryo. The first rests upon the relationship between the size of the hen's egg and the growth of its embryo. Some workers have considered that 'chick size was limited significantly by the space in the eggshell during the last two or three days of incubation' (Wiley, 1950); however, there are data suggesting that the correlation between egg size and embryo weight is detectable by the tenth day of incubation (Byerly, 1932; Romanoff, 1960). At that stage, a restraint on embryo growth by the physical confinement of the eggshell seems unlikely. However, the surface area of the eggshell is directly and uniquely determined by the size of the egg. Using data from several different strains of *Gallus domesticus*, we can show that DCO at the eighteenth day of incubation is directly related to the surface area of the shell

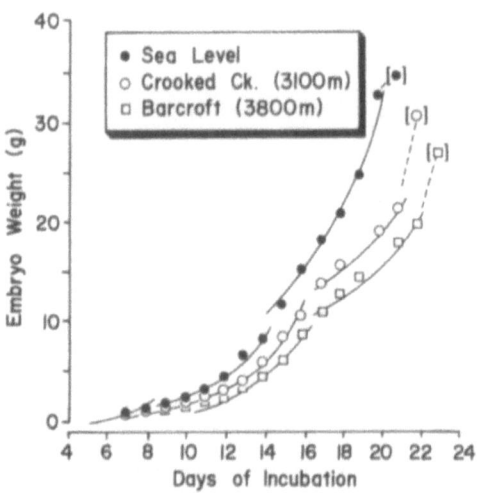

Fig. 3. Wet weight of chicken embryos during incubation at sea level, 3100 meters, and 3800 meters, plotted against incubation time. The data are from Smith et al. (1969). The figure is taken from Wangensteen et al. (1974) with permission.

(Fig. 4). Calculations of the surface area of the chorioallantoic network of eggs exposed to room air give a value at 18 days of incubation that closely approximates the surface area of the eggshell, suggesting that the chick embryo utilizes the entire surface area of its shell for gas exchange by the eighteenth day of incubation (Bissonnette and Metcalfe, 1978).

As would be expected, when embryo weight is plotted against the surface area of the egg (rather than the weight of the egg), the correlation between egg size and embryo weight persists (Fig. 5; Metcalfe et al., 1981). When eggs are half-covered throughout incubation by a membrane that is relatively impermeable to O_2, DCO is reduced by approximately 20%, and a significant reduction in embryo weight at 18 days can be demonstrated. Thus, decreasing the O_2 availability by decreasing the diffusing capacity without changing the space within the eggshell limits embryo size (Fig. 6). The most powerful argument for O_2 availability as the determinant of embryo growth is shown by the augmented growth of embryos when the O_2 availability is increased by incubation in 60% O_2 (Fig. 7). Here, embryo weight at 18 days increases, without an increase of space within the eggshell.

Additional supportive evidence for the concept that O_2 availability normally limits the growth rate of embryo can be drawn from the growth curve of the embryonic chick. In precocial species of birds embryonic growth slows as the time of hatching approaches (Vleck et al., 1980). The sigmoid curve of growth during the period of chorioallantoic respiration is paralleled by the curve of total O_2 consumption and also by the curve relating carbon monoxide diffusing capacity to incubation time (Temple and Metcalfe, 1970; Vleck et al., 1980). All of these

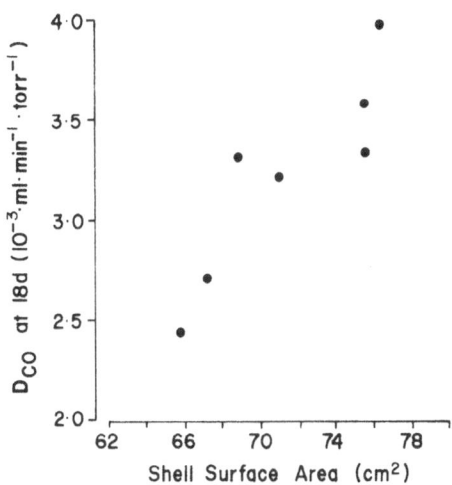

Fig. 4. Relationship of total carbon monoxide diffusing capacity (DCO) at 18 days of incubation to the surface area of chicken eggs. Mean data from seven experimental groups, ranging in size from eight to 20 eggs, are shown. All were incubated in 21% O_2. Taken from Bissonnette and Metcalfe (1978) with permission.

curves flatten perceptibly at the time when the P_{O_2} in oxygenated allantois venous blood is declining. Our central hypothesis, that the chick's growth rate is limited by O_2 availability, would suggest that when oxygenation declines because the chick outgrows the DCO of its egg, growth is slowed and O_2 consumption plateaus. Fig. 8 is taken from a paper by Wangensteen and his colleagues (1974) and raises an intriguing speculation. The solid line shows O_2 consumption of the chick embryo throughout incubation near sea level and was calculated from the data of several authors, some of it unpublished. An increase in O_2 consumption occurs following internal pipping and the onset of air breathing, but before complete hatching. As shown in Table 1, the O_2 tension in arterialized blood from the chorioallantoic vein falls progressively from the fourteenth through the seventeenth days of incubation, then rises with the onset of pulmonary respiration. Since we have attributed the declining growth rate from the sixteenth through the eighteenth days of incubation to relative hypoxemia, can we show that the increased oxygenation that begins with the onset of pulmonary respiration coincides not only with an increase in O_2 consumption, but also with an increase in growth rate? We have not measured embryonic weights after the eighteenth day of incubation. To study the effects of internal pipping upon growth, we turn to the data from Romanoff (1960) (Fig. 9). Between the eigh-

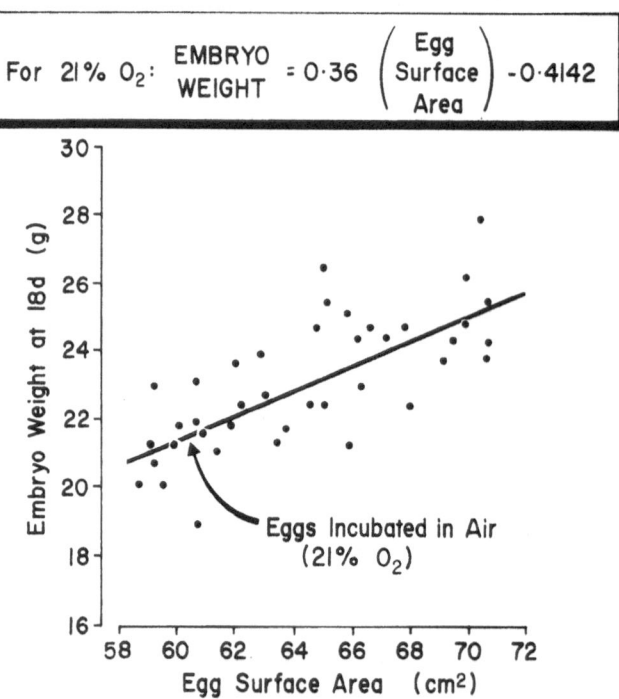

Fig. 5. The data and the line of best fit relating chick embryo wet weight at 18 days to shell surface area (calculated from weight of the freshly laid egg) of eggs incubated in 21% O_2. The linear regression equation is given at the top of the figure. Reproduced from Metcalfe et al. (1981) with permission.

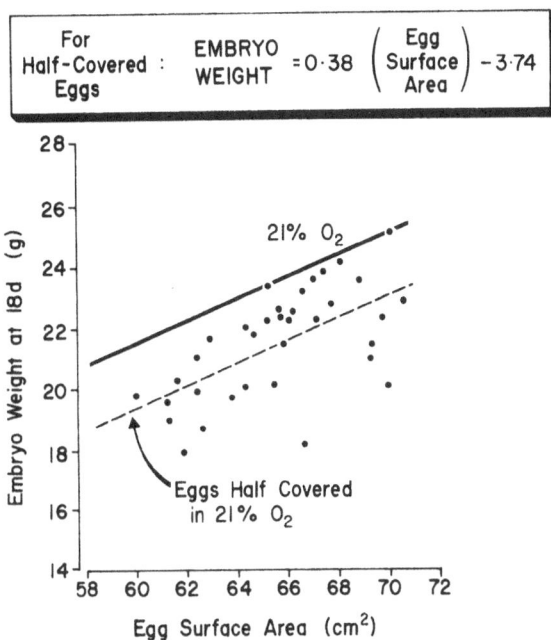

Fig. 6. The solid line is identical with the line in Fig. 5, derived from the data obtained from eggs incubated (uncovered) in 21% O_2. The dashed line is the line of best fit of the data shown by circles and obtained from half-covered eggs incubated in 21% O_2 for 18 days. The linear regression equation is given at the top of the figure. Reproduced from Metcalfe et al. (1981) with permission.

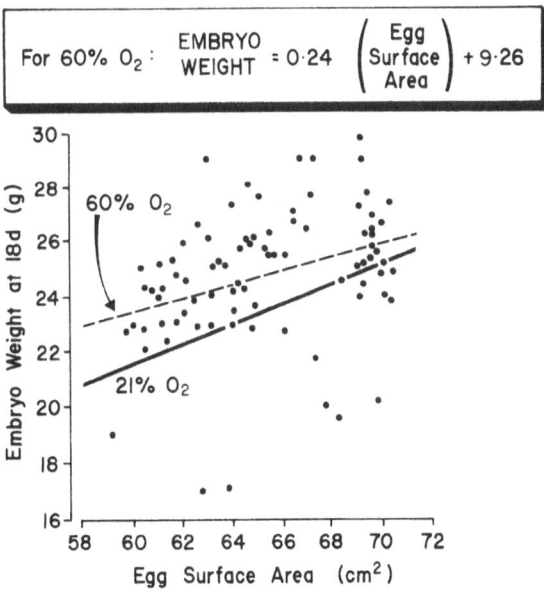

Fig. 7. The solid line is identical with the line in Fig. 5, derived from data obtained from eggs incubated in 21% O_2. The dashed line is the line of best fit of the data shown by the dots and obtained from eggs incubated in 60% O_2 for 18 days. The linear regression equation for the data is given at the top of the figure. Reproduced from Metcalfe et al. (1981) with permission.

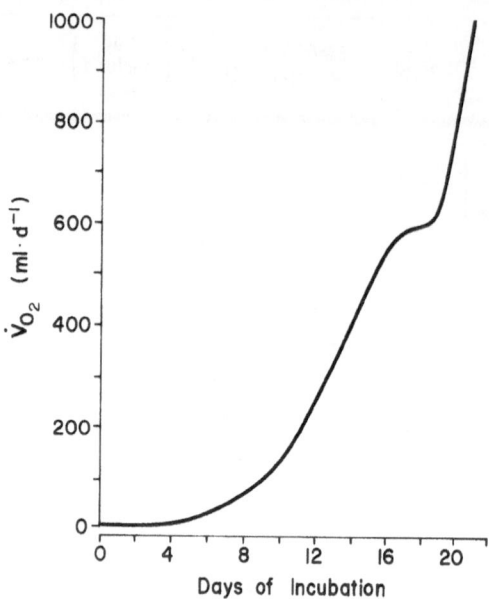

Fig. 8. The O_2 consumption ($\dot{V}_{O_2}$) of the chicken embryo during incubation near sea level. The line was calculated from the data of Lokhorst and Romijn (1965), Romanoff (1967), Visschedijk (1968), and Paganelli and Rahn (unpublished). The figure is modified from Wangensteen et al. (1974) with permission.

Table 1. Values for O_2 tension (P_{O_2}) in oxygenated blood of the chorioallantoic vein ('Venous') and deoxygenated blood of the chorioallantoic artery ('Arterial') during the last week of incubation. Statistical comparisons were made between the data for one days and those of the preceeding day. Reproduced from Freeman and Misson (1970) with permission.

Days of incubation	P_{O_2} (torr)	
	Venous	Arterial
14	71.70 ± 1.292†	29.48 ± 1.743
15	60.21 ± 1.651***	24.67 ± 1.437*
16	55.09 ± 1.756	24.75 ± 0.977
17	48.68 ± 1.582***	21.19 ± 0.820**
18	49.79 ± 1.952	22.63 ± 1.061
19 chorioallantoic respiration	52.19 ± 2.225	20.65 ± 0.687
+ lung respiration	68.24 ± 1.791***	37.19 ± 2.435***
+ pipped shell	82.91 ± 2.410***	40.85 ± 2.275
1 day after hatching	34.25 ± 1.967	109.08 ± 5.843

* P<0.05

** P<0.01

*** P<0.001

† Mean ± s.e. of twenty-five observations, excepting data for the 1-day-old chick where eight (arterial) and fifteen (venous) observations were made.

teenth and nineteenth days of incubation the weight gain of the chick embryo is nearly twice that measured from the seventeenth to the eighteenth day of incubation. It seems fair to suggest that whatever restraint is limiting embryonic growth late in incubation is relieved after internal pipping occurs. Oxygen availability seems the most prominent candidate for the controlling variable.

It has recently been reported that the O_2 consumption of fertile hen's eggs increases with acute exposure to 100% O_2 after the ninth day of incubation and that a detectable increase in O_2 uptake occurs when one square centimeter of the eggshell above the air cell is removed experimentally (Hoiby et al., 1983). These observations also suggest that O_2 availability limits the metabolic rate of the chick embryo. It is likely that some of the increase in metabolism that occurs with exposure to increased O_2 concentrations is utilized for growth.

There is evidence that development, as well as growth, is accelerated by increased O_2 availability. Exposure of early embryos to different O_2 tensions for 48–72 h results in developmental acceleration of the embryo with increasing O_2 pressures (Bartels et al., 1973). Recent work in our laboratory shows that when eggs are incubated in 70% O_2 accelerated development is detectable by the third day of incubation.

The effect of altered O_2 availability on the growth of individual organs of the embryo is not in uniform (McCutcheon et al., 1982). The heart of an embryo incubated for 18 days in 60% O_2 exhibits a 22% increase in wet weight above controls, compared to a 12% increase for the hyperoxic embryo as a whole and an increase in brain growth of only 3% (Fig. 10). In contrast, when O_2 availability is reduced during incubation in air, by covering half of the shell with a membrane impermeable to O_2, the wet weight of the heart at 18 days is insignificantly affected, brain growth is reduced by 5% and the weight of the whole embryo is reduced by 9%. This pattern of growth retardation with sparing of the brain and heart relative to the muscles is reminiscent of the intra-uterine growth retardation seen in human newborns after placental infarction (Naeye, 1966), a pattern that

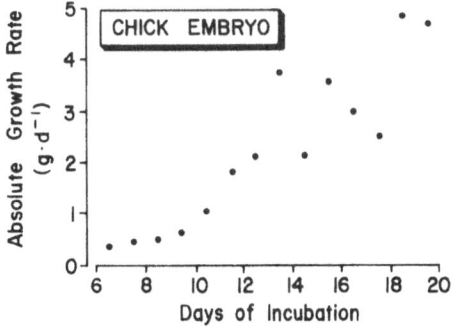

Fig. 9. The average daily gain in wet weight of chicken embryos from eggs of approximately 60 grams incubated at 37.5 °C in 21% O_2. Data from Romanoff (1960).

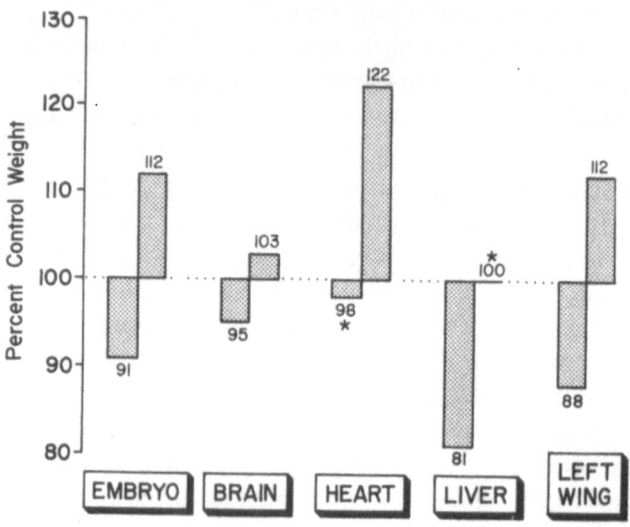

Fig. 10. Percentage deviation of tissue wet weights from control values (= 100%) on day 18. Tissues from chick embryos incubated in 60% O_2 are represented by the bars in the upper half of the figure. Tissues of embryos from half-covered eggs incubated in 21% O_2 are represented by bars in the lower half. The number with each bar gives the value represented by the length of that bar. All values are significantly different from control (P<0.05), except those indicated by an asterisk. Reproduced from McCutcheon et al. (1982) with permission.

can be reproduced experimentally by limitation of uterine blood flow in sheep (Creasy et al., 1972). The data shown in Fig. 10 are not substantially altered when comparisons are made on the basis of dry weights of the embryo and its component tissues, showing that only minor changes in tissue water content occur, either with hyperoxic growth acceleration or hypoxic growth retardation.

More recently, we have demonstrated a dose-related effect of increased O_2 concentration in accelerating growth (Stock et al., 1983). When eggs are incubated in 40% O_2, their embryos have a pattern of growth enhancement which is qualitatively similar to, but quantitatively less than, that resulting from incubation in 60% O_2. When eggs are incubated in 70% O_2, growth is accelerated above that characteristic of 60% O_2 until the twelfth day of incubation (Fig. 11). Subsequently, the 70% O_2 embryos show growth retardation. We attribute the growth retardation observed late in incubation in the 70% O_2 embryos to a limitation of nutrient supply. It has been demonstrated that tissue growth, i.e. protein synthesis, is a more expensive process for the avian embryo than is tissue maintenance (Vleck et al., 1980). When the finite energy stores of the egg are utilized early in incubation for accelerated protein synthesis, the cost of tissue maintenance rises as a consequence, and growth later in incubation is necessarily limited.

What is the mechanism by which O_2 stimulates embryonic growth and develop-

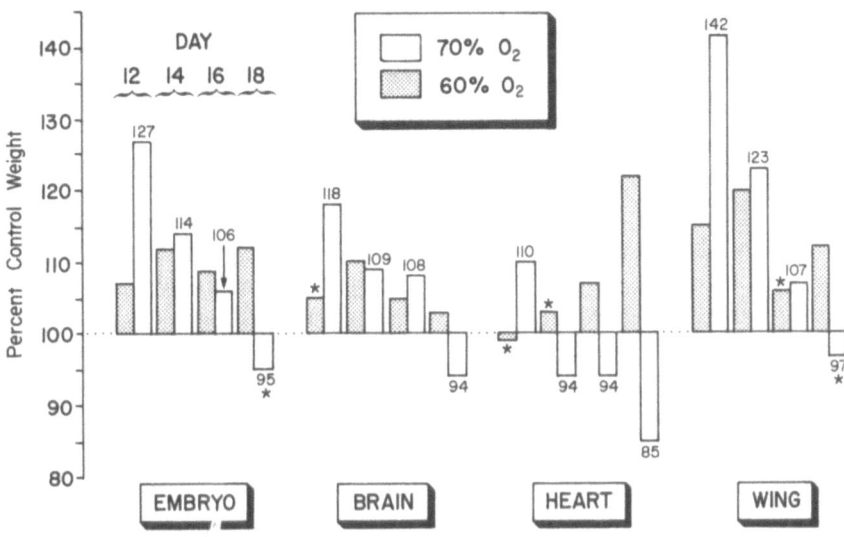

Fig. 11. Percentage deviation of tissue wet weights from controls (= 100%) on incubation days 12, 14, 16 and 18. Results from embryos incubated in 70% O_2 (clear bars) are compared with data from 60% O_2 (stippled bars). The number with each bar gives the value represented by the length of that bar. All values are significantly different from control (P<0.05), except those indicated by an asterisk. Reproduced from Stock et al. (1983) with permission.

ment? Two general hypotheses may be advanced. According to the first, some particular cell or group of cells in the embryo controls the embryonic growth rate and the activity of these cells is regulated by O_2 availability. For example, if growth hormone synthesis were sensitive to arterial O_2 tension, that sensitivity would be revealed by growth outside the areas of growth hormone production. Because growth acceleration can be detected very early in embryonic development, before many of the known anabolic factors have been identified, this alternative seems unlikely; however its lack of support may derive from methodological limitations, so it cannot be discarded.

The alternative hypothesis to explain the mechanism of oxygen's effect on growth is that increased P_{O_2} stimulates energy production directly in embryonic tissues, and the increased concentrations of adenosine triphosphate (ATP) are used, at least in part, for growth. There is considerable evidence that the oxidative metabolic machinery of organisms, including that of mammals, responds to both acute and chronic changes in O_2 availability. A decrease in the activity of several of the cytochromes has been demonstrated to result from hypoxia both in vitro and in vivo (Pious, 1970; Park et al., 1973; Mela et al., 1976; Hare and Hodges, 1982). Conversely, the increased O_2 availability that follows parturition is correlated with increases both in the number of mitochondria per unit weight of tissue (Hommes and Richters, 1969; Jakovcic et al., 1971) and the concentration of cytochromes per unit weight of mitochondrial protein (Mela et al., 1975, 1976). This hypothesis also draws support from studies of erythrocyte metabolism in hypoxic and hyperoxic eggs. The concentration of ATP in red cells of chick embryos incubated in 21% O_2 declines after the fourteenth day of incubation (Fig. 12; Ingermann et al., 1983). In contrast, the ATP concentration in cells of embryos incubated in 70% falls significantly less. Furthermore, when eggs incubated in 21% O_2 are switched to a hyperoxic incubator on the seventeenth day of incubation, the fall in ATP is reversed (Table 2). Finally, the erythrocyte ATP concentration of hyperoxic eggs can be made to fall by switching them to 21% O_2 on the seventeenth day of incubation. If the nucleated erythrocyte of this species can be considered characteristic of other embryonic tissues, with respect to the mechanism of ATP generation, the availability of high energy phosphate bonds of ATP appears to be enhanced by increased O_2 availability; in other tissues some of the extra ATP may be used for growth.

Control of growth by O_2 availability is manifest in species other than *Gallus domesticus*. The embryos of several species of fish show an increase in developmental rate with increasing concentrations of dissolved O_2 up to the point of complete equilibration with air at 5 °C (Hamor and Garside, 1976). Hyperoxic conditions were not studied to our knowledge. Experimental methods of inducing chronic hyperoxia in mammalian fetuses have not been developed. However, the growth rate of the human fetus slows in the last two weeks before parturition (Babson et al., 1970), then rises within one month after the onset of air breathing (Babson and Benda, 1976). An O_2 limitation of growth during normal fetal development may thus be a general phenomenon.

We have summarized evidence suggesting that O_2 availability limits embryonic growth and development in the chicken. This limitation appears to exist under

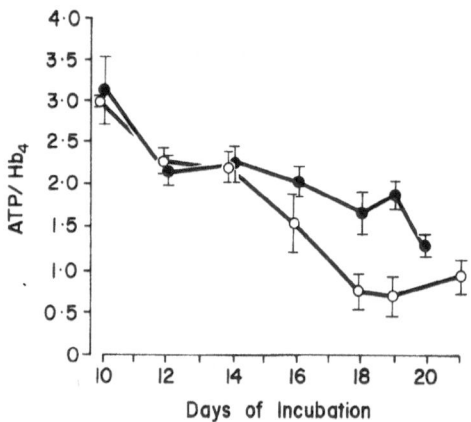

Fig. 12. Mean erythrocytic ATP concentration (relative to hemoglobin tetramer concentration) in hyperoxic (closed circles) and normoxic (open circles) chick embryos as a function of incubation age in days. Vertical bars represent plus or minus one standard deviation. Reproduced from Ingermann et al. (1983) with permission.

Table 2. Effect of acute changes in ambient O_2 concentration on erythrocytic ATP levels in the chick embryo. Day 17 experimental eggs were switched from normoxic to hyperoxic or from hyperoxic to normoxic conditions; unswitched eggs served as controls (C). Values are mean mole ATP per mole hemoglobin tetramer plus or minus one standard deviation. Reproduced from Ingermann et al. (1983) with permission.

Protocol	ATP, Day 18
21% O_2(C)	0.56 ± 0.18
	(n = 3)
21% O_2-70% O_2	* 0.99 ± 0.12
	(n = 8)
70% O_2(C)	1.37 ± 0.08
	(n = 3)
70% O_2-21% O_2	* 0.96 ± 0.17
	(n = 8)

* P<0.01, switched vs. unswitched (unpaired t-test)

normal conditions. The growth restraint can be accentuated by hypoxic exposure and can be reduced or eliminated by hyperoxic exposure. A slower rate of growth in the face of reduced O_2 availability may permit the embryo to conserve O_2. Conversely, when O_2 is in excess, accelerated growth and O_2 utilization may benefit the embryo by maintaining tissue P_{O_2} within tolerable limits.

References

Allen, S.C. (1963). A comparison of the effects of nitrogen lack and hyperoxia on the vascular development of the chick embryo. *Aeros. Med.* 34: 897–899.

Babson, S.G., Behrman, R.E. and Lessel, R. (1970). Fetal growth: Liveborn birth weights for gestational age of white middle class infants. *Pediatrics* 45: 937–944.

Babson, S.G. and Brenda, G.I. (1976). Growth graphs for the clinical assessment of infants of varying gestational age. *J. Pediatr.* 89: 814–820.

Bartels, H., Bartels, R., Brilmayer, Th. and Peters, W.-W. (1973). Frühentwicklung von Hühnerembryonen bei Normoxie und Hyperoxie. *Pneumonologie* 149: 67–73.

Bissonnette, J.M. and Metcalfe, J. (1978). Gas exchange of the fertile hen's egg: Components of resistance. *Respir. Physiol.* 34: 209–218.

Byerly, T.C. (1932). Growth of the chick embryo in relation to its food supply. *J. Exp. Biol.* (London) 9: 15–44.

Creasy, R.K., Barrett, C.T., de Swiet, M., Kahanpää, K.V. and Rudolph, A.M. (1972). Experimental intrauterine growth retardation in the sheep. *Am. J. Obstet. Gynecol.* 112: 566–573.

Flemister, L.J. and Cunningham, B. (1940). The effect of increased atmospheric pressure on the allantoic vascular bed and the blood picture of the developing chick. *Growth* 4: 63–65.

Freeman, B.M. and Misson, B.H. (1970). pH, P_{O_2} and P_{CO_2} of blood from the foetus and neonate of *Gallus domesticus*. *Comp. Biochem. Physiol.* 33: 763–772.

Hamburger, V. and Hamilton, H.L. (1951). A series of normal stages in the development of the chick embryo. *J. Morphol.* 88: 49–92.

Hamor, T. and Garside, E.T. (1976). Developmental rates of embryos of Atlantic salmon, *Salmo salar* L., in response to various levels of temperature, dissolved oxygen, and water exchange. *Can. J. Zool.* 54: 1912–1917.

Hare, J.F. and Hodges, R. (1982). Oxygen-stimulated cytochrome oxidase assembly in hepatocyte monolayer cultures. *J. Cell. Physiol.* 113: 23–27.

Hoiby, M., Aulie, A. and Reite, O.B. (1983). Oxygen uptake in fowl eggs incubated in air and pure oxygen. *Comp. Biochem. Physiol.* 74A: 315–318.

Hommes, F.A. and Richters, A.R. (1969). Mechanism of oxidation of cytoplasmic reduced nicotinamide adenine dinucleotides in the developing rat liver. *Biol. Neonat.* 14: 359–364.

Ingermann, R.L., Stock, M.K., Metcalfe, J. and Shih, T.-B. (1983). Effect of ambient oxygen on organic phosphate concentrations in erythrocytes of the chick embryo. *Respir. Physiol.* 51: 141–152.

Jakovcic, S., Haddock, J., Getz, G.S., Rabinowitz, M. and Swift, H. (1971). Mitochondrial development in liver of foetal and newborn rats. *Biochem. J.* 121: 341–347.

Lokhorst, W. and Romijn C. (1965). Some preliminary observations on barometric pressure and incubation. In: Energy Metabolism. K.L. Blaxter, ed., Academic Press, London, pp. 419–424.

McCutcheon, I.E., Metcalfe, J., Metzenberg, A.B. and Ettinger, T. (1982). Organ growth in hyperoxic and hypoxic chick embryos. *Respir. Physiol.* 50: 153–163.

Mela, L., Goodwin, C.W. and Miller, L.D. (1975). Correlation of mitochondrial cytochrome concentration and activity to oxygen availability in the newborn. *Biochem. Biophys. Res. Commun.* 64: 384–390.

Mela, L., Goodwin, C.W. and Miller, L.D. (1976). In vivo control of mitochondrial enzyme concentrations and activity by oxygen. *Am. J. Physiol.* 231: 1811–1816.

Metcalfe, J. (1967). The oxygen supply of the foetus. In: CIBA Foundation Symposium on Development of the Lung. A.V.S. de Reuck and R. Porter, eds., J. &A. Churchill, Ltd., London, pp. 258–271.

Metcalfe, J., McCutcheon, I.E., Francisco, D.L., Metzenberg, A.B. and Welch, J.E. (1981). Oxygen availability and growth of the chick embryo. *Respir. Physiol.* 46: 81–88.

Naeye, R.L. (1966). Abnormalities in infants of mothers with toxemia of pregnancy. *Am. J. Obstet. Gynecol.* 95: 276–283.

Needham, J. (1963). Chemical Embryology, Vol. 2. Hafner, New York, p. 696.

Park, C.D., Mela, L., Wharton, R., Reilly, J., Fishbein, P. and Aberdeen, E. (1973). Cardiac mitochondrial activity in acute and chronic cyanosis. *J. Surg. Res.* 14: 139–146.

Pious, D.A. (1970). Induction of cytochromes in human cells by oxygen. *Proc. Natl. Acad. Sci.* 65: 1001–1008.

Remotti, E. (1933). Adattamento respiratorio e suo substrato morfologico nello sviluppo embrionale degli Uccelli. *Boll. Mus. Lab. Zool. Anat. Comp.* 13: 3–21.

Robertson, T.B. (1923). The Chemical Basis of Growth and Senescence. Lippincott, Philadelphia. Cited in : Needham, A.E., 1964. The Growth Process in Animals. D. Van Nostrand Co., Inc., Princeton, New Jersey, p. 13.

Romanoff, A.L. (1960). The Avian Embryo. Macmillan Co., New York, pp. 1147–1148.

Romanoff, A.L. (1967). Biochemistry of the Avian Embryo. A Quantitative Analysis of Prenatal Development. John Wiley & Sons, New York, p. 287.

Romijn, C. and Roos, J. (1938). The air space of the hen's egg and its changes during the period of incubation. *J. Physiol.* (London) 94: 365–379.

Smith, A.H., Burton, R.R. and Besch, E.L. (1969). Development of the chick embryo at high altitude. *Fed. Proc. Fed. Am. Soc. Exp. Biol.* 28: 1092–1098.

Stock, M.K., Francisco, D.L. and Metcalfe, J. (1983). Organ growth in chick embryos incubated in 40% or 70% oxygen. *Respir. Physiol.* 52: 1–11.

Tazawa, H. (1971). Measurement of respiratory parameters in blood of chicken embryo. *J. Appl. Physiol.* 30: 17–20.

Tazawa, H., Ar, A., Rahn, H. and Piiper, J. (1980). Repetitive and simultaneous sampling from the air cell and blood vessels in the chick embryo. *Respir. Physiol.* 39: 265–272.

Tazawa, H., Mikami. T. and Yoshimoto. C. (1971). Effects of reducing the shell area on the respiratory properties of chicken embryonic blood. *Respir. Physiol.* 13: 352–360.

Temple, G.F. and Metcalfe, J. (1970). The effects of increased incubator oxygen tension on capillary development in the chick chorioallantois. *Respir. Physiol.* 9: 216–233.

Visschedijk, A.H.J. (1968). The air space and embryonic respiration. I. The pattern of gaseous exchange in the fertile egg during the closing stages of incubation. *Br. Poult. Sci.* 9: 173–184.

Vleck, C.M., Vleck, D. and Hoyt. D.F. (1980). Patterns of metabolism and growth in avian embryos. *Am. Zool.* 20: 405–416.

Wangensteen, O.D., Rahn, H., Burton. R.R. and Smith, A.H. (1974). Respiratory gas exchange of high altitude adapted chick embryos. *Respir. Physiol.* 21: 61–70.

Wiley, W.H. (1950). The influence of egg weight on the pre-hatching and post-hatching growth rate in the fowl. I. Egg weight-embryonic development ratios. *Poult. Sci.* 29: 570–574.

15. Regulation of oxygen affinity of embryonic blood during hypoxic incubation

Rosemarie Baumann

Zentrum Physiologie, Medizinische Hochschule, 3000 Hannover 61, Federal Republic of Germany

Abstract

Chronic incubation in normobaric hypoxia (13.5% O_2) leads to distinct changes of the O_2 affinity of embryonic chicken blood. The basis for these changes is a premature appearance of definitive red cells in the circulation of hypoxic embryos. In consequence the transition from embryonic to adult hemoglobin, from high to low ATP levels. is initiated at an earlier stage, so that hypoxic embryos of 9 days of incubation have an O_2 affinity comparable to that of 14 day old normal embryos. The striking increase of the O_2 affinity of hypoxic blood is best correlated to the fall of the ATP concentration and a significant increase of red cell pH.

Introduction

Changes of whole blood O_2 affinity are a prominent feature of adaptation to hypoxia in adult animals, as first shown in the classic study of Hall et al. (1936), who first demonstrated that animals native to high altitude had an increased O_2 affinity. While in adult mammals and birds this increase can in general be attributed to the properties of the hemoglobins, which either have an elevated intrinsic O_2 affinity or show decreased interaction with organic phosphates (cf. Petschow et al., 1977). lower vertebrates often respond by changing the organic phosphate concentration. In several fishes a decrease of the environmental P_{O_2} causes a fall of the red cell ATP concentration (Wood and Johansen, 1972; Wood et al., 1975; Greany and Powers, 1978) with a concomitant increase of the O_2 affinity.

It is long known that hypoxic episodes during embryonic and fetal development can lead to retarded growth and increased embryonic mortality in the chicken or other embryos (Beattie and Smith, 1975). but very little is known about the ability of the embryo to adapt to hypoxia. This question is especially pertinent with regard to the regulation of embryonic blood O_2 affinity. It has been shown that the O_2 extraction of embryonic chicken blood is up to 80% or more (Tazawa and Mochizuki, 1977) and that the drastic increase of embryonic blood O_2 affinity

Seymour, R.S., (ed.) Respiration and metabolism of embryonic vertebrates.
© *1984, Dr W. Junk Publishers, Dordrecht/Boston/London. ISBN-13:978-94-009-6538-6*

during the second half of incubation (Bartels et al., 1966; Lomholt, 1975; Tazawa et al., 1976) keeps arterial O_2 saturation constant despite a decrease of arterial P_{O_2} from 90 torr to about 50 torr shortly before hatching (Tazawa and Mochizuki, 1977). The adaption of hemoglobin O_2 affinity during normal development involves changes in hemoglobin type as well as alterations in the concentrations of allosteric effectors that regulate embryonic or adult hemoglobin function. While ATP is the dominant organic phosphate of early development (Bartlett and Borgese, 1976) 2,3 DPG concentrations peak during the last week of embryonic development (Isaacks and Harkness, 1975). In early embryos regulation of blood O_2 affinity seems to involve additional, but not yet identified red cell metabolites (Baumann et al., 1982).

In the chicken embryo, the change from embryonic to adult hemoglobin is due to an exchange of two cell populations (Burns and Ingram, 1973; Chapman and Tobin, 1979). The so-called primitive erythrocyte produces only embryonic hemoglobin types and lacks the capacity for self renewal. From the 6th day of incubation it is replaced by immature precursors of the definitive red cell series that produce adult hemoglobin, but can also synthesize small quantities of embryonic hemoglobin at first (Chapman and Tobin, 1979). The mechanism responsible for the switch from embryonic to adult hemoglobin production is unknown, but there are indications that extragenetic influences are involved (Schalekamp et al., 1982).

We have investigated the influence of normobaric hypoxia (13.5% O_2) on the O_2 affinity of embryonic blood between 4 to 14 days of development. The results show that in hypoxia there are changes in embryonic erythropoiesis, leading to a premature appearance of definitive red cells with adult hemoglobin. There are also significant differences in the red cell organic phosphate pattern of hypoxic embryos, that become first apparent after about 1 week of incubation. These alterations are reflected in the elevated O_2 affinity of blood from hypoxic embryos. The results suggest that the ontogenetic pattern of erythropoietic and O_2 affinity changes is under partial control of the environmental O_2 pressure.

Material and methods

Fertilized eggs from White-Leghorn chicken were incubated either in air or in 13.5% O_2 at 37.5°C and 60% relative humidity. The O_2 concentration was checked periodically with a Beckman O_2 analyzer. Timing of the eggs started when they entered the incubator. Blood was taken from the embryo by puncture of a larger blood vessel with a glass capillary or syringe mounted on a micromanipulator. Oxygen binding curves of red cell suspensions and hemoglobin solutions was obtained with a stepwise photometric method as described previously (Baumann et al., 1982). Adult chicken hemoglobins A and D were separated by column chromatography on CM 52-cellulose using a linear pH gradient of 0.01 M phosphate buffer pH 6.5 and 0.02 M phosphate buffer pH 8.0. Determinations of red cell 2.3 DPG, ATP and plasma lactate concentration were made with test-kits (Boehringer Co. Mannheim). Gas mixtures for equilibration were provided by Wösthoff mixing pumps. Hematological data were determined with standard methods. pH was measured with pH-meter 64 and MI 410 combination microelectrode (Microelectrodes Inc., Londonderry, NH, U.S.A.). Red cell pH was measuréd as described elsewhere (Baumann et al., 1982). Data on total embryonic hemoglobin and blood volume are taken from a previous publication (Baumann et al., 1983).

Results

The O_2 affinity changes in the blood of hypoxic and normoxic embryos are shown in Fig. 1. The half-saturation pressure (P_{50}) at pH 7.4 has its maximum at day 5 in the hypoxic embryo (86.3 torr) and at day 6 in the normal embryo (90.8 torr). Between 5 to 9 days it decreases by 37.5 torr in the hypoxic embryo to 48.8 at day 9, when P_{50} in the normal embryo is 73.5 torr. From 9 to 14 days there is only a comparatively small further increase of the O_2 affinity of hypoxic blood (P_{50} is 42.8 torr at day 14). The corresponding value in normal blood is 49.5 torr. Similar impressive differences are seen when one compares the red cell ATP and 2,3 DPG concentration of normal and hypoxic embryos (Fig. 2). Most prominent is a rapid decrease of the ATP concentration in hypoxic blood after 7 days of incubation. The largest difference exists at day 10 where the hypoxic blood has 1.3 mol ATP/mol Hb_4 compared to 3.1 mol ATP/mol Hb_4 in normoxic blood. The reverse is seen for 2,3 DPG; there the hypoxic red cell contains 0.42 mol 2,3 DPG/

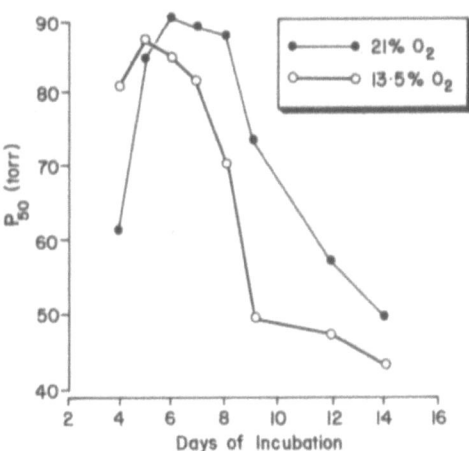

Fig. 1. Changes in O_2 affinity (P_{50}) in dependence of incubation for hypoxic and normoxic embryos. P_{50} is measured at constant pH of 7.4 and 37 °C.

mol Hb_4 at day 10, when the normal red cell has about 0.1 mol 2,3 DPG/mol Hb_4. In the hypoxic embryos the 2,3 DPG concentration increases steadily to 0.95 mol 2,3 DPG/mol Hb_4 at day 14 compared to 0.18 mol/mol Hb_4 in the normoxic embryo.

In order to get an estimate of the blood pH at early stages of incubation (e.g. between 6 to 9 days of development) we equilibrated whole blood with different concentrations of CO_2 and measured whole blood pH. The results are shown in Table 1 and demonstrate that there are no significant differences between blood from hypoxic and normoxic embryos, so that one can assume that arterial and venous pH will not be very different in both groups. At 6 to 7 days the pH of the arterialized blood of hypoxic and normoxic embryos should be around pH 7.7 if one takes the P_{CO_2} of the arterialized blood to be close to the one measured in the air cell of early embryos (Romijn and Roos, 1938; Wangensteen et al., 1974).

The pH of red cell lysates was also determined. Throughout embryonic development there are drastic changes in intracellular pH. While the earliest

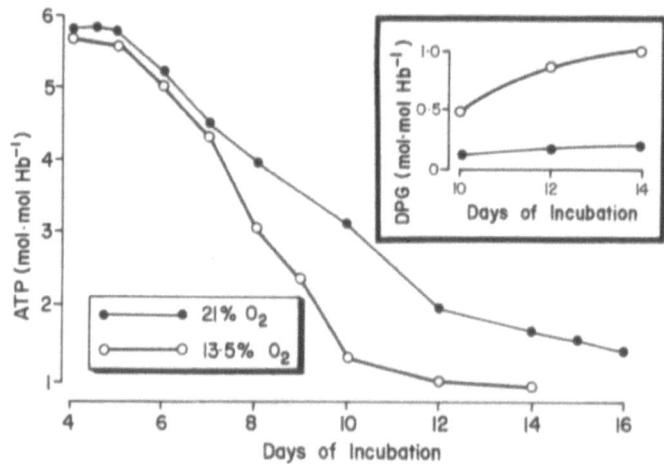

Fig. 2. Changes of red cell ATP and 2,3 DPG concentration in hypoxic and normoxic embryos between 4 to 14 days of incubation.

Table 1. pH of blood from hypoxic and normoxic embryos in dependence of P_{CO_2}.

	day 6	day 7		day 8		day 9	
P_{CO_2} (torr)	21% O_2	21% O_2	13.5% O_2	21% O_2	13.5% O_2	21% O_2	13.5% O_2
7	7.776	7.742	7.696	7.662	7.78	7.693	7.68
10	7.676	7.633	7.587	7.575	7.67	7.598	7.577
20	7.482	7.423	7.377	7.405	7.457	7/413	7.366

investigated embryos (6 day) have a low red cell pH of about 6.8 at pH_e 7.4, which is found in both hypoxic an normoxic embryos, there is a pronounced difference in the rate of change thereafter. Fig. 3 shows measurements of red cell pH at day 10, 12 and 14 in hypoxic embryos compared to normal embryos. In the normal embryos the pH_c increases by about 0.25 pH units (at constant pH_e of 7.4) between 6 to 14 days, the increase is much more pronounced in the hypoxic red cells where at day 12 and 14 one finds a cell pH of about 7.2 to 7.25 at pH_e of 7.4 and the pH_c of erythrocytes from 10 day hypoxic embryos is already higher than that of 14 day old normal embryos.

Discussion

The present study supplements our previous results on the O_2 affinity and organic phosphate concentration changes of blood from hypoxic embryos (Baumann et al., 1983) and allows an extended discussion of the mechanisms that are responsible for the increased O_2 affinity of hypoxic embryos. Since the hypoxic embryo is unable to increase its total hemoglobin or blood volume (Figs. 4, 5), a mechanism common in adult animals subjected to hypoxia, the O_2 affinity changes represent the principal adaptive mechanism. The major influence of hypoxia is directed on the onset of definitive erythropoiesis (Baumann et al., 1983) with the result that definitive red cells and adult hemoglobin appear at an earlier stage in the circulation of hypoxic embryos (Fig. 6). Compared to definitive erythropoiesis, primitive red cell production and function is much less influenced (Baumann et al., 1983). The fact that hypoxia is unable to increase the red cell mass, indicates

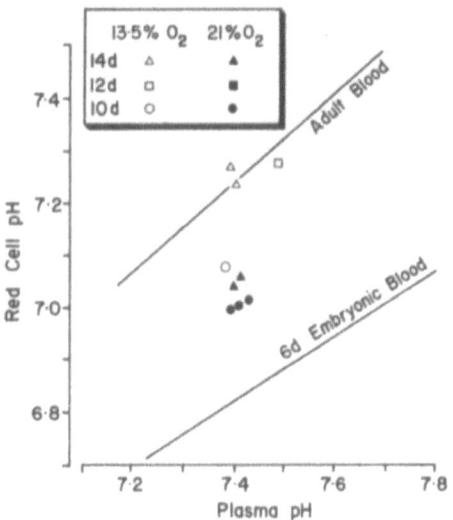

Fig. 3. Developmental changes in red cell pH.

that red cell production is at maximum rate under normal conditions of incubation. However it is possible to change the relative amount of primitive and definitive red cells in the population. In this regard it is noteworthy that Samarut (1979) has isolated from early chicken embryos a substance with properties similar to that of avian erythropoetin, which raises the possibility that early definitive erythropoiesis is at least in part under hormonal control.

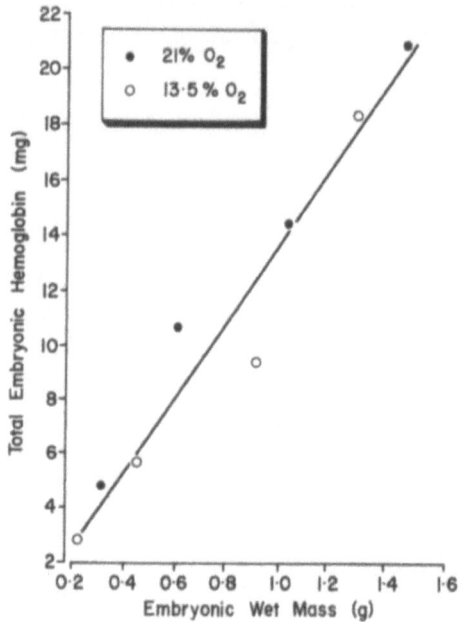

Fig. 4. Total embryonic hemoglobin measured in hypoxic and normoxic embryos.

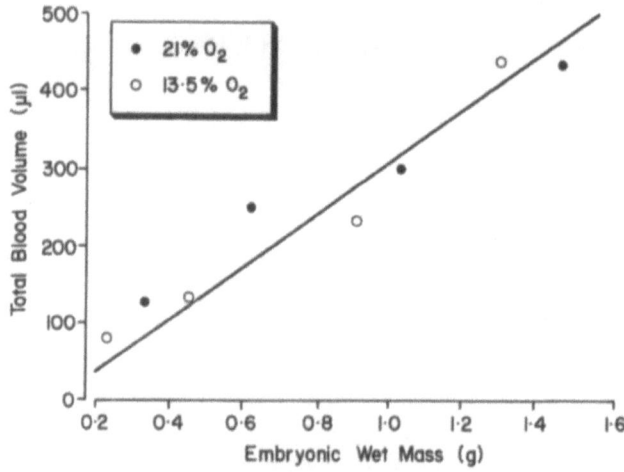

Fig. 5. Total blood volume of hypoxic and normoxic embryos.

226

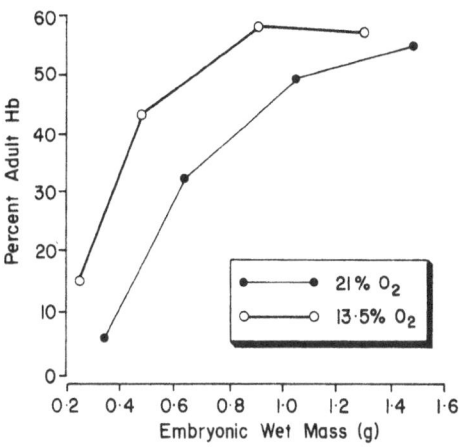

Fig. 6. Circulating adult hemoglobin (percent) in hypoxic and normoxic embryos.

We now turn to the mechanisms responsible for the increased O_2 affinity of hypoxic blood. For normal embryonic blood it has been demonstrated first by Misson and Freeman (1972), that the increase of blood O_2 affinity before hatching is best correlated with the fall of the red cell ATP concentration. Excepting the earliest stages (i.e. prior to day 7) a fair correlation is seen between the lower ATP concentration of hypoxic red cells and their increased O_2 affinity. It is unknown at present how the ATP or the 2.3 DPG concentration of avian embryonic blood are regulated. It cannot at the present be decided if the fall of the ATP concentration reflects the release of definitive cell populations with different, but inherently fixed organic phosphate concentrations or if the individual definitive red cell is able to regulate the ATP concentration in response to the ontogenetic changes of the P_{O_2} of arterialized blood. If one compares the time course of the change in red cell ATP and O_2 affinity in normal and hypoxic embryos it is evident that hypoxia accelerates the reduction of red cell ATP. Nevertheless, the general pattern of change is very similar, which suggests that to a large extent the ontogenetic changes of red cell ATP reflect the different metabolite pattern of the various definitive cell populations. These are released earlier in hypoxia hence the transition from high to low ATP starts earlier. In addition results of Ingermann et al. (1983) indicate that individual cells may respond to in vivo hypoxia by a reduction of their ATP concentration, a response also seen in some fish erythrocytes (cf. Tetens and Lykkeboe, 1981). It is however hardly probable that the fall of the ATP concentration is caused by cellular hypoxia, which does not exist when eggs are incubated in 13.5% O_2. In vitro we could deplete definitive and primitive red cells of ATP only when incubating them in anoxia and there was no difference in the response of both cell types (Baumann and Haller, unpublished observation). During in vivo hypoxic incubation, on the other hand, there is no change in ATP in the primitive red cells. This

227

suggests that extraerythrocytic mechanisms may be involved in the regulation of definitive red cell ATP concentration. A similar conclusion was reached by Tetens and Lykkeboe (1981) for fish erythrocytes. Indeed it has been shown that intracellular ATP concentrations can be changed in response to hormonal action, e.g. growth factors (Whetton and Dexter, 1983).

A second factor responsible for the changes of O_2 affinity is the red cell pH. While one can predict an increase of red cell pH due to lowered organic phosphate concentration, when the pH is determined through the Donnan-distribution (Duhm, 1971), this effect does not sufficiently explain the results obtained in the present study.

In hypoxic blood the sum of the organic phosphate concentration (ATP and 2,3 DPG) remains nearly constant between day 10 to day 14, yet cell pH at day 10 is about 0.2 units lower and a similar result is obtained when one compares hypoxic and normoxic blood at day 14, where the hypoxic red cell pH is increased by 0.25 units and total organic phosphate is the same. This result supports previous data indicating that embryonic red cell pH is in part regulated through active processes (Baumann and Haller, 1983).

A third factor that might possibly influence the O_2 affinity in vivo is the changing ratio of the adult hemoglobin components A and D. Early definitive red cells have a HbA/HbD ratio of less that 1, which increases to the adult value of about 2.5 during late ontogeny (Bruns and Ingram, 1973, Chapman and Tobin, 1979), and between 6 to 9 days the ratios of hypoxic and normoxic embryos are different (Baumann et al., 1983). However this seems of little importance, since variation of the HbD content from 30% to 70% of total hemoglobin causes only small changes of O_2 affinity (Fig. 7) in the absence or presence of ATP or 2,3 DPG.

Conclusion

Hypoxic exposure leads to adaptive changes of blood O_2 affinity in chicken embryos, caused by premature release of definitive red cells. The increase of O_2 affinity in hypoxic blood is correlated with a fall of the red cell ATP concentration and an increase of red cell pH. Furthermore, the present results support the conclusion that positive selection has had a dominant influence in establishing the ontogenetic pattern of hemoglobin types and red cell metabolites, which best allow a constant adjustment of blood O_2 affinity to changing environmental or arterial P_{O_2}.

References

Bartels, H., Hiller, G. and W. Reinhardt (1966). Oxygen affinity of chicken blood before and after hatching. *Respir. Physiol.* 1: 345–356.

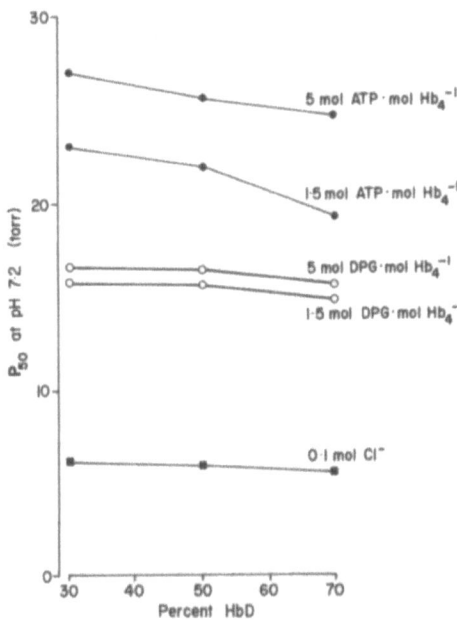

Fig. 7. Influence of amount of HbD on O_2 affinity in the absence and presence of organic phosphates. Temperature 37 °C, hemoglobin concentration 40 g · l⁻¹.

Barlett, G.R. and Borgese, T.A. (1976). Phosphate compounds in red cells of the chicken and duck embryo and hatchling. *Comp. Biochem. Physiol.* A55: 207–210.

Baumann, R., Padeken, S. and Haller, E.A. (1982). Functional properties of embryonic chicken hemoglobins. *J. Appl. Physiol.* 53: 1439–1448.

Baumann, R., Padeken, S., Haller, E.A. and Brilmayer, T.H. (1983). Effects of hypoxia on oxygen affinity, hemoglobin pattern, and blood volume of early chicken embryos. *Am. J. Physiol.* 244: R733–R741.

Baumann, R. and Haller E.A. (1983). Chloride distribution and intracellular pH of primitive red cells from chicken blood. *Naunyn-Schmiedeberg's Arch. of Pharm. Suppl.* Vol. 322R, 23.

Beattie, J. and Smith, A.H. (1975). Metabolic adaption of the chick embryo to chronic hypoxia. *Am. J. Physiol.* 228: 1346–1350.

Bruns, G.A.P. and Ingram, V.M. (1973). The erythroid cells and hemoglobins of the chick embryo. *Philos. Trans. R. Soc. London* 266: 225–305.

Chapman, B.S. and Tobin, A.J. (1979). Distribution of developmentally regulated hemoglobins in embryonic erythroid populations. *Dev. Biol.* 69: 375–387.

Duhm, J. (1971). Effects of 2.3 diphosphoglycerate and other organic phosphate compounds on oxygen affinity and intracellular pH of human erythrocytes. *Pflügers Arch.* 32: 341–356.

Greany, G.S. and Powers, D.A. (1978). Allosteric modifiers of fish hemoglobins: in vitro and in vivo studies of the effect of ambient oxygen and pH on erythrocyte ATP concentrations. *J. Exp. Zool.* 203: 339–349.

Hall, F.G., Dill, D.B. and Guzman-Barron, E.S. (1936). Comparative physiology in high altitudes. *J. Cell. Comp. Physiol.* 8: 301–313.

Ingermann, R.L., Stock, M.K., Metcalfe, J. and Shih, T.B. (1983). Effect of ambient oxygen on organic phosphate concentrations in erythrocytes of the chick embryo. *Respir. Physiol.* 51, 141–152.

Isaacks, R.E. and Harkness, D.R. (1975). 2.3 diphosphoglycerate in erythrocytes of chick embryos. *Science* 189: 393–394.

Lomholt, J.P. (1975). Oxygen affinity of bird embryo blood. *J. Comp. Physiol.* 99: 339–343.

Misson, D.H. and Freeman, B.M. (1972). Organic phosphates and oxygen affinity of chick blood before and after hatching. *Respir. Physiol.* 14: 323–352.

Petschow, D., Würdinger, I., Baumann, R. Duhm, J., Braunitzer, G. and Bauer, C. (1977). Causes of high blood oxygen affinity in animals living at high altitude. *J. Appl. Physiol.* 42: 139–143.

Romijn, C., and Roos, J. (1938). The air space of the hen's egg and its changes during the period of incubation. *J. Physiol.* (London) 94: 365–379.

Samarut, J. (1979). Erythropoetin-like factor production by young chick blastoderms. *Dev. Biol.* 70: 278–282.

Schalekamp, M., de Jonge, P. and van Goor, D. (1982). Is erythroid differentiation a matter of all-or-none transcription only? In: Cell function and differentiation, Part A. Alan R. Liss. New York, 1982. pp. 25–33.

Tazawa, H. and Mochizuki, M. (1977). Oxygen analyses of chicken embryo blood. *Respir. Physiol.* 31: 203–215.

Tazawa, H., Ono, T. and Mochizuki, M. (1976). Oxygen dissociation curve for chorioallantoic capillary blood of chicken embryo. *J. Appl. Physiol.* 40: 393–398.

Tentens, V. and Lykkeboe, G. (1981). Blood respiratory properties of rainbow trout, *Salmo gairdneri:* responses to hypoxic acclimation and anoxic incubation of blood in vitro. *J. Comp. Physiol.* 145: 117–125.

Wangensteen, O.D., Rahn, H. Burton, R.R. and Smith, A.H. (1974). Respiratory gas exchange of high altitude adapted chick embryos. *Respir. Physiol.* 21: 61–70.

Whetton, A.D. and Dextor, M. (1983). Effect of haematopoietic growth factor on intracellular ATP levels. *Nature* 303: 629–631.

Wood, S.C. and Johansen, K. (1972). Adaption to hypoxia by increased HbO_2 affinity and decreased red cell ATP concentration. *Nature New Biol.* 237: 278–279.

Wood, S.C., Johansen, K. and Weber, R.E. (1975). Effects of ambient P_{O_2} on hemoglobin-oxygen affinity and red cell ATP concentration in a benthic fish *Pleuronectes platessa*. *Respir. Physiol.* 25: 259–267.

16. Blood oxygen affinity in relation to yolk-sac and chorioallantoic gas exchange in the developing chick embryo

Robert Blake Reeves

Department of Physiology, School of Medicine, State University of New York, Buffalo, New York 14214, U.S.A.

Abstract

A complete profile of blood O_2 affinity throughout chick development from the inception of blood formation to adulthood shows three clearly marked phases. In Phase I from incubation day 4 to 8, P_{50} increases from 38 to 52 torr; in Phase II from day 8 to day 18 (pipping) P_{50} falls from 52 to 30 torr; in Phase III P_{50} rises from 30 torr at pipping to 47 torr in the adult chicken. Phase I, concurrent with the replacement of red cells that derive from yolk-sac hematopoesis to definitive red cells formed in the embryo, is not associated with hypoxic stress to the embryo. High P_{50} of Phase I is an adaptation that assures higher mixed venous O_2 partial pressures in the presence of significant mixed venous shunt fractions. Phase II is coincident with chorioallantoic gas exchange, definitive red cells, and progressive hypoxia as the O_2 consumption of the embryo increases. The increase in blood O_2 affinity, due to progressive decrease in red cell [ATP], is adaptive in ensuring adequate O_2 saturation of arterialized blood. In Phase III, the transition to convective air-breathing and removal of the P_{O_2} restriction on loading, the increase in P_{50} is an advantage for O_2 delivery to tissues. This increase in P_{50} is due to the rise in intra-erythrocyte [IPP]. P_{50} changes during chick development illustrate the interplay between two major factors affecting the O_2 pressure at which delivery of O_2 to tissue occurs: large shunt fractions and a diffusive limitation on hemoglobin loading. When hypoxia threatens adequate loading, lower P_{50} secures the required arterial O_2 saturations. In the absence of an hypoxic limitation on loading, as in Phases I and III, blood O_2 affinity decreases in favor of maximizing tissue O_2 delivery.

For blood to fulfill its function in O_2 transport blood O_2 affinity must reflect a compromise between an affinity high enough to assure efficient O_2 loading and an affinity low enough the preserve an adequate O_2 partial pressure gradient to deliver O_2 to the tissues. A regulated blood O_2 affinity is characteristic of each species but a wide range of affinities has been found even among air-breathing vertebrates. The ultimate physiological determinants of a particular affinity set-

Seymour, R.S., (ed.) Respiration and metabolism of embryonic vertebrates.
© 1984, Dr W. Junk Publishers, Dordrecht/Boston/London. *ISBN-13: 978-94-009-6538-6*

point are yet to be clarified for any species; we lack hypotheses that define how a specific regulated affinity is suited to one species and not another. Many studies have explored the proximate molecular mechanisms that govern blood O_2 affinity and are utilized in its regulation; i.e., genetically determined specific hemoglobins have their intrinsic O_2 affinity modulated by metabolic control systems that determine red cell concentrations of hemoglobin allosteric effectors. But answers are as yet obscure to the more general question of how a specific regulated blood O_2 affinity is adapted to the respiratory function of the whole organism.

The chicken (*Gallus domesticus*) is the only species in which whole blood O_2 affinity has now been measured under in vivo physiological conditions through all stages of development from immediately after commencement of embryonic hematopoesis through full adult status (Lapennas and Reeves, 1983a, b). Fig. 1 illustrates in a very schematic way two of the principal limiting features of O_2 delivery to avian embryonic tissues. One is the presence of a fixed diffusion resistance for respiratory gas exchange imposed by the presence of the shell and shell membranes. The second is the extensive but as yet poorly measured right to left shunting that occurs throughout embryonic development. This paper reviews the profound ontogenetic excursions in blood O_2 affinity recently reported in this species, inquires to what extent known genetic and/or physiological controls are operating, and explores how these changes in O_2 affinity relate to patterns of diffusive gas exchange and central venous shunting.

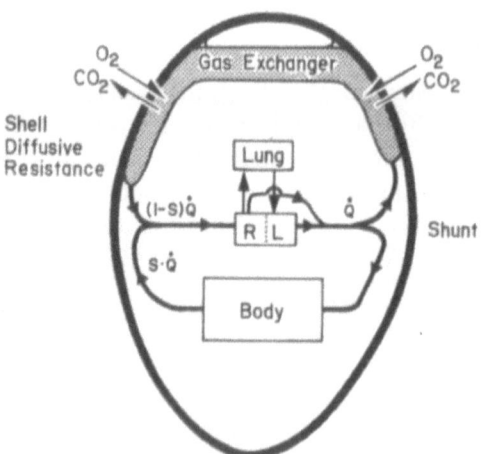

Fig. 1. Schematic of chick embryo circulation in ovo. Arterialized blood returning from the gas exchanger is mixed with mixed venous blood from the tissues of the embryo. The relative flow fractions of the cardiac output ($\dot{Q}$) are designated $(1-S)\dot{Q}$ for arterialized blood and $S \cdot \dot{Q}$ for the shunt fraction.

232

Usefulness of new micromethods

Early studies on avian blood O_2 affinity foundered on two technical problems. Adult avian erythrocytes have a metabolic rate many times that found in non-nucleated mammalian red cells; metabolic rates of avian embryonic red cells may exceed adult values as much as ten-fold (Grima et al., 1983). These properties gave rise to significant errors in methods that relied on tonometry, because tono-metered sample O_2 tension was less than the equilibrating gas mixture. Further-more, as study of earliest incubation stages was approached the tiny blood sample volume that could be collected became limiting. Even micro techniques for determining O_2 equilibrium data were inadequate if relevant acid-base studies could not be run on the same sample. To circumvent these problems and yet work with whole blood we applied to the chick embryo and adult chicken a new dynamic blood film technique for obtaining O_2 equilibrium curves (O_2EC) (Reeves, 1980; Reeves et al., 1982). This method needs less than one microliter of whole blood spread as a thin film one red cell in thickness between thin Teflon membranes to measure O_2 saturation by dual-beam spectrophotometry; the O_2 tension in equilibrium with the film is registered with an O_2 electrode. The O_2 permeability of the Teflon membrane is sufficiently great that error arising from red cell metabolism is negligible. To circumvent the need for a blood sample from small embryos on which to determine acid-base data so that blood film pH could be computed, we utilized the observation that air cell CO_2 partial pressure is essentially the same as arterialized blood (Tazawa et al., 1980). Hence by carrying out our blood film equilibrium curve measurements at the same CO_2 partial pressure as was measured in each egg's air cell we were assured that physiological blood pH was attained.

Blood oxygen affinity during chick development

The half-saturation partial pressures of O_2 (P_{50}) of physiological O_2 equilibrium curves of adult and embryonic blood as a function of incubation age are shown in Fig. 2. Blood O_2 affinity changes significantly with age; indeed the magnitude of the variation is the largest yet reported for any vertebrate in vivo. Three phases can be distinguished: (I) in the period 4 to 8 days, P_{50} increases from 38 to 52 torr; (II) the remainder of development in the egg, days 8 to 18, shows a sustained decrease in P_{50} from 52 to 30 torr and (III) from pipping until adulthood is achieved there is an increase in P_{50} from 30 to 47 torr. Compared with mammals, these blood O_2 affinities are remarkably low. The CO_2 Bohr coefficient also changed during development increasing from values of -0.29 at 6 days incubation age to -0.54 at pipping and in adult birds.

233

Phase I. The yolk-sac gas exchange period

These three phases of altered blood O_2 affinity correlate closely with functionally identifiable modes of respiratory exchange. Phase I, coincident with the use of the yolk-sac area vasculosa for gas exchange, concludes with the period in which the chorioallantois becomes established as the principal gas exchange site (ca. days 8–9). The striking blood O_2 affinity feature of Phase I is the decrease in blood O_2 affinity that occurs during incubation age 3–8 days. This trend suggests that the embryo, replete with a circulation to assist in gas exchange since day 2, is not obliged to increase blood O_2 affinity to foster O_2 loading.

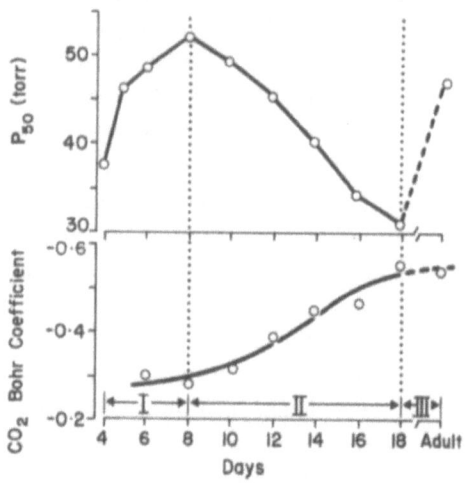

Fig. 2. The variation during chick development of blood O_2 affinity measured at 38 °C and the CO_2 tension of air cell gas is shown above; the corresponding time course of the CO_2 Bohr coefficient is depicted below.

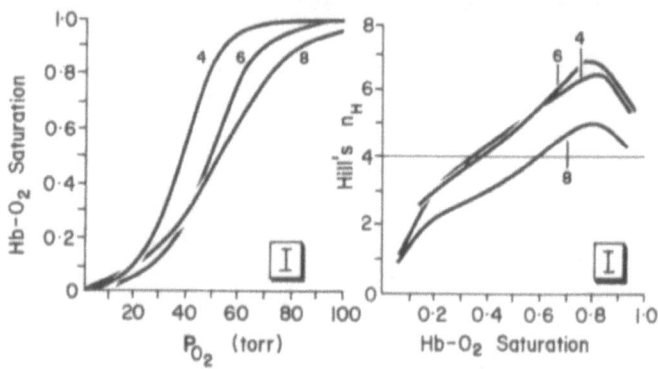

Fig. 3. O_2EC of chick embryo blood during Phase I are shown at left. The magnitude of Hill's n_H versus saturation for each O_2EC indicated is given on right. Numbers besides curves indicate incubation age in days.

Phase I is also distinguished from the other phases by a different shape to the O_2 equilibrium curve as is illustrated in Fig. 3. Curves from days 4 to 6 are conspicuously steeper, more sigmoid, than are other curves. The Hill plot ($\log S/(1 - S) = n_H \log P_{O_2} + K$) defines a constant n_H, the slope; when the Hill plot is non-linear, the first derivative at any point ($\Delta \log S/(1 - S)/\Delta \log P_{O_2}$) is also termed Hill's n_H. The Hill's n_H of the chick O_2EC in Phase I is markedly saturation dependent as is also shown in Fig. 3; at low saturation, n_H is between 1 and 2 while at $S = 0.8$, n_H reaches 6.5. Hill's n_H values at higher saturations diminish as development proceeds but even in the adult bird exceeds 3.0 (see Fig. 10).

Utility of low blood oxygen affinity when shunt exists

Is there an advantage to having blood of low affinity? Some answer to this question can be gotten by first examining the effect of low blood O_2 affinity on unloading in the tissues of the embryo and then by examining the consequences of low affinity for O_2 loading. Fig. 4 presents two O_2 equilibrium curves appropriate to Phase I of differing affinity and identifies on each three points: a loading point indicating the saturation achieved in the ideal gas exchange organ (P_c, S_c) where gas and blood tensions approach equilibrium; an arterial point (P_a, S_a) of lower saturation and O_2 partial pressure because of the admixture of shunt venous blood; and a mixed venous point ($P_{\bar{v}}$, $S_{\bar{v}}$), a measure of tissue oxygenation. Each curve represents the same metabolic state except for a difference in blood O_2 affinity: i.e., hemoglobin concentration, O_2 consumption, cardiac output, and hence a-v O_2 extraction, as well as shunt fraction are constant, as is the loading O_2 tension in the gas exchange organ. The curve with the higher P_{50} has significantly greater arterial and mixed venous O_2 pressures. A right shifted curve in the

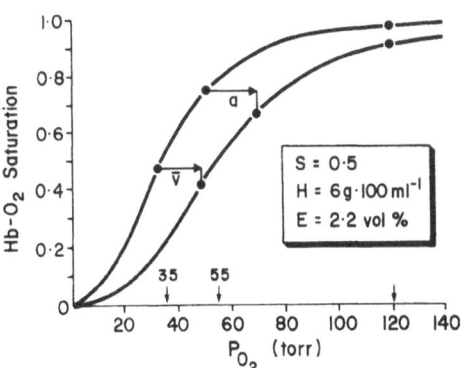

Fig. 4. Calculated effect of the increase in P_{50} occurring during Phase I on arterial and mixed venous O_2 pressures for a shunt fraction of 0.5, hemoglobin $60 \, g \cdot l^{-1}$ and extraction of 2.2 vol. % if metabolism and cardiac output are kept constant.

presence of significant shunt delivers O_2 to the tissues at a higher O_2 pressure (Rossoff et al., 1980).

Some measure of the sensitivity of this phenomenon to blood O_2 affinity and to the magnitude of the shunt fraction can be gauged from the calculations shown in Fig. 5. For a shunt fraction of 0.5, mixed venous P_{O_2} increases by 7 torr for every 10 torr increment in P_{50}. For arterial P_{O_2} the increment is greater, 10 torr per 10 torr increase in P_{50}. It is necessary to emphasize strongly that this advantage obtains only so long as loading O_2 tension remains high; as loading tensions falls a crossover point is reached where the right shifted curve arterial and mixed venous tensions now become less than the corresponding reference curve tensions (Aberman, 1977). However these computations demonstrate that chick blood with as low an O_2 affinity as can be fully loaded best serves the needs of the tissues in

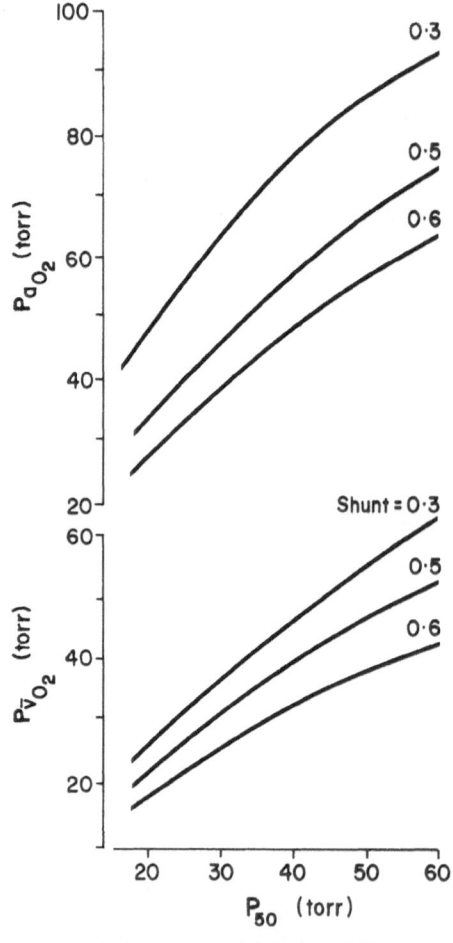

Fig. 5. Calculated sensitivities of arterial (Pa_{O_2}) and mixed venous $P\bar{v}_{O_2}$) blood O_2 tensions to variation in blood O_2 affinity (P_{50}). Three shunt fractions are indicated; other conditions as in Fig. 4.

terms of O_2 delivery in an embryonic circulation involving venous shunt admixtures of significant magnitude.

Is oxygen loading of blood adequate with yolk-sac gas exchange?

Phase I gas exchange is mediated by the area vasculosa of the yolk sac membrane. Little is known about the yolk-sac membrane as a gas exchanger, however, because of the great difficulty in obtaining blood samples for analysis. Nonetheless, O_2 consumption increments are impressive in this period inasmuch as each day of incubation sees nearly a doubling of the previous day's O_2 demand (Freeman and Vince. 1974).

Significant differences in the O_2 diffusion path from atmosphere to red cell between yolk-sac versus chorioallantoic gas exchanger are illustrated in Fig. 6. Although resistance to O_2 diffusion through the shell and outer shell membrane is identical, early in Phase I the inner shell membrane is hydrated and its resistance to O_2 diffusion is large (Kutchai and Steen, 1971; Lomholt, 1976). The time course of drying of the inner shell membrane is not fully established as yet, but by day 6 this resistance is decreased significantly and approaches the values for later incubation stages.

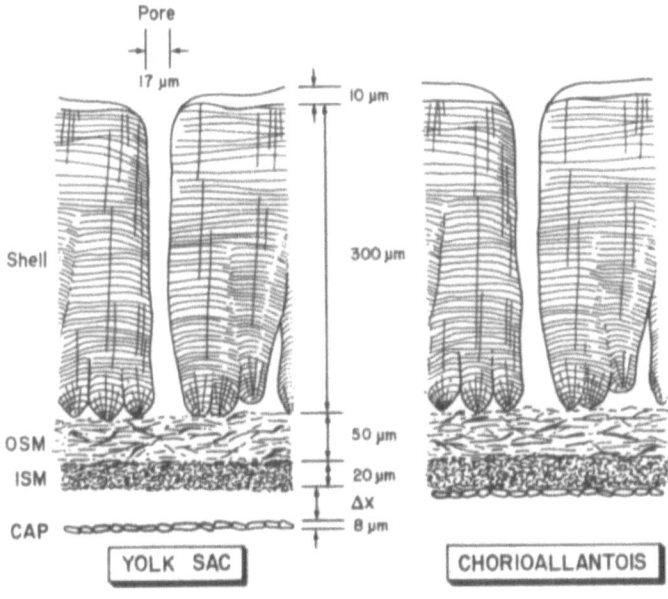

Fig. 6. Comparison of diffusion pathway through shell and associated membranes during yolk-sac gas exchange stage (4–8 days) (left) with later chorioallantoic gas exchange stage (10–19 days) (right) (after Freeman and Vince. 1974, p. 125). (OSM, outer shell membrane; ISM, inner shell membrane; CAP, capillary blood).

An additional resistance to O_2 diffusion, a layer of albumen (labelled Δx in Fig. 6), also unique to Phase I is present. In the 3 to 7 day embryo the yolk sac membrane growth over the surface of the yolk is rapid, achieving by 7 days a maximal area of 50 cm², practically 75% of the total shell surface area (Romanoff, 1960). The vitelline membranes surrounding the yolk have ruptured or disappeared, in part through uptake of water from the albumen, conspicuously altering the shape of the yolk mass. A fertile egg candled at this stage shows the embryo lying on top of the yolk sac at the uppermost pole of the egg whatever its position. The yolk mass, like a large oil drop, floats over the albumen assuming the dome-shape of the inner shell membrane above it. When the egg is turned, the yolk-mass covered superiorly by the area vasculosa is also free to turn; in an egg with a portion of the shell and shell membrane removed, the yolk mass covered by the embryo and yolk-sac membrane can be shown to move freely with respect to the inner shell membrane. Hence there must exist an albumen layer of variable thickness, Δx, (see Fig. 6) between the inner shell membrane and the vascular yolk-sac membrane through which diffusive exchange of respiratory gases os obliged to occur.

Fig. 7 estimates the magnitude of the added ΔP_{O_2} arising from the presence of the albumen layer of thickness Δx in the diffusion path. Using Fick's first law we can calculate the ratio $\Delta P_{O_2}/\Delta x$ for any specific day of incubation where the O_2 consumption and area of the yolk sac membrane area vasculosa are known; calculations extending over period of days 3–8 are shown. At the highest consumptions occurring in Phase I the ΔP_{O_2} from the flux across the shell, outer shell,

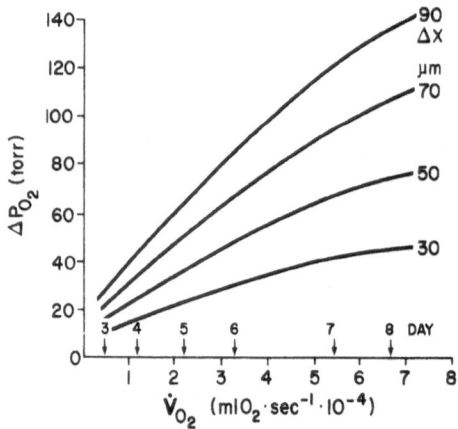

Fig. 7. Oxygen partial pressure gradient (ΔP_{O_2} torr) required to deliver O_2 consumption ($\dot{V}_{O_2}$), a function of thickness (Δx) of albumen solution (water), calculated from Fick's first law,

$$\dot{V}_{O_2} = D_{O_2} \cdot A_v \cdot a_{O_2} \, (\Delta P_{O_2}/\Delta x).$$

$\dot{V}_{O_2}$ values indicated are for incubation days 3 to 8 (Freeman and Vince, 1974); note that areas (A_v) for diffusion (area vasculosa of yolk-sac), a function of embryo age, are required (Romanoff, 1960). The diffusion coefficient of O_2 in water at 37 °C ($D_{O_2} = 3.3 \cdot 10^{-5} \cdot cm^2 \cdot sec^{-1}$, Grote 1967) and O_2 solubility in water at 38 °C ($a_{O_2} = 3.3 \cdot 10^{-5} ml\, O_2 \cdot torr^{-1}$) are also utilized in this calculation.

dry inner shell and capillary membranes is of the order of 5 torr (Paganelli, 1980). The air cell P_{O_2} in this phase is 140 torr and P_{50} values are 40–50 torr. Hence the ΔP_{O_2} across the albumen layer can be as large as (140–5–50) or circa 85 torr and still provide an O_2 flux sufficient to achieve saturations of 50%. This calculation overestimates the maximal permissible ΔP_{O_2} because saturations greater than 50% seem likely. With 85 torr as the upper bound, the albumen layer thickness compatible with adequate oxygenation according to Fig. 7 can be as great as 50 micrometers. The smallest conceivable albumen layer thickness might be that of an unstirred layer at high shear velocities, one of several micrometers. A more likely estimate for albumen layer thickness in vivo is 20–30 micrometers; this range of estimates appears fully compatible with normal O_2 uptakes and high (ca. 100 torr) loading O_2 tensions. Tentatively one can conclude that despite the limitations of the yolk-sac membrane for gas exchange, there appears to be no serious limitation on O_2 loading pressure in the capillaries of the yolk-sac membrane. Were it not for the speedy replacement of yolk-sac gas exchange with the chorioallantois over days 8–10, however, yolk-sac gas exchange limitations would inevitably appear before days 10–12.

Erythroid cell lines in Phase 1

The origin of the decrease in blood O_2 affinity in Phase I is most likely attributable to the radical change in erythrocyte populations that commences mid-way in this phase (Ingram, 1981). Initial chick red cell populations derive as a cohort of cells formed in the area vasculosa of the yolk-sac. These primitive red cells lack a self-perpetuating stem cell population and are characterized by major embryonic hemoglobins P and P' and minor hemoglobins E and M; the major and minor hemoglobins are thought to reside separately in two sub-populations of the primitive red cell series (Cirotto et al., 1977). A second definitive red cell population arises within the embryo from true erythroid stem cells in the liver and marrow; definitive red cells containing hemoglobins A and D first appear at day 5 but very rapidly outnumber the primitive erythrocytes; by day 10 the ratio of definitive to primitive erythrocytes is 3–4 (Ingram, 1980). The progressive increase in P_{50} over days 4 to 8 mirrors this event as does the change in O_2 equilibrium curve shape (see Fig. 3). It is the primitive yolk-sac erythrocyte line which is characterized by the steeply sigmoid ($n_H > 6$) O_2 equilibrium curve with a P_{50} of 38 torr. The blood O_2 equilibrium curves from chicks of 10–12 day incubation age (see Fig. 9) largely reflect definitive red cell characteristics. Thus the alterations in O_2 affinity of Phase 1 are principally attributable to the genetically determined switch in erythroid cell lines. Recent evidence suggests that experimentally induced hypoxia hastens the appearance of the definitive red cell line (Baumann et al., 1983).

Phase II. The chorioallantoic gas exchange period

Phase II is concomitant with gas exchange through the chorioallantois and the period in which the greatest metabolic rate and O_2 consumption are achieved within the egg (see Fig. 8). Establishment of the chorioallantoic contribution to gas exchange commences when the budding allantois makes contact with the inner shell membrane on day 6; the chorioallantoic membrane quickly becomes the principal gas exchange organ and completely covers the shell membrane by day 12 (Fitze-Gschwind, 1973). Fig. 8 shows the time course of the very large increase in O_2 consumption that reaches a maximum about day 16 and cannot increase further until the onset of pipping. As the O_2 demand increases the O_2 tension in the air cell falls from an initial value near 140 torr to a level of about 100 torr immediately before pipping.

Evidence for oxygen limitation on growth in Phase II

There is now a growing body of evidence that the O_2 consumption of the embryo toward the end of Phase II, from day 14 onward to pipping, is increasingly limited by the rate of O_2 delivery to the tissues (Metcalfe et al., 1981). Observations of

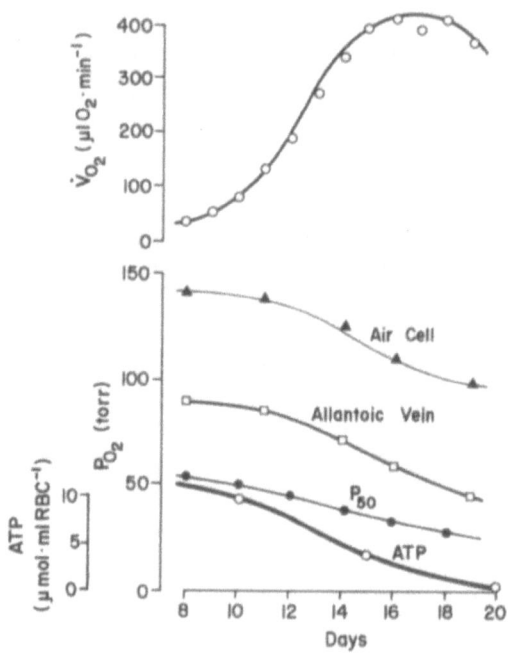

Fig. 8. Time course during Phase II of whole egg O_2 consumption ($\dot{V}_{O_2}$), O_2 tension in air cell gas and allantoic venous (arterialized) blood, blood O_2 affinity (P_{50}) and red cell [ATP].

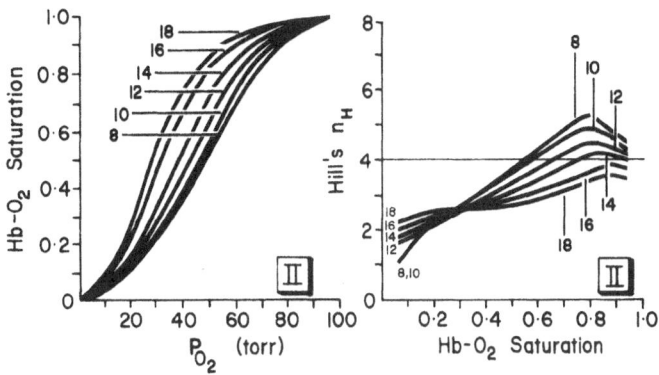

Fig. 9. O$_2$EC and Hill's n$_H$ versus saturation plots characteristic of Phase II.

embryo growth under hyperoxic conditions clearly establish that the normoxic incubated embryo in ovo is hypoxic. This situation is also reflected in the large decrease observed in P$_{50}$ during Phase II. Fig. 9 shows the progressive shift to the left of the O$_2$EC from a P$_{50}$ of 52 torr at day 8 to the lowest P$_{50}$ in the whole of the life-cycle of the chicken, a value of 31 at day 18.

Uncertainty concerning red cell loading oxygen tension

Falling in parallel with the diminishing air cell O$_2$ tension are the allantoic vein O$_2$ tension values (see Fig. 8) reported by Piiper et al. (1980). These workers suggested that the large air cell to allantoic venous blood O$_2$ tension difference is principally attributable to venous shunt in the chorioallantoic circulation. This conclusion however intimates that O$_2$ loading occurs at O$_2$ tensions close to those recorded for the air cell which does not fall below 100. If the P$_{O_2}$ gradient across the structures of the inner shell membrane and chorioallantoic capillaries is as small as has been estimated, about 10 torr (Paganelli, 1980), then the loading tension for blood throughout Phase II would exceed 80 torr. Inspection of Fig. 9, however, demonstrates that no limitation on O$_2$ loading would obtain if that were the case. Even blood at day 8 with a P$_{50}$ of 52 torr would be 90% saturated. Instead it would appear, judging from the progressive increase in blood O$_2$ affinity, that the allantoic vein blood is affected by shunt only to a small degree. In order for the progressive increase in blood affinity to represent an adaptive response of the embryo to developing hypoxia, it is necessary to conclude that the allantoic vein O$_2$ tensions reflect O$_2$ loading conditions in the chorioallantoic capillaries and not shunt admixture in the gas exchanger itself. This conclusion suggests a larger true A-a gradient than heretofore suspected and that some diffusion resistance has been seriously underestimated.

The increase in blood O_2 affinity during Phase II is the resultant of at least two processes. A decrease in intra-erythrocyte ATP concentration over the time course of Phase II has been well documented (Isaacks et al., 1976; Bartlett and Borgese, 1976). Fig. 8 shows how closely the change in P_{50} with red cell ATP level is correlated. Recently it has been demonstrated that the fall in red cell ATP can be attenuated by incubation under hyperoxic conditions (Ingermann et al., 1983). Furthermore there is also at this time a progressive decrease in the ratio Hb D/Hb A from 0.85 to 0.45 over this period (Ingram, 1981). Since Hb D has a higher affinity for O_2 than Hb A (Vandecasserie et al., 1973), the overall effect on blood affinity must be principally attributed to the alterations in ATP levels. Recent work has made evident that the decrease in embryo blood P_{50} during Phase II is the expression of a regulatory system but precisely how a diminished chorioallantoic venous blood O_2 tension is translated into a decrease in red cell ATP levels is not known. However, without this important adaptation in blood O_2 affinity to falling loading tensions, assuring reasonable O_2 saturation values for blood returning to the embryo, attainment of normal embryo weights at pipping could not be achieved.

The pattern of progressively increasing O_2 affinity as embryonic O_2 demand accelerates followed by a marked decrease in affinity once air-breathing has become fully established is also characteristic of mammalian development (Jelkmann and Bauer, 1977). This suggests that increased affinity to secure adequate loading can be accomplished without serious encroachment upon O_2 delivery during development. But when the period of hypoxic stress engendered by the limitations of diffusive conductance in the eggshell or of placental gas exchange in utero is alleviated, the organism reverts to a lowered blood O_2 affinity to promote O_2 delivery to tissues during periods of maximal physical activity.

Phase III. Blood O_2 affinity after commencement of air-breathing

Starting with Phase III, the commencement of lung breathing at pipping for adult life markedly increases the O_2 tension at O_2 loading. Having removed the hypoxic restriction to O_2 loading of a limiting diffusive shell conductance, the advantages of lower affinity for increasing the driving force for O_2 delivery to tissue can be achieved. Fig. 10 depicts the right shift in the O_2EC that occurs in the weeks after hatching. This process is effected by increased levels of inosinepentaphosphate (IPP) within the erythrocytes (Isaacks et al. 1976). IPP is the most potent allosteric modifier of hemoglobin O_2 affinity known. How this modification is controlled and what sensors initiate the affinity change are as yet undetermined.

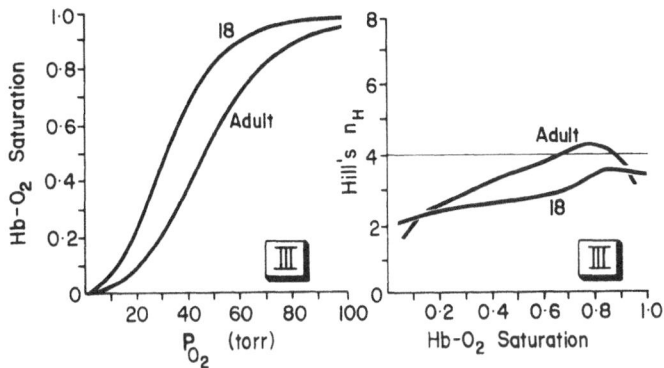

Fig. 10. O$_2$EC and Hill's n_H versus saturation data for chick embryo blood immediately prior to pipping at day 18 compared with data from full-grown adult chicken.

References

Aberman, A. (1977). Crossover P_{O_2}, a measure of the variable effect of increased P_{50} on mixed venous P_{O_2}. *Am. Rev. Respir. Dis.* 115: 173–175.

Bartlett, G.R. and Borgese, T.A. (1976). Phosphate compounds in red cells of the chicken and duck embryo and hatchling. *Comp. Biochem. Physiol.* 55A: 207–210.

Baumann, R., Padeken, S., Haller, E. and Brilmayer, T. (1983). Effects of hypoxia on oxygen affinity, hemoglobin pattern, and blood volume of early chicken embryos. *Am. J. Physiol.* 244: R733–R741.

Cirotto, C., Panera, F. and Geraci, G. (1977). Two different populations of primitive erythroid cells in the chick embryo. *Dev. Biol.* 61: 384–387.

Fitze-Gschwind, V (1973). Zur Entwicklung der Chorioallantoismembran des Huhnchens. *Ergeb. Anat. Entwicklungsgesch.* 47: 7–52.

Freeman, B.M. and Vince, M.A. (1974). *Development of the Avian Embryo.* Wiley, New York, p. 121.

Grima, M., Girard, H. and Dejours, P. (1983). Blood oxygen consumption and erythrocyte types during embryonic and post-natal growth in chicken. *Am. J. Physiol.* 244: C32–C36.

Grote, J. (1967). Die Sauerstoffdiffusionskonstanten in Lungengewebe und Wasser und ihre Temperatureabhangigkeit. *Pfluegers Arch.* 295: 245–254.

Ingermann, R., Stock, M., Metcalfe, J. and Shih, T. (1983). Effect of ambient oxygen on organic phosphate concentrations in erythrocytes of the chick embryo. *Respir. Physiol.* 51: 141–152.

Ingram, V.M. (1981). Hemoglobin switching in amphibians and birds. In: *Hemoglobins in Development and Differentiation.* G. Stamatoyannopoulos and A.W. Nienhuis, eds., Liss, Now York, pp. 147–160.

Isaacks, R., Harkness, D., Froeman, G., Goldman, P., Adler, J., Sussman, S. and S. Roth (1976). Relationship between the major phosphorylated metabolic intermediates and oxygen affinity of whole blood in chick embryos and chicks. *Comp. Biochem. Physiol.* 53A: 151–156.

Jelkmann, W. and Bauer, C. (1977). Oxygen affinity and phosphate compounds of red blood cells during intrauterine development in rabbits. *Pfluegers Arch.* 372: 149–156.

Kutchai, H. and Steen, J. (1971). Permeability of the shell and shell membranes of hen's eggs during development. *Respir. Physiol.* 11: 265–278.

Lapennas, G. and Reeves, R.B. (1983a). Oxygen affinity and equilibrium curve shape in blood of chicken embryos. *Respir. Physiol.* 52: 13–26.

Lapennas, G. and Reeves, R.B. (1983b). Oxygen affinity of blood of adult domestic chicken and red jungle fowl. *Respir. Physiol.* 52: 27–39.

Lomholt, J.P. (1976). The development of the oxygen permeability of the avian egg shell and its membranes during incubation. *J. Exp. Zool.* 198: 177–184.

Metcalfe, J., McCutcheon, I., Francisco, D., Metzenberg, A. and J. Welch (1981). Oxygen availability and growth of the chick embryo. *Respir. Physiol.* 46: 81–88.

Paganelli, C.V. (1980). The physics of gas exchange across the avian eggshell. *Am. Zool.* 20: 329–338.

Piiper, J., Tazawa, H., Ar, A. and Rahn, H. (1980). Analysis of chorioallantoic gas exchange in the chick embryo. *Respir. Physiol.* 39: 273–284.

Reeves, R.B. (1980). A rapid micro method for obtaining oxygen equilibrium curves on whole blood. *Respir. Physiol.* 42: 229–315.

Reeves, R.B., Park, J., Lapennas, G. and Olszowka, A. (1982). Oxygen affinity and Bohr coefficients of dog blood. *J. Appl. Physiol.* 53: 87–95.

Romanoff, A.L. (1960). *The Avian Embryo.* Macmillan, New York.

Rossoff, L., Zeldin, R., Hew, E. and Aberman, A. (1980). Changes in blood P_{50}: Effects on oxygen delivery when arterial hypoxemia is due to shunting. *Chest* 77: 142–146.

Tazawa, H. (1980). Oxygen and CO_2 exchange and acid-base regulation in the avian embryo. *Am. Zool.* 20: 395–404.

Tazawa, H., Ar, A., Rahn, H. and Piiper, J. (1980). Repetitive and simultaneous sampling from the air cell and blood vessels in the chick embryo. *Respir. Physiol.* 39: 265–272.

Vandecasserie, C., Paul, C., Schnek, A. and Leonis, J. (1973). Oxygen affinity of avian hemoglobins. *Comp. Biochem. Physiol.* 44A: 711–718.

17. Permeability of eggshells of native chickens in the Peruvian Andes

Fabiola Leon-Velarde[1], Jose Whittembury[1], Cynthia Carey[2] and Carlos Monge[3]

Abstract

Introduction of Castilian chickens to the Andes about 400 years ago and the presence of human settlements up to 4000 m which raise possible descendants of these chickens afforded an opportunity to study long-term acclimatization of the avian eggshells over a wider altitudinal range than previously possible. Eggs were collected at four locations between sea level and 3900 m in the Andes of Peru. Eggshell permeability (P) at intermediate altitudes was lower than the sea level value, but P of two of the three groups collected at the highest altitude increased sizeably. The curvilinear relation between P and barometric pressure is interpreted as adaptations for water conservation at intermediate altitudes and for improving O_2 availability to the embryo at the highest altitude.

Introduction

The conductance of avian eggshell to gases (G) has apparently been under selective pressure to meet mutually antagonistic requirements. The conductance must be large enough to permit diffusion of sufficient O_2 into the egg to support metabolic requirements, but it must also be restrictive enough to prevent excessive losses of CO_2 and H_2O vapor from the egg. The reduction of barometric pressure (P_B) at high altitude presents a special challenge for optimal eggshell design for two reasons. First, O_2 availability to embryos decreases as P_{O_2} drops

[1] Laboratorio de Biofisica, Universidad Peruana Cayetano Heredia, Lima, Peru.
[2] Department of EPO Biology and Institute of Arctic and Alpine Research, University of Colorado, Boulder, Colorado, 80309, U.S.A.
[3] Laboratorio de Biofisica, Universidad Peruana Cayetano Heredia, Lima, Peru.
[3] Department of Medicine, School of Medicine, University of Miami, Miami, Florida, U.S.A.

Seymour, R.S., (ed.) Respiration and metabolism of embryonic vertebrates.
© 1984, Dr W. Junk Publishers, Dordrecht/Boston/London. ISBN-13:978-94-009-6538-6

with P_B. Second, the increase in the diffusion coefficient of gases (D) at high altitude could enhance diffusion of CO_2 and H_2O vapor from the egg because of the inverse relation between D and P_B (Paganelli et al., 1975; Erasmus and Rahn, 1976; Rahn, 1984).

Several recent studies have addressed the questions concerning how eggshell conductances to water vapor (G_{H_2O}) of eggs laid at high altitudes vary in comparison with those laid by conspecific populations at sea level and if these differences are associated with the change in D with P_B (see Carey, 1980). These studies have shown that mean G_{H_2O} of eggs of wild birds (Packard et al., 1977; Rahn et al., 1977; Sotherland et al., 1980; Carey et al., 1983) and of domesticated fowl (Wangensteen et al., 1974; Rahn et al., 1977) breeding in the mountains are significantly lower than that of eggs of lowland populations, though one exception has been identified (Taigen et al., 1980). Since the observed decrease in G_{H_2O} generally parallelled the increase in D_{H_2O}, these adjustments have been interpreted as an adaptation to prevent excessive water loss which could result from the increase in D_{H_2O} at low P_B (Wangensteen et al., 1974; Rahn et al., 1977, Packard et al., 1977; Sotherland et al., 1980; Carey et al., 1983). The observation that daily water loss of naturally incubated eggs is independent of variation in D_{H_2O} over a 3050 m altitudinal gradient supports this interpretation (Carey et al., 1983). Therefore, it appears that conservation of water and/or CO_2 has proven more important for embryonic survival than maximization of O_2 availability, at least to certain altitudes.

Since wild birds breed successfully up to 6500 m (Rahn, 1977), it is important to ask whether the progressive decline in eggshell conductance to gases observed in previous studies continues to the highest altitudes or whether O_2 availability becomes limiting at some critical altitude and selects for an increase in conductance, which could be achieved by an enlargement of the functional pore area of the shell (A_p) and/or by a reduction in shell thickness (L_p) (Rahn and Ar, 1974). Conductance is derived from a combination of the terms $(D/RT) \cdot (A_p/L_p)$. These terms originate in a modified expression of Fick's first law of diffusion (Ar et al., 1974), which describes the factors that determine gas flux ($\dot{V}_g$) through an eggshell:

$$\dot{V}_g = (D_g/RT) \cdot (A_p/L_p) \cdot \Delta P_g \tag{1}$$

The descriptions of terms, units, and theory for this equation are presented by Wangensteen et al. (1970/71), Paganelli et al. (1975), and Rahn (1984).

It was our goal to investigate how eggshell construction has been selected at high altitude by the multiple and conflicting necessities of conserving H_2O vapor and CO_2 while maximizing O_2 availability. Since it was desirable to gather eggs over the widest possible altitudinal gradient, we studied eggs of chickens (*Gallus*) in the Peruvian Andes. Several reasons justified this choice. First, some information already exists on eggs of chickens acclimatized for about 15 years to 3800 m in the White Mountains of California and for approximately 40 years at 3500 m in the mountains of India (Wangensteen et al., 1974; Rahn et al., 1977). Further, the introduction of chickens to the Peruvian Andes by the Spaniards over 400 years ago (Murra, 1981), the ease of transportation in the Andes, and the presence of human settlements up to 4000 m afford an important opportunity to study acclimatization of avian eggs over a wider range of altitudes than previously attempted and to evaluate effects of long-term acclimatization on eggshell structure.

Material and methods

Since successful breeding of chickens at high altitudes is problematical (see Discussion), it is common for many Andean communities to obtain eggs for food by transporting hens bred from commercial stock at lower altitudes. It was essential for this study that we obtain eggs laid at high altitudes by chickens originating from stock acclimatized to altitude over many generations. Therefore, we obtained the eggs described here directly from the natives who raised their own flocks. These eggs, which we designated 'native', differ distinctly in size and color from eggs of commercial breeds obtained in Lima and other low altitude locations. We also avoided eggs of special breeds.

Eggs were collected at Pachacamac (altitude = 100 m, P_B = 751 torr), a town about 30 km from Lima on the coastal plain; Huanuco (altitude = 1900 m, P_B = 604 torr), a semi-tropical town on the eastern slopes of the Andes; the Urubamba Valley (altitude = 2800 m, P_B = 540 torr), a fertile farming region near Cuzco; and Puno (altitude = 3900 m, P_B = 469 torr), a town located by Lake Titicaca on the Peruvian altiplano.

Because of the importance of eggs collected at Puno for this study, three different groups of eggs were gathered: one at a street market (designated 'market eggs'), another group from a native woman who lived in a hut about 8 km from Puno ('Chucuito eggs'), and the last from a small farm close to Puno ('Puno eggs').

Eggs were transported by car or by air to Lima (P_B = 754 torr). Eggshell conductance to water vapor was measured by placing eggs over dry silica gel in a 26×21 cm desiccator located in a constant temperature cabinet at 25 °C. Eggs were weighed daily for 6 days to establish the average daily water loss for each egg. This value was divided by the saturation water vapor pressure at 25 °C to obtain G_{H_2O} according to the equation given by Ar et al. (1974). The surface area was determined by covering the egg with a polyvinylchloride-stained film. After the film dried, it was peeled off the egg, duplicated, and weighed according to methods described by Wangensteen et al. (1970/71). The accuracy of this technique, checked by using a sphere of known surface area, was approximately 94% (Paganelli et al., 1974). The eggs were then emptied, the shells were dried, and shell thickness was measured with an appropriate micrometer caliper.

The values for shell thickness and G_{H_2O} were used to calculate the effective pore area of the shell (A_p) and the permeability of the eggshell (P, in ml STP $\cdot$ sec^{-1} $\cdot$ cm^{-2} $\cdot$ torr^{-1}) using the appropriate equations given by Rahn et al. (1977). Krogh's permeation coefficient (K, in ml STP $\cdot$ sec^{-1} $\cdot$ cm^{-1} $\cdot$ torr^{-1}) was derived by multiplying P by shell thickness. This value provides a direct expression of Fick's first law of diffusion.

Average values for each measurement were analyzed by one-way analysis of variance in conjunction with Scheffe's procedure. This procedure grouped statistically indistinguishable means into homogenous subsets. Levels of significance were accepted at $P<0.05$. The relation of G_{H_2O}, P, and K to P_B was described by least-squares regression equation of progressively higher orders until no further increase in significance was obtained.

Results

Eggshell conductance to H_2O vapor varied significantly ($P<0.0001$) among the six groups (Table 1). The average G_{H_2O} of the sea level (Pachacamac) eggs and one of the Puno groups (Chucuito) formed one homogeneous subset; those two means differed significantly from those of eggs collected at intermediate altitudes and at the other two locations in Puno. The general trend illustrated by these data in reference to the relation of G_{H_2O} to altitude was that the average G_{H_2O} decreased from sea level to the intermediate altitudes, and then increased in some, but not all, of the groups collected at Puno.

Eggshell conductance to gases is a function of both pore area (A_p) and eggshell thickness (L_p) (Wangensteen et al., 1970/71; Ar et al., 1974). Both A_p and L_p varied significantly among groups ($P<0.001$ for both; Table 1). Average eggshell thickness of the sea level eggs was significantly greater than the means of the other groups; the thinnest eggshells comprise a homogeneous subset of values from Huanuco and two Puno groups (Puno and Chucuito). Average A_p of the sea level eggs was also significantly higher than all other means, which together formed one homogeneous subset.

Average fresh egg mass and surface area differed significantly ($P<0.0001$ for both) among groups (Table 1). No altitudinal relationships were evident for either of these values, since average masses and surface areas of the sea level, Urubamba Valley, and Puno eggs fell into one homogeneous subset. Due to the significant variation in these parameters, however, calculation of eggshell permeability, P, which corrects for changes in surface area (Rahn et al., 1977), is essential to evaluate variation in G_{H_2O} among groups. Permeability varied significantly among groups ($P<0.001$; Table 1), as did Krogh's permeation coefficient (K) ($P<0.001$), which corrects for variation in shell thickness. Again, it should be noted that no direct linear relation was obvious between K or P and P_B.

The lack of a direct linear relation between G_{H_2O}, P, or K with altitude or P_B

Table 1. Physical properties of native chicken eggs collected at various locations in the Peruvian Andes. Values are means ± standard errors. Results of analysis of variance are at bottom of columns.

Location	Egg		Shell				
	Mass g	Area cm²	Thickness mm	G_{H_2O} mg·day⁻¹·torr⁻¹	Pore Area mm²	P ml(STP)·cm⁻²·sec⁻¹·torr⁻¹·10⁻⁶	K ml(STP)·cm⁻¹·sec⁻¹·torr⁻¹·10⁻⁶
Pachacamac n = 18	57.96 ± 1.56	62.49 ± 1.13	0.35 ± 0.005	13.61 ± 0.58	2.14 ± 0.10	3.11 ± 0.11	11.0 ± 0.43
Huanuco n = 11	49.13 ± 1.17	58.11 ± 0.83	0.30 ± 0.011	10.13 ± 0.69	1.33 ± 0.10	2.51 ± 0.16	7.39 ± 0.53
Urubamba n = 12	52.05 ± 1.67	59.13 ± 1.32	0.31 ± 0.009	9.59 ± 0.74	1.34 ± 0.13	2.36 ± 0.15	7.34 ± 0.64
Puno-Puno n = 6	48.18 ± 1.86	54.23 ± 1.26	0.29 ± 0.006	8.71 ± 0.86	1.09 ± 0.11	2.39 ± 0.25	6.39 ± 0.65
Puno-Market n = 7	48.14 ± 1.77	56.42 ± 1.85	0.31 ± 0.016	10.26 ± 0.68	1.40 ± 0.07	2.62 ± 0.16	8.07 ± 0.44
Puno-Chucuito n = 7	55.90 ± 1.51	63.65 ± 1.17	0.25 ± 0.007	11.25 ± 0.52	1.27 ± 0.07	2.55 ± 0.12	6.44 ± 0.28
df	5,60	5,60	5,60	5,60	5,60	5,60	5,60
F-ratio	6.820	6.556	16.869	9.070	15.890	4.852	12.991
F-probability	0.0001	0.0001	0.0001	0.0001	0.0001	0.001	0.0001

suggested that a curvilinear regression line could best describe the relationship. This suggestion was supported by an analysis of r^2 values of linear and second-order regression lines calculated for these variables (Table 2). A clear improvement in fit and significance was obtained by use of a curvilinear regression line. The equation for the relation of P $(cm^3STP \cdot sec^{-1} \cdot cm^{-2} \cdot torr^{-1})$ to P_B (torr) is:

$$P \cdot 10^{-6} = 8.05 - (0.0205\,P_B) + (1.86 \cdot 10^{-5}\,P_B^2) \qquad (2)$$

The r^2 values calculated with previously published data on eggs of red-winged blackbirds (*Agelaius*) and robins (*Turdus*) breeding between sea level and 3050 and 3450 m, respectively, (Carey et al., 1983) showed similar improvement (Table 2). These r^2 values in Table 2 are higher than those published by Carey et al. (1983), since averages were used for the analysis in Table 2 rather than the data for individual eggs.

The relation of P to P_B is best visualized when the means for each group, presented as a fraction of the sea level value (Table 3) are plotted as a function of P_B (Fig. 1). This procedure illustrates the decrease in P at intermediate altitudes, followed by an increase at the highest altitude. Similar fractional values of P are plotted for previously published data on eggs laid between sea level and over 3000 m (Table 3). It is of interest that the single means for eggs of chickens laid at 3800 m in California and at 3500 m in India fall close to the curve generated from data for Andean chicken eggs, despite differences in acclimatization time.

Discussion

The pore area (A_p) and shell thickness (L_p) are features of the eggshell that govern gaseous exchange between the embryo and its environment (Wangensteen et al., 1970/71; Wangensteen and Rahn 1970/71). The functional performance of these features has been variously quantified as eggshell conductance $(mg \cdot day^{-1} \cdot torr^{-1})$ and permeability $(cm^3 \cdot sec^{-1} \cdot cm^{-2} \cdot torr^{-1})$. The use of each term has advantages and disadvantages; a full review of these expressions is provided by Piiper et al. (1971). Conductance is used most frequently and is possibly the most accurate since it depends on only one measurement, daily mass loss (Ar et al., 1974). Calculations of P and K are subject to errors in the measurement of surface area and shell thickness. The advantage of K is that it is a direct expression of Fick's first law of diffusion, but it results in larger statistical variability than does P. We have emphasized use of P in this study (Table 3, Fig. 1) because it shows the least amount of statistical variation and because it is probably the preferred method when dealing with intraspecific comparisons involving variation in egg mass.

Use of chicken eggs to describe adaptations of avian eggs to high altitude might appear inappropriate, judging from observations that chicken embryos are remarkably sensitive to hypoxia (Visschedijk et al., 1980), that embryos of commer-

Table 2. Correlation coefficients (r^2) for linear and parabolic regression lines. *Gallus* data from this study: *Turdus* and *Agelaius* from Carey et al. (1983). P, K, and G_{H_2O} are defined in the text.

	Gallus			*Turdus*			*Agelaius*		
	P	K	G_{H_2O}	P	K	G_{H_2O}	P	K	G_{H_2O}
Linear	0.608	0.748	0.567	0.726	0.687	0.787	0.804	0.686	0.740
Parabolic	0.891	0.864	0.772	0.963	0.988	0.983	0.850	0.784	0.832

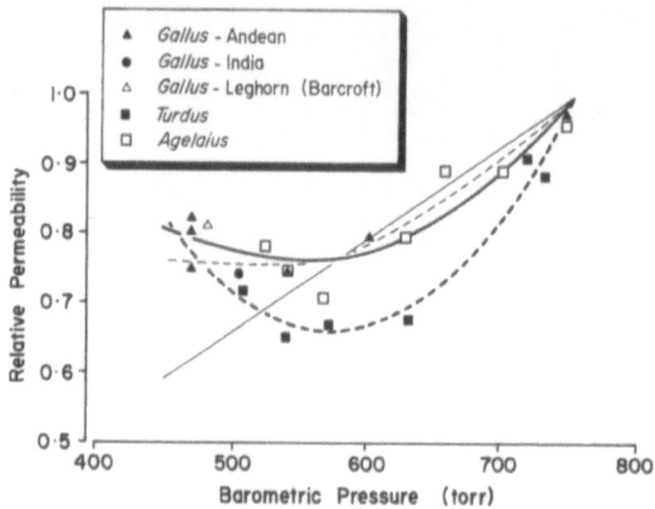

Fig. 1. The permeability of avian eggs to gases, presented as a fraction of the sea level average for each group, as a function of barometric pressure. The straight, continuous line represents the predicted values that would occur at any altitude if permeability were to be reduced in exact proportion to the reduction in barometric pressure. The solid curve represents the second-order regression equation (Eq. 2) calculated for the Andean chicken eggs. The interrupted lines are the regression equations:

$$P \cdot 10^{-6} = 7.426 - 0.02175 \, P_B + 1.88 \cdot 10^{-5} \, (P_B)^2$$
$$P \cdot 10^{-6} = 3.154 - 0.00717 \, P_B + 7.11 \cdot 10^{-6} \, (P_B)^2$$

for *Turdus* and *Agelaius*, respectively.

cial turkeys and chickens in the United States must be provided with supplemental O_2 above 1600 m for optimal hatchability and that commercial breeds rarely hatch at all above 3000 m (Moreng, 1983). We did consider study of the small, green eggs of possible descendants of a fowl domesticated by pre-Columbian natives which are periodically available in markets of a few montane villages. But, because the status and origin of this fowl, tentatively designated *Gallus inauris*, are unclear (Wilhelm, 1978) and because it is impossible to obtain eggs laid by this group at sea level and intermediate altitudes, we focussed this study on eggs of probable descendants of Castilian chickens which have apparently reproduced successfully in the Andes since their introduction by the Spanish about 1537 A.D. (Murra, 1981). Study of these chicken eggs is further justified by findings of past studies indicating that eggshell structure and hatchability of chicken eggs can be modified by long-term acclimatization (Wangensteen et al., 1974; Rahn et al., 1977).

Ar and Rahn (1980) have proposed that regulation of the rate of water loss by formation of the appropriate shell structure is essential for optimal hatchability. Although it appears that tolerances of avian embryos may be much greater than formerly appreciated (Carey, 1984), the reduction in G_{H_2O} observed in previous

Table 3. Physical characteristics of avian eggs presented as a fraction of the sea level value.

Location	P_B	Egg		Shell				Reference
		Mass	Area	G_{H_2O}	A_p	P	K	
Gallus								
Huanuco	0.80	0.85	0.93	0.74	0.62	0.79	0.66	This study
Urubamba	0.72	0.90	0.95	0.70	0.63	0.74	0.65	
Puno	0.62	0.88	0.93	0.74	0.58	0.79	0.62	
Ladakh	0.69	0.94	0.96	0.72	0.65	0.74	0.68	Rahn et al. 1977
Barcroft	0.63	0.78	0.85	0.68	0.68	0.81	0.81	Wangensteen et al. 1974
Turdus								
Kansas	0.98	0.93	0.98	0.99	0.90	0.90	0.86	Carey et al. 1983
Boulder, CO	0.84	1.05	1.01	0.77	0.77	0.67	0.70	
Gunnison, CO	0.75	1.07	1.02	0.76	0.76	0.66	0.69	
Gothic, CO	0.73	1.05	1.02	0.76	0.76	0.65	0.68	
Bellv., CO	0.69	1.07	0.98	0.78	0.78	0.71	0.74	
Agelaius								
Kansas	0.94	0.96	0.91	0.83	0.84	0.89	0.90	Carey et al. 1983
Bonny, CO	0.88	1.02	0.96	0.88	0.87	0.89	0.88	
Boulder, CO	0.84	1.07	1.02	0.82	0.84	0.79	0.81	
Gunnison, CO	0.75	1.10	1.03	0.74	0.74	0.70	0.70	
Gothic, CO	0.73	1.06	1.02	0.77	0.82	0.74	0.79	
Bellv., CO	0.70	1.02	0.98	0.78	0.81	0.78	0.82	

studies that roughly corresponds to the increase in D_{H_2O} at high altitudes supports the possibility that these adjustments maximize embryonic viability at intermediate altitudes (Wangensteen et al., 1974; Rahn et al., 1977; Carey et al., 1983). Three other existing studies that have not shown a good correspondence between G_{H_2O} and D_{H_2O} are influenced by problems concerning the effects of age on G_{H_2O} (see Carey, 1980). However, all studies, including this one, which have tested eggs laid above 3000 m, uniformly have shown an undercompensation of G_{H_2O} relative to the value predicted if G_{H_2O} were to be reduced in exact proportion to the change in P_B (Wangensteen et al., 1974; Rahn et al., 1977; Carey et al., 1983).

Our interpretation of the undercompensation of G_{H_2O} or P and eggshell permeability above 3000 m is that it serves to improve O_2 availability to the embryo, at the cost of increasing rates of diffusion of CO_2 and H_2O vapor. While O_2 consumption of embryos of wild birds appears to be unaffected by hypoxia to altitudes around 2900–3600 m (Carey et al., 1982), the increase in permeability may become selectively advantageous to meet metabolic requirements at a critical altitude above 3000 m.

These adjustments may contribute to embryonic survival because the increase in D_{O_2} at low P_B cannot compensate for the decrease in ambient P_{O_2}. If the increase in D_{O_2} at low P_B did succeed in increasing G_{O_2} sufficiently to overcome the decrease in ambient P_{O_2}, the air cell P_{O_2} (P_{AO_2}) would be P_{IO_2}. However, D_{O_2} cannot increase G_{O_2} sufficiently to prevent a fall in P_{AO_2} at low P_B. Therefore, dP_{AO_2}/dP_B has a positive slope. Here is an explanation of this phenomenon:

The P_{IO_2} is the ambient partial pressure of O_2, corrected for the water vapor tension inside the egg at incubation temperature (see Wangensteen and Rahn, 1970/71). Therefore

$$P_{IO_2} = 0.209 \, (P_B - P_{H_2O}) \tag{3}$$

If D_{O_2} is inversely proportional to P_B, the conductance equation given by Ar et al. (1974) can be expressed:

$$\dot{V}_{O_2} = G_{O_2} \cdot \frac{760}{P_B} \cdot (P_{IO_2} - P_{AO_2}) \tag{4}$$

where $\dot{V}_{O_2}$ is O_2 flux.
From Eqs. 3 and 4 we obtain:

$$\dot{V}_{O_2} = \frac{G_{O_2}}{P_B} \cdot 760 \cdot [0.209(P_B - P_{H_2O}) - P_{AO_2}] \tag{5}$$

and from Eq. 5:

$$P_{AO_2} = P_B(0.209 - \frac{\dot{V}_{O_2}}{G_{O_2} \cdot 760}) - 0.209 \, P_{H_2O} \tag{6}$$

If P_{AO_2} is to remain independent of P_B, then

$$dP_{AO_2}/dP_B = 0 \tag{7}$$

Derivation of Eq. 6 gives:

$$d\,P_{AO_2}/d\,P_B = 0.209 - \frac{\dot{V}_{O_2}}{G_{O_2} \cdot 760} = 0.209 - \frac{\Delta P_{O_2}}{760} \tag{8}$$

where $\Delta P_{O_2} = P_{IO_2} - P_{AO_2}$.
Equating Eqs. 7 and 8:

$$d\,P_{AO_2}/d\,P_B = 0 = 0.209 - \frac{\Delta P_{O_2}}{760} \tag{9}$$

Thus $\Delta P_{O_2} = 159$ torr which is an impossibly high physiological figure. Therefore, there exists a positive slope for the relationship between P_{AO_2} and P_B and the increase in D_{O_2} does not compensate for the fall in P_{IO_2}.

It is important to note that even if increased eggshell P does enhance O_2 delivery to the embryo above 3000 m, it clearly does not solve all the problems associated with O_2 diffusion. Chicken embryos of stock acclimatized for 15 years at 3800 m exhibited depressed metabolic rates at all developmental stages, low growth rates, low hatchling masses, prolonged incubation periods, and low levels of hatchability (Wangensteen et al., 1974). We talked with several Peruvian native women who raised native chickens at 3000 m in Tarma. They said that hatchability was less than 50% and egg-laying was sporadic. They believed that low air temperature was the cause of low hatchability; they had improved hatching somewhat by keeping brooding hens in the house near the stove or by hand-making a deep, well-insulated nest for the hens. Fidel C. Puchoc Ollero, a zootechnical engineer in Tarma, confirmed that reproduction of native chickens at 3000 m and above was so problematical that only a few natives had the patience to continue it; most purchased hens brought to Tarma from Lima. He made no attempt to breed these transported hens, since their eggs never hatched at that altitude. Transport of birds bred at sea level is apparently the only commercially feasible method for providing eggs for large numbers of people in montane towns.

Since the eggshell and dry outer membrane contribute only about 1/3 of the total resistance to O_2 diffusion into a chicken egg (Piiper et al., 1980), it is perhaps understandable that modification of P may not be totally successful in maximizing hatchability at high altitudes. In the absence of any hard data on the subject, we might hypothesize that eggs of wild birds breeding successfully to 6500 m differ from those of chickens not only in the resistance to O_2 diffusion posed by both the shell and the inner shell and chorioallantoic membranes, but also in other features, such as pattern of blood flow (Piiper et al., 1980), O_2 carrying capacity of the blood (Black and Snyder, 1980), and other unknown adaptations. We can conclude that adjustment of shell structure has occurred during long-term acclimatization of Andean chickens and has probably contributed to their ability to reproduce at high altitude. However, up to 400 years of acclimatization has not selected for other features of eggs that make reproduction of wild birds successful at very high altitudes.

Acknowledgements

We thank Dr Hermann Rahn for his stimulating ideas and advice. Preparation of the manuscript was supported in part by NSF DEB 79–23403 to C.C.

C.M. is grateful for the continuous support received from the University of Miami School of Medicine.

References

Ar, A., Paganelli, C.V., Reeves, R.B., Greene, D.G. and Rahn, H. (1974). The avian egg: water vapor conductance, shell thickness, and functional pore area. *Condor* 76: 153–158.

Black, C.P. and Snyder, G.K. (1980). Oxygen transport in the avian egg at high altitude. *Am. Zool.* 20: 461–468.

Carey, C. (1980). Adaptation of the avian egg to high altitude. *Am. Zool.* 20: 449–459.

Carey, C. (1984). Tolerance of avian embryos to modification of eggshell conductances to gases. MS submitted.

Carey, C., Garber, S.D., Thompson, E.L. and James, F.C. (1983). Avian reproduction over an altitudinal gradient – II. Physical characteristics and water loss of eggs. *Physiol. Zool.* 56: 340–352.

Carey, C., Thompson, E.L., Vleck, C.M. and James, F.C. (1982). Avian reproduction over an altitudinal gradient – Incubation period, hatchling mass, and embryonic oxygen consumption. *Auk* 99: 710–718.

Erasmus BD. and Rahn, H. (1976). Effects of ambient pressures, He and SF_6 on O_2 and CO_2 transport in the avian egg. *Respir. Physiol.* 27: 53–64.

Moreng, R.E. (1983). Incubation and growth of fowls and turkeys in high altitude environments. *World Poult. Sci. J.* 39: 47–51.

Murra, J.V. (1981). Las etnocategorias de un khipu estatal. In: La Tecnología en el Mundo Andino. Subsistencia y mensuración. Vol. I. H. Lechtman and A.M. Soldi, eds., Universidad Autónoma de Mexico, Mexico, pp. 433–442.

Packard, G.C., Sotherland, P.R. and Packard, M.J. (1977). Adaptive reduction in permeability of avian eggshells to water vapor at high altitudes. *Nature* 266: 255–256.

Paganelli, C.V., Ar, A., Rahn, H. and Wangensteen, O.D. (1975). Diffusion in the gas phase: the effects of ambient pressure and gas composition. *Respir. Physiol.* 25: 247–258.

Paganelli, C.V., Olszowka, A. and Ar, A. (1974). The avian egg: surface area, volume, and density, *Condor* 76: 319–325.

Piiper, J., Dejours, P., Haab, P. and Rahn, H. (1971). Concepts and basic quantities in gas exchange physiology. *Respir. Physiol.* 13: 292–304.

Piiper, J., Tazawa, H., Ar, A. and Rahn, H. (1980). Analysis of chorioallantoic gas exchange in the chick embryo. *Respir. Physiol.* 39: 273–285.

Rahn, H. (1977). Adaptation of the avian embryo to altitude: the role of gas diffusion through the egg shell. In: Respiratory Adaptations, Capillary Exchange, and Reflex Mechanisms. A.S. Paintal and P. Gill-Kumar, eds., Vallabhbhai Patel Chest Institute, University of Delhi, India, pp. 94–105.

Rahn, H., Carey, C., Balmas, K., Bhatia, B., and Paganelli, C.V. (1977). Reduction of pore area of the avian eggshell as an adaptation to altitude. *Proc. Natl., Acad. Sci., U.S.A.* 74: 3095–3098.

Rahn, H. (1984). Altitude adaptation: organisms without lungs. Proceedings of the Third Banff International Hypoxia Symposium. (in press).

Rahn, H. and Ar, A. (1974). The avian egg: incubation time and water loss. *Condor* 76: 147–152.

Rahn, H. and Ar, A. (1980). Gas exchange of the avian egg: time, structure, and function. *Am. Zool.* 20: 477–484.

Sotherland, P.R., Packard, G.C., Taigen, T.L. and Boardman, T.J. (1980). An altitudinal cline in conductance of cliff swallow (*Petrochelidon pyrrhonota*) eggs to water vapor. *Auk* 97: 177–185.

Taigen, T.L., Packard, G.C., Sotherland, P.R., Boardman, T.J. and Packard, M.J. (1980). Water-vapor conductance of black-billed magpie (*Pica pica*) eggs collected along an altitudinal gradient. *Physiol. Zool.* 53: 163–169.

Visschedijk, A.H.J., Ar, A., Rahn, H. and Piiper, J. (1980). The independent effects of atmospheric pressure and oxygen partial pressure on gas exchange of the chicken embryo. *Respir. Physiol.* 39: 33–44.

Wangensteen, O.D. and Rahn, H. (1970/71). Respiratory gas exchange by the avian embryo. *Respir. Physiol.* 11: 31–45.

Wangensteen, O.D., Rahn, H., Burton, R.R. and Smith, A.H. (1974). Respiratory gas exchange of high altitude adapted chick embryo. *Respir. Physiol.* 21: 61–70.

Wangensteen, O.D., Wilson, D. and Rahn, H. (1970/71). Diffusion of gases across the shell of the hen's egg. *Respir. Physiol.* 11: 16–30.

Wilhelm, O.E. (1978). The pre-columbian araucanian chicken (*Gallus inauris*) of the Mapuche Indians. In: Advances in Andean Archeology. D.L. Browman, ed., Mouton Publ., The Hague, pp. 189–196.

18. Eggshell conductances of avian eggs at different altitudes

Cynthia Carey[1], Donald F. Hoyt[2] and Theresa L. Bucher[3], Diane L. Larson[1]

Abstract

Previous studies have documented that the conductance to water vapor (G_{H_2O}) of eggs collected at high altitudes from wild and domesticated birds is decreased below sea level values in approximate proportion to the decrease in barometric pressure (P_B) at the montane locations. This study tested whether this reduction could result from physiological adjustment of shell structures by the female in response to detection of a change of P_B or a correlate thereof. The effect of a 28% reduction in P_B on G_{H_2O} shell thickness, shell permeability to gases, and egg mass was tested by comparisons of eggs laid by *Coturnix* quail and Bengalese finches at sea level and after transport to 2900 m. The mean G_{H_2O} of the eggs laid by the finches and 5 of 7 quail did not differ significantly from averages of eggs laid by the same birds at sea level. The G_{H_2O} of the two remaining quail either increased above pre-transport levels or fluctuated in a manner that was not correlated with the direction of change of P_B. Average egg mass of one finch and several quail increased significantly following transport, but since this effect was also noted in control quail remaining at sea level for the entire period, the change is probably associated with aging. We conclude that the reduction in G_{H_2O} observed in eggs laid by wild and domesticated birds at high altitudes most likely reflects long-term selection for females which lay genetically fixed shell structures appropriate for that habitat.

[1] *Department of EPO Biology and Institute of Arctic and Alpine Research, University of Colorado, Boulder, Colorado, 80309, U.S.A.*
[2] *Department of Biological Sciences, California State Polytechnic University, Pomona, California, U.S.A.*
[3] *Department of Biology, University of California, Los Angeles, California, 90024, U.S.A.*

Seymour, R.S., (ed.) Respiration and metabolism of embryonic vertebrates.
© *1984, Dr W. Junk Publishers. Dordrecht/Boston/London. ISBN-13: 978-94-009-6538-6*

Introduction

Gases diffuse between the avian embryo and the environment through pores in the eggshell (Wangensteen et al., 1970/71; Wangensteen and Rahn, 1970/71; Paganelli et al., 1975). The pore area of the eggshell (A_p) and the shell thickness (L_p), together forming the conductance of the shell to gases (G), contribute to the control of rates of gaseous flux (Ar et al., 1974; Paganelli et al., 1978). Since embryonic survival to hatching depends importantly on adequate gas exchange, natural selection has apparently favored particular combinations of sizes and numbers of pores and shell thicknesses appropriate for the egg mass, incubation period, and habitat conditions of each species (Rahn and Ar, 1974; Ar and Rahn, 1980).

When the average conductance to water vapor (G_{H_2O}) of species breeding in habitats that are stressful for optimal embryonic gas exchange is compared either to allometrically predicted values or to the average G_{H_2O} of conspecific populations breeding in less stressful habitats, sizeable differences have been observed (see Carey, 1983, for summary). These differences have been interpreted as adjustments in shell structure that maximize survival of the embryo. For example, the increase in G_{H_2O} above allometrically predicted levels in eggs laid in hypoxic, hypercarbic, and humid nests is thought to facilitate optimal gas exchange (Lomholt, 1976; Birchard and Kilgore, 1980; Seymour and Ackerman, 1980). Similarly, the significant reduction in G_{H_2O} of eggs laid at high altitudes may compensate for the increased tendency of gases to diffuse at low barometric pressure (P_B) and may prevent excessive water loss from those eggs (see Carey, 1980, for summary). It is unknown whether these differences in G_{H_2O} reflect long-term, evolutionary selection for females producing genetically fixed eggshells appropriate for these environmental conditions or whether they result from short-term physiological modification of shell structures in response to detection by the female of environmental gas concentrations or other related variables known to affect embryonic gas exchange.

Since the ability of birds to colonize new habitats clearly depends on successful reproduction, it is important to establish how adjustments in shell structure which could contribute to embryonic survival are achieved. This study addresses the question concerning whether short-term manipulation of P_B can cause a change in G_{H_2O} similar in direction and magnitude to those known for eggs laid by wild and domesticated birds between sea level and at least 3000 m (Wangensteen et al., 1974; Rahn et al., 1977; Carey et al., 1983; Leon-Velarde et al., 1984a). Barometric pressure is an ideal environmental variable with which to test the ability of females to vary eggshell G_{H_2O} because: 1) the theoretical and experimental effects of P_B on gaseous diffusion through eggshells have been well documented (Paganelli et al., 1975; Erasmus and Rahn, 1976), 2) the G_{H_2O} of eggs of wild and domesticated birds breeding in montane areas has been more fully studied than in any other problematical (humid, hypoxic or hypercarbic) gaseous habitat, and 3) a previous study suggests that domesticated fowl (*Gallus domesticus*) can modify G_{H_2O} in response to variation in P_B (Rahn et al., 1982).

Material and methods

Coturnix quail ('D-1 Pharoah Quail') and Bengalese finches (*Lonchura striata*) were obtained from local breeding stock in the Los Angeles, California (designated CA) area and were housed at approximately 50 m above sea level at California Polytechnic University in Pomona and the University of California at Los Angeles, respectively. Female quail were housed individually, but female finches required the presence of a courting male for egg-laying.

Eggs were collected in CA from each female quail and finch for periods of 4 and 8 weeks, respectively. The difference in time periods reflected the unpredictability of clutch production by the finches. Then, three pairs of finches and five quail were retained in CA as controls for the effect of aging on shell conductance, while nine quail and six pairs of finches were shipped by air express from Los Angeles to Denver, Colorado (designated CO) and then transported immediately to the Mountain Research Station (MRS) of the University of Colorado. The MRS is located at 2900 m near Nederland, CO. The duration of the entire transport process from Los Angeles to the MRS was about 6 h.

Eggs were collected from the control quail in CA and from the birds at the MRS for 6 weeks. The finches remaining as controls in CA stopped laying eggs following the departure of the experimental birds, presumably due to disruption of social interactions necessary for breeding. Only two of the six pairs of finches laid eggs during their stay at the MRS. Therefore, we report here only the results from eggs of these two pairs. Two of the nine quail ceased laying within two weeks of transfer to the MRS; results from their eggs laid both before and after the move to high altitude were discarded.

We attempted to match as closely as possible the environmental and nutritional conditions at the MRS with those the birds encountered in CA, even though no evidence exists that these factors affect shell synthesis or structure. Quail food, 'Turkey Grow Crumbles' purchased from Kruse Grain and Milling company, El Monte, CA, was shipped to CO to ensure that the quality and composition were identical in both locations. In both locations the finches received a commercial finch mix which was matched for the appropriate proportions and kinds of seeds. Vitamins, wheat germ oil, and health grit for finches and oyster shell grit for quail were provided ad libitum in CA and shipped to CO with the birds. Finches received romaine lettuce and baked chicken eggshells daily in both locations. Since the unique character of Los Angeles area tap water could not be duplicated at the MRS, deionized water was provided for both species at both locations. Photoperiod was 14 and 12 h for the quail and finches, respectively in CA. Both groups were held in 14-h photoperiods at the MRS. Ambient temperatures were

261

maintained at about 25 °C. Nest boxes, identical to those in CA, and dead grasses were provided for the finches at the MRS.

Quail eggs were collected on the date they were laid but finch eggs were incubated for at least 4 days following the completion of the clutch before they were collected. This delay avoided the complication that G_{H_2O} of eggs of certain small passerines changes in the early stages of incubation (Carey, 1983). Eggs were individually marked, wrapped in Saran wrap, and stored temporarily in a refrigerator. Eggshell G_{H_2O} of eggs laid in CA was measured at the respective universities where the birds were housed. Eggs laid in CO were measured at the University of Colorado in Boulder at 1600 m. Values obtained at the latter location were corrected to standard P_B (760 torr). To ensure that no errors were introduced into the results for G_{H_2O} by differences in procedures in CO and CA, the G_{H_2O} of 15 quail eggs was measured in CO and then the eggs were shipped to CA for remeasurement. The values obtained for the eggs in CO averaged 8% lower than the values for the same eggs in CA.

After G_{H_2O} of eggs was measured in CA, the eggs were shipped to CO where the fresh egg mass and shell thickness (quail eggs only) of all eggs laid in CA and CO were determined. Procedures for these measurements and equations for calculation of eggshell permeability, which corrects for variation in surface area, are detailed in Carey et al. (1983). Shells of finch eggs were too fragile for accurate determination of shell thickness.

Data from eggs were analyzed statistically for each female separately so that the eggs laid at sea level by a given female served as the controls for her eggs laid at high altitudes. One-way analysis of variance, in conjunction with Scheffe's procedure to identify homogenous subsets of means that are not statistically distinguishable, was used to compare averages of G_{H_2O}, fresh egg mass, permeability, and shell thickness (quail only). Means from 5 (3 sea level and 2 montane) and 6 (3 sea level and 3 montane) clutches were compared for the two finch females. Since *Coturnix* lay continuously under conditions of captivity and long photoperiod, data from eggs of each female were grouped into 5 'clutches' (2 sea level and 3 montane). Each 'clutch' represented the eggs laid by a female within a two week period.

If mean egg mass of a particular female varied significantly among clutches according to one-way analysis of variance, the variation among average G_{H_2O} was retested using analysis of variance with clutch as the major effect and egg mass as a covariate.

Results

If birds have indeed developed mechanisms for sensing P_B and for reducing G_{H_2O} in approximate proportion to the decrease in P_B at any montane location, we would predict that the G_{H_2O} of finches and quail housed at the MRS would have been reduced approximately 28%, corresponding to the 28% decrease in P_B at 2900 m. This predicted decrease is large enough to afford statistical detection of any differences between sea level and montane eggs. It also overcomes a possible 8% decrease in G_{H_2O} in eggs laid in CO which could be expected from differences between methods for measurement of G_{H_2O} between CA and CO (see Material and Methods). However, the results of this study indicate that the average G_{H_2O} of quail and finches did not decrease as predicted.

Coturnix quail

Mean G_{H_2O} of 5 clutches from 5 out of 7 quail transported to the MRS and from 4 of 5 quail maintained in CA did not vary significantly (Table 1). The mean G_{H_2O} of one control bird in CA (# 15) and 1 transported bird (# 1) increased significantly by the end of the 10-week extent period (Table 1, Fig. 1). Mean G_{H_2O} of the other quail (# 9) fluctuated significantly but exhibited no directional trend (Fig. 1).

Egg mass increased significantly in 5 of the 7 quail transported to altitude (Table 1, Fig. 2). It is doubtful that this increase resulted from the change in P_B, since the average egg mass of 1 control quail (# 12) in CA also increased significantly and that of 3 other control quail increased gradually, but not significantly, during the same time period (Fig. 2). It is most probable that the change in egg mass resulted from maturational or aging effects.

Variation in G_{H_2O} among clutches can be analyzed without the effects of egg mass in two ways: by calculation of eggshell permeability (P, in $cm^3 STP \cdot sec^{-1} \cdot cm^{-2} \cdot torr^{-1}$) which corrects for variation in surface area, and by subjecting mean G_{H_2O} of each bird to analysis of variance with egg mass as a covariate. Eggshell P did not differ significantly among clutches of 4 out of 5 control quail and 4 out of 7 transported quail (Table 1). The mean P of 1 control quail (# 15) and 1 experimental quail (# 1) significantly increased during the test period and that of # 9 fluctuated significantly with no directional trend. Average P of quail # 7 decreased about 15% between the first and fifth clutches. Since egg mass of that bird increased about 11% during the experiment and since G_{H_2O} did not vary significantly among clutches (Table 1), the bird appears to have adjusted the numbers of pores per square area to compensate for the increased egg size.

When variation in mean G_{H_2O} was analyzed with egg mass as a covariate, significant differences were found among clutches of 3 transported birds but no controls (Table 2). These results again reflect the significant increase in G_{H_2O} in eggs of quail # 1, the significant fluctuation in mean G_{H_2O} of quail # 9, and the change in the number of pores per square area in quail # 7.

Eggshell G_{H_2O} is a function of both the functional pore area and shell thickness (Ar et al., 1974). Since shell thickness did not vary significantly among clutches of any quail (Table 1), the few significant differences in G_{H_2O} found in this study must result from changes in the pore numbers or sizes.

Table 1. Results of one-way analysis of variance of physical properties of eggs laid by *Coturnix* quail. 'Control' birds were maintained for the entire experiment at sea level and 'Transported' birds were shipped to 2900m after 4 weeks at sea level.

Bird	Conductance			Egg mass			Shell thickness			Permeability		
	df	F-Ratio	F-Prob.	df	F-Ratio	F-Prob.	df	F-Ratio	F-Prob.	df	F-Ratio	F-Prob.
Control												
# 10	4,29	.967	.443	4,21	2.332	.097	4,21	.369	.827	4,21	.939	.465
# 11	4,48	1.116	.361	4,41	2.374	.070	4,40	.292	.881	4,42	1.457	.234
# 12	4,28	1.107	.376	4,20	16.146	.000	4,20	.150	.960	4,20	1.276	.320
# 15	4,26	3.069	.376	4,20	16.146	.000	4,20	.150	.960	4,20	1.276	.320
# 16	4,34	2.514	.062	4,27	2.284	.091	4,27	.286	.884	4,27	1.596	.209
Transported												
# 1	4,62	8.022	.000	4,51	9.909	.000	4,54	1.212	.317	4,54	7.295	.001
# 2	4,54	1.399	.248	4,52	4.420	.005	4,52	.307	.872	4,52	1.697	.166
# 3	4,51	.352	.842	4,45	1.517	.215	4,45	.970	.434	4,45	.349	.844
# 4	4,47	1.498	.220	4,46	1.327	.275	4,46	.876	.486	4,47	1.534	.209
# 6	4,52	2.170	.086	4,52	4.165	.006	4,52	1.388	.252	4,52	1.921	.122
# 7	4,54	1.125	.355	4,44	21.441	.000	4,53	.359	.836	4,52	2.801	.036
# 9	4,63	3.325	.016	4,57	27.129	.000	4,58	.136	.968	4,57	4.698	.003

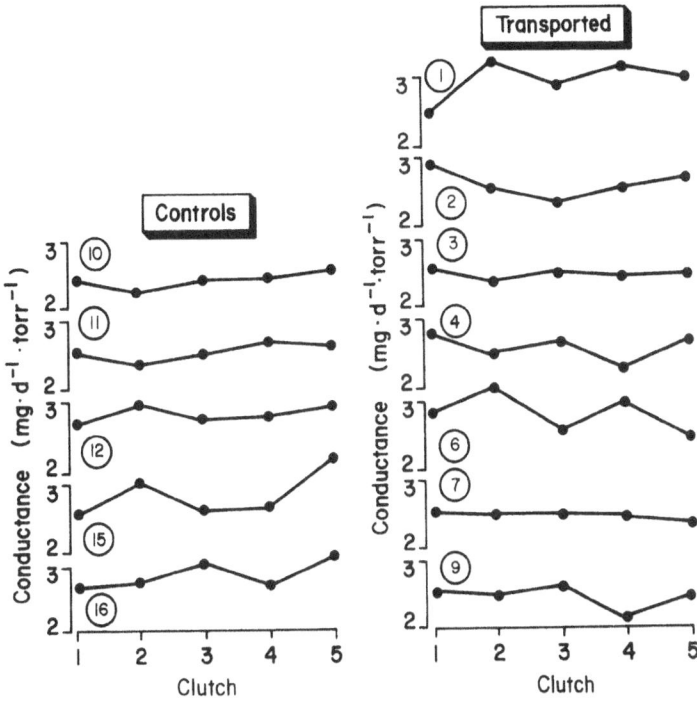

Fig. 1. Mean conductance (mg · day⁻¹ · torr⁻¹) for 5 clutches of *Coturnix* quail held for 10 weeks at sea level (Controls) or transported to 2900 m for 6 weeks (Transported). Transported birds laid the first two clutches at sea level and the last 3 clutches at high altitude. Control birds laid all 5 clutches at sea level.

Bengalese finches

Mean G_{H_2O} and P did not vary significantly among clutches of the two finch hens (Table 3). Egg mass of finch #2 varied significantly among clutches, but exhibited no directional trend. Analysis of variance for G_{H_2O} of finch #2 with mass as a covariate produced no significant results (Table 3).

Discussion

The diffusion coefficient for gases increases inversely with P_B (Paganelli et al., 1975; Erasmus and Rahn, 1976). Therefore, if all other factors affecting embryonic gas exchange are held equal, water vapor and CO_2 will diffuse out and O_2 will diffuse into an egg more rapidly at high altitude than at sea level. The decrease in G_{H_2O} in eggs laid by wild and domesticated birds at high altitude is thought to counteract the increased tendency of water vapor and CO_2 to diffuse from the egg (Wangensteen et al., 1974; Rahn et al., 1977; Carey et al., 1983;

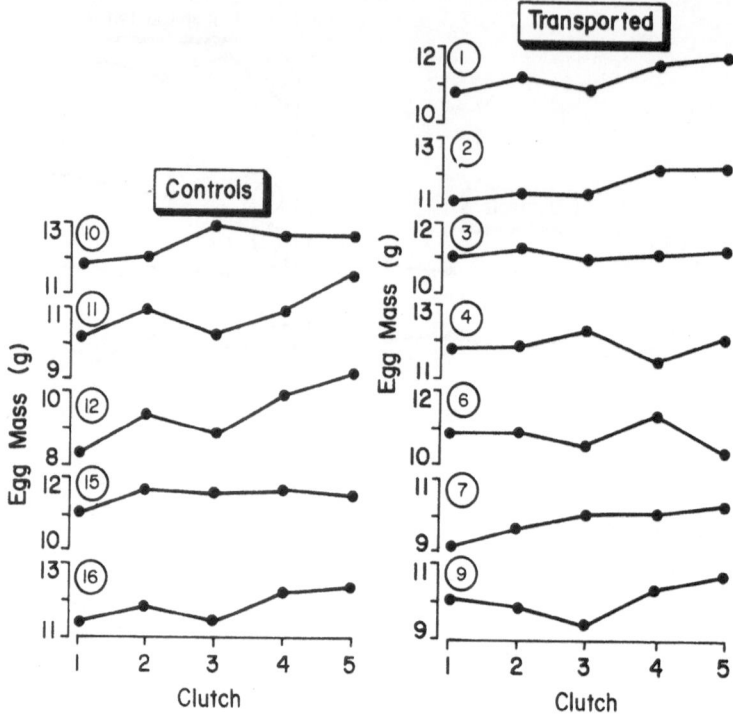

Fig. 2. Mean egg mass (g) for 5 clutches of *Coturnix* quail held for 10 weeks at sea level (Controls) or transported to 2900 m for 6 weeks (Transported). Transported birds laid the first two clutches at sea level and the last 3 clutches at high altitude. Control birds laid all 5 clutches at sea level.

Table 2. Results of analysis of variance of conductance (G_{H_2O}) of *Coturnix* quail eggs with egg mass as a covariate. This test was used only when egg mass differed significantly among clutches in Table 1.

Bird		F-Ratio	F-Prob.
Control			
# 12	Mass	6.865	.019
	Clutch	.735	.582
Transported			
# 1	Mass	17.663	.001
	Clutch	5.750	.001
# 2	Mass	.015	.902
	Clutch	1.380	.255
# 6	Mass	1.090	.302
	Clutch	1.853	.135
# 7	Mass	.050	.824
	Clutch	4.140	.007
# 9	Mass	1.567	.216
	Clutch	3.005	.026

Table 3. Results of one-way analysis of variance and analysis of variance (egg mass as covariate) of physical properties of eggs laid by two finches.

One-way analysis of variance

Bird	Conductance			Egg Mass			Permeability		
	df	F-Ratio	F-Prob.	df	F-Ratio	F-Prob.	df	F-Ratio	F-Prob.
# 2	4.24	.908	.478	4.23	4.039	.016	4.24	.941	.461
# 3	5.24	2.391	.076	5.15	2.813	.076	5.16	1.210	.366

Analysis of variance of G_{H_2O} with egg mass as a covariate

Bird # 2	df	F-Ratio	F-Prob.
Mass	1	.247	.625
Clutch	4	.918	.475

267

Leon-Velarde et al., 1984a). Since regulation of water loss by shell structures has been postulated to be essential for optimal hatchability (Ar and Rahn, 1980), it might be hypothesized that the observed reduction in G_{H_2O} of montane eggs could have contributed to the success of these species in colonizing montane habitats.

An ability to modify eggshell conductance in relation to the P_B on the breeding grounds would require both sensory ability to detect P_B and physiological capacities to modify the number or sizes of pores, since alteration of shell thickness does not appear to be a feature of adaptation of eggs to high altitude (Carey, 1980). A very acute ability to detect small changes in P_B evidently exists in some birds (Kreithen and Keeton, 1974). Since detection of changes in P_B associated with weather fronts may be crucial for species migrating over water or deserts (Kreithen and Keeton, 1974), it is perhaps not surprising that this sense has been developed by at least some birds. Unfortunately, practically nothing is currently known about how pores are formed and whether the number of pores can be changed markedly during the life-time of a hen (see Carey, 1983).

The two species used in this study, however, failed to demonstrate that they have the ability to modify G_{H_2O} in response to variation in P_B. One possible explanation for this result is that we were experimenting with species that have been domesticated for centuries (see Sossinka, 1982). Selection by humans for various characteristics associated with breeding may have abolished previously existing abilities to adjust shell features in response to environmental changes. However, Rahn et al. (1982) have shown that mean G_{H_2O} of a group of chickens acclimatized to 3800 m and transported to 1200 m did change in the appropriate direction. Since chickens have undoubtedly been under more intense selection for eggshell characteristics than have these quail and finches, the influence of domestication might be ruled out.

Another possible explanation for our findings is that the 6-week period at high altitude was too short for the birds to adjust eggshell characteristics. However, the alteration of G_{H_2O} by chickens occurred well within a 6-week period (Rahn et al., 1982). Further, breeding seasons at high altitudes are short relative to those at lower elevations and breeding usually takes place as soon as weather and snow conditions permit following the arrival of the birds from their lowland wintering grounds (Morton, 1976). It is doubtful that montane species could afford an adjustment period as long as 6 weeks prior to egg-laying since young must be fledged and prepared for migration to lower altitudes before cold temperatures and snow occur in late summer.

Our findings that eggshell G_{H_2O} of quail and finches do not change in response to manipulation of P_B contrast with the study of Rahn et al. (1982) on eggs of chickens. It is difficult to compare their results to our own because of differences in experimental design and statistical treatment of the data. The differences in findings might solely be due to the acclimatization of the chickens to higher altitudes (3800 m) than in this study. However, the altitudinal gradient over which the birds were transported was fairly similar (2600 m in Rahn et al., 1982 and 2800 m in this study). Further, our findings are supported by a recently completed study involving chicken eggs that also found no significant change in G_{H_2O} associated with variation in P_B (Leon-Velarde et al., 1984b). Chickens hatched at sea level were transported to 2800 m. The G_{H_2O} of eggs laid at that altitude by those birds after maturation did not differ significantly from sea level controls. These results of Leon-Velarde and co-workers and our own findings lead us to conclude that the observed reduction in G_{H_2O} of eggs laid by wild birds and domesticated fowl reproducing at high altitude for generations most likely results from long-term selection for females laying genetically fixed shell structures appropriate for that particular altitude rather than from short-term physiological changes. Since considerable variation in G_{H_2O} exists in wild populations of birds at any altitude (Carey et al., 1983), it is probable that movement and successful breeding of a population up or down an altitudinal gradient is not necessarily limited by shell structure.

Acknowledgements

This study was supported in part by NSF DEB79-23403. We thank Dr James Halfpenny for use of the facilities of the Mountain Research Station and Ms Bonnie Zetta for her conscientious care of the birds.

References

Ar, A., Paganelli, C.V., Reeves, R.B., Greene, D.G. and Rahn, H. (1974). The avian egg: water vapor conductance, shell thickness, and functional pore area. *Condor* 76: 153–158.

Ar, A. and Rahn, H. (1980). Water in the avian egg: overall budget of incubation. *Am. Zool.* 20: 373–384.

Birchard, G.F. and Kilgore, D.J. (1980). Conductance of water vapor in eggs of burrowing and nonburrowing birds: implications for embryonic gas exchange. *Physiol. Zool.* 53: 284–292.

Carey, C. (1980). Adaptation of the avian egg to high altitude. *Am. Zool.* 20: 449–459.

Carey, C. (1983). Structure and function of avian eggs. In: Current Ornithology. R.F. Johnston, ed., Plenum Press, New York, pp. 69–103.

Carey, C., Garber, S.D., Thompson, E.L. and James, F.C. (1983). Avian reproduction over an altitudinal gradient – II. Physical characteristics and water loss of eggs. *Physiol. Zool.* 56: 340–352.

Erasmus, B.D. and Rahn, H. (1976) Effects of ambient pressures, He and SF_6 on O_2 and CO_2 transport in the avian egg. *Respir. Physiol.* 27: 53–64.

Kreithen, M.L. and Keeton, W.T. (1974). Detection of changes in atmospheric pressure by the homing pigeon, *Columba livia. J. Comp. Physiol.* 89: 73–82.

Leon-Velarde, F., Whittembury, J., Carey, C. and Monge, C. (1984a). Permeability of eggshells of native chickens in the Peruvian Andes. In: Respiration and metabolism of embryonic vertebrates. R.S. Seymour, ed., Junk, The Hague.

Leon-Velarde, F., Whittembury, J., Carey, C. and Monge, C. (1984b). Shell characteristics of eggs laid at 2800 m by hens transported from sea level 24 hours after hatching. *J. exp. Zool.* 230: 137–139.

Lomholt, J.P. (1976). Relationship of weight loss to ambient humidity of bird eggs during incubation. *J. Comp. Physiol.* 105: 189–196.

Morton, M.L. (1976). Adaptive strategies of *Zonotrichia* breeding at high latitude or high altitude. In: Proceedings of the XVI International Ornithological Congress, Australian Academy of Science, Canberra, pp. 323–336.

Paganelli, C.V., Ackerman, R.A. and Rahn, H. (1978). The avian egg: in vivo conductances to oxygen, carbon dioxide, and water vapor in late development. In: Respiratory function in birds, adult and embryonic. J Piiper ed., Springer-Verlag, Berlin, pp. 212–218.

Paganelli, C.V., Ar, A., Rahn, H. and Wangensteen, O.D. (1975). Diffusion in the gas phase: the effects of ambient pressure and gas composition. *Respir. Physiol.* 25: 247–258.

Rahn, H. and Ar, A. (1974). The avian egg: incubation time and water loss. *Condor* 76: 147–152.

Rahn, H., Carey, C., Balmas, K., Bhatia, B. and Paganelli, C.V. (1977). Reduction of pore area of the avian eggshell as an adaptation to altitude. *Proc. Natl. Acad. Sci. U.S.A.* 74: 3095–3098.

Rahn, H., Ledoux, T., Paganelli, C.V. and Smith, A.H. (1982). Changes in eggshell conductance after transfer of hens from an altitude of 3,800 to 1,200 m. *J. Appl. Physiol.* 53: 1429–1431.

Seymour, R.S. and Ackerman, R.A. (1980). Adaptations to underground nesting in birds and reptiles. *Am. Zool.* 20: 437–447.

Sossinka, R. (1982). Domestication in birds. In: Avian biology, Vol. VI. D.S. Farner, J.S. King, and K.C. Parkes, eds., Academic Press, New York, pp. 373–403.

Wangensteen, O.D. and Rahn, H. (1970/71). Respiratory gas exchange by the avian embryo. *Respir. Physiol.* 11: 31–45.

Wangensteen, O.D., Rahn, H., Burton, R.R. and Smith, A.H. (1974). Respiratory gas exchange of high altitude adapted chick embryos. *Respir. Physiol.* 21: 61–70.

Wangensteen, O.D., Wilson, D. and Rahn, H. (1970/71). Diffusion of gases across the shell of the hen's egg. *Respir. Physiol.* 11: 16–30.

19. Factors controlling the rate of incubation water loss in bird eggs

Hermann Rahn

Department of Physiology, State University of New York at Buffalo, Buffalo, New York, 14214, U.S.A.

Abstract

The relative water content of freshly laid eggs is essentially the same as that of eggs at the end of incubation (Ar and Rahn, 1980). To achieve this state of hydration at the end of incubation, presumably needed for optimal hatching success, an egg must lose a total amount of H_2O equal to 18% of its initial mass, given an overall O_2 uptake of $102 \, ml \cdot g^{-1}$ (Hoyt and Rahn, 1980). Field observations show that on average 15% of initial egg mass is lost as water vapor by diffusion across the eggshell (Ar and Rahn, 1980), while additional water is lost by other routes after pipping.

Several factors determine the daily rate of water loss from incubating eggs. This rate is directly proportional to the total water vapor pressure difference between the egg and the ambient environment surrounding the nest, and inversely proportional to the total diffusive resistance to water vapor of the eggshell plus that of the nest. A loss of 15% of initial egg mass during incubation implies a daily loss equal to $0.15 \, W/I$, where W = initial egg mass, g, and I = incubation time, days. The general equation relating these facts is

$$\dot{M}_{H_2O} = (0.15 \, W/I) \times 10^3 = (P_e - P_a)/(R_e + R_n)$$

where $\dot{M}_{H_2O}$ = rate of water loss, $mg \cdot d^{-1}$; 10^3 = conversion of g to mg; P_e and P_a = water vapor pressure of egg and ambient environment, respectively, torr; R_e and R_n = diffusive resistance to water vapor of eggshell and nest, respectively, $torr \cdot d \cdot mg^{-1}$.

Introduction

During development the metabolism of an avian embryo is responsible for an increase in the relative water content of the whole egg. However, this is offset by a continuous loss of water vapor across the pores of the eggshell so that by the end of incubation the relative water content of the pipped egg, or hatchling, is essentially the same as that of the freshly laid egg. Water vapor is lost not only

Seymour, R.S., (ed.) Respiration and metabolism of embryonic vertebrates.
© *1984, Dr W. Junk Publishers, Dordrecht/Boston/London. ISBN-13: 978-94-009-6538-6*

from the egg to the microclimate of the nest but also from the nest to the final sink, the ambient atmosphere. The rate of this diffusive flux is proportional to the water vapor difference between the egg and the ambient atmosphere and inversely proportional to the total resistance offered by the egg plus the nest. Because the magnitude of the existing water vapor gradient depends in large measure upon the absolute humidity of the air outside the nest, the total resistance of the egg plus the nest must be adjusted to fit this imposed constraint, if a given rate of water loss is to be realized. Here I review briefly the rate of egg hydration at the end of incubation, the changes in relative egg water content attributable to metabolism, and the reported water loss of eggs during natural incubation. I then examine in detail the rate of water loss of eggs of 14 species incubated in various climates where the total water vapor difference varied from 21 to 44 torr.

State of hydration

In Table 1 are shown the states of hydration (expressed as percent water content) of eggs of 37 species just prior to hatching, or their hatchlings (Ar and Rahn, 1980). It can be presumed that these values represent the normal state of hydration for optimal hatching success. Not only does the relative water content vary between altricial and precocial eggs, but more important is the fact that the relative water content at the end of incubation is essentially the same as that of the fresh egg. One might thus generalize that an incubating egg must lose a given amount of water to achieve a state of hydration at the end of incubation which is the same as that of the fresh egg. How large is this amount of water?

Predicted incubation water loss

If no water were lost during incubation it can be shown that the normal metabolism of the embryo would increase the relative water content of the egg above initial levels not only because the relative solid content would be reduced, but also because metabolic water would be formed by oxidation of lipids. In the example described below the relative water content of a freshly laid precocial egg would, by the end of incubation, increase from 75% to 80% if water loss were prevented.

Loss of solids

A typical freshly laid precocial egg of 110 g has a shell mass of 10 g (Paganelli et al., 1974). This leaves 100 g of egg contents, of which 74.7 g are water and 25.3 g are solids (Carey et al., 1980). During development 102 ml $O_2 \cdot g^{-1}$ of initial egg mass

Table 1. Comparison of state of hydration (expressed as percent water content and standard deviation) of fresh egg contents, hatchlings or pipped eggs in 37 species arranged according to maturity at hatching (Ar and Rahn, 1980).

	n	Fresh egg	Hatchling
Altricial	12	82.4 (1.8)	83.8 (2.2)
Semi-altricial	8	82.0 (1.1)	83.5 (2.1)
Semi-precocial	7	77.1 (1.1)	78.1 (1.3)
Precocial	10	72.2 (2.9)	72.2 (2.6)

are consumed (Hoyt and Rahn, 1980), or 11.2 l O_2, which has a mass of 15.9 g. Because the overall respiratory quotient during development is close to 0.73, where the mass of O_2 consumed is equal to the mass of CO_2 produced, the total mass of CO_2 eliminated is also equal to 15.9 g. Assuming that all CO_2 is derived from oxidation of lipids and, because C represents 0.27 of the mass of CO_2, (0.27 × 15.9 g) or 4.3 g of C is removed from the solid fraction.

To calculate the total mass of lipids oxidized we must add the mass of H bound to carbon in the form of fatty acids lost by oxidation. Assuming long-chain fatty acids with approximately 2 H for each C, (2/12 × 4.3 g) or 0.72 g of H must be added to 4.3 g C, which yields 5.0 g of lipids as a reasonable estimate of the total solids oxidized. This is equal to 20% of the initial 25.3 g of solids. Therefore, a 20% loss of solids has to be balanced by a 20% loss of water, if the relative water content is to remain the same. This would require a loss of 0.2 × 74.7 g or 14.9 g of water, which is equal to 13.5% of the initial egg mass of 110 g.

Metabolic water

The oxidation of 5.0 g of lipids will produce (5.0 × 1.07) or 5.35 g of metabolic water (Schmidt-Nielsen, 1975). This amount must also be eliminated if the relative water content of an egg is to remain the same. Thus the total amount of water that must be eliminated during the course of incubation is equal to (14.9 + 5.4) or 20.3 g, which equals (20.3/110) or 18.5% of the initial egg mass of 110 g.

Observed egg water loss

Before discussing the observed rate of water loss of incubating eggs it is important to comment upon its measurement. The fact that the normally-occurring mass loss is equivalent to water loss only was well described by Drent (1970). Not only did he cite earlier experimental observations of others which supported this view, but he pointed out that the overall respiratory quotient during development was close to 0.73 where the mass of O_2 taken up equals the mass of CO_2 eliminated. Thus, changes in egg mass reflected only the loss of water vapor. More recent experiments with artificially and naturally incubated eggs have verified this concept (Ar and Rahn, 1980; Grant et al., 1982c). Here measurements were made by weighing freshly laid eggs and then at various advanced stages of incubation reweighing the eggs after their air cells had been replaced by water injections. Eggs treated in such manner regained exactly their original mass.

According to the previous calculation a typical egg would have to lose about 18% of its original mass if at the end of incubation it were to attain a relative water content equal to that of its initial concentration. What are the mass changes that have been observed in the field? A recent survey of the literature showed that if the reported daily mass loss is multiplied by the incubation time, the total mass or water loss averages 15% (s.d. = 2.5) of the intial egg mass in 83 species (Ar and Rahn, 1980). As previously emphasized, this is the amount lost by diffusion of water vapor across the pores of the shell and does not represent the additional amount which is lost once the eggshell is pipped. Few measurements of this additional loss are available, but field measurements in four species of Procellariiformes (Pettit and Whittow, 1983) and in the Black-necked Stilt (Grant, 1982) give a range from 1.4 to 6% of the initial egg mass with an average value of 3.5%. When this post-pipping water loss is added to that lost by diffusion across the intact part of the shell, the best estimate is that the total water loss for the average egg is about 18% of the initial egg mass. This agrees well with the prediction based upon the average oxygen uptake of eggs and the assumption that the metabolic energy is mainly derived from oxidation of lipids. Lastly, field observations cited above (Ar and Rahn, 1980) allow one to predict the daily rate of water loss (limited to diffusion of water vapor through the pores of the shell) as:

$$\dot{M}_{H_2O} = 0.15 \, W/I$$

where $\dot{M}_{H_2O}$ = rate of water loss, $g \cdot d^{-1}$; W = initial egg mass, g; and I = incubation time, d.

Factors determining rate of water loss

During incubation water vapor is lost from egg to nest and from nest to ambient environment. In his analysis of factors contributing to water loss from the nest Walsberg (1980) provides estimates of additional water vapor released from the skin or brood patch. The magnitude of such rates is difficult to assess and has not been considered. Thus in this model we assume that in the steady state the two rates are the same and can be defined as follows:

$$\dot{M}_{H_2O} = G_c(P_c - P_n) \tag{2}$$

$$\text{and } \dot{M}_{H_2O} = G_n(P_n - P_a) \tag{3}$$

where $\dot{M}_{H_2O}$ = rate of water loss from egg or nest, $mg \cdot d^{-1}$; G_e = conductance of eggshell to water vapor, $mg \cdot d^{-1} \cdot torr^{-1}$; G_n = conductance of nest to water vapor, $mg \cdot d^{-1} \cdot torr^{-1}$; P_e = water vapor pressure of egg, torr; P_n = water vapor pressure of the nest, torr; P_a = water vapor pressure of the ambient environment, torr.

Thus water vapor flux from the egg to the final sink of the ambient atmosphere must overcome two resistance in series, that of the eggshell and that of the nest. Because $1/R = G$, where R = resistance to water vapor flux, Eqs. 1 and 2 can be arranged as follows:

$$\dot{M}_{H_2O} \cdot R_e = (P_e - P_n) \tag{4}$$

$$\text{and } \dot{M}_{H_2O} \cdot R_n = (P_n - P_a) \tag{5}$$

Rearranging Eqs. 4 and 5 yields:

$$\dot{M}_{H_2O} = \frac{(P_e - P_a)}{(R_e + R_n)} = \frac{(P_e - P_a)}{(1/G_e) + (1/G_n)} \tag{6}$$

A brief description of the methods employed for measuring the various factors in equations 2 to 6 follows.

Egg water vapor pressure

P_e is the saturation vapor pressure at a given egg temperature. The egg temperatures shown in Table 2 varied from 38.6 °C in the Ring-necked Pheasant to about 33 °C in Leach's Storm Petrel, corresponding to saturation vapor pressures of 51.3 and 38.6 torr, respectively.

Nest water vapor pressure

P_n is the absolute humidity of the nest microclimate surrounding an egg. It was measured by placing calibrated egg hygrometers in the nest and calculating the mean vapor pressure from the measured increase in mass (Rahn et al., 1977).

Ambient water vapor pressure

P_a is calculated from ambient air temperature and relative humidity. In many other cases it was determined using the egg hygrometers described above, by placing them near the nest. In the warm and humid Marshall Islands, with an air temperature of 28 °C and a R.H. of 78%, the calculated $P_a = 22$ torr, whereas in

Table 2. Measurements of egg mass, incubation time, egg water loss, egg and ambient temperature, and absolute humidities (P) of egg, nest, and environment of 14 species incubating in various climates.

Nest type	Location	Species	W Egg mass g	I Incuba-tion days	$\dot{M}_{H_2O}$ Water loss mg·d^{-1}	G_{H_2O} Conduct. mg·d^{-1}·torr^{-1}	T_c Egg temp. °C	P_c Egg torr	P_n Nest torr	P_a Ambient torr	T_a Ambient °C	Ref.
Burrow	Bay of Fundy, Atlantic	Leach's Storm Petrel *Oceanodroma leucorhoa*	10.5	42	47	1.55	33.4	38.6	7.9	7.9*	8.3*	1
	Midway Isl., Pacific	Bonin Petrel *Oceanodroma hypoleuca*	39.5	48	110	5.2	33.8	39.4	18.1	14.9*	20.8*	2
	Hawaii, Pacific	Wedge-tailed Shearwater *Puffinus pacificus*	60.0	54	155	6.1	35.0	42.2	19.6	17.3*	29.3*	3
Ground	Midway Isl., Pacific	Laysan Albatross *Diomedea immutabilis*	285.0	65	674	32.0	36.0	44.6	17.3	13.7	22.0	4
	Midway Isl., Pacific	Black-footed Albatross *Diomedea nigripes*	305.0	66	707	32.5	35.6	43.6	19.1	13.7	22.0	4
	Gulf of California	Heermann's Gull *Larus heermanni*	53.4	23	278	11.0	36.8	46.6	18.4	14.4	23.0	5
	Buffalo, New York	Ring necked Pheasant *Phasianus colcichus*	33.0	28	190	6.6	38.6	51.3	20.0	13.7	22.0	6
	Salton Sea, California	Black-necked Stilt *Himantopus mexicanus*	21.0	25	153	5.0	37.9	49.3	22.9	16.0	26.0	7

Nest	Location	Common name	Scientific name									
	Spitsbergen, Norway	Eider	*Somateria mollissima*	108.0	26	590	21.6	37.3	47.8	19.2	4.0	2.8	8
	Spitsbergen, Norway	Barnacle Goose	*Branta leucopsis*	115.0	24.5	572	22.3	35.9	44.2	17.3	4.0	2.8	8
Cliff	Spitsbergen, Norway	Black-legged Kittiwake	*Rissa tridactyla*	49.5	27	326	9.1	37.4	48.1	14.6	6.0	2.8	8
Tree	Beaufort. N. Carolina	Great Egret	*Casmerodius albus*	48.6	27	231	7.6	35.6	44.0	15.0	12.0	16.7	9
	Beaufort. N. Carolina	White Ibis	*Eudocimus albus*	50.8	22	290	7.8	37.1	47.0	14.0	11.0	20.5	9
Bare Branch	Marshall Isl., Pacific	White Tern	*Gygis alba*	21.4	36	74	3.5	35.4	43.1	22.0	22.0	28.3	10
	Midway Isl., Pacific	White Tern	*Gygis alba*	23.3	36	79	2.5	35.3	42.9	14.6	14.0	20.0	11

* Vapor pressure and temperature in burrow air

References: (1) Rahn and Huntington, unpublished, (2) Grant et al., 1982a, (3) Whittow et al., 1982, (4) Grant et al., 1982b, (5) Rahn and Dawson, 1979, (6) Rahn et al., 1977, (7) Grant, 1982, (8) Rahn et al., 1983, (9) Vleck et al., 1983, (10) Rahn et al., 1976, (11) Pettit et al., 1981.

Spitsbergen, with a mean air temperature of 2.8°C and a R.H. of 71%, the calculated $P_a = 4$ torr (Table 2). Thus the maximum difference in $(P_c - P_a)$ to which the egg and nest in our series were exposed ranged from 21 to 44 torr.

Eggshell resistance

R_e is the inverse of eggshell conductance, G, which is conventionally measured by dividing the daily mass loss of an egg by the water vapor pressure difference across the pores when the egg is kept at a constant temperature in a desiccator (Ar et al., 1974). This resistance is determined by the number of pores, their mean cross-sectional area, and their length (shell thickness) (Paganelli, 1980) and has been found to be constant during incubation in most non-passerine birds.

Nest resistance

As defined in Eq. 5, R_n is the water vapor difference across the nest $(P_n - P_a)$ divided by the water loss rate $\dot{M}_{H_2O}$. The assumption that this transfer is only by diffusive transport is only for convenience. Some of the vapor will diffuse through the feathers and the nest material, while some will also be transported by convection whenever the incubating parent moves off the egg, stands up, turns the egg, or leaves the nest. Thus R_n depends not only upon the type of material used in nest construction and the way incubated eggs are covered by the feathers, but also by the incubation behavior of the parent, such as incubation bouts, stand-up periods, egg-turning, and overall nest attendance.

Discussion

Equations 1 and 6 can be combined to evaluate the various factors which influence the daily rate of water loss.

$$\dot{M}_{H_2O} = \frac{0.15 \, W}{I} \times 10^3 = \frac{P_c - P_a}{R_e + R_n} \tag{7}$$

The first equation predicts the average daily rate of water loss based on the initial egg mass and incubation time of 83 species (Ar and Rahn, 1980). When multiplied by 10^3 it expresses the rate in $mg \cdot d^{-1}$ instead of $g \cdot d^{-1}$. The second equation defines the conditions which will meet this requirement. For a given rate of water loss the total resistance $(R_e + R_n)$ must conform to the magnitude of $(P_c - P_a)$, which as noted below can vary from ca. 21 to 44 torr, depending in large part on the ambient temperature and humidity and to a lesser degree on the egg temperature. For example, for a given rate of water loss the total rsistance $(R_e + R_n)$, or

278

$(1/G_e + 1/G_n)$, must double if $(P_c - P_a)$ increases from 20 to 40 torr.

In Table 2 are shown measurements which either represent or allow one to calculate all the variables of Eq. 7 in 14 species and in addition the values for the mean ambient temperatures, type of nest site, and geographic location of nesting area. Among these species the egg mass varies from 10 to 305 g, incubation time from 22 to 65 d, daily rate of water loss from 47 to 707 mg, egg temperatures from 33 to 38.6 °C, ambient temperatures from 4 to 29 °C, and $(P_e - P_a)$ from 21 to 44 torr. These heterogeneous conditions provide a data base for testing the general validity of Eq. 7.

Fractional water loss

In Table 3, column 1, are shown F values which represent the fraction of initial egg mass lost during incubation. These were calculated as the product of the observed daily water loss rate and incubation time divided by initial egg mass. F varies from 0.12 to 0.19, with a mean value of 0.14 (s.d. = 0.02), which is similar to the mean value of 0.15 (s.d. = 0.025) reported previously (Ar and Rahn, 1980).

Application of G_e measurements to field conditions

In the method of Ar et al. (1974) for measurement of eggshell conductance, the whole shell surface is exposed to the desiccator atmosphere. When an egg is incubated, part of its surface may be covered by the brood patch, thus reducing the *effective* conductance (or increasing the *effective* resistance). This consideration brings into question the quantitative application of shell conductance values obtained under laboratory conditions. How can this question be resolved?

Rearrangement of Eq. 2 allows one to calculate *effective* G_e during natural incubation from the following relation:

$$G_e = \dot{M}_{H_2O}/(P_e - P_n) \tag{8}$$

where $\dot{M}_{H_2O}$, P_e and P_n were taken from Table 2. This value divided by the *measured* G_e (Table 2) for each species given a ratio as shown in Table 3, column 5. The overall mean ratio is statistically not different from 1.0 and has a coefficient of variation of 12%. The agreement between calculated *effective* and *measured* G_e is all the more remarkable because the *effective* G_e was calculated from three independent field measurements and suggests that the directly measured value can be applied under field conditions with some degree of confidence.

Table 3. Fractional water loss during incubation, differences between measured and calculated nest humidity and egg temperature, resistance of the nest as % of the total resistance to the water vapor transport, and ratio of calculated to measured shell conductance.

	(1) Fractional water loss	(2) Nest humid. meas.-calc.	(3) Egg temp. meas.-calc.	(4) Nest resist. Tot. Resist.	(5) Calc. G_{H_2O} / Meas. G_{H_2O}
	F	torr	°C	%	
Leach's Storm Petrel	0.19	0.0	0.2	0	0.99
Bonin petrel	0.13	− 0.1	0.1	13	1.16
Wedge-tailed Shearwater	0.14	2.8	− 1.2	9	1.13
Laysan Albatross	0.15	− 6.2	2.7	12	0.77
Black-footed Albatross	0.15	− 2.8	1.1	18	0.89
Heermann's Gull	0.12	− 2.9	1.1	12	0.90
Ring-necked Pheasant	0.16	− 2.5	0.9	17	0.92
Black-necked Stilt	0.18	4.2	− 1.4	21	1.16
Eider	0.14	− 1.3	0.5	35	0.95
Barnacle Goose	0.12	− 1.3	0.6	33	0.96
Black-legged Kittiwake	0.18	2.3	− 1.0	20	1.07
Great Egret	0.13	1.4	− 0.7	9	1.05
White Ibis	0.13	4.1	− 1.4	8	1.12
White Tern	0.12	0.0	0.0	0	1.00
White Tern	0.12	3.3	− 1.4	0	1.12
Mean	0.14	0.07	0.01	–	1.01
s.d.	0.2	3.1	1.2		0.12

(1) Fractional water loss = $(\dot{M}_{H_2O} \cdot I)/W$
(2) Calculated nest humidity. $P_n = P_e - (\dot{M}_{H_2O}/G_{H_2O})$
(3) Calculated egg temperature derived from saturation vapor pressure of egg, $P_e = P_n + (\dot{M}_{H_2O}/G_{H_2O})$
(4) Relative nest resistance = $(P_n - P_a)/\dot{M}_{H_2O}$ divided by $(P_e - P_a)/\dot{M}_{H_2O}$
(5) Calculated G_{H_2O} from equation (2) where $G_{H_2O} = \dot{M}_{H_2O}/(P_e - P_n)$

Calculated vs measured nest humidity

The question whether the brood patch in contact with the egg surface reduces the *effective* conductance of the eggshell also applies to egg hygrometers. However, one might argue that because the *effective* conductance of intact eggs, as shown above, is reasonably close to the *measured* conductance, the same principle applies to egg hygrometers. One can compare the value of nest humidity, P_n, derived by egg hygrometry (Table 2) with the value of P_n calculated by subtracting from P_e the $(P_e - P_n)$ difference obtained from the ratio $(\dot{M}_{H_2O}/G_e)$ according to Eq. 2. The difference between directly measured and calculated values is shown for each species in Table 3, column 2. The average difference is essentially zero.

Egg temperature prediction

As pointed out previously (Rahn et al., 1977), one can theoretically predict the average egg temperature on the basis of three independent gravimetric measurements, namely, the rate of water loss, $\dot{M}_{H_2O}$, the eggshell conductance for water vapor, G_e, and the nest humidity, P_n. Because $\dot{M}_{H_2O}/G_e = (P_e - P_n)$, Eq. 2, the difference can be added to P_n, yielding P_e, the saturation water vapor pressure of the egg from which the egg temperature is then derived. Shown in Table 3, column 3, are the differences between the measured and derived egg temperatures, which range from $+2.7$ to $-1.4\,°C$, with an average difference of $0.07\,°C$ (s.d. $= 1.2$).

Nest resistance vs egg resistance

Walsberg (1980) raised the question whether a significant difference exists between the absolute humidity of the microclimate of the nest and that of the environment and whether one can speak of a nest resistance to water vapor. As shown in Table 2, the difference $(P_n - P_a)$ is very small or even absent in some species, but can be as high as 13 or 15 torr in the Eider and the Barnacle Goose. The careful measurements by Chattock (1925) throughout the incubation of the hen (reproduced in Rahn et al., 1977) showed an average $(P_n - P_a)$ difference of ca. 9 torr. More recently Grant et al. (1982b) followed the P_n and P_a values in the Laysan and Black-footed Albatross over a 56-day period and observed a consistent difference of 4–5 torr between P_n and P_a. The data in Table 2 suggest that nest resistance to water vapor may be absent in some species such as the White Tern, which nests on bare branches. However, in other species such as the Eider, the Barnacle Goose, and the Kittiwake nesting in the extremely dry region of Spitsbergen, the rather well-constructed nests offer considerable resistance to the flux of water vapor. Table 3, column 4, gives calculated nest resistance, $(P_n - P_a)/\dot{M}_{H_2O}$, as a percentage of total resistance, $(P_e - P_a)/\dot{M}_{H_2O}$. For three species this is zero. For stick nesters, such as the Great Egret and the White Ibis, the relative resistance offered by the nest is obviously small. The same applies to the shearwater nesting in a burrow. All remaining species are ground nesters were the nest structure (and incubating behavior) provide a significant fraction of the overall resistance to water loss. Furthermore, if such a nest resistance did not actually exist, then according to Eq. 7 the eggshell resistance, R_e, would have to increase by the amount indicated in column (4) for the nest resistance. For example, in the case of the Eider, R_e would have to increase by 35%. On the other hand, in the case where nest resistance is zero, the entire resistance to water vapor loss resides in the eggshell.

An example of two types of nesting conditions is illustrated in Fig. 1, which shows the water vapor pressure and temperature gradients between the egg and

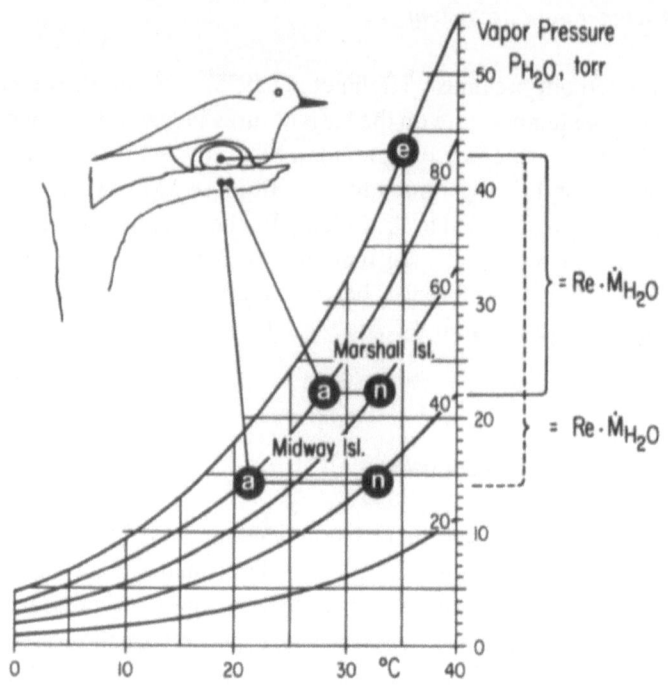

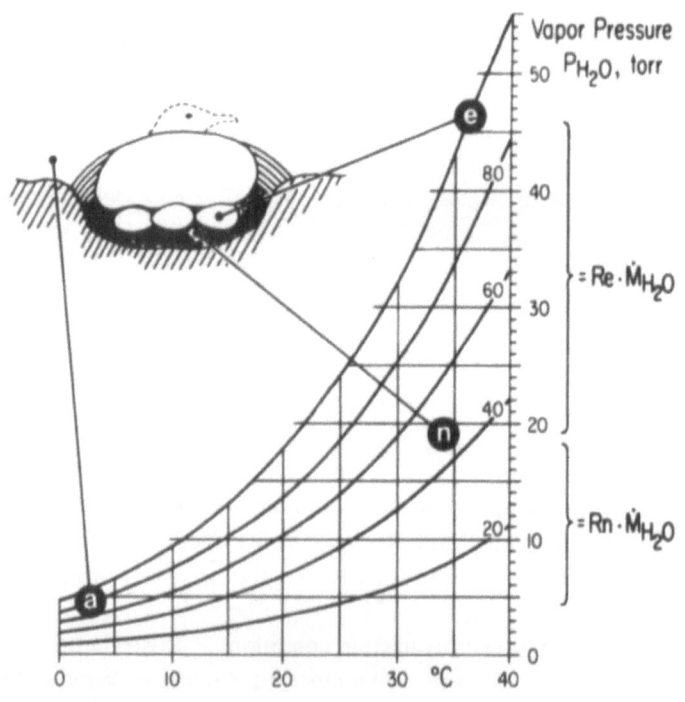

the ambient environment. The Eider nests in the unusually cold and dry environment of Spitsbergen where $(P_e - P_a)$ is 44 torr, whereas the White Tern of the Marshall Islands nests in a warm and humid environment where $(P_e - P_a)$ is 21 torr. In the latter case the total resistance is offered by the eggshell since $P_n = P_a$, while in the Eider the total resistance is divided between eggshell and nest in a ratio of 65 to 35 respectively.

Table 4. Water vapor pressure in the egg, P_e, in the nest, P_n, and in the ambient atmosphere, P_a, for species in which P_n was not measured, but calculated. From these values the nest resistance $(P_n - P_a/\dot{M}_{H_2O})$ was calculated as % of the total water vapor resistance $(P_e - P_a/\dot{M}_{H_2O})$, as shown in Fig. 2.

	P_e Egg	P_n Nest	P_a Ambient	$P_n - P_a$	$P_e - P_a$	$\dfrac{P_n - P_a}{P_e - P_a}$	Refer.
	torr	torr	torr	torr	torr	%	–
Sooty Tern (Hawaii)							
Sterna fuscata	49.7	21.8	14	7.8	35.7	22	(1)
White-capped Noddy (Marshall Isl.)							
Anous tenuirostris	47.6	24	23	1.0	24.6	0	(1)
Cattle Egret (Beaufort, NC)							
Bubulcus ibis	46.6	16.4	16	0.4	30.6	0	(2)
Royal Tern (Beaufort, NC)							
Sterna maxima	49.2	25.1	15	10.1	34.2	30	(2)
American Avocet (Salton Sea, CA)							
Recurvirostra americana	45.3	24.2*	16.9	7.3	28.4	26	(3)
Killdeer (Salton Sea, CA)							
Charadrius vociferus	47.3	25.2*	13.4	11.8	33.9	35	(3)
Glaucous-winged Gull (Alaska)							
Larus glaucescens	44.6	20	6.4	13.6	31.0	44	(4)
Adelie Penguin (Antarctica)							
Pysgoscelis adeliae	44.3	10.2	3.0	7.2	41.3	17	(5)
South-Polar Skua (Antarctica)							
Catharacta maccormicki	44.6	9.1	3.0	6.1	41.6	15	(6)

* P_n was measured

References: (1) Rahn et al., 1976, (2) Vleck et al., 1983, (3) Grant, 1982, (4) Morgan et al., 1978, (5) Rahn and Hammel, 1982, (6) Rahn and Hammel, unpublished.

Fig. 1. Stimultaneous water vapor and temperature differences from the egg, e, to the nest, n, to the ambient environment, a, in the Eider incubating in the Arctic and the White Tern incubating in the warm Pacific, with isopleths of relative humidity of 20–100%. In the Eider $(P_e - P_a) = 44$ torr which is divided between $(P_e - P_n)$ of the egg and $(P_n - P_a)$ of the nest. The brackets show the relative magnitudes of the egg resistance, R_e, and the nest resistance, R_n. In the White Tern of the Marshall Islands $(P_e - P_a) = 21$ torr, and since $P_n = P_a$, the total resistance is reflected by the egg, R_e. The White Tern of Midway Island incubates at a lower ambient temperature and vapor pressure and a larger $(P_e - P_a)$ of 29 torr. Because $\dot{M}_{H_2O}$ was essentially the same in both localities, R_e of Midway Island was proportionally higher (or shell conductance proportionally lower).

Nest resistance in other species

In addition to the relative nest resistances shown for the species in Table 3, similar computations were made for 9 other species where egg temperature, $\dot{M}_{H_2O}$, G_e and P_a were measured, but P_n was calculated as noted above (Table 4). In Fig. 2 the relative nest resistance is plotted as a function of the total water vapor pressure difference that existed between the egg and the ambient environment $(P_e - P_a)$. This pressure difference, imposed by the egg temperature on the one hand and the ambient humidity on the other, dictates the resistance that egg and nest must offer to achieve a given rate of daily water loss. In general, as $(P_e - P_a)$ increases from a warm, humid climate to a cold, dry climate, the relative contribution of nest resistance increases from zero to values above 30%, and that offered by the eggshell decreases from 100% to ca. 70%. The special cases of the Pied-billed Grebe and the Egyptian Plover are indicated by square symbols at $(P_e - P_a)$ values of 13 and 17 torr and are discussed below.

Nests of the Grebe and the Egyptian Plover

Two unusual nesting situations deserve attention: eggs in the wet, floating nests of the Grebe and eggs of the Egyptian Plover buried in hot, dry sand. What they are share in common is that these eggs are covered part or most of the time and

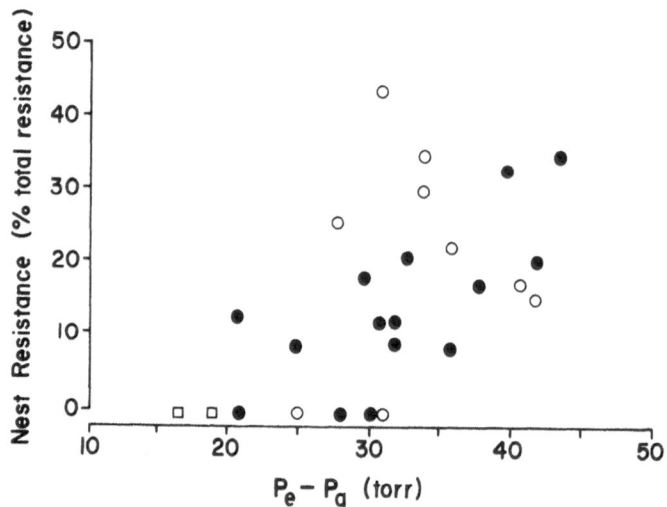

Fig. 2. Nest resistance (R_n) as percent of total water vapor resistance ($R_e + R_n$) for 25 species plotted as a function of the total imposed water vapor gradient. $(P_e - P_a)$. In a warm, humid environment this gradient is small and generally associated with little or no nest resistance. In cold, dry areas $(P_e - P_a)$ is large and nest resistance becomes a greater fraction of the total resistance. Solid circles from Table 2, open circles from Table 4. Square symbols represent the Pied-billed Grebe and the Egyptian Plover – see text.

that the final sink for water vapor loss from the egg is not the ambient atmosphere but the immediate surroundings in the nest. Thus nest conductance, in the sense used above, is absent and the entire resistance to water loss is in the eggshell. Therefore, the entire water vapor transport is described by Eq. 2 where $\dot{M}_{H_2O} = G_c(P_e - P_n)$.

Eggs of the Pied-billed Grebe (*Podilymbus podiceps*) are laid in wet, floating nests which are covered by wet plant material when the bird is not on the nest. The 22 g eggs lose $156\ mg \cdot d^{-1}$ (Davis and Ackerman, 1983). For such a normal rate of water loss to occur into the very humid air of a wet, floating nest, a very high shell conductance (or very low resistance) is required. The conductance was $12.3\ mg \cdot d^{-1} \cdot torr^{-1}$, 2.5 times larger than predicted on the basis of egg mass and incubation time. $(P_e - P_n)$ was estimated to be (156/12.3), or 13 torr (Davis and Ackerman, 1983) (see Fig. 2).

The eggs of the Egyptian Plover (*Pluvianus aegyptius*), on the other hand, are covered by a few mm of sand in the hot, arid region along the banks of the Baro River, Ethiopia. (Howell, 1979). During the hot part of the day the parent carries water to the nest and repeatedly moistens the area over the egg. These 9.5 g eggs lose $35\ mg \cdot d^{-1}$, and have an average egg temperature of 37.5 °C, equivalent to a $P_e = 48$ torr. P_n fluctuates depending on the degree of nest wetting, and is estimated by Howell to have a mean value of 31.3 torr throughout a 24 h period. Thus $P_e - P_n = 17$ torr, shown in Fig. 2.

Is nest humidity regulated?

It was originally proposed that the behavior of the parent could modify nest humidity by changes in incubating behavior, e.g., frequency of egg turning, stand-up periods, altering the tightness with which the brood patch was applied to the egg, or ventilation of the nest (Rahn et al., 1976, 1977). However, simultaneous measurements of P_n and P_a over the long incubation period of the Laysan Albatross have shown that changes in ambient P_a were followed by nearly identical changes in P_n (Grant et al., 1982b). Chattock (1925) observed similar changes during the incubation of chickens. More recently Walsberg (1983) demonstrated that by acutely exposing the natural nests of the House Finch (*Carpodacus mexicanus*) and the Phainopepla (*Phainopepla nitens*) to an increase in nest humidity he could reduce the rate of egg water loss. All these observations suggest that an incubating bird is unable to cope with changes in nest humidity and is therefore at the mercy of large changes in ambient humidity. Under certain conditions this could have a profound effect upon the overall water loss.

The data in Table 1 and Eq. 7 suggest that the total water vapor gradient $(P_e - P_a)$ imposes certain constraints upon the total water vapor resistance $(R_e + R_n)$ if a given rate of water loss is to be achieved. Of the total resistance, that offered by the eggshell is the major one. What is the evidence that long-term adjustments of R_e occur in response to change in climate and humidity?

The best examples are the changes in pore area of species that incubate at various altitudes. Because the diffusion coefficients of gases increase with altitude, the pore area of a shell should be reduced proportionally so that the normal water vapor flux from the egg to the environment is preserved. Recent experiments in chickens transferred from 3800 m to 1500 m revealed changes in shell conductance over a 2 month period which approximately compensated for the altered barometric pressure (Rahn et al., 1982). Another example is the difference between G_c of eggs of the White Tern nesting in the warm Marshall Islands (28 °C) and those nesting in the relatively cooler climate of Midway Island (20 °C) (see Table 2 and Fig. 1). In each case the egg temperature, P_c, and the daily water loss were the same. However, $(P_c - P_a)$ was 29 torr on Midway Island and 21 torr in the Marshall Islands. Because $(P_n - P_a)$ was zero in both cases, the (29/21) or 38% difference in $(P_c - P_a)$ should, according to Eq. 6, be reflected in a similar difference in R_c. R_c on Midway Island = (1/2.5) or 0.400, and in the Marshall Islands, (1/3.5) or 0.286 torr·d·mg^{-1}, resulting in a 40% difference (0.40/0.29). The G_c values were based on an average of 14 eggs for Marshall Islands terns and 12 eggs for Midway Island terns. While this is an isolated example, it does suggest that an adaptation of eggshell conductance to a climate which imposed a different $P_c - P_a$ has occurred in conformance to Eq. 6.

Summary

If one assumes that F = 0.15, or some other given fraction of the initial egg mass that must be lost as water to provide the proper state of hydration prior to hatching, then Eq. 7 can be rearranged as follows:

$$(R_c + R_n) = 10^3 (I/FW)(P_c - P_a) \tag{9}$$

The right side of the equation gives the independent variables for each species. For a given value of I, F and W the ambient environment will in large part determine $(P_c - P_a)$ and therefore the magnitude of the dependent variable $(R_c + R_n)$. The relative magnitude of R_n depends upon the particular species and its nesting site. Generally, the larger $(P_c - P_a)$ the greater the role of R_n in the total vapor resistance $(R_c + R_n)$.

Acknowledgement

The author is very grateful and indebted to C.V. Paganelli, P.R. Sotherland, and G.S. Grant for their invaluable criticisms and comments.

References

Ar, A., Paganelli, C.V., Reeves, R.B., Greene, D.G. and Rahn, H. (1974). The avian egg: water vapor conductance, shell thickness, and functional pore area. *Condor* 76: 153–158.

Ar, A. and Rahn, H. (1980). Water in the avian egg: overall budget of incubation. *Am. Zool.* 20: 373–384.

Carey, C., Rahn, H. and Parisi, P. (1980). Calories, water, lipid and yolk of avian eggs. *Condor* 82: 335–343.

Chattock, A.P. (1925). On the physics of incubation. *Phil. Trans. Roy. Soc. London. Ser. B* 213: 397–450.

Davis, T.A. and Ackerman, R.A. (1983). Water loss by avian eggs in wet nests. *Fed. Proc.* 42: 616 (Abstract).

Drent, R. (1970). Functional aspects of incubation in the Herring gull, *Behav. Suppl.* 17:1–132.

Grant, G.S., Pettit, T.N., Rahn, H., Whittow, G.C. and Paganelli, C.V. (1982a). Regulation of water loss from Bonin Petrel (*Pterodroma hypoleuca*) eggs. *Auk* 99: 236–242.

Grant, G.S., Pettit, T.N., Rahn, H., Whittow, G.C. and Paganelli, C.V. (1982b). Water loss from Laysan and Black-footed Albatross eggs. *Physiol. Zool.* 55: 405–414.

Grant, G.S., Paganelli, C.V., Pettit, T.N., Whittow, G.C. and Rahn, H. (1982c). Determination of fresh egg mass during natural incubation. *Condor* 84: 121–122.

Grant, G.S. (1982). Avian incubation: Egg temperature, nest humidity, and behavioral thermoregulation in a hot environment. Ornithol. Monogr. No. 30, Amer. Ornithol. Union, Washington, D.C. pp. 75.

Howell, T.R. (1979). Breeding biology of the Egyptian Plover, *Pluvianus aegyptius*. Univ. Calif. Publ. Zool. Vol. 113, Univ. of California Press, Los Angeles, pp. 76.

Hoyt, D.F. and Rahn, H. (1980). Respiration of avian embryos – a comparative analysis. *Respir. Physiol.* 39: 255–264.

Morgan, K.R., Paganelli, C.V. and Rahn. H. (1978). Egg weight loss and nest humidity during incubation in two Alaskan gulls. *Condor* 80: 272–275.

Paganelli, C.V., Olszowka, A. and Ar, A. (1974). The avian egg: surface area, volume and density. *Condor* 76: 319–325.

Paganelli, C.V. (1980). The physics of gas exchange across the avian eggshell. *Am. Zool.* 20: 329–338.

Pettit, T.N., Grant, G.S., Whittow, G.C., Rahn, H. and Paganelli, C.V. (1981). Respiratory gas exchange and growth of White Tern embryos. *Condor* 83: 355–361.

Pettit, T.N., and Whittow, G.C. (1983). Water loss from pipped Wedge-tailed Shearwater eggs. *Condor* 85: 107–109.

Rahn, H., Paganelli, C.V., Nisbet, I.C.T. and Whittow, G.C. (1976). Regulation of incubation water loss in seven species of terns. *Physiol. Zool.* 49: 245–259.

Rahn, H., Ackerman, R.A. and Paganelli, C.V. (1977). Humidity in the avian nest and egg water loss during incubation. *Physiol. Zool.* 50: 269–283.

Rahn, W. and Dawson, W.R. (1979). Incubation water loss in eggs of Heermann's and Western Gulls. *Physiol. Zool.* 52: 451–460.

Rahn, H. and Hammel, H.T. (1982). Incubation water loss, shell conductance, and pore dimensions in Adelie Penguin eggs. *Polar Biol.* 1: 91–97.

Rahn, H., Ledoux, T., Paganelli, C.V. and Smith, A.H. (1982). Changes in eggshell conductance after transfer of hens from an altitude of 3,800 to 1,200 m. *J. Appl. Physiol.: Respirat. Environ. Exercise Physiol.* 53: 1429–1431.

Rahn, H., Krog, J. and Mehlum, F. (1983). Microclimate of the nest and egg water loss of the Eider and other waterfowl in Spitsbergen. *Polar Res.* 1: 171–183.

Schmidt-Nielsen, K. (1975). Animal Physiology. Cambridge University Press, Cambridge, p. 211.

Vleck, C.M., Vleck, D., Rahn, H. and Paganelli, C.V. (1983). Nest microclimate, water-vapor conductance, and water loss in heron and tern eggs. *Auk* 100: 76–83.

Walsberg, G.E. (1980). The gaseous microclimate of the avian nest during incubation. *Am. Zool.* 20: 363–372.

Walsberg, G.E. (1983). A test for regulation of nest humidity in two bird species. *Physiol. Zool.* 56: 231–235.

Whittow, G.C., Ackerman, R.A., Paganelli, C.V. and Pettit, T.N. (1982). Prepipping water loss from the eggs of the Wedge-tailed Shearwater. *Comp. Biochem. Physiol.* 72A: 29–34.

20. A preliminary study of local oxygen tensions inside bird eggs and gas exchange during early stages of embryonic development

J.P. Lomholt

Department of Zoophysiology, University of Aarhus, DK-8000 Aarhus C, Denmark

Abstract

Measurements of O_2 tension (P_{O_2}) have been done in chicken eggs aged 1 to 6 days by means of a needle O_2 electrode. The steepness of the drop in P_{O_2} seen as the electrode is moved into the egg depends on the age of the embryo and on the site of insertion relative to the position of the vascularized portion of the yolk sac.

The steepest gradient is found when insertion is through the part of the shell membrane in direct contact with the vascular yolk sac. The minimal values of P_{O_2} decline with increasing age up to about 4 d, where values of 0–10 torr are found. These near anoxic values disappear after the 4th day as a result of an increase in the O_2 permeability of the inner shell membrane.

Observations on eggs in which a part of the shell is laquered indicate that only the fraction of the shell in direct contact with the yolk sac vessels is important to gas exchange of the four-day-old embryo.

Introduction

The physiology of gas exchange and gas transport of the avian embryo has been studied in much detail, and the role of the chorioallantoic membrane as an organ of gas exchange has long been appreciated. The literature is extensive, but reviews can be found in the books by Bartels (1970) and Freeman and Vince (1974) and much information is included in the symposium reports edited by Piiper (1978) and Carey (1980).

Nearly all this work is concerned with that part of the period of incubation where the chorioallantois is well developed, that is with the latter two-thirds of embryonic life. This wealth of information contrasts sharply with the scarcity of data dealing with the early stages of development before the chorioallantois assumes a prominent role.

Apart from the fact that the later stages of development are more accessible to experimentation, the reason for directing most of the effort towards the study of the later stages of development has been the interesting situation faced by the

Seymour, R.S., (ed.) Respiration and metabolism of embryonic vertebrates.
© 1984, Dr W. Junk Publishers, Dordrecht/Boston/London. *ISBN-13:978-94-009-6538-6*

embryo, namely a steadily increasing demand for O_2 and at the same time a fixed area available for gas exchange determined by the porosity of the egg shell. As a consequence of this the embryo experiences an increasing degree of hypoxia towards the end of the incubation period. Much of the work on avian embryonic respiration has been undertaken to understand how the embryo copes with this situation.

The O_2 supply of the early embryo might seem to present no major problems. The embryo is small and so is the O_2 consumption, although on a wet mass-specific basis it is high (Romijn and Lokhorst, 1951). The P_{O_2} of the air space gas is high, above 140 torr (Romijn and Roos, 1938; Wangensteen and Rahn, 1970/71). The O_2 affinity of embryonic blood is low at the early stages of development. Half saturation tensions of about 40–80 torr have been reported (Lomholt, 1975; Bartels and Baumann, 1977; Baumann et al., 1983; Lapennas and Reeves, 1983). There are, however, factors which may tend to complicate the oxygenation of the early embryo and which should be taken into account when considering the conditions for gas exchange during early embryonic stages. The O_2 permeability of the shell membranes is low (Kutchai and Steen 1971; Lomholt, 1976; Tullett and Board, 1976). Before the proliferation of the chorioallantois, the only vascularized extraembryonic membrane is the yolk sac. Although this is brought into contact with the inner shell membrane through the formation of the subgerminal fluid (New, 1956), this contact is probably not as close as is the case for the chorioallantois. Furthermore the yolk sac makes contact with only a part of the shell. Hence the entire shell area may not be involved in gas exchange at these stages.

In an attempt to gain insight into the conditions for O_2 supply of the early avian embryo, P_{O_2} has been recorded in the immediate surroundings of the embryo by means of micro O_2 electrodes. In addition, experiments in which a part of the egg shell was laquered have been used as a first approach to determine the relative importance of different parts of the shell to gas exchange.

Material and methods

Domestic chicken eggs were incubated for periods ranging from 24 h to 6 d. One hour before measurements were to be done, the eggs were placed broad end up in the incubator to ensure that the embryonic area of the yolk was located towards the broad end and thus made contact with the air space of the egg. The egg shell and the outer shell membrane was broken away over the air space, while the inner membrane, which lines the inner aspect of the air space, was left intact. Since at these early stages of development the O_2 concentration of the air space gas does not deviate significantly from that of atmospheric air, this procedure will not alter the conditions for gas exchange of the embryo. Immediately after opening, the egg was mounted in a thermostatted egg chamber to maintain the incubation temperature during the measurement of P_{O_2}. A hole in the lid of the egg chamber permitted insertion of the O_2 electrode (cathode and reference electrode) as well as visual inspection through a stereo microscope.

P_{O_2} was measured by a platinum needle electrode which is basically a simplified version of the design described by Baumgärtl and Lübbers (1973, 1983) and a separate silver/silver chloride reference electrode, which was inserted about 1 cm away from the point of insertion of the cathode. The platinum electrode had a tip diameter of about 5 μm and produced a measuring current at air saturation (P_{O_2} about 149 torr) of 100–500 pA. The polarizing voltage was 700 mV. Before being used, the performance of an electrode was checked in a thermostatted vessel containing a Ringer solution (26 mM $NaHCO_3$, 2.6 mM $CaCl_2$, 1 mM glucose, 124 mM NaCl, 4.96 mM KCl, 1.3 mM KH_2PO_4, 13 mM $MgSO_4$). Electrodes which showed a stable signal in a pure NaCl solution often showed a downward drift in the Ringer solution. When the different components of the Ringer solution were added to the test vessel one at a time, the drift was found to begin as soon as calcium and bicarbonate were both present. The signal slowly recovered upon tranfer to a pure NaCl solution, whereas a brief rinse in dilute HCl immediately brought the signal back to the original level. This observation is in accordance with what has recently been described by Baumgärtl and Lübbers (1983) and is probably an indication that $CaCO_3$ is deposited on the platinum cathode. The stability may be improved by increasing the thickness of the membrane covering the cathode, although it was found very difficult to control adequately the application of the membrane. Not all of the electrodes showing a stable calibration in the Ringer solution were found to be adequate for measurements in the egg. Data were taken only from electrodes showing less than 10% decline in the signal at air saturation after the 10–15 min taken to make a series of measurements. The drift was checked by calibration immediately before and after a run in the egg. The inevitable small drift was corrected for by assuming it to be linear between these two points. The needle electrode was mounted on a motor driven micro manipulator and inserted in steps of 20 or 50 μm. After finishing the measurements, the inner shell membrane was removed and the points of insertion located in relation to the position of the embryo and the area vasculosa of the yolk sac membrane.

Because of the difficulties in producing electrodes which are sufficiently stable when exposed to the egg contents, it has only been possible to make measurements in from 3–5 eggs of each of the different stages of incubation. Consequently, it is not at present possible to give average values of O_2 tensions at specific sites. Instead, the graphs presented bring examples of the measurements obtained from a single egg of each of the different stages of development.

For the shell laquering experiment, the large circumference of the egg was drawn with a soft pencil on the shell, and one half of the shell was covered with a thick layer of a fast setting glue. The eggs were either turned every 3 h or they were not turned at all. After 4 d of incubation, the eggs were opened and the embryos classified as being either alive and of normal size or dead and/or retarded in growth.

Results

Examples of P_{O_2} profiles obtained at various stages of development are given in Figs. 1–5. Inserts on the figures represent the opening in the shell over the air space and indicate the approximate site of electrode insertion relative to the sinus terminalis which is the circular vessel surrounding the vascular part of the yolk sac, or in the case of Fig. 1, the edge of the germinal disc. In the embryo incubated for 28 h, the minimal P_{O_2} is between 90 and 100 torr when insertion is 2 mm outside the edge of the germinal disc, whereas the tension drops to about 45 torr when the electrode is inserted through the germinal disc about 2 mm from the edge (Fig. 1).

In the two-day-embryo, O_2 tension drops to between 20 and 60 torr when insertion is into the vascular part of the yolk sac. Penetrating farther results in a rise in P_{O_2} as the electrode has passed through the tissue of the yolk sac membrane and into the underlying subgerminal fluid. When the electrode is inserted a few millimeters outside the sinus terminalis, the drop in O_2 tension is much less steep (Fig. 2).

In the three-day-embryo the pattern is similar but the drop in O_2 tension with depth of insertion is much steeper (Fig. 3). The minimal values are extremely low, less than 10 torr and sometimes indistinguishable from zero.

In the four-day-embryo, depicted in Fig. 4, the situation is somewhat similar to that in Fig. 3. In both cases the inner shell membrane looks moist and transparent to the yellow color of the yolk. In contrast to this the P_{O_2} drop shown in Fig. 5,

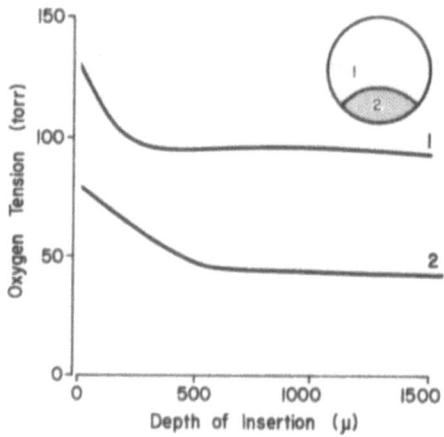

Fig. 1. P_{O_2} profiles in a chicken egg incubated for 28 h. The insert shows the approximate site of electrode insertion relative to the germinal disc (stipple) and hole in the shell (circle). Distance of insertion from edge of germinal disc is (1): 2 mm. (2): 2 mm.

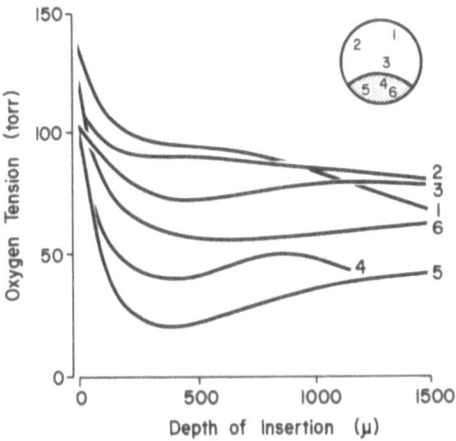

Fig. 2. P_{O_2} profiles in a chicken egg incubated for 51 h. The insert shows the approximate site of electrode insertion relative to the vascular part of the yolk sac (stipple). Distance of insertion from the sinus terminalis is (1): 6 mm, (2): 5 mm, (3): 1 mm, (4): 1 mm, (5): 4 mm, (6): 4 mm.

which is from another embryo of the same age, is much less steep and the near anoxic values seen in Figs. 3 and 4 are absent. In this embryo the inner shell membrane looks dry, whitish and opaque. This can be taken as an indication that the O_2 permeability of the membrane has increased, whereas this has not yet happened in the example in fig. 4. In embryos incubated for 5–6 d, the pattern is very similar to that seen in Fig. 5.

Results from the shell laquering experiment are given in Table 1. Eggs were treated in 5 different ways as indicated in the table.

The effect on development and hatchability of turning the eggs during incubation is well known (Lundy, 1969; New 1957). The effect of turning is, however, relatively small, and a comparison of group 1 and 2 in Table 1 reveals the differences to be non-significant. The eggs of group 3 have the lower half of the shell laquered and are not turned. Development in this group does not differ from that of the non-laquered eggs of group 1 and 2, indicating that the lower half of the shell is of little importance to gas exchange during the first 4 d of incubation.

In contrast to this, development is seriously disturbed in group 4 which have the upper half of the shell laquered and are not turned. This indicates that gas exchange takes places primarily across the part of the shell in contact with the yolk sac membrane, and suggests that the yolk sac membrane functions as a gas exchanger during early stages of development, as does the chorioallantois of the older stages.

The eggs of group 5 are laquered as is the case for group 3 and 4, but in contrast to those, they have been turned every 3 h. Development is normal in this group. Because the embryo is free to rotate inside the egg and because the density of the subgerminal fluid immediately beneath the embryo is smaller than that of the rest

293

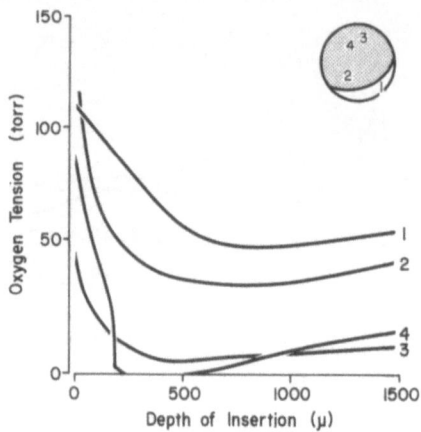

Fig. 3. As Fig. 2. Egg incubated for 70 h. Distance of insertion from the sinus terminalis is (1): 2 mm, (2): 2 mm, (3): 10 mm, (4): 10 mm.

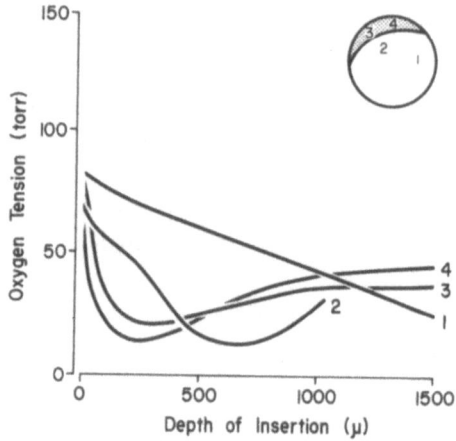

Fig. 4. As Fig. 2. Egg incubated for 100 h. Distance of insertion from the sinus terminalis is (1): 10 mm, (2): 5 mm, (3): 3 mm, (4): 3mm.

of the egg contents, the embryo always faces upward. It thus appears that intermittent exposure to a laquered shell does not seriously interfere with development.

Discussion

The technical problems in getting reliable measurements of P_{O_2} inside the eggs with the micro O_2 electrodes have been considerable. As a result of this the data obtained are limited and must be considered of a preliminary nature. Recently a modified micro O_2 electrode suitable for recording O_2 tensions in acid microbial

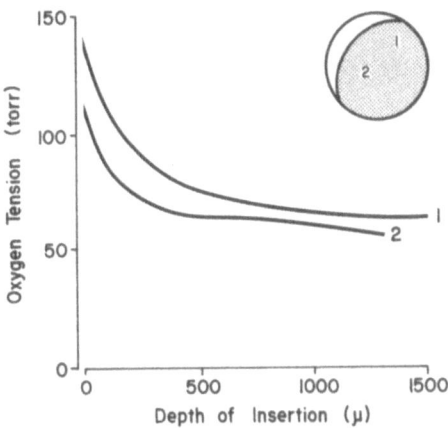

Fig. 5. As Fig. 2. Egg incubated for 96 h. Distance of insertion from the sinus terminalis is (1): 8 mm, (2): 10 mm.

Table 1. The effect of turning and laquering of one half of the egg shell on development after 4 d of incubation.

Treatment	Dead or retarded	Alive, normal size
1 Not laquered/turned	1	9
2 Not laquered/not turned	5	14
3 Lower half laquered/not turned	1	9
4 Upper half laquered/not turned	15	4
5 Laquered/turned	3	15

The difference between any two groups has been tested (Chi-square test). Group 4 is significantly different from any other (P<0.005). None of the other groups are significantly different from each other (P>0.05).

mats and other harsh environments has been described (Revsbech and Ward, 1983). If this type of electrode proves to be resistant to chemical interference from the egg contents, a more systematic and detailed study of O_2 tensions in the chicken egg will be possible.

Samples of subgerminal fluid, which is the extraembryonic fluid in most direct contact with the embryo at the earliest stages of development, show a rather high P_{O_2} of about 70–90 torr (Dawes, 1972; personal observations). This suggests that the early embryo is far from being subjected to very hypoxic conditions. However, such samples of necessity will show some sort of an average P_{O_2} of the relatively large volume of subgerminal fluid. In contrast, the highly localized in situ measurements with the micro O_2 electrode indicate that within about 3 d of development the P_{O_2} gradient across the inner shell membrane becomes very steep and P_{O_2} in the immediate surroundings of the embryo comes close to zero. P_{O_2} may in some cases become zero in the tissue of the yolk sac membrane. This must indicate that at this stage the embryo is in a situation of sparse O_2 availability.

The finding of very low P_{O_2} substantiates the functional significance of the large rise in O_2 permeability of the inner shell membrane which takes place around the 4th–6th day of incubation (Kutchai and Steen, 1971; Lomholt, 1976; Tullett and Board, 1976). The increase in permeability is the result of a dehydration of the membrane brought about by an increase in the colloid osmotic pressure of the albumen (Lomholt, 1976). As a result of this dehydration the appearance of the inner shell membrane changes from semitransparent, yellowish to being whitish and opaque. From the present measurements it appears that the near anoxic values seen while O_2 permeability is still low, disappear as a result of the rise in permeability.

The tolerance of the chicken embryo to a lowered O_2 content in ambient air has been reported to be highest at the earliest stages (0–4 d) of development (Taylor et al., 1956; Grabowski and Paar, 1958). This seems puzzling in the light of the very large drop of P_{O_2} across the shell membrane. If P_{O_2} immediately inside the membrane is close to zero, any drop in ambient P_{O_2} should cause a reduction in the diffusive gradient and thus in the supply of O_2 across the shell. This might be suggestive of a capacity in the early embryo for anaerobic metabolism, a possibility discussed by Kucera et al. (present volume).

The finding that the P_{O_2} gradient is much steeper across the part of the shell membrane overlying the vascular part of the yolk sac suggests that the major part of the gas exchange of the embryo will take place across the corresponding part of the shell. The fact that laquering the lower half of the egg shell is of no consequence to development, whereas laquering the upper half of the shell in the unturned egg results in high mortality and retardation of growth can be taken as direct evidence for this (Table 1).

The significance to heat transfer of an intimate contact between the defeathered skin of the brood patch and the egg shell has often been stated (Drent, 1975). The part of the shell in contact with the brood patch will, at the early stages of development, also be the part in contact with the yolk sac vessels. It is tempting to ask whether this contact may have any consequences for the respiratory gas exchange of the embryo. Exactly how large a part of the shell is in contact with the brood patch is not known. Especially in species with a clutch of only one egg, such as in various sea birds, it may be a considerable fraction of the shell.

Surface P_{O_2} of the skin of normal humans has been found to be close to zero (Evans and Naylor, 1967; personal observations), whereas higher values of up to 50 torr were found on the normal skin as well as on the brood patch of the chicken (personal observations). Whether this higher skin surface P_{O_2} in the chicken may be of any significance to gas exchange across the part of egg shell in contact with this skin must at present remain speculative.

The observation that the deleterious effect on development of covering the upper half of the egg shell in the unturned egg is offset by regular turning, may suggest an additional respiratory significance of the well known behavior of frequent turning of the eggs observed in wild birds.

References

Bartels, H. (1970). Prenatal Respiration. North Holland Publishing Company, Amsterdam.

Bartels, H. and Baumann, R. (1977). Respiratory function of hemoglobin. In: Respiratory Physiology II, International Review of Physiology. J.C. Widdicombe ed., University Park Press, Baltimore.

Baumann, R., Padeke, S., Haller, E.A. and Brilmayer, T. (1983). Effects of hypoxia on oxygen affinity, hemoglobin pattern, and blood volume of early chicken embryos. Am. J. Physiol. 244 (Regulatory Integrative Comp. Physiol. 13): R733–R741.

Baumgärtl, H. and Lübbers, D.W. (1973). Platinum needle electrode for polarographic measurement of oxygen and hydrogen. In: Oxygen Supply. M. Kessler, ed., Urban and Schwarzenberg, München, pp. 130–136.

Baumgärtl, H. and Lübbers, D.W. (1983). Microcoaxial needle sensor for polarographic measurement of local O_2-pressure in the cellular range of living tissue. Its construction and properties. In: Polarographic Oxygen Sensors. E. Gnaiger and H. Forstner, eds., Springer, Berlin, pp. 37–65.

Carey, C. (1980). Introduction to the symposium: Physiology of the avian egg (and a series of accompanying papers). Am. Zool. 20: 325–484.

Dawes, C.M. (1972). The oxygen tensions of the extraembryonic fluids of the chick embryo. Comp. Biochem. Physiol. 43A: 119–124.

Drent, R. (1975). Incubation. In: Avian Biology, Vol. 5. D.S. Farner and J.R. King, eds., Academic Press, New York, pp. 333–420.

Evans, N.T.S. and Naylor, P.F.D. (1967). The systemic oxygen supply to the surface of human skin. Respir. Physiol. 3: 21–37.

Freeman, B.M. and Vince, M.A. (1974). Development of the Avian Embryo. Chapman and Hall, London.

Grabowski, C.T. and Paar, J.A. (1958). The teratogenic effects of graded doses of hypoxia on the chick embryo. Am. J. Anat. 103: 313–347.

Kutchai, H. and Steen, J.B. (1971). Permeability of the shell and shell membranes of hen's eggs during development. Respir. Physiol. 11: 265–278.

Lapennas, G.N. and Reeves, R.B. (1983). Oxygen affinity and equilibrium curve shape in blood of chicken embryos. Respir. Physiol. 52: 13–26.

Lomholt, J.P. (1975). Oxygen affinity of bird embryo blood, J. Comp. Physiol. 99: 339–343.

Lomholt, J.P. (1976). The development of the oxygen permeability of the avian egg shell and its membranes during incubation. J. Exp. Zool. 198: 177–184.

Lundy, H.L. (1969). A review of the effects of temperature, humidity, turning and gaseous environment in the incubator on the hatchability of the Hen's Egg. In: The Fertility and Hatchability of the Hen's Egg. T.C. Carter and B.M. Freeman, eds., Oliver & Boyd, Edinburg.

New, D.A.T. (1956). The formation of sub-blastodermic fluid in hens' eggs. J. Embryol. Exp. Morphol. 4: 221–227.

New, D.A.T. (1957). A critical period for the turning of hens' eggs. J. Embryol. Exp. Morphol. 5: 293–299.

Piiper, J. (1978). Respiratory Function in Birds, Adult and Embryonic, J. Piiper, ed., Springer-Verlag, Berlin.

Revsbech, N.P. and Ward, D.M. (1983). Oxygen micro electrode that is insensitive to medium chemical composition: Use in an acid microbial mat dominated by Cyanidium caldarium. Applied and Environmental Microbiology 45: 755–759.

Romijn, C. and Ross, J. (1938). The air space of the hen's egg and its change during the period of incubation. J. Physiol. 94: 365–379.

Romijn, C. and Lokhorst, W. (1951). Foetal respiration in the hen. *Physiol. Comp. Oecol.* 2: 187–197.

Taylor, L.W., Sjodin, R.A. and Gunns, C.A. (1956). The gaseous environment of the chick embryo in relation to its development and hatchability. *Poult. Sci.* 35: 1206–1215.

Tullett, S.G. and Board, R.G. (1976). Oxygen flux across the integument of the avian egg during incubation. *Br. Poult. Sci.* 17: 441–450.

Wangensteen, O.D. and Rahn, H. (1970/71). Respiratory gas exchange by the avian embryo. *Respir. Physiol.* 11: 31–45.

21. Oxygen and glucose uptakes in the early chick embryo

Pavel Kučera, Eric Raddatz and Anne Baroffio

Institute of Physiology, Medical Faculty of the University of Lausanne, Switzerland

Abstract

The uptake of O_2 and glucose and the production of CO_2 and lactate were measured using non-invasive techniques in the intact in vitro developing chick blastodisc during gastrulation-neurulation (18 to 24 h of incubation).

Glucose represents the primary source of energy at this period of development. Under normoxic conditions, about 30 and 50% of glucose are oxidized, respectively, at stages 4 and 7. The rest is converted to lactate. The oxidation rate in the embryo is as high as in the most active differentiated tissues and the additional aerobic glycolysis is comparable to that found in malignant tumors. The total ATP produced is estimated at 7–10 $nmol \cdot h^{-1} \cdot (\mu g\ protein)^{-1}$. The anaerobic glycolysis can account for only 20% of this production.

The O_2 and glucose uptakes vary according to regions of the blastodisc. In the area pellucida, the metabolism increases in parallel to the growth but the capacity to oxidize glycose remains constant. In the extraembryonic area opaca, the metabolism increases more rapidly than the growth parameters and the efficiency of the glucose utilization seems to increase as well.

Introduction

Previous studies of embryonic metabolism have demonstrated that the morphogenetic activities of the young chick embryo depend on the oxidative metabolism of carbohydrates.

As far as the metabolism of the whole embryo is concerned, the following has been shown:

(1) The oxidative activity (the O_2 uptake related to the nitrogen content or to the mass) rises rapidly during the first 17 h of incubation, reaches a plateau during the next 10 to 20 h (Philips, 1941, 1942) and then may slowly decrease (Bartels and Baumann, 1972). Thus it seems that a peak of oxidative activity occurs probably at the end of gastrulation and the onset of neurulation.

(2) The main sugar utilized seems to be glucose, although other sugars can also

Seymour, R.S., (ed.) Respiration and metabolism of embryonic vertebrates.
© *1984, Dr W. Junk Publishers, Dordrecht/Boston/London. ISBN-13:978-94-009-6538-6*

be metabolized to some extent (see Spratt, 1952, for review) and although the embryo does not lack the capacity to metabolize other substrates, e.g. proteins (Needham, 1932). The importance of glucose has been inferred from studies of survival of the embryos in presence of different substrates, but direct measurements of glucose uptake by the embryo have not been done.

As far as the patterns of metabolic activity within the embryo are concerned, qualitative as well as quantitative regional differences of metabolic activity have been suggested on the basis of indirect observations of differential localization and activity of some reducing enzymes (Spratt, 1951, 1952).

Recently, we have developed a new system of in vitro culture that respects the structural and functional integrity of the embryo. Using this culture system, the oxidative metabolism within the gastrulating and neurulating embryo has been mapped by means of sensitive and non-invasive techniques (Kucera and de Ribaupierre, 1980; Kucera and Raddatz, 1980).

In this paper, the spatio-temporal pattern of the oxidative activity as demonstrated by these techniques (Raddatz and Kucera, 1983) is compared to the glucose uptake as determined using the 2-deoxyglucose method.

Material and methods

The embryonic stages 4 to 7, according to Hamburger and Hamilton (1951), were obtained by a preincubation of fertilized eggs (White Leghorn) at 37.5 °C and 80% relative humidity. The blastodiscs, still attached to the vitelline membrane, were carefully cleaned of yolk (technique of New, 1955) and then mounted either onto a special metallic support allowing a rapid exchange of incubation media or into a closed transparent chamber (Fig. 1) allowing development under perfusion or stopflow conditions. The first procedure was adopted for the glucose uptake measurements, the second for the measurements of O_2 uptake and CO_2 production. The incubation medium was a modified Tyrode solution containing 7.5 mmol $\cdot$ l^{-1} of glucose and bicarbonate as principal buffer.

As the degree of expansion of the embryo varies to a great extent, the dimensions of blastodiscs were systematically measured.

Oxygen uptake

The O_2 consumption of the incubated embryos has been determined using a computer-controlled scanning microspectrophotometer (Kucera et al., 1979). For this purpose, a purified human hemoglobin has been added to the perfusion medium and saturated at the $P_{50} = 54$ torr. When the perfusion is stopped for a short time, the adjacent respiring tissue deoxygenates the hemoglobin and the concomitant spectral change is recorded. The hemoglobin thus serves at the same

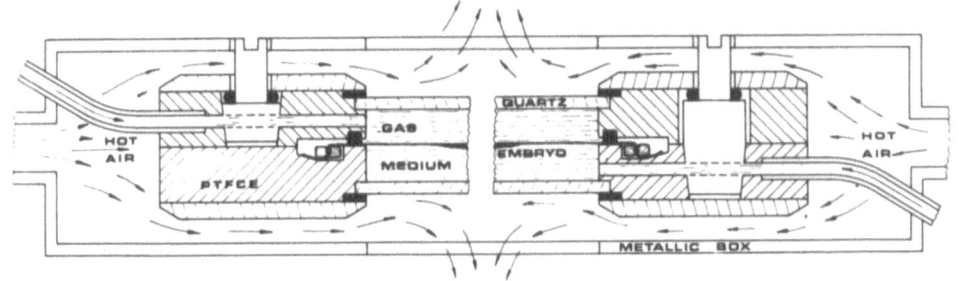

Fig. 1. The 'artificial egg': the embryonic disc adhering to the vitelline membrane is suspended in the central part of the incubation chamber (Diameter 2 cm. volume 1.89 ml) tightly closed by two windows and thermo-stabilized with hot air. Two sections corresponding to different radii are represented (hence the broken aspect) to show the system of stopcocks allowing closed or open system conditions.

time as O_2 donor as well as indicator of tissue respiration (Barzu and Borza, 1967). The conditions to be respected and the calculation of O_2 uptake from such recordings have been described in detail (Kucera and Raddatz, 1980). This technique allows the determination of the O_2 consumption of minute embryonic regions and the total O_2 uptake by the embryo is then obtained by integration of the local measurements over the scanning area. Alternatively, the difference between the O_2 concentrations in the inflowing and outflowing perfusate is measured and multiplied by the perfusion rate (Fick's principle). Both spectrophotometric methods give comparable results (Raddatz and Poretti, unpublished).

Glucose uptake

The glucose uptake of the incubated embryos, has been determined using the 2-deoxyglucose (2-DG) method originally developed for measurements of the local cerebral glucose utilization (Reivich et al., 1971; Sokoloff et al., 1977). The 2-DG is an analogue of glucose which is transported into the cells and then phosphorylated like the glucose. The deoxyglucose-6-phosphate is not further metabolized and, if present in traces, accumulates in the cells at a rate proportional to the utilization of glucose. It has been shown that this is also true for the embryonic cells (Baroffio and Kucera, 1983).

The blastodiscs at the desired stages were preincubated at 37 °C in the Tyrode solution for 30 min and then transferred to a warm Tyrode solution containing traces of C^{14} labeled 2-DG ($0.6-1 \, \mu Ci \cdot ml^{-1}$) and incubated for 15 min. They were then loosened from the vitelline membrane, washed and reincubated for additional 30 min in non-radioactive, DG-free Tyrode solution to let any free, non-metabolized DG diffuse out of the cells. The metabolism was then stopped by immersion in an ice-cold Tyrode solution.

The area pellucida and area opaca were carefully separated by microdissection, each part of the embryo was solubilized and its radioactivity determined by liquid scintillation counting and converted into nmols of glucose taken up per hour.

Alternatively, contact autoradiographs were made using the whole blastodiscs spread and dried onto a glass slide. The autoradiographs representing maps of glucose consumption have been evaluated by scanning densitometry. The mean optical density of the autoradiograph corresponding to the area pellucida was determined and calibrated in terms of the mean glucose consumption determined from the radioactivity counting. Thus knowing the density value for a given region of the area pellucida, the corresponding glucose uptake can be estimated.

The carbon-dioxide and lactate productions

These two parameters have been estimated indirectly from the pH variations in the nutritive medium under stopflow condition (2 h for the stage 4, 1 h for the stage 6–7, the O_2 content being sufficient for 4 hours of normal development). The bicarbonate concentration of the medium was adjusted to give a pH value of about 7.4 when equilibrated with the calibration gas (air) at ATPS conditions, and the CO_2 titration curve (the log P_{CO_2}-pH line) of this solution was determined experimentally. In each experiment, three values of pH have been measured at room temperature: the first one, in the perfusing medium saturated with the calibration gas, the second one, at the end of the incubation, in the medium withdrawn from the incubation chamber under gas-tight conditions and the third one after the re-equilibration of this sample with the calibration gas. Knowing the volume of the chamber and using the equation of Henderson-Hasselbalch and that of the CO_2 titration line, the volatile acidity change corresponding to the CO_2 produced by respiration and the non-volatile acidity change, reflected by the HCO_3^- decrease and corresponding mainly to the lactate produced, can be calculated (see e.g. Siggaard-Andersen, 1974).

Results

Normoxic conditions

Thirty-four blastodiscs at stage 4–5 (definitive primitive streak, intermediate head process) and 23 blastodiscs at stage 6–7 (headfold, beginning of neurulation, 1 somite) have been studied. The parameters measured in the whole disc, the area pellucida and the area opaca are represented in the first part of Table 1.

All the parameters characterizing the growth and the metabolic rates of the embryo increase considerably during the 6 h of development. Despite a great interindividual variability all the differences are significant (F-test, t-test).

302

Table 1. The parameters characterizing the growth and metabolism of gastrulating-neurulating chick embryos (18–24 h of incubation) under various experimental conditions. The mean values and standard deviations are given. n: number of embryos, HCO_3^-: bicarbonate decrease in the incubation medium.

	n	Blastodisc	A. Pellucida	A. Opaca
NORMOXIA				
Stage 4–5				
Protein content (μg)	5	79	21 ± 5	58 ± 2
Area (mm²)	14	51 ± 11	6 ± 1	43 ± 11
$\dot{V}_{O_2}$ (nmol·h^{-1})		87 ± 49	16 ± 7	71 ± 49
Area (mm²)	5	49 ± 12		
$\dot{V}_{CO_2}$ (nmol·h^{-1})		94 ± 27	?	?
HCO_3^- (nmol·h^{-1})		59 ± 24		
Area (mm²)	4	61 ± 7	5 ± 1	57 ± 7
Glucose (nmol·h^{-1})		53 ± 13	8 ± 5	44 ± 9
Stage 6–7				
Protein content (μg)	5	145	36 ± 5	109 ± 13
Area (mm²)	8	117 ± 19	10 ± 1	104 ± 18
$\dot{V}_{O_2}$ (nmol·h^{-1})		272 ± 125	27 ± 11	245 ± 125
Area (mm²)	5	115 ± 22		
$\dot{V}_{CO_2}$ (nmol·h^{-1})		202 ± 54	?	?
HCO_3^- (nmol·h^{-1})		110 ± 32		
Area (mm²)	5	86 ± 12	9 ± 2	77 ± 12
Glucose (nmol·h^{-1})		78 ± 2	15 ± 3	63 ± 3
ANOXIA				
Stage 4–5				
Area (mm²)	3	52 ± 12		
$\dot{V}_{CO_2}$ (nmol·h^{-1})		16 ± 14	?	?
HCO_3^- (nmol·h^{-1})		166 ± 15		
Area (mm²)	3	44 ± 3	4 ± 1	39 ± 3
Glucose (nmol·h^{-1})		81 ± 8	18 ± 2	63 ± 7
KCN				
Stage 4–5				
Area (mm²)	3	61 ± 16	5 ± 1	57 ± 16
Glucose (nmol·h^{-1})		96 ± 27	21 ± 4	75 ± 23

When the O_2 and glucose uptakes are related to the protein content the comparison between the area pellucida (AP) and area opaca (AO) gives the following values for O_2 and glucose respectively in nmols·h^{-1}·μg^{-1}: at stage 4–5, AP: 0.76 and 0.38, AO: 1.22 and 0.76; at stage 6–7, AP: 0.75 and 0.41, and AO: 2.25 and 0.58. Thus the metabolic activity is high and continues to increase in the area opaca whereas it remains lower and constant in the area pellucida. At each stage, the mean O_2 consumption ($\dot{V}_{O_2}$) and CO_2 production ($\dot{V}_{CO_2}$) values do not differ significantly, suggesting a respiratory exchange ratio near unity. The glucose uptake as measured, exceeds that calculated from the $\dot{V}_{O_2}$ (p<0.01) but does not differ significantly from that calculated from the $\dot{V}_{CO_2}$ and lactate production. Thus only a fraction of glucose is fully oxidized by the embryo, the rest being converted to lactate. At stage 4–5, this fraction is the same, about 30%, in both regions. At stage 6–7, it stays at 30% in the area pellucida but increases up to 65% in the area opaca.

The intraindividual variations of the glucose uptake are presented in more detail in Fig. 2. The pictures of the optical density of the embryos (upper third) and the corresponding autoradiographs (middle third) were obtained using a TV image processor allowing a quantitative analysis of images, e.g. averaging, filtering and colour display (Kucera, to be published). The values of optical density at each image point are represented according to a calibrated grey scale (minimum black, maximum white). The area pellucida is delimited from the peripheral area opaca by an optically dense ring corresponding to the yolk granules tightly adhering to the cells.

It can be seen that the area opaca, although constituted of essentially one layer of flattened cells (notice the lower optical density), shows an intense labelling. In the area pellucida, the radioactivity seems to be modulated according to the cell packing density, for example, the anterior axial part of the embryo where the cell number is high shows also a higher 2-DG accumulation.

When the glucose uptake estimated in different regions is compared to that calculated from the O_2 uptake found in the same regions (Raddatz and Kucera, 1983) the pattern shown at the bottom of Fig. 2 is obtained. Both parameters seem to vary in a roughly similar manner but a closer inspection reveals that the glucose could be oxidized to a different extent even within the area pellucida itself. In the presented comparison, the ratio of the glucose oxidized to the total glucose taken up varies from 0.16 to 0.64 at stage 4–5 and from 0.20 to 0.51 at stage 6–7. The mean value is however the same, about 35%, for the both stages and comparable to the results obtained by counting.

Anoxic conditions

The second part of Table 1 summarizes the results obtained in 9 embryos (stage 4–5) incubated either without O_2 or in the presence of 10^{-3} molar KCN.

As compared to the normoxic situation (P_{O_2} = 140 torr) the glucose uptake of the blastodisc increases by 53%, that of the area pellucida by 125% and that of the area opaca by 43%. This increase corresponds to the loss of bicarbonate (lactate production) indicating that all the glucose was anaerobically glycolyzed. Thus the anaerobic glycolysis is about 2.8 times higher than the 'aerobic glycolysis'. The results obtained with the cyanide are very similar.

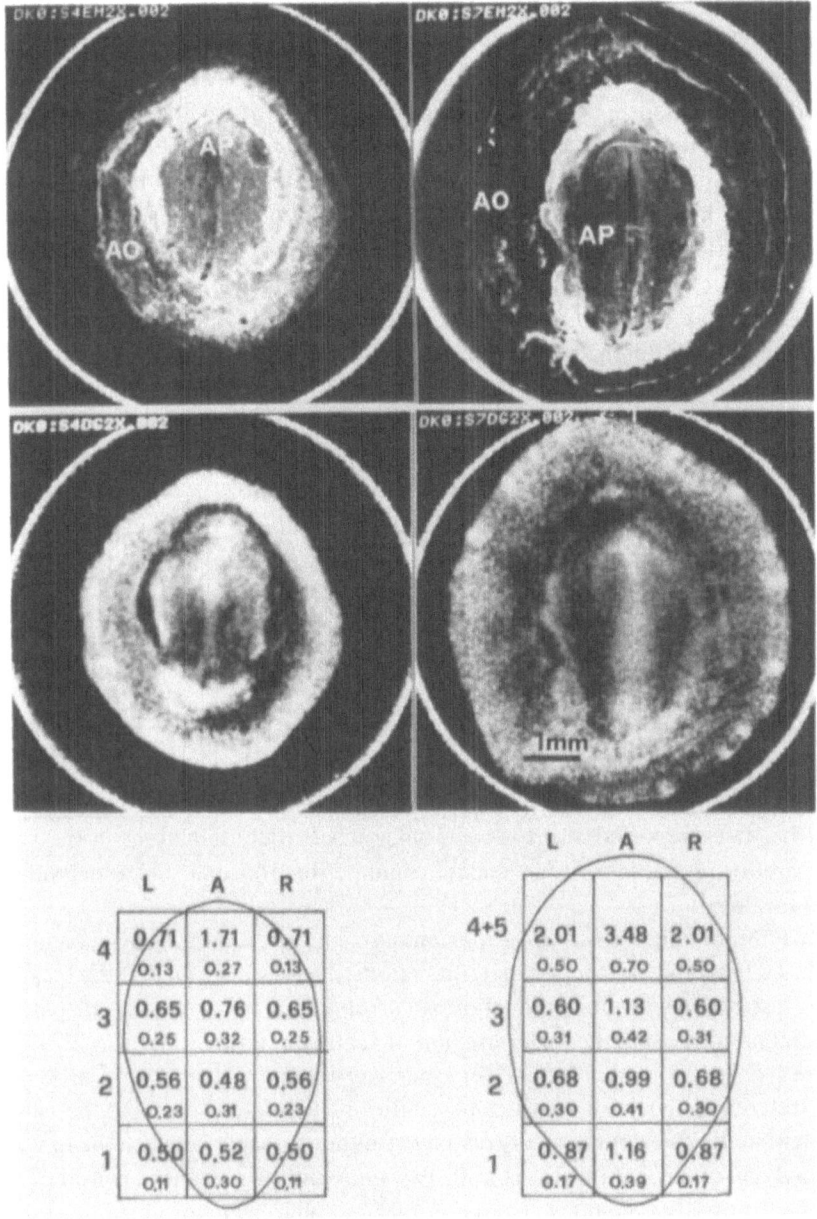

Fig. 2. Patterns of the optical density in the whole blastodiscs (top) and in the corresponding autoradiographs (center). The intensity of white is proportional to the optical density. Bottom: comparison of the uptakes of glucose (nmol·h⁻¹) as averaged within 12 regions of area pellucida (labels). Big figures are total glucose, small figures are glucose oxidized.

Left side: stage 4–5, right side: stage 6–7. AO: area opaca, AP: area pellucida (regions: L-left, A-axial, R-right; 1-posterior, 4(5)-anterior).

Discussion

All the measured parameters show a high coefficient of variation (from 16 to 50%). Our methods show the following errors: ±5% for the O_2 consumption, up to ±8% for the CO_2 and up to 12% for the HCO_3^- determinations. Thus, as the results were obtained in relatively small and different series of embryos, the main reason for the observed scatter is the (well-known) interindividual and inter-seasonal variability (1) in the true developmental age, (2) in the size of blastodiscs and (3) in the relative development of the area pellucida and area opaca. The error of glucose determination is difficult to evaluate as, for the moment, we have no comparable references. Nevertheless, the data are coherent enough to allow the evaluation of the oxidative and glucose metabolism during the gastrulation-neurulation period.

The embryos consume between 24 and $44 \, nl \cdot h^{-1} \cdot (\mu g \, protein)^{-1}$ of O_2 and between 0.4 and $0.8 \, nmol \cdot h^{-1} \cdot (\mu g \, protein)^{-1}$ of glucose. The $\dot{V}_{O_2}$ values are the highest yet found in the embryo, certainly due to the excellent in vitro culture system. The values of glucose uptake are the first reported. These metabolic rates compare well with those found in the most active differentiated tissues. For example, values between 9 and $50 \, nl \cdot h^{-1} \cdot \mu g^{-1}$ for O_2 (Fedoroff and Hertz, 1977) and between 0.3 and $1.2 \, nmol \cdot h^{-1} \cdot \mu g^{-1}$ for glucose (Sokoloff et al., 1977) have been found in active cerebral tissues.

The respiratory exchange ratio as obtained from the mean $\dot{V}_{CO_2}$ and $\dot{V}_{O_2}$ values (different series of embryos) is 1.08 for stage 4–5 but 0.74 for stage 6–7. However, as the $\dot{V}_{CO_2}$ and $\dot{V}_{O_2}$ do not differ significantly and as a respiratory exchange ratio of 1 has been reported also for even older stages (determinations in the same preparation, Needham, 1932) glucose could be the principal substrate oxidized also in stage 6–7.

In addition, glucose is to a great extent also converted aerobically to lactate as deduced from the bicarbonate determinations. The assumption that the bicarbonate loss reflects lactate production is corroborated by the results of parallel determinations using the LDH method of Gutmann and Wahlefeld (1974). At stage 4–5, 46 ± 6 nmol of lactate were found against 59 ± 24 nmol of bicarbonate, the difference being not statistically significant. It is interesting that the lactate content in the egg albumen and yolk increases most steeply between the first and second day of incubation (Romanoff, 1967), i.e. during the period considered in the present study. The rates of aerobic and anaerobic lactate production, respectively 0.75 and $2.1 \, nmol \cdot h^{-1} \cdot (\mu g \, protein)^{-1}$, are comparable to those found in several malignant tumors (see e.g. Racker, 1976) and in the rat embryos of 11–13 days (Neubert et al., 1971).

As the $\dot{V}_{CO_2}$ and lactate production were determined in a closed system in which the final pH usually decreased to about 6.8, it may be argued that the metabolism had been affected by the combined respiratory and metabolic acidosis. Comparison of these parameters with the O_2 and glucose uptakes, determined in an open and efficiently buffering system, shows that this was not the case. The $\dot{V}_{O_2}$ and $\dot{V}_{CO_2}$ on one hand, and the glucose uptake and the sum of CO_2 and lactate produced on the other hand, match rather well. Moreover, all the embryos developed at the normal speed.

It is quite probable that glucose, on which the whole blastodisc so closely depends, is supplied essentially from exogenous stores, i.e. from the albumen and yolk. An intense glycogen breakdown does occur in the embryo but at the stages of early formation of the area pellucida. Thereafter no glycogen seems to be present in the cells (Eyal-Giladi et al., 1979).

The area pellucida where the embryo develops and the area opaca from which develop the extraembryonic adnexa seem to differentiate from one another just within the period studied. In the area pellucida, metabolism roughly follows growth. Fig. 2 (bottom) indicates that the regions lying postero-lateral to Hensen's node (located in region A3) oxidize glucose more than the anterior regions where neurulation begins. More experiments are needed to confirm this pattern. In the area opaca, the metabolism is higher already at stage 4 and continues to increase. This increase, greater than the growth parameters would predict, takes place in parallel with a better efficiency of the glucose utilization found at stage 6–7. The rather moderate value of anaerobic glycolysis in the area opaca supports this idea as well.

It can be concluded that the one-day chick embryo belongs to the most active energy consuming living systems. Both oxidation and aerobic glycolysis are necessary to cover the energy demand which can be estimated at about 7–10 nmol ATP per hour and per microgram of protein.

The important question, namely how this energy is used by the developing embryo is completely open. It has been calculated that the oxidation of glucose itself can account for the whole synthesis of proteins in the area pellucida and that it liberates even more energy than necessary in the area opaca (Raddatz and Kucera, 1983). It is possible that in the latter region, a great deal of the extra energy is spent on mechanical work. As the periphery of the blastodisc expands rapidly during the studied period, the inner cells, especially in the area opaca, must withstand the resulting tensions. This would necessitate a constant and energy-dependent modulation of cytoskeletal elements. The fact that glucose uptake falls to about 50% when the blastodisc is loosened from the vitelline membrane (unpublished) strongly supports this hypothesis.

Acknowledgements

We thank Mrs. M.-B. Burnand for the expert technical aid and Dr M. Markert for the lactate determinations. This work was in part supported by the grant 3.243–0.82 from the Swiss National Science Foundation.

References

Baroffio, A. and Kucera, P. (1983). The deoxyglucose method considered for studies of glucose metabolism in the chick embryo. *Experientia* 39: 629.

Bartels, H. and Baumann, F. (1972). Metabolic rate of early embryos (4–22 somites) at varying oxygen pressures. *Respir. Physiol.* 16: 1–15.

Barzu, O. and Borza, V. (1967). Spectrophotometric method for assay of mitochondrial oxygen uptake using oxyhemoglobin as indicator and oxygen donor. *Anal. Biochem.* 21: 344–357.

Eyal-Giladi, H., Raveh, D., Feinstein, N. and Friedlander, M. (1979). Glycogen metabolism in the prelaid chick embryo. *J. Morphol.* 161: 23–38.

Fedoroff, S. and Hertz, L. (1977). Cell, Tissue, and Organ Cultures in Neurobiology. Academic Press, New York, pp. 39–71.

Gutmann, I. and Wahlefeld, A.W. (1974). L-(+)-lactate. Determination with lactate dehydrogenase and NAD. In: Methods of Enzymatic Analysis, Vol. 3. H.U. Bergmeyer, ed., Verlag Chemie Weinheim; New York, Academic Press, pp. 1464–1468.

Hamburger, V. and Hamilton, H.L. (1951). A series of normal stages in the development of the chick embryo. *J. Morphol.* 88: 49–92.

Kucera, P., de Ribaupierre, Y. and de Ribaupierre, F. (1979). Computer-controlled double-beam scanning microspectrophotometry for rapid microscopic image reconstructions. *J. Microsc.* 116: 173–184.

Kucera, P. and de Ribaupierre, Y. (1980). Evaluation of the NAD redox states in a living chick embryo using a computer-controlled double-beam scanning microfluorometry. *Microsc. Acta.* Suppl. 4: 283–287.

Kucera, P. and Raddatz, E. (1980). Spatio-temporal micro-measurements of the oxygen uptake in the developing chick embryo. *Respir. Physiol.* 39: 199–215.

Needham, J. (1932). On the true metabolic rate of the chick embryo and the respiration of its membranes. *Proc. R. Soc. London.* 110: 46–74.

Needham, J. (1932). A manometric analysis of the metabolism in avian ontogenesis. II. The effects of fluoride, iodoacetate, and other reagents on the respiration of blastoderm, embryo, and yolk-sac. *Proc. R. Soc. London.* 112: 114–138.

Neubert, D., Peters, H., Teske, S., Köhler, E. and Barrach, H.-J. (1971). Studies on the problem of 'aerobic glycolysis' occurring in mammalian embryos. *Naunyn-Schmiedebergs Arch. Pharmakol.* 268: 235–241.

New, D.A.T. (1955). A new technique for the cultivation of the chick embryo in vitro. *J. Embryol. Exp. Morphol.* 3: 326–331.

Philips, F.S. (1941). The oxygen consumption of the early chick embryo at various stages of development. *J. Exp. Zool.* 86: 257–289.

Philips, F.S (1942). Comparison of the respiratory rates of different regions of the chick blastoderm during early stages of development. *J. Exp. Zool.* 90: 83–100.

Racker, E. (1976). A New Look at Mechanisms in Bioenergetics. Academic Press, New York, pp. 162–167.

Raddatz, E. and Kucera, P. (1983). Mapping of the oxygen consumption in the gastrulating chick embryo. *Respir. Physiol.* 51: 153–166.

Reivich, M., Sano, N. and Sokoloff, L. (1971). Development of an autoradiographic method for the determination of regional glucose consumption. In: Brain and Blood Flow. Ross and Russel, eds. Pitman, London, pp. 397–400.

Romanoff, A.L. (1967). Biochemistry of the Avian Embryo. J. Wiley, New York, pp. 210 and 222.

Siggaard-Andersen, O. (1974). The Acid-Base Status of the Blood. Munksgaard, Copenhagen. pp. 29–44.

Sokoloff, L., Reivich, M., Kennedy, C., Des Rosiers, M.H., Patlak, C.S., Pettigrew, K.D., Sakurada, O. and Shinohara, M. (1977). The (14C) deoxyglucose method for the measurement of local cerebral glucose utilization: theory, procedure, and normal values in the conscious and anesthetized albino rat. *J. Neurochem.* 28: 897–916.

Spratt, N.T. (1951). Demonstration of spatial and temporal patterns of reducing enzyme systems in early chick blastoderms by neotetrozolium chloride, potassium tellurite and methylene blue. *Anat. Rec.* 109: 384–385.

Spratt, N.T. Jr. (1952). Metabolism of the early embryo. *Ann. N.Y. Acad. Sci.* 55: 40–50.

22. Influence of L-thyroxine and thiourea on metabolism and lung respiration in embryonic chicks

J. Wittmann, W. Kugler and H. Kolb[1]

Abstract

L-thryoxine (18.9 nmol) injected into chicken eggs accelerates the development of lung function, the decrease of liver metabolites, and the onset of hatching. A low dose of thiourea (13.1 μmol) exerts an opposite effect on these parameters. Under the influence of a high dose of thiourea (32.8 μmol) hatching, lung ventilation and the normal increase of somatic activity are suppressed. Furthermore, the increase of O_2 consumption and the changes of O_2 and CO_2 concentration in the air cell occurring normally towards the end of incubation are also prevented under these conditions. The results suggest a decisive role for pulmonary gas exchange in initiating hatching.

Introduction

The mechanism of thyroid hormone has been a subject of research for almost a century, but as yet there is no unifying concept describing thyroid hormone function. Thyroid hormones are known to modulate energy metabolism (Martius and Hess, 1951; Lardy, 1954; Sterling, et al., 1980) and to influence growth and differentiation (Greenberg et al. 1974). Obviously, these two effects are independently mediated.

To demonstrate the developmental effect of thyroid hormones, amphibians and avian embryos provide sensitive models because their development is directly influenced by them. As shown in previous reports (Beyer, 1952; Romanoff and Laufer, 1956; Balaban and Hill, 1971), hatching of chick embryos is accelerated under the influence of thyroid hormones, whereas antithyroid drugs like thiourea

Institut für Physiologie, Physiologische Chemie und Ernährungsphysiologie, Tierärztliche Fakultät, Ludwig-Maximilians-Universität München, Veterinärstrasse 13, D-8000 München 22, F.R.G.
[1] Klinisch-Chemisches Institut am Städtischen, Krankenhaus Harlaching, München, F.R.G.

Seymour, R.S., (ed.) Respiration and metabolism of embryonic vertebrates.
© *1984, Dr W. Junk Publishers, Dordrecht/Boston/London. ISBN-13: 978-94-009-6538-6*

retard hatching. The mechanisms underlying these effects are poorly understood. Therefore, we have undertaken a study of the mechanism leading to the acceleration or retardation of hatching under the influence of L-thyroxine and thiourea, respectively.

Visschedijk (1968) showed first, that O_2 supply plays a role in regulating the hatching process. A few years later Wangensteen and Rahn (1970/71) and Wangensteen et al. (1970/71) were able to demonstrate that in avian eggs gas exchange is determined by the egg shell permeability, its surface area and the partial pressure gradient across the egg shell. In chicken eggs the maximal capacity of O_2 uptake is nearly reached about the 16th day of incubation. During the following days the O_2 demand of the embryo is increased, especially with the onset of lung ventilation. Because the egg shell limits gas exchange, a progressive hypoxic situation develops, which may be the trigger for external pipping (Visschedijk, 1968).

As the hatching process seems to be related to the limited gas exchange, gas metabolism and lung ventilation have been studied in chicks, whose hatching has been influenced by L-thyroxine and thiourea.

Material and methods

Fertile eggs from a commercial supplier were incubated conventionally in a forced draught incubator. The air temperature was maintained at 37–38 °C and humidity at 63.7%. In order to treat the embryo with thyroid hormones and antithyroid drugs, a small hole was made into the egg shell on day 17 of incubation and L-thyroxine (18.9 nmol) or thiourea (13.1 μmol or 32.8 μmol) were injected into the eggs, avoiding direct injection of the embryo. The hole was covered with glue after the treatment.

The method used for the determination of hepatic glycogen and ATP were essentially those of Keppler and Decker (1970) and Lamprecht and Trautschold (1970). The methods have been described in more detail in a previous report (Wittmann and Weiss, 1981). Frequencies of respiration and somatic activity were recorded from pressure changes released in the air cell and at the egg shell surface. Technical details were reported in previous papers (Kugler et al., 1982; Wittmann et al., 1983).

To measure partial pressures of O_2 and CO_2 in the air cell and the O_2 consumption of the embryo, methods described by Tazawa et al. (1980) and Scholander and Edwards (1942), respectively, were used.

Blood samples were collected by heart puncture and centrifuged for 10 min at 2000 xg. The plasma was stored at -20 °C. Plasma L-thyroxine was determined by radioimmunoassay using kits from Clinical Assay.

Results and discussion

In chicken liver, glycogen and ATP levels decrease at the end of incubation (Freeman, 1965; Freeman, 1971; Wittmann and Weiss, 1981). On the basis of our previous experiments (Wittmann and Weiss, 1981) it seems very likely that the changes in the levels of these metabolites are produced by hypoxia formed at this developmental stage.

Injected L-thryoxine (18.8 nmol) into the eggs on day 17 of incubation accelerates hatching by 24 h and decreases hepatic glycogen and ATP levels (Table 1). The reverse is observed when eggs are treated with 13.1 μmol thiourea. Furthermore a premature onset of lung ventilation is induced by treating the embryos with L-thyroxine, whereas thiourea exerted a retarding effect on lung respiration (Wittmann et al., 1983).

These findings suggest that the premature onset of lung ventilation leads to a premature formation of hypoxia within the egg, which may be the trigger for the accelerated hatching and for the changes of hepatic glycogen and ATP levels. On the other side, the retarding effect of thiourea on metabolism and hatching may be caused by the delayed formation of hypoxia due to delayed lung ventilation.

It is reasonable to suggest that these effects of thiourea are brought about by the inhibition on thyroid hormone secretion. As shown in Fig. 1, in controls, L-thyroxine levels are increased at the end of incubation. Treating the embryos with thiourea prevented this thyroid hormone increase.

The retarding effect on metabolism and development was evoked with a comparatively low dose of thiourea (13.1 μmol). In previous experiments (Romanoff and Laufer, 1956; Balaban and Hill, 1971), treating the embryo with higher amounts of thiourea was found to prevent chick embryos from hatching. In our experiments about 70–80% of the embryos were prevented from hatching by

Table 1. The influence of L-thyroxine (18.8 nmol) and thiourea (13.1 μmol), injected in chicken eggs on day 17, on hepatic glycogen and ATP levels on day 20.

Treatment	Incubation (d)			
	17	20	17	20
	glycogen (μmol glucose/g wet mass)		ATP (μmol/g wet mass)	
Control	11.2 ± 25.2	78.8 ± 11.2	1.65 ± 0.17	1.04 ± 0.14
	(7)	(6)	(7)	(8)
Thiourea		143.18 ± 20.45***		1.39 ± 0.25*
		(6)		(6)
L-thyroxine		34.8 ± 9.8***		0.82 ± 0.14**
		(6)		(12)

* P<0.05; ** P<0.01; *** P<0.001

Results are given as means ± s.d. Significance of changes was tested by Student's t-test. The number of determinations is given in parentheses.

treating them with 32.8 µmol thiourea on day 17 of incubation, although almost 40% survived until day 25 or longer. As a rule, internal but not external pipping was performed.

If external pipping is induced by active breathing, then it would be reasonable to suggest that in chick embryos prevented from hatching, the development of lung function is disturbed. In the following experiments, therefore, ventilation activity was recorded in controls and thiourea-treated embryos. Because lung ventilation may exert an influence on somatic activity (Wittmann et al., 1983), somatic movements were also registered.

In control embryos, lung respiration is initiated at day 20 of incubation with a continuous increase in ventilation frequency (Fig. 2). The development of lung function is followed by a marked increase in somatic activity. However, under the influence of thiourea, lung ventilation was prevented (Fig. 3). Concomitantly, the significant increase of somatic activity as recorded in controls was also prevented. In a few cases of thiourea-treated chicks active breathing started, but ceased again several hours later. Like lung ventilation, the initial increase of somatic activity was also followed by a decrease in somatic activity in those animals.

In contrast to the controls no increase in O_2 consumption under the influence of thiourea was observed on day 21 of incubation (Table 2). Furthermore, the increase in CO_2 and the decrease in O_2 in the air cell due to lung respiration did not occur in thiourea treated chicks (Table 2). However, there was no significant difference in the respiratory exchange ratio (RE) between controls and the

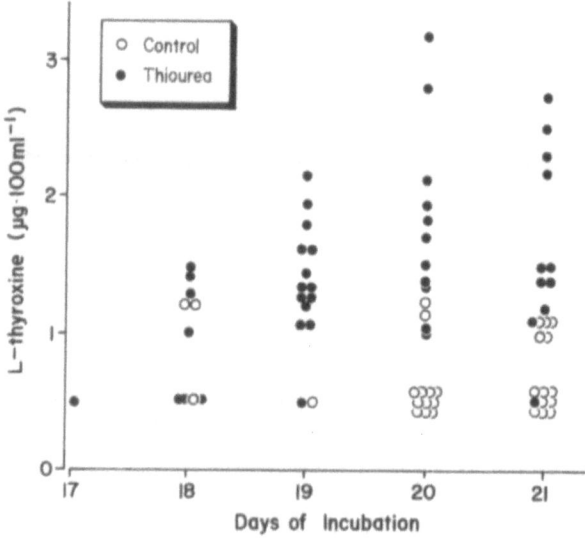

Fig. 1. The influence of thiourea on plasma concentration of L-thyroxine. In the experimental group 13.1 µmol thiourea was injected on day 17 of incubation.

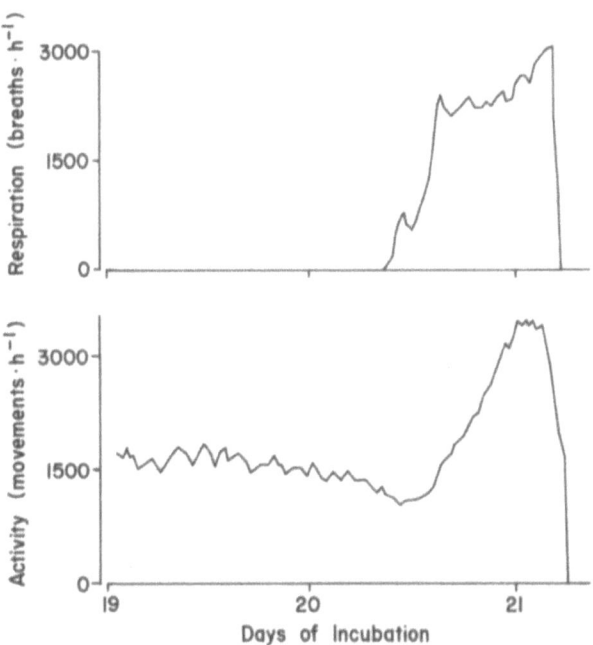

Fig. 2. Activity and respiratory pattern of embryonic chicks at the end of incubation.

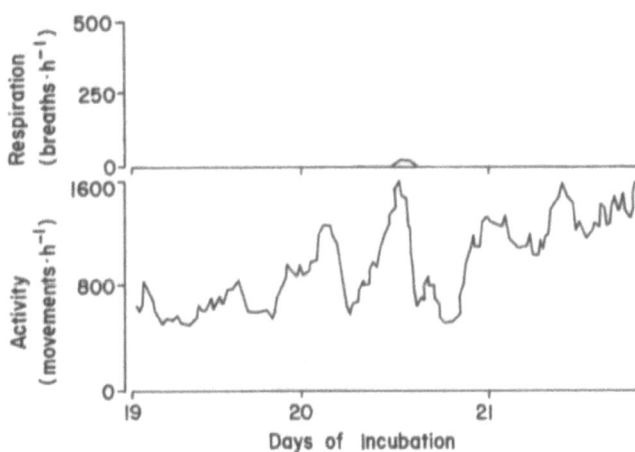

Fig. 3. Activity pattern of a chick embryo with suppressed lung respiration. In the experimental group 32.8 μmol thiourea was injected into the egg on day 17 of incubation.

315

Table 2. The influence of thiourea (32.8 μmol), injected on day 17, on O_2 consumption and on O_2 and CO_2 content in the air cell of embryonic chickens.

| | Incubation (d) | | | |
	17	21	18	20
	O_2 consumption (ml·h^{-1} STPD)			
Control	17.1 ± 3.5	31.1 ± 7.2	O_2% 18.22 ± 0.76	16.21 ± 1.02
	(7)	(11)	(13)	(7)
			CO_2% 2.38 ± 0.63	4.16 ± 0.72
			(13)	(7)
			RE 0.69	0.69
Thiourea		10.0 ± 3.74***	O_2% 18.53 ± 0.50	18.35 ± 1.08***
		(29)	(7)	(10)
			CO_2% 2.39 ± 0.46	2.59 ± 0.90***
			(7)	(10)
			RE 0.78	0.78

$P < 0.001$ RE = respiratory exchange ratio
Results are given as means ± s.d. Significance of changes was tested by Student's t-test. The number of determinations is given in parentheses.

experimental group. These findings suggest a decisive role for pulmonary gas exchange in initiating hatching in chicken embryos. Because active breathing is suppressed in embryos treated with the high dose of thiourea, no hypoxic situation (Table 2) develops and, as a consequence, the chick does not pip externally. In the intact egg presumably insufficient O_2 is available for the vigors of hatching. This conclusion may be drawn from the low somatic activity in embryos prevented from hatching (Fig. 3).

As pointed out above, the effects produced by the low dose of thiourea, may be attributed to its antithyroid action. However, the suppressive effect on hatching and lung ventilation, due to high doses of thiourea is possibly not or only partially caused by the inhibition of thyroid hormone secretion. In preliminary studies we observed an increased incorporation of ^{14}C-leucine into lung protein from embryos treated with the high doses of thiourea. Due to the anabolic effects of thyroid hormones, a decrease rather than an increase in labeled protein would have been expected, if the antithyroid action of thiourea is prevailing. Hollinger et al. (1974) observed the same phenomenon in rat lung and suggested that edemagenic changes of the lung that are caused by thiourea are involved in the enhanced incorporation of labeled amino acid into protein. Finally, we found that pulmonary hemoglobin extinction (ΔE_{546nm}/100 mg dry weight) increases from 104.4 ± 16.7 on day 17/18 of incubation to 190.2 ± 25.6 on day 21 of incubation. In thiourea treated lungs no increased of hemoglobin extinction has been observed. This finding supports the contention of a restricted lung circulation. In this context it is interesting to note that more recent studies of the influence of thiourea (Dauber et al., 1983) suggest an enhancement of lung leucotrienes, which are potent vasoconstrictors. More work is needed to elucidate the mechanism leading to the prevention of hatching by high doses of thiourea.

Acknowledgements

We are indebted to Prof. H. Rahn, State University of New York at Buffalo, for his advice and instructions. This research was supported by the Deutsche Forschungsgemeinschaft.

References

Balaban, M. and Hill, J. (1971). Effects of thyroxine level and temperature manipulations upon the hatching of chick embryo (*Gallus domesticus*). *Dev. Psychobiol.* 4: 17–35.

Beyer, R.F. (1952). The effect of thyroxine upon the general metabolism of the intact chick embryo. *Endocrinology* 50: 497–503.

Dauber, J., Pluss, W., Stenmark, K., Weil, J. and Voelkel, N. (1983). Leukotrienes in thiourea-induced lung injury edema. *Ann. Rev. Resp. Dis.* 127: 304.

Freeman, B.M. (1965). The importance of glycogen at the termination of the embryonic existence of *Gallus domesticus*. *Comp. Biochem. Physiol.* 14:217–222.

Freeman, B.M. and Manning, A.C. (1971). Glycogenolysis and lipolysis in *Gallus domesticus* during the perinatal period. *Comp. Gen. Pharmac.* 2: 198–204.

Greenberg, A.H., Najjar, S. and Blizzard, R.M. (1974). Effects of thyroid hormone on growth differentiation and development. In: handbook of Physiology, Vol. III. Monte A. Green and D.H. Soleman, eds., American Physiological Society, Washington, D.C., pp. 373–390.

Hollinger, M.A., Giri Shri, N., Hwang, F. and Budd, E. (1974). Effect of thiourea on rat lung protein Synthesis. *Res. Commun. Chem. Pathol. Pharmacol.* 8(2): 319–326.

Keppler, D. and Decker, A.K. (1970). Glykogen. In:Methoden der enzymatischen Analyse. H.U. Bergmeyer, ed., pp. 1089–1093.

Kugler, W., Wittmann, J. and Petry, H. (1982). Kontinuierliche Registrierung von Motilität und Lungenatmung in der Pränatalen Phase des Kükens. *Zbl. Vet. Med.* A 29: 395–402.

Lamprecht, W. and Trautschold, D. (1970). ATP-Bestimmung. In: Methoden der enzymatischen Analyse, H.U. Bergmeyer, ed., pp. 2024–2033.

Lardy, H.A. (1954). Effect of thyroid hormone on enzyme systems. *Brookhaven Symp. Biol.* 7: 90–97.

Martius, C. and Hess, B. (1951). The mode of action of thyroxine. *Arch. Biochem. Biophys.* 33: 468–487.

Romanoff, A.L. and Lauffer, H. (1956). The effect of injected thiourea on the development of some organs of the chicken embryo. *Endocrinology* 56: 611–619.

Scholander, P.F. and Edwards, G.A. (1942). Micro-respiration apparatus. *Rev. Sci. Instru.* 13: 292–295.

Sterling, K., Brenner, M.A. and Sakurado, T. (1980). Rapid effect of triiodo-thyronine on the mitochondrial pathway in rat liver in vivo. *Science* 210: 340–343.

Tazawa, T., Ar, A., Rahn, H. and Piiper, J. (1980). Repetitive and simultaneous sampling from the air cell and blood vessels in the chick embryo. *Respir. Physiol.* 39: 265–272.

Visschedijk, A.H.J. (1968). The air space and embryonic respiration. *Br. Poult. Sci.* 9: 173–210.

Wangensteen, O.D. and Rahn, H. (1970/71). Respiratory gas exchange by the avian embryo. *Respir. Phys.* 11: 31–45.

Wangensteen, O.D., Wilson, D. and Rahn, H. (1970/71). Diffusion of gases across the shell of the hen's egg. *Respir. Physiol.* 11: 16–30.

Wittmann, J. and Weiss, A. (1981). Studies on the metabolism of glycogen and adenine nucleotides in embryonic chick liver at the end of incubation. *Comp. Biochem. Physiol.* 69C: 1–6.

Wittmann, J., Kugler, W. and Petry, H. (1983). Motility pattern and lung respiration of embryonic chicks under the influence of L.-thyroxine and thiourea. *Comp. Biochem. Physiol.* 75A: 379–384.

23. Patterns of lung aeration in the perinatal period of domestic fowl and Brush Turkey

Roger S. Seymour

Department of Zoology, University of Adelaide, Adelaide, S.A., 5001, Australia

Abstract

During the transition from chorioallantoic to pulmonary respiration in domestic fowl, the embryonic lungs are aerated by displacement of fluid and lung expansion due to growth. Initial lung aeration may occur initially by convection of fluid from the parabronchial lumina. The total lung volume increases almost twice as fast as the rest of the body in this period, but the air-free volume changes little. Thus the proliferation of parabronchial air capillaries appears to be responsible for aspiration of air to the gas exchange surfaces. Aeration is most rapid during the interval between internal and external pipping, but continues through hatching; at hatching, the total lung organ volume, relative to yolk-free embryo mass, is $17.7 \, ml \cdot kg^{-1}$ and air represents 44% of this volume.

Eggs of the Brush Turkey, a megapode, lack an air cell because they are incubated in mounds where high humidity restricts evaporative water loss. Therefore the embryos neither pip internally nor ventilate the lungs prior to hatching. There is no air in the lungs before the chorioallantois is destroyed and the first breath taken. Hatching is rapid and the lungs become aerated during the first hours after hatching; O_2 consumption rises in parallel with lung aeration. Megapodes may be able to hatch 'viviparously' because the abnormally thin shell makes hatching quick and energetically inexpensive. Other birds probably require the development of higher capacity for pulmonary gas exchange before they can successfully hatch from thicker shells.

Introduction

A transition from chorioallantoic to pulmonary gas exchange occurs during the last days of incubation in domestic fowl (Visschedijk, 1968a, b). After pipping internally (IP), the chick begins to breathe within the air cell created by evaporative water loss during the course of incubation. Internal pipping begins the

Seymour, R.S., (ed.) Respiration and metabolism of embryonic vertebrates.
© *1984, Dr W. Junk Publishers, Dordrecht/Boston/London. ISBN-13: 978-94-009-6538-6*

perinatal* period. During the first 12 h, lasting until external pipping (EP), there is a gradual rise in pulmonary O_2 and CO_2 exchange while chorioallantoic respiration is constant. For about 16 h between external pipping and hatching (H), pulmonary O_2 and CO_2 exchange increase further and chorioallantoic gas exchange progressively drops. At hatching, the chorioallantois is non-functional and the lung assumes total importance. Thus domestic fowl maintain the chorioallantoic gas exchange while developing pulmonary function.

This gradual shift in mode of gas exchange is thought to be unique to birds and necessary because of the structure of the adult respiratory system. Air flow is unidirectional through a relatively non-distensible lung (Duncker, 1972; Macklem et al., 1979; Scheid, 1979). A constant-volume lung appears to be essential in birds because collapsed air capillaries, with diameters of 3–10 μm (Duncker, 1971), could not be reinflated against high surface tension existing within them, despite highly developed surfactant film (Pattle, 1978).

Similarly, the fluid-filled embryonic lungs develop within an apparently inexpansible pleural space, deeply incised by the ribs and bounded ventrally by the horizontal and oblique septa.

Therefore ventilation by lung expansion, which results in the major part of alveolar aeration in mammals, does not appear to be effective in hatchling birds. Nevertheless, the parabronchi and their air capillaries must be aerated before the avian lung can become an effective gas exchanger.

When the embryo pips internally into the air cell of the egg, the first breaths aerate the primary bronchi and air sacs. It is then possible to aerate the lung by one, or a combination, of hypothetical mechanisms: (1) reabsorption of the pulmonary fluid into the blood or lymph capillaries, (2) convection of fluid from the parabronchial lumina as air is forced through them from the air sacs, and (3) increase in total lung volume with growth and proliferation of air capillaries which thereby aspirate air.

To assess the importance of these mechanisms, this study documents changes in lung mass and density during the perinatal period in domestic fowl (*Gallus domesticus*) and correlates the data with the anatomy of the parabronchi and the development of pulmonary gas exchange. The state of lung aeration in embryonic domestic fowl has been observed in a qualitative way by Vince and Tolhurst (1975). The present study extends their investigation on a volumetric basis.

The data from domestic fowl, which begin breathing long before hatching, are then compared with data from the Brush Turkey (*Alectura lathami*). The embryos of these magapode birds develop in an incubation mound where high humidity reduces evaporative water loss to levels where no useful air cell forms (Seymour and Ackerman, 1980). Therefore pulmonary respiration is prevented until the chick breaks through the shell and destroys the chorioallantois. There

---

* The 'perinatal' period includes both the 'paranatal' period (between IP and H) and an undefined neonatal period after H.

320

appears to be practically no overlap in chorioallantoic and pulmonary gas exchange. This 'viviparous' pattern of lung development and aeration may represent a further adaptation associated with the unique form of incubation in megapode birds.

Material and methods

Fertile domestic fowl eggs were obtained from a local hatchery and incubated at 37.5 °C in a forced convection incubator. From the onset of day 18 until 2 d after hatching, eggs were taken at selected stages of incubation as judged from the presence of pip holes in the air cell membranes or the shell. Lung air volume was determined with the method of Burton and Smith (1968). Through the chorioallantoic membrane, the neck was broken and the trachea clamped simultaneously with a hemostat. The yolk sac was carefully removed intact and the embryo weighed. The skin, extremities, and internal organs were removed from the thorax, leaving the lungs in situ behind the horizontal and oblique septa. The thorax (with lungs) was quickly weighed (to 0.1 mg) with a Mettler AE 163 balance, first in air and then in a cage suspended in container of water. The water level was maintained constant by zeroing the balance with the thorax in the water but not in the cage. Then the lungs were carefully separated from the thorax, leaving extraneous tissue attached to the thorax, and the lungs were weighed in air and the thorax in water. The density of air-free lung tissue was determined by weighing pre-IP and degassed lungs in water. Degassing occurred in a glass 10 ml saline-filled syringe by pulling the plunger back several times and ejecting the bubbles.

In another series of experiments, eggs were incubated synchronously and embryos were selected at 2–3 h intervals. Because Burton and Smith (1968) noted that air could enter the vasculature in lungs removed in air, the lungs in this series were removed under water before determination of density. Although care was taken to avoid compression and loss of gas, this procedure resulted in the loss of about 15% of total lung volume, and about 36% of the gas, in late stage embryos. These data, therefore, were useful only in demonstrating the trends within each stage.

Both lungs in the latter experiment were fixed in 4% formalin. Right lungs were paraffin embedded, sectioned (7 μm) roughly perpendicular to the parabronchi at three locations (i.e. near the mediodorsal secondary bronchi, middle of the parabronchi, and near the medioventral secondary bronchi), H–E stained and examined. Measurements of the parabronchial lumen diameter and the distance between adjacent parabronchi were made with an optical micrometer.

Lungs of five Brush Turkey embryos were removed under water because it was important to determine whether any air existed in the lungs before hatching. The eggs were collected on Kangaroo Island and incubated in the laboratory in mound material at 34 °C. The embryos were aged from a series of measurements of O_2 consumption, calibrated against a standard curve generated from 11 embryos allowed to hatch (Vleck et al., 1984). The ages of unhatched embryos were estimated to be 2 d before hatching and the day of hatching. The age was substantiated by the mass of their yolk sacs (Table 2). In Mallee Fowl (*Leipoa ocellata*) the yolk sac shows a decline of $3.86 \text{ g} \cdot \text{d}^{-1}$ during the last few days of incubation (Vleck et al., 1984). Two hatched embryos were sampled, one at about 5–10 min, and the other 4 h, after the onset of pulmonary respiration. The lungs of 6 Brush Turkey embryos and hatchlings were examined histologically.

Rates of O_2 consumption ($\dot{V}_{O_2}$) were measured by open-flow respirometry (Vleck et al., 1984) in three Brush Turkey embryos during and after hatching. Meanwhile, the birds were observed through a clear acrylic chamber.

Results

Aeration of domestic fowl lungs

An increase in total lung organ volume during the perinatal period was accounted for by increases in both air-free and air-filled volume (Table 1). The data show that the lung was increasing in size almost twice as fast as the rest of the body. The progressive increase in air was correlated with a drop in lung density.

Because the embryos were growing, it is more useful to express lung volumes relative to yolk-free embryo mass (Fig. 1). Expressed this way, the air-free volume changed little during the perinatal period, except for a statistically significant drop of $2.53 \text{ ml} \cdot \text{kg}^{-1}$ (95% CI of difference = ±1.79), or 21%, between pre-IP and IP embryos. The decrease in air-free volume during the IP stage was accompanied by an even larger increase in air volume of $4.70 \text{ ml} \cdot \text{kg}^{-1}$ (95% CI = 3.10), thus producing a net increase in total lung volume.

The density of pre-IP and degassed lung tissue was $1.03 \text{ g} \cdot \text{cm}^{-3}$ (Table 1), a

Table 1. Lung volume and density in domestic fowl embryos before and during the perinatal period. Stages are described in the text. Data ($\bar{x} \pm 95\%$ CI) are from lungs within the thorax.

Stage	n	Yolk-free embryo mass (g)	Volume (μl) Air	Air-free	Total	% Air	Density ($g \cdot cm^{-3}$)
Pre–IP	13	26.50 ± 2.29	-5 ± 2	314 ± 37	309 ± 37	-1.6 ± 0.8	1.03 ± 0.01
IP	7	29.71 ± 4.04	142 ± 102	275 ± 47	418 ± 116	30.5 ± 17.8	0.72 ± 0.18
EP	9	36.61 ± 3.26	262 ± 65	352 ± 47	615 ± 85	42.0 ± 6.9	0.60 ± 0.07
Hatch <1 d	7	35.65 ± 4.04	279 ± 41	349 ± 42	628 ± 75	44.4 ± 2.8	0.57 ± 0.03
Hatch 2 d	3	46.04 ± 9.03	466 ± 243	492 ± 248	958 ± 450	48.7 ± 10.8	0.53 ± 0.11

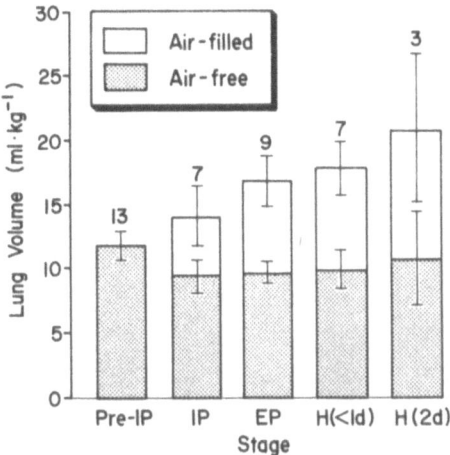

Fig. 1. Lung volumes (ml · kg⁻¹ yolk-free embryo) of domestic fowl during the perinatal period. Total lung volume is divided into air-filled (gas space) and air-free (tissues plus fluids) volumes. Stages are before internal pipping (pre-IP), internally pipped (IP), externally pipped (EP), and during the first or second day after hatching [H(<1d), H(2d)]. Data are means and 95% CI.

value similar to data of King and Payne (1962) and Burton and Smith (1968) for adult domestic fowl. At hatching time, the mass-specific total organ volume was $17.7 \, \text{ml} \cdot \text{kg}^{-1}$ for embryos with the thorax intact (Table 1). Two days after hatching, the total organ volume was $20.7 \, \text{ml} \cdot \text{kg}^{-1}$. These values are similar to data for 5- and 9-week-old domestic fowl (18.4 and $15.5 \, \text{ml} \cdot \text{kg}^{-1}$) but are considerably above the average for adults ($12.2 \, \text{ml} \cdot \text{kg}^{-1}$; Burton and Smith, 1968). Thus the mass-specific total lung volume is maximum shortly after hatching and decreases as the bird approaches adulthood. At hatching, the chick lung volume is 44% air and the fraction appears to increase after hatching (Table 1). Lungs of older domestic fowl (5 weeks to adult) hold about 70% air (Burton and Smith, 1968).

Air first appeared in the lungs of domestic fowl embryos about 35 h before predicted hatching (Fig. 2). All embryos prior to this time were pre-IP. Two living embryos, however, were pre-IP at 30 and 16 h before predicted hatching time. After IP, the lungs became positively buoyant and there was a progressive increase in air fraction throughout the perinatal period. A regression line, fitted by the least squares method to fractional air content of aerated lungs and incubation time, showed a significant increase ($r = 0.77$, $N = 32$) (Fig. 2). This line extrapolates to zero air volume at 50 h before predicted hatching, or 15 h before any air appeared in the lung.

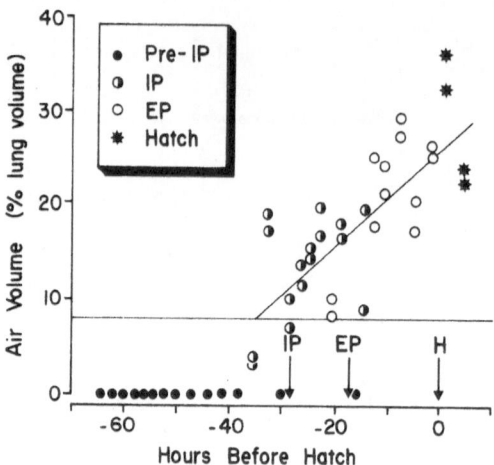

Fig. 2. Air volume in lungs removed from embryonic or hatchling domestic fowl which were incubated synchronously. The nominal times of internal pipping (IP), external pipping (EP) and hatching (H) are from Visschedijk (1968a). The horizontal line represents the volume of the parabronchial lumina taken from histological sections (Fig. 4). Air volume is underestimated because an unknown amount of gas was lost from lungs removed from the thorax under water.

Histology of parabronchi of domestic fowl

Sections of parabronchi of domestic fowl embryos showed the development of the gas exchange surface in agreement with descriptions and photomicrographs presented by Duncker (1978). In pre-IP lungs, the parabronchial mantle appeared dense because there were few infundibula and air capillaries (Fig. 3a, b). The number and size of infundibula increased in older embryos regardless of the state of aeration of the lung. For example, in the first embryo examined at 64 h before hatch, there were about 5–8 infundibula visible in cross section, but in the embryo which was still pre-IP 16 h before hatch, there were 11–15 wide infundibula visible. The infundibula were widest at the EP and H stages and the mantle appeared very porous (Fig. 3d, e). There were very few air capillaries in pre-IP lungs (Fig. 3c) but they were abundant at the EP stage (Fig. 3f).

The histological sections of domestic fowl lungs showed a constant rate of increase in size of the parabronchi (Fig. 4). Assuming that the mean cross-sectional dimensions represent the entire lung, one can estimate volume. At IP, the volume of the parabronchial lumina was 8.1% of the total parabronchial volume. Total parabronchial volume increased 35% between IP and H. This is about half of the gross increase in organ volume observed over the same period in intact lungs within the thorax (Table 1) and points to partial collapse of lungs removed from the thorax.

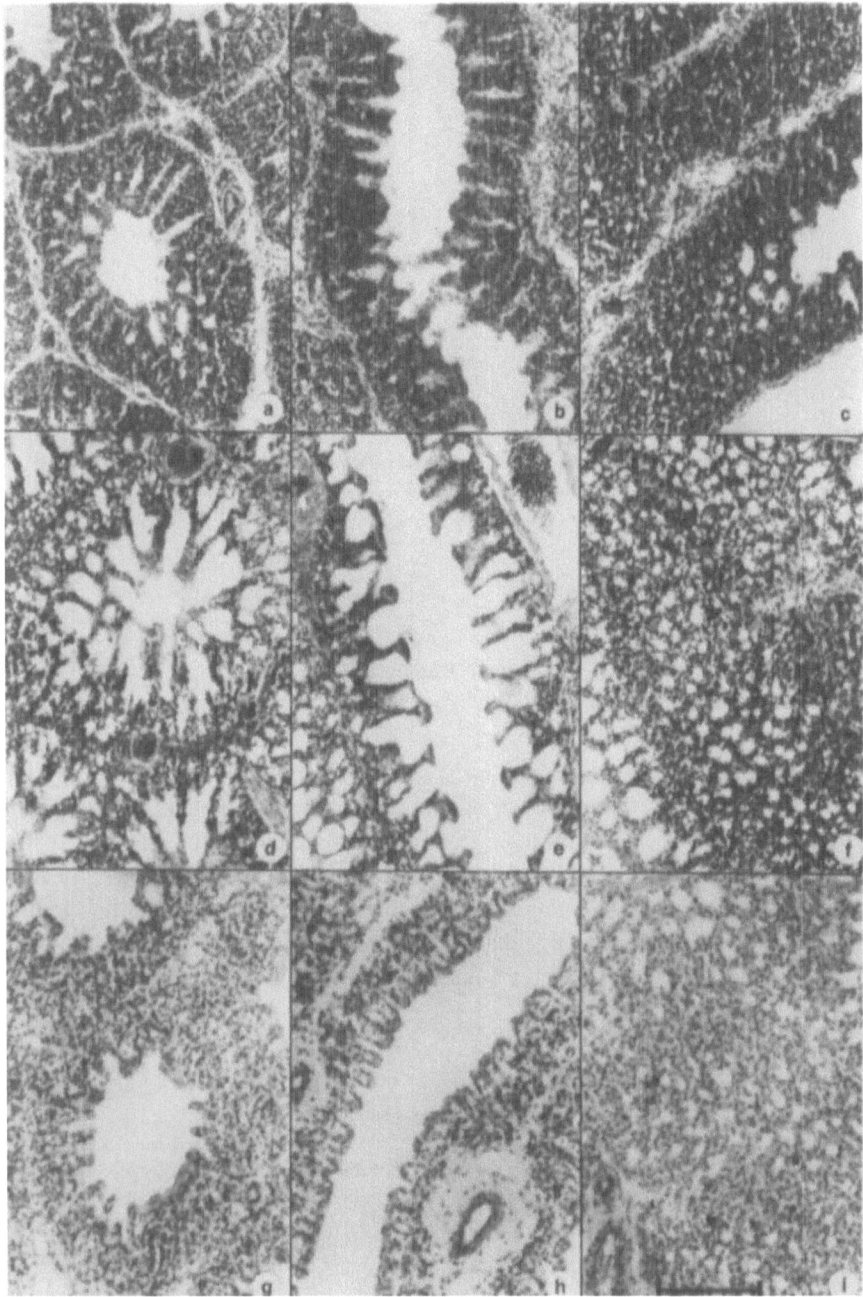

Fig. 3. Histological sections of parabronchii of embryonic domestic fowl (DF) and Brush Turkeys (BT). a: DF, cross section, pre-IP, ca. 60 h before hatch. b: DF, long section, pre-IP, ca. 58 h before hatch. c: DF, oblique section, pre-IP, ca. 50 h before hatch. d: DF, cross section, EP, ca. 1 h before hatch. e: DF, long section, EP, ca. 1 h after hatch. f: DF, oblique section, ca. 10 h before hatch. g: BT, cross section, ca. 4 d before hatch. h: BT, long section, ca. 7 d before hatch. i: BT, oblique section, ca. 4 h after hatch. All formalin fixed, wax embedded, sectioned 7 μm thick and stained with hematoxylin and eosin. The scale bar is 100 μm.

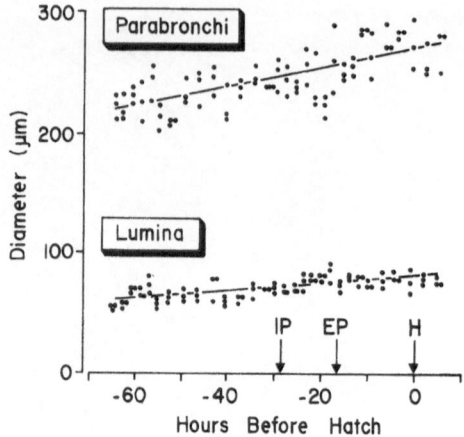

Fig. 4. Diameters of parabronchi and their lumina during with the final stages of incubation of the domestic fowl. Lines were fitted by least squares regression.

Aeration of Brush Turkey lungs

Three pairs of lungs removed before hatching were negatively buoyant. One of these was removed in the field approximately 7 d before hatching and could not be weighed, but the other two each had densities of $1.023 \, g \cdot cm^{-3}$ (Table 2), similar to the values in domestic fowl (Table 1). Within 5–10 min after the fluid drained from another egg at hatching, lung density was 0.987 which corresponds to an air volume of 2.8% (Table 2). In the embryo 4 h post hatch, the density was 0.840, or 19.2% air. These results demonstrate that there is no aeration of the lung within the egg prior to hatching and that aeration can occur rapidly after the first breath. The rate of aeration, calculated from these two birds, is $4.2\% \cdot h^{-1}$, or approximately eight times faster than in domestic fowl (Fig. 2).

Table 2. Lung aeration in Brush Turkey embryos and hatchlings.

Chick	Stage*	Yolk-free body mass (g)	Yolk mass (g)	Lung mass (g)	Lung air vol (μl)	Lung total vol (μl)	% Air	Density $(g \cdot cm^{-3})$
1	≈2 DBH	118.7	22.6	1.15	0	1.12	0	1.023
3	<1 DBH	106.9	16.9	1.29	0	1.26	0	1.023
8	5–10							
	MAH	108.6	14.9	1.20	34	1.22	2.8	0.987
2	4 HAH	111.2	12.4	1.14	261	1.36	19.2	0.840

* DBH = days before hatch   MAH = minutes after hatch   HAH = hours after hatch

326

Histology of Brush Turkey lungs

The four Brush Turkey lungs examined at stages near hatching were similar. The dimensions of both the lumen and mantle of the parabronchi were slightly larger, and the mantle denser, in Brush Turkeys (Fig. 3g, h) than in hatchling domestic fowl. In Brush Turkey lungs ($N = 6$), the parabronchial lumen volume was 7.3% of the total lung volume. Duncker (1978) shows that the lungs of altricial birds are less well developed than those of precocial species. In altricial species, the parabronchi are small and the mesenchyme mantle is thin with poorly developed infundibula and air capillaries. On the other hand, parabronchi of precocial species are larger and possess thicker mantles, deeper infundibula and well developed air capillaries (Fig. 3i). One sees a precocious development of air capillaries in Brush Turkey lungs beginning at about 4 days before hatching.

Hatching in Brush Turkeys

$\dot{V}_{O_2}$ averages $61 \, \text{ml} \cdot \text{h}^{-1}$ in Brush Turkey embryos immediately before hatching (Vleck et al., 1984). However $\dot{V}_{O_2}$ is markedly unstable during the last 3 d of incubation, presumably because of periodic movements of the embryo. Baltin (1969) ascribed these movements to attempts to hatch, but there is no evidence to support this notion. He is correct, however, in reporting that the shell is broken suddenly by pressure exerted by the legs and the back. Shell fragments often break off of the egg before the membranes are broken because the shell separates from the single shell membrane in late development. When the shell membrane is exposed, either naturally or artificially in eggs about to hatch, $\dot{V}_{O_2}$ drops to about $40–50 \, \text{ml} \cdot \text{h}^{-1}$ for a variable period. When the feet eventually rip long slits through the chorioallantois and shell membrane, about 25 ml of fluid drains from the egg and the chick immediately begins deep breathing. On some occasions, a little blood is lost from the chorioallantois but on others, almost none appears in the fluid. There is no blood flow in the chorioallantois when it is first adequately visible, as little as 3 min after it is ruptured.

$\dot{V}_{O_2}$ has been measured in three Brush Turkey embryos during and after hatching. The onset of pulmonary respiration marks the beginning of a slow rise in $\dot{V}_{O_2}$ in the first hours after hatching; most of the change occurs during the first hour (Fig. 5). During this period, the chick's movements are limited to ventilation (ca. $68–112 \, \text{min}^{-1}$) and occasional struggling from the remains of the shell. It requires about 1 h for the chick to free itself completely from the membranes and shell. Chicks buried in acrylic cylinders filled with mound material rest for 3–4 h after hatching and then begin to burrow up to the surface. Bouts of digging alternate with long periods of rest when the chicks may be observed to ventilate heavily. In the field, emergence from the mound takes about 2 d (Baltin, 1969; Vleck et al., 1984).

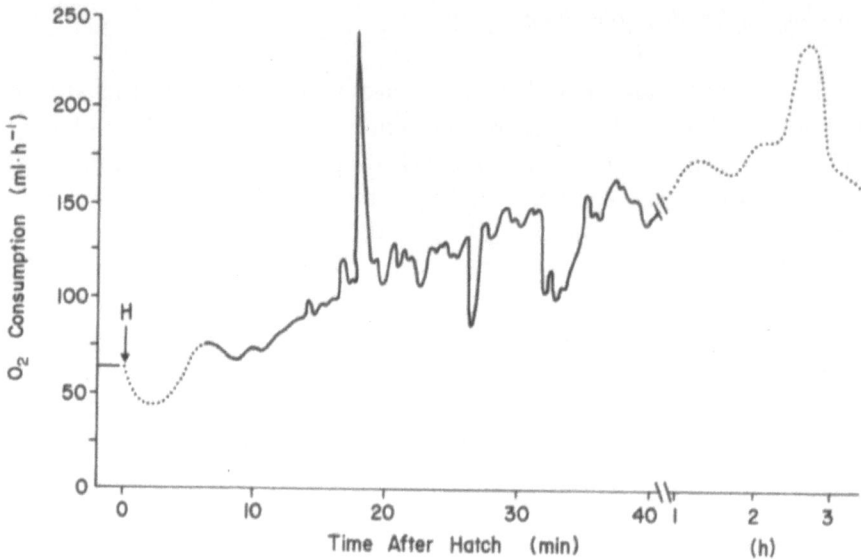

Fig. 5. Oxygen uptake during and after hatching in a Brush Turkey. The time of hatching was not exactly known but the chorioallantois was ruptured during the first 5 min after the point H. Although O_2 uptake was not recorded during this time, an indication of the drop observed in other hatchlings is presented. Although the lower pole of the egg was broken off, the chick's body and head remained in the upper part of the shell; when the upper part was broken at 18 min, the stale gas around the chick was released. Ignoring this artifact, maximum O_2 uptake is seen to increase gradually in the first few hours after hatching.

Discussion

Aeration of domestic fowl lungs

During the perinatal period, air apparently enters the domestic fowl lung by fluid replacement and total lung growth. Reabsorption of fluid is indicated by the 21% decrease in air-free lung volume during the period between IP and EP (Table 1, Fig. 1). However, less fluid appears to leave the lung than the volume of air entering during this period, because total lung volume increases (Table 1). After EP, the air-free volume begins to increase again, possibly due to a combination of at least three factors: (1) a slowing of fluid removal from the lung, (2) continued proliferation of pulmonary parenchyma and (3) an increase in pulmonary blood volume associated with the decrease in chorioallantoic gas exchange that begins at EP (Visschedijk, 1968a).

Removal of parabronchial fluid and its replacement with air can occur by forced convection and by reabsorption according to Starling forces. Some degree of fluid convection is a possibility in the flow-through parabronchi of the avian lung. The fluid in the lumina might be flushed out by air forced through from the

328

air sacs. One aspect of the present data consistent with forced convection is the scarcity of aerated lungs containing a small volume of air (Fig. 2). Only one of 15 IP embryos in both series of measurements had an air volume less than the parabronchial lumen volume of 8.1%.

However, it is difficult to conceive how air from the air sacs could be directed through the parabronchi when the primary bronchi and medioventral secondary bronchi are so much wider. There are no anatomical valves in these large bronchi (Scheid, 1979), and the surface tension should be much lower there. Air should leave the air sacs preferentially via the larger bronchi. However, the parabronchi of the neopulmo may be aerated by convection, especially in passerine and fowl-like birds in which the neopulmo is interposed between the abdominal air sacs and the primary bronchus (Duncker, 1971, 1972).

Even if the fluid in the parabronchi is convected away, this mechanism cannot aerate the air capillaries because they form a meshwork that nevertheless is functionally blind-ended in the mantle of individual parabronchi (Brackenbury and Akester, 1978; Scheid, 1979). This fluid must ultimately be reabsorbed into the body. Reabsorption probably continues throughout the perinatal period, but is most evident during the IP–EP interval (Table 1, Fig. 1). The decrease in air-free volume during this interval is about 2.5 times the parabronchial lumen volume and occurs despite proliferation of pulmonary tissue.

Reabsorption of alveolar fluid into the blood or lymph channels occurs in the lungs of mammals. In foetal lambs, for example, the process occurs largely during the first 5 h of pulmonary respiration and is evident by a progressive decrease in lung mass approximately equivalent to the volume of alveolar fluid (Strang, 1968).

Aeration by lung expansion is contrary to the concept of a constant-volume avian lung. However, the embryonic lung is growing and the increase in total organ space necessarily enlarges the air spaces. Although the relative mass of the lung does not change during the perinatal period, the absolute increase in total lung volume of about 50% pulls gas into the air capillaries from the air sacs and aerated parabronchial lumina (Table 1). Because the number of parabronchi in the late-developed embryo does not differ from that of the hatchling or the adult, increase in lung size is accomplished mainly by proliferation of air and blood capillaries and by enlargement of the parabronchial lumina (Duncker, 1978). This growth therefore increases both air-free and air-filled lung volumes.

Vince and Tolhurst (1975) noted that the slow development of lung aeration in domestic fowl reflects the use of the lung for O_2 uptake (Visschedijk, 1968a). Except during the time between IP and EP, when pulmonary gas exchange is limited by resistance across the unbroken shell, pulmonary O_2 uptake and lung air volume rise in parallel (Fig. 2). This correlation is interpreted as a progressive increase in the surface area available for gas exchange through aeration of the air capillaries.

How much lung aeration is necessary for survival on pulmonary gas exchange alone? Visschedijk (1968a) helped answer this question when he put paraffin oil on the shell over the chorioallantois and thereby forced the embryos to rely almost entirely on the lung. Survival progressively increased as the oil was applied later during the perinatal period. Survival just after internal pipping was 20%, increasing to 100% immediately before external pipping. All eggs hatched normally when the whole shell was coated with paraffin oil after external pipping. External pipping corresponds to an air fraction of 30–42% (Table 1). Therefore the lung is capable of keeping the embryo alive long before hatching and with a relatively small amount of gas in it. Moreover, pulmonary gas exchange was restricted in Visschedijk's experiments because the IP embryo was forced to breathe within the air cell where the P_{O_2} was certainly depressed. The normal air cell P_{O_2} at this stage is about 100 torr when the entire shell is available for gas exchange (Wangensteen and Rahn, 1970/71). Oiling of perhaps 75% of the shell would drastically decrease shell conductance and lower the P_{O_2} considerably further. It is significant that, despite this low air cell P_{O_2}, some of Visschedijk's embryos survived when the shell covering the chorioallantois was oiled just after internal pipping. This observation leads to the hypothesis that domestic fowl embryos could survive a 'viviparous birth' if they could hatch without pipping. There are reported cases of domestic fowl hatching upside down in the shell where they could not have pipped into the air cell, but it is not clear that breathing had not occurred before hatch. Duncker (1978) states that the amnion becomes aerated 2–3 d before hatching, whether or not the embryo internally pips.

Aeration of Brush Turkey lungs

There is no air in the Brush Turkey lung until the first breath is taken when the fluids drain from the ruptured chorioallantois (Table 2). The two post-hatch birds suggest that aeration progresses within the few hours after hatching in parallel with the rise in O_2 consumption (Fig. 5). Thus the development of pulmonary gas exchange is similar in Brush Turkeys and in domestic fowl, except that it occurs about 8 times faster in Brush Turkeys. Whereas a great proportion of lung aeration occurs by lung growth in domestic fowl embryos, there is less time for this in Brush Turkeys and initial aeration must occur largely by fluid displacement.

Brush Turkeys clearly demonstrate a 'viviparous birth' because chorioallantoic and pulmonary gas exchange do not occur simultaneously in a perinatal period. These birds survive a period of several minutes when there is so little air in the lungs that the air capillaries must be non-functional (Table 2). Diffusion of O_2 from the parabronchial lumen may be sufficient. However, the extent of anaerobic metabolism and the degree of extrapulmonary gas exchange deserves further investigation. Although O_2 consumption drops during hatching, it does not drop much. Gas exchange in the air sacs, amounting to about 4.5% of the total gas exchange in ducks (Magnussen et al., 1976), may be higher in Brush Turkeys.

Hatching in megapodes is facilitated by an abnormally thin and fragile shell (Seymour and Ackerman, 1980). The metabolic cost of hatching is doubtless less in these birds than in typical birds with thicker shells. Thus megapodes may be able to survive a 'viviparous birth', accompanied by a quick burst of anaerobic metabolism, whereas most other birds require the development of a higher capability for pulmonary gas exchange during their prolonged escape from their shells.

Acknowledgements

This study was supported by the Australian Research Grants Scheme. I thank David Bradford, Sandi Poland, David Booth and Cathie Buddle for help with the measurements and Ruth Hughes for drafting the figures. Appreciation is due to A.H.J. Visschedijk and H.-R. Duncker for comment on the manuscript.

References

Baltin, S. (1969). Zur Biologie und Ethologie des Talegalla-Huhns (*Alectura lathami Gray*) unter besonderer Berücksichtigung des Verhaltens während der Brutperiode. *Z. Tierpsychol.* 26: 524–572.

Brackenbury, J. and Akester, A.R. (1978). A model of the capillary zone of the avian tertiary bronchus. In: Respiratory Function in Birds, Adult and Embryonic. J. Piiper, ed., Springer-Verlag, Berlin, Heidelberg, New York, pp. 125–128.

Burton, R.R. and Smith, A.H. (1968). Blood and air volume in the avian lung. *Poult. Sci.* 47: 85–91.

Duncker, H.-R. (1971). Die Festlegung des Bauplanes der Vogellunge beim Embryo und das Problem ihres postnatalen Wachstums. *Verh. Anat. Ges. Jena.* 66: 273–277.

Duncker, H.-R. (1972). Structure of avian lungs. *Respir. Physiol.* 14: 44–63.

Duncker, H.-R. (1978). Development of the avian respiratory and circulatory systems. In: Respiratory Function in Birds, Adult and Embryonic. J. Piiper, ed., Springer-Verlag, Berlin, Heidelberg, New York, pp. 260–273.

King, A.S. and Payne, D.C. (1962). The maximum capacities of the lungs and air sacs of *Gallus domesticus. J. Anat.* 96: 495–503.

Macklem, P.T., Bouverot, P. and Scheid, P. (1979). Measurement of the distensibility of the parabronchi in duck lungs. *Respir. Physiol.* 38: 23–35.

Magnussen, H., Willmer, H. and Scheid, P. (1976). Gas exchange in air sacs: contribution to respiratory gas exchange in ducks. *Respir. Physiol.* 26: 129–146.

Pattle, R.E. (1978). Lung surfactant and lung lining in birds. In: Respiratory Function in Birds, Adult and Embryonic. J. Piiper, ed., Springer-Verlag, Berlin, Heidelberg, New York, pp. 23–32.

Scheid, P. (1979). Mechanisms of gas exchange in bird lungs. *Rev. Physiol. Biochem. Pharmacol.* 86: 137–186.

Seymour, R.S. and Ackerman, R.A. (1980). Adaptations to underground nesting in birds and reptiles. *Am. Zool.* 20: 437–447.

Strang, L.B. (1968). Uptake of liquid from the lungs at the start of breathing. In: Development of the Lung. A.V.S. de Reuck, R. Porter, eds., Ciba Foundation Symposium, Churchill, London, pp. 348–361.

331

Vince, M.A. and Tolhurst, B.E. (1975). The establishment of lung ventilation in the avian embryo: the rate at which lungs become aerated. *Comp. Biochem. Physiol.* 52A: 331–337.

Visschedijk, A.H.J. (1968a). The air space and embryonic respiration. 1. The pattern of gaseous exchange in the fertile egg during the closing stages of incubation. *Br. Poult. Sci.* 9: 173–184.

Visschedijk, A.H.J. (1968b). The air space and embryonic respiration. 2. The times of pipping and hatching as influenced by an artificially changed permeability of the shell over the air space. *Br. Poult. Sci.* 9: 185–196.

Vleck, D., Vleck, C.M. and Seymour, R.S. (1984). Energetics of embryonic development in the megapode birds, Mallee Fowl (*Leipoa ocellata*) and Brush Turkey (*Alectura lathami*). *Physiol. Zool.* (in press).

Wangensteen, O.D., Rahn, H. (1970/71). Respiratory gas exchange by the avian embryo. *Respir. Physiol.* 11: 31–45.

24. Carbon dioxide transport and acid-base balance in chickens before and after hatching

Hiroshi Tazawa

Department of Electronic Engineering, Muroran Institute of Technology, Muroran, Japan

Abstract

During the late stages of incubation, when the CO_2 increases within the shell due to higher metabolic rate and a fixed shell conductance, the increases in standard CO_2 content, Haldane effect, CO_2 capacitance coefficient and hematocrit augment CO_2 transport and buffer capacity. The buildup of CO_2 results in respiratory disturbances of acid-base balance, but the decrease in pH is mitigated by an increase in non-respiratory bicarbonate. In the final stages of the prenatal period, a metabolic change in pH, partially due to lactate buildup, becomes dominant over the respiratory change.

After the chick pips internally and begins pulmonary ventilation that promotes CO_2 loss, standard bicarbonate of Van Slyke decreases. Blood P_{CO_2} varies widely among embryos and the average P_{CO_2} is not significantly different from the final stages of prenatal period. Hence, the average pH decreases further due to a fall in non-respiratory bicarbonate. The contribution of pulmonary ventilation to the acid-base status is variable, probably in conformity with an individual established degree of ventilatory movement.

Finally the chick pips externally and undergoes respiratory compensation and a rise in pH. After hatching, the blood P_{CO_2} further decreases by the large convective conductance of pulmonary ventilation but the pH change is small due to a corresponding fall in non-respiratory bicarbonate. This is immediately followed by a gradual increase in blood P_{CO_2} and precipitous rise in non-respiratory bicarbonate synchronizing with the initiation of feeding. The postnatal pH results in a light positive overswing during 2–3 d and stabilizes within 1–2 weeks.

Introduction

The ontogeny of avian respiration is similar to the phylogeny of respiration among vertebrates insofar as both include a transition from an 'aquatic' existence with an external gas exchanger, to a 'terrestrial' mode of gas exchange with internal lungs. The transition in birds occurs with a shift from chorioallantoic to

Seymour, R.S., (ed.) Respiration and metabolism of embryonic vertebrates.
© *1984, Dr W. Junk Publishers, Dordrecht/Boston/London. ISBN-13:978-94-009-6538-6*

pulmonary respiration during the final stages of incubation. Initially the chorio-allantois is separated from the atmosphere by a protective barrier, the eggshell and its membranes, and therefore is ventilated mainly by diffusion. The diffusive conductance of the eggshell is nearly constant and progressively imposes a restriction on embryonic gas exchange during development. Once the embryo externally pips, however, the lung is used as an additional gas exchanger and the respiratory restriction is released. During the perinatal period, chorioallantoic and pulmonary functions change reciprocally until the lung assumes the entire role at hatching. Thus the mode of ventilation of the two exchangers changes from one dominated by diffusion to a combination of diffusion and convection and finally to convection alone. Because CO_2 exchange differs in diffusive and convective systems, it is of interest to examine the changes in blood CO_2 transport and acid-base status that occur during the perinatal period in chickens. This study also deals with technical details associated with obtaining reliable blood samples from newly hatched chicks.

Material and methods

Blood sampling

In embryos, the blood was collected directly from the allantoic vein or artery into a heparinized glass syringe as stated previously (Tazawa, 1971). Blood from the allantoic vein is 'arterialized' blood draining the gas exchanger of the embryo and therefore is the functional counterpart of the blood from the pulmonary vein of hatchlings. However, the blood in the allantoic artery is a mixture of the systemic venous blood and the chorioallantoic arterialized blood due to the central and extracardiac shunt (Tazawa and Mochizuki, 1977; Tazawa, 1978). Despite the mixture, this 'mixed' blood from the allantoic artery is referred to as 'venous' blood to compare with the mixed venous blood of the hatchling chick.

While there are several reports of anaerobic blood sampling from avian em-bryos (Dawes and Simkiss, 1969; Freeman and Misson, 1970; Erasmus et al., 1970/71; Tazawa, 1971; Girard, 1971; Boutilier et al., 1977; Tazawa et al., 1980), there are only a few technical descriptions of blood sampling from hatchling chicks (Freeman and Misson, 1970). It is desirable to sample blood from unanesthetized and undisturbed birds to avoid possible artefacts in blood respiratory variables. Because birds become quiet in the dark, I initially blindfolded the chicks but abandoned this technique when it was discovered that it produced respiratory and metabolic acidosis. Chicks usually became calm, and often closed their eyes, when held on their backs in the palm. Blood was collected by direct cardiac puncture by inserting a needle of a syringe into the heart from the ventro-cranial side of the thorax. Either arterialized or mixed venous blood was obtained by selecting the puncture site and angle, both of which vary with development of the chick. In newly hatched chicks, the left ventricle was reached via a puncture through the median line and the right ventricle was reached from slightly right of center. With development, the puncture sites were shifted leftward. The blood volume of chicks is greater than that of embryos so it was often possible to obtain samples from the same chicks at different ages.

After hatching, the chicks were kept in the incubator for about half a day and then transferred to a circular cage warmed by a lamp. Data were obtained from chicks which were not fed and watered until after blood sampling on day 1 and from those fed and watered immediately after entering the cage.

Blood CO_2 dissociation curves

Fertile eggs of White Leghorn hens were incubated at 38 °C and were used for blood CO_2 dissociation curves. Blood was sampled from embryos aged from 10 to 20 d. The 20-day embryos pipped externally. The first blood samples from chicks were taken within 12 h after hatching and these were referred to as day 0. Further samples were obtained on days 1–4 and 6 and from the wing veins of 8-month hens.

Bloods from 10- and 12-day embryos were pooled for determination of CO_2 dissociation curves on oxygenated and deoxygenated samples. Thereafter, samples came from individual animals. The curves were derived from total CO_2 content, measured with a Natelson blood microanalyzer (Natelson, 1951), of samples equilibrated with ten gas mixtures (ca. 1.5%, 3.5%, 5.5%, 7.5% and 10% CO_2; 25% or 0% O_2; balance N_2) in an Instrumentation Laboratories tonometer, model 273. From five pairs of P_{CO_2} and C_{CO_2}, the regression equation expressed by C_{CO_2}–A · $P_{CO_2}{}^B$ was calculated by the least square method. The correlation coefficient for all relations was larger than 0.995.

Acid-base status

Blood was collected anaerobically from chickens of the Warren breed at days 14, 16, 18, 19 of incubation (prenatal period) and during internal and external pipping (perinatal period). Blood was sampled from chicks every day from 0 to 6 d and at 14, 15 and 17 d after hatching. P_{CO_2} and pH were measured with a Radiometer blood micro system (BMS 3) immediately after collection. The bicarbonate concentration ($[HCO_3^-]$) was calculated from the Henderson-Hasselbalch equation using pK' = 6.1.

The pH was measured in blood equilibrated with selected CO_2 in 25% O_2. The regression equation for pH and calculated $[HCO_3^-]$, $[HCO_3^-] = C - b \cdot pH$, was determined by the least square method. The constant, b, is the buffer value in mmol $[HCO_3^-]/l \cdot pH$. Simultaneously, the equation, $\log P_{CO_2} = D - E \cdot pH$, was derived. With these equations, standard bicarbonate ($[HCO_3^-]$ at $P_{CO_2} = 40$ torr) and the bicarbonate of Van Slyke ($[HCO_3^-]$ at pH = 7.4, referred to as $[HCO_3^-]_{7.4}$) were calculated. The plasma CO_2 at $P_{CO_2} = 40$ torr was obtained as the sum of standard bicarbonate and dissolved CO_2 in mmol $[HCO_3^-]/l$ plasma.

Other blood variables

Hematocrit (Hct) was measured throughout the transition from embryos to chicks in both series of experiments.

Blood lactate concentrations in some experiments were measured enzymatically with lactate test kits (Boehringer, Mannheim, FRG).

Results

Blood CO_2 dissociation curves

Fig. 1 shows the CO_2 dissociation curves of oxygenated and deoxygenated bloods of chicks ranging from 10-day embryos to 8-month hens. For the individual developmental stages, the CO_2 content of oxygenated blood at $P_{CO_2} = 40$ torr ($C_{CO_2(40)}$) is plotted in Fig. 2. These values are calculated from dissociation curves of individual animals. The solid line connects the average values. The plots show the range of scattering of individual CO_2 dissociation curves among animals of the same age. Two plots at day 10 and day 12 are quite similar to each other. Each value represents pooled blood of 8 embryos.

The difference between deoxygenated and oxygenated blood demonstrates that the Haldane effect in whole blood becomes large in newly hatched chicks compared with embryos.

Averaged values of $C_{CO_2(40)}$ in plasma and red blood cells are also shown in Fig. 2. The plasma $C_{CO_2(40)}$ is the sum of standard bicarbonate and dissolved CO_2 which is multiplied by (1-Hct/100). The difference of plasma CO_2 from the oxygenated whole blood CO_2 is the value in the RBC.

Blood P_{CO_2}, pH and $[HCO_3^-]$

The P_{CO_2}, pH and $[HCO_3^-]$ of arterialized and venous blood of embryos and hatchlings are shown in Figs. 3–5. The values for arterial blood of hens cited from published reports for domestic fowl are also depicted (Chiodi and Terman, 1965; Frank and Burger, 1965; Frankel, 1965; Calder and Schmidt-Nielsen, 1968; Frankel and Frascella, 1968; Kawashiro and Scheid, 1975). A change in venous variables is almost in parallel with the arterialized blood.

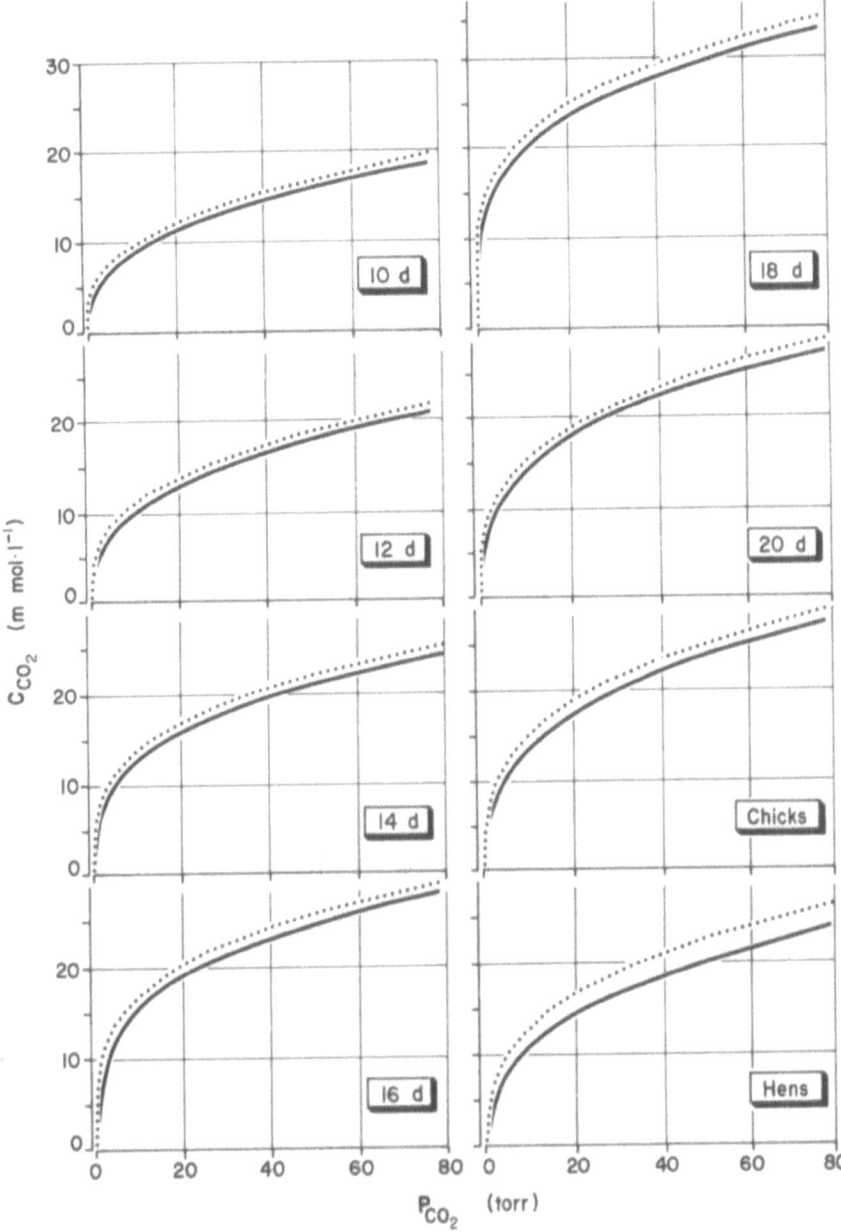

Fig. 1. Average CO_2 dissociation curves of oxygenated (solid curve) and deoxygenated blood (broken curve) of chicks during prenatal (days 10–18), perinatal (day 20) and postnatal period (days 0–6) and of 8-month-old hens.

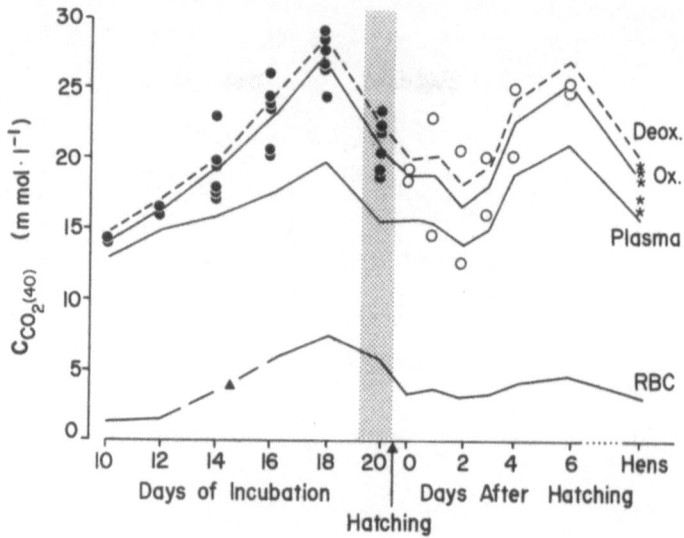

Fig. 2. CO₂ content at $P_{CO_2} = 40$ torr in whole blood (Deox.; deoxygenated blood, Ox.; oxygenated blood), plasma (Plasma) and erythrocytes (RBC) during development from embryos to hens. The closed circles (embryos) and open circles (newly hatched chicks) and the cross (hens) show the C_{CO_2} value in the oxygenated blood. Shaded area indicates the perinatal period.

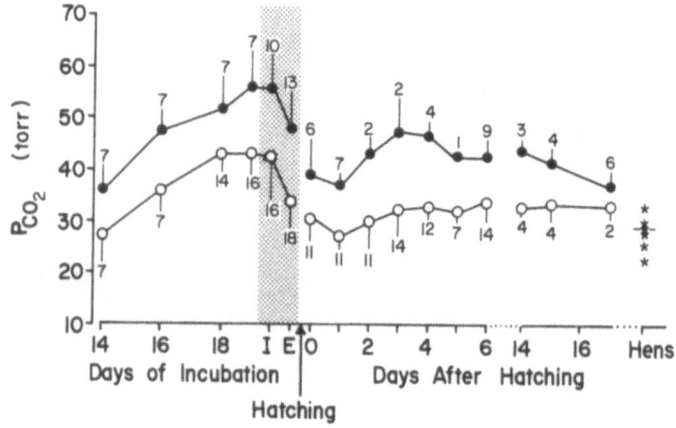

Fig. 3. Changes in P_{CO_2} of arterialized (open circle and cross) and venous blood (closed circle) before, during and after pipping and hatching. The number of determinations and the standard deviation are indicated.

Bicarbonate of Van Slyke ([HCO₃⁻]₋₄)

The bicarbonate concentration of oxygenated plasma at pH = 7.4, is shown in Fig. 6 for two breeds. This value is an index for non-respiratory bicarbonate, and the line for 24 mM is drawn to show the base excess.

338

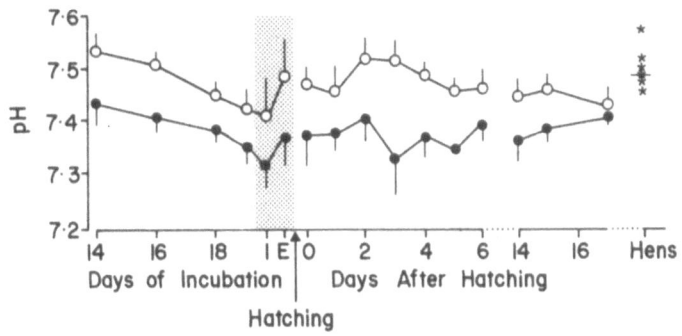

Fig. 4. Changes in pH of arterialized and venous blood during development. Symbols as in Fig. 3.

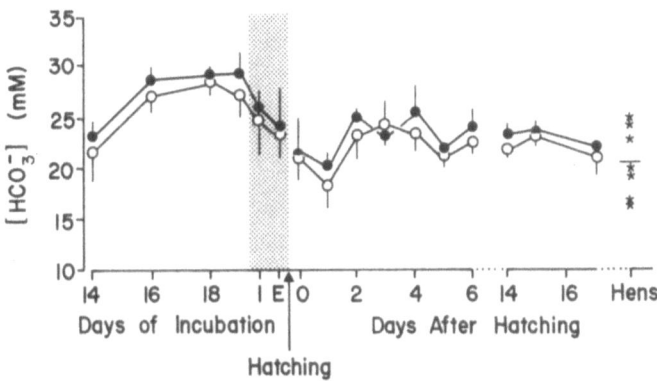

Fig. 5. Changes in [HCO$_3^-$] of arterialized and venous blood during development. Symbols as in Fig. 3.

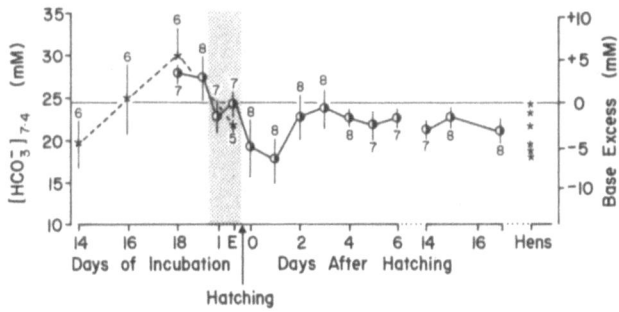

Fig. 6. Changes in the bicarbonate at standard pH ([HCO$_3^-$]$_{7.4}$) during development. Results are from two series of experiments with two breeds (solid line; Warren breed, broken line; White Leghorn breed).

339

Hematocrit

Results from two series of experiments are presented in Fig. 7. Besides the tendency for the Hct value to be slightly larger in the White Leghorn breed than in the Warren breed, a marked difference between two experimental results appears during the perinatal period. Namely, while the Hct of the White Leghorn breed continues increasing up to hatching, that of the Warren breed diminishes during internal pipping and further after hatching. In the postnatal period, the Hct begins increasing after 2–3 d in the Warren breed. This difference in Hct during perinatal period would be rather due to unknown substantial causes than attributed to the different breed.

Effects of repeated sampling

There were no significant differences between values of P_{CO_2}, pH, $[HCO_3^-]$, lactate or Hct in samples taken from the same chicks on day 6 and day 14 (Table 1). In another set of experiments, however, pH and $[HCO_3^-]$ were significantly different between blood samplings on day 3 and 4 and samplings on day 17 (Table 2). The difference is not caused by the repeated samplings, but is attributed to the physiological changes in pH and $[HCO_3^-]$ shown in individual chicks (Figs. 4 and 5). This is substantiated by the lack of significance between samples taken from the same chicks on days 6 and 15 (Table 3). This suggests that the acid-base status in early postnatal periods becomes stable after the first week to two weeks.

Comparison of Hct between days 3–4 and 6, both of which came from the first sampling, indicates that the Hct increases during development (Tables 2 and 3). However, the Hct on day 17 was markedly low compared with day 15 (Tables 2 and 3). Both values were determined on the second sampling. Because the Hct

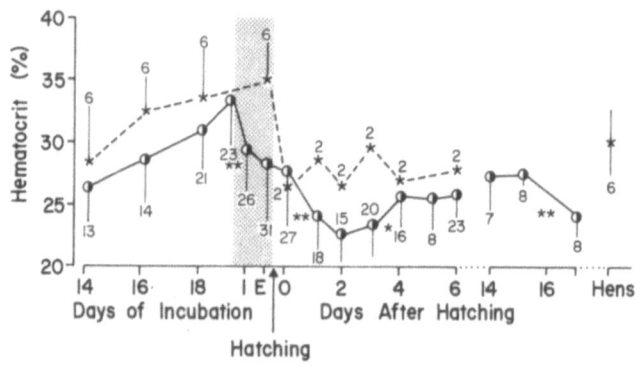

Fig. 7. Changes in the Hct during development. Results from two series of experiments (solid line, Warren; broken line, White Leghorn). Double asterisk (**) indicates that the difference between two adjacent days is significant at P<0.01 and single asterisk (*) at P<0.05.

Table 1. Comparison of selected blood properties between day 6 (left side of columns) and day 14 (right side) of the same chicks.

No.		P_{CO_2} (torr)		pH		[HCO$_3$] (mM)		Lactate (mM)		Hct (%)	
1	a	36	31	7.471	7.475	24	22	3.1	3.9	28	26
2	a	35	35	7.450	7.440	23	22	2.9	2.4	27	27
3	a	36		7.414		21		4.2		29	
	v̄		46		7.323		24		5.3		29
4	v̄	44	43	7.375	7.364	24	23	2.6	2.6	25	27
5	v̄	39	42	7.405	7.386	23	23	2.3	2.3	28	27
6	a		33		7.420		21		2.8		33
	v̄	47		7.346		24		2.1		28	
7	a		32		7.450		22		4.8		24
	v̄	41		7.432		25		4.0		26	

a, arterialized blood; v̄, mixed venous blood.

Table 2. Comparison of selected blood properties between the first sampling at day 3 or 4 and the second sampling at day 17. Blood is arterialized.

	Age (days)	No.	P_{CO_2} (torr)	pH	[HCO$_3$] (mM)	Lactate (mM)	Hct (%)
1st	3	4	35 ± 3	7.492[a] ± 0.019	25[a] ± 2	3.7 ± 1.0	23 ± 1
	4	6	35 ± 2	7.487[b] ± 0.030	24[a] ± 2	4.0 ± 1.1	24 ± 2
2nd	17	6	33 ± 2	7.433 ± 0.030	21 ± 1	3.3[c] ± 1.2	24[c] ± 2

[a] $P<0.01$; [b] $P<0.02$; [c] N = 8 (6 arterialized bloods + 2 venous bloods).

increases with development, the decrease in Hct during days 15 and 17 is not physiological, but attributed to the repetitive sampling. That is, the first sampling, especially from young chickens such as 3–4-day, diminishes the Hct due to hemodilution, and during 1–2 weeks of development the Hct resumes the value of the first sampling.

During embryonic stages, the repetitive samplings have no effect on acid-base balance but decrease the Hct (Tazawa et al., 1980; Tazawa, 1982).

Effect of feeding and watering

Differences in the blood respiratory properties are not significant in fed and unfed

1-day chicks (Table 4). However, the P value was only slightly larger than 0.05 and there is a tendency for pH, Hct, $[HCO_3^-]$ and $[HCO_3^-]_{7.4}$ to be higher in fed and watered chicks. Moreover, values of Hct, combining arterialized and mixed venous blood, are significantly different (P<0.01). Perhaps the variability in the present experiments obscured real differences that may be demonstrated with greater sample sizes.

Table 3. Comparison of selected blood properties between the first sampling at day 6 and the second sampling at day 15. Samples are arterialized and mixed venous bloods.

	Age (days)	No.	P_{CO_2} (torr)	pH	$[HCO_3^-]$ (mM)	Lactate (mM)	Hct (%)
a		5	36 ± 2	7.451 ± 0.039	23 ± 1	4.2 ± 0.7	27 ± 1
1st	6						
v̄		3	44 ± 4	7.397 ± 0.044	25 ± 2	4.0 ± 0.9	27 ± 0.2
a		4	34 ± 3	7.462 ± 0.030	23 ± 1	3.5 ± 1.3	28 ± 1
2nd	15						
v̄		4	42 ± 3	7.380 ± 0.018	23 ± 1	3.5 ± 0.9	28 ± 1

Table 4. Comparison of blood respiratory variables between unfed and fed chicks (1-day-old).

	No.	P_{CO_2} (torr)	pH	$[HCO_3^-]$ (mM)	Lactate (mm)	Hct (5)	$[HCO_3^-]_{7.4}$ (mM)
unfed	11	27 ± 3	7.460 0.046	18 ± 2	3.2 ± 0.6	25 ± 3	17.4 ± 2.8
a							
fed	4	28 ± 3	7.482 ± 0.032	20 ± 2	3.0 ± 0.6	27 ± 1	20.8 ± 2.0
unfed	7	38 ± 2	7.375 ± 0.030	20 ± 1	4.2 ± 1.3	24 ± 3	
v̄							
fed	3	34 ± 9	7.414 ± 0.026	22 ± 1	4.1 ± 0.8	28 ± 2	

a, arterialized blood; v̄, mixed venous blood.

Discussion

Blood CO$_2$ transporting properties before and after hatching

Individual CO$_2$ dissociation curves at the same ages exhibit considerable variability in whole blood $C_{CO_2(40)}$ (Fig. 2). This is attributed to natural variations in diffusive conductance of the shell which produce predictable differences in acid-base status of the embryo (Tazawa et al., 1983b). Despite the variability, however, there are obvious trends during the transition from embryos to chicks. The curves tend to move upward with development during the prenatal stages (day 10 to day 18) and to drop during the perinatal and postnatal periods. The changes in CO$_2$ dissociation curve, indicated in terms of $C_{CO_2(40)}$, parallel changes in blood P_{CO_2} during the commensurate stages (Fig. 3), thus indicating a metabolic compensation to respiratory changes in pH.

Changes in whole blood C_{CO_2} are attributed to not only plasma C_{CO_2} but also erythrocyte (Fig. 2). The participation of both plasma and erythrocytes in acid-base regulation increase toward the end of the prenatal stages. The difference in C_{CO_2} between oxygenated and deoxygenated whole bloods increases during embryonic development. Thus the Haldane effect progressively increases CO$_2$ transport. The slope of the CO$_2$ dissociation curves (capacitance coefficient, Piiper et al., 1971) at $P_{CO_2} = 40$ torr increases toward the end of the prenatal period and decreases during hatching. This is attributed mainly to a decrease in blood CO$_2$ because the capacitance coefficient is a function of blood buffering and C_{CO_2} (Tazawa and Piiper, unpublished).

During the late stages of incubation, when the CO$_2$ increases within the shell due to higher metabolic rate and a fixed shell conductance, the increase in $C_{CO_2(40)}$, Haldane effect, CO$_2$ capacitance coefficient and Hct augment CO$_2$ transport and buffer action. At this time, moreover, there is a corresponding increase in allantoic blood flow which further favors transport (Tazawa 1978, 1980). After pipping and pulmonary ventilation that promote CO$_2$ loss, however, the values decrease again, concommitant with an increased contribution of pulmonary ventilation to CO$_2$ transport.

Acid-base status

During development up to internal pipping and pulmonary ventilation, the buildup of CO$_2$ results in respiratory disturbances of acid-base balance. However, as indicated in the pH-[HCO$_3^-$] diagram (Fig. 8), the decrease in pH is mitigated by an increase in non-respiratory bicarbonate (Fig. 6). At the final stages of the prenatal period (days 18–19), metabolic changes in pH becomes dominant over respiratory change. This occurs because, at days 18 to 19, prior to internal pipping, the air cell P_{O_2} becomes minimum and the O$_2$ demand of the embryo is maximum.

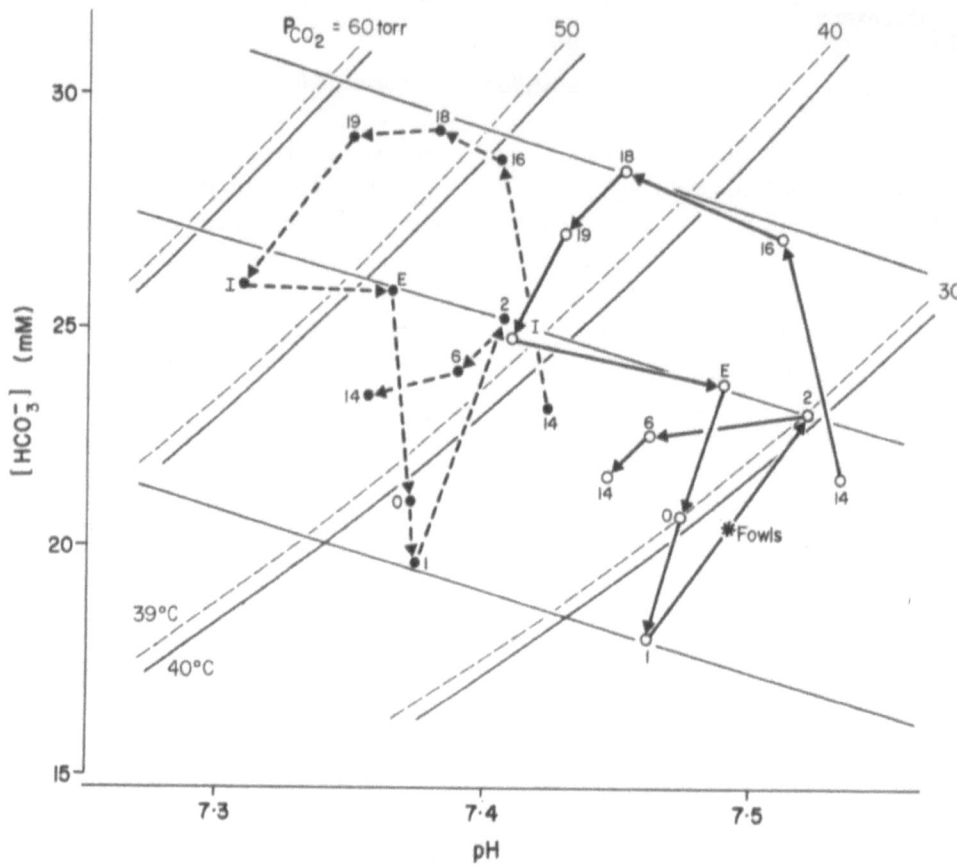

Fig. 8. pH-[HCO₃] diagram showing the acid-base statuses of arterialized (open circle) and venous blood (closed circle) during prenatal (ages are indicated by 14, 16, 18 and 19), perinatal (indicated by I and E for internal and external pipping embryos, respectively), and postnatal period (ages are indicated by 0, 1, 2, 6 and 14). The cross shows the averaged acid-base status of adult fowls (Chiodi and Terman, 1965; Frank and Burger, 1965; Frankel, 1965; Calder and Schmidt-Nielsen, 1968; Frankel and Frascella, 1968; Kawashiro and Scheid, 1975).

Blood lactate concentration begins to increase in this period (Tazawa et al., 1983a). Even after internal pipping, the respiratory restraint is not reduced enough to diminish blood P_{CO_2} but on the contrary, the pH decreases further due to a fall in non-respiratory bicarbonate (Fig. 6). By piercing the eggshell and breathing atmospheric air, however, embryos undergo respiratory increase in pH.

After hatching, the blood P_{CO_2} is decreased by the large convective conductance of pulmonary ventilation but the pH change is small due to a corresponding fall in non-respiratory bicarbonate. On day 2, non-respiratory bicarbonate of the hatchling returns to the level of that in pipping embryos. With an increase in blood P_{CO_2}, the postnatal pH decreases from a maximum of 7.52 on day 2 to about 7.45 after two weeks. The averaged status of adult hens so far reported is slightly on the side of respiratory alkalosis compared with 1- to 2-week chicks.

References

Boutilier, R.G., Gibson, M.A., Toews, D.P. and Anderson, W. (1977). Gas exchange and acid-base regulation in the blood and extraembryonic fluids of developing chicken embryo. *Respir. Physiol.* 31: 81–89.

Calder, W.A. and Schmidt-Nielsen, K. (1968). Panting and blood carbon dioxide in birds. *Am. J. Physiol.* 215: 477–482.

Chiodi, C. and Terman, J.W. (1965). Arterial blood gases of domestic hen. *Am. J. Physiol.* 208: 798–800.

Dawes, C. and Simkiss, K. (1969). The acid-base status of the blood of developing chick embryo. *J. Exp. Biol.* 50: 79–86.

Erasmus, D.W., Howell, B.J. and Rahn. H. (1970/71). Ontogeny of acid-base balance in the bullfrog and chicken. *Respir. Physiol.* 11: 46–53.

Frank, F.R. and Burger, R.E. (1965). The effect of carbon dioxide inhalation and sodium bicarbonate ingestion on egg shell deposition. *Poult. Sci.* 44: 1604–1606.

Frankel, H.M. (1965). Blood lactate and pyruvate and evidence for hypocapnic lacticacidosis in the chicken. *Proc. Soc. Exp. Biol. Med.* 11: 261–263.

Frankel, H.M. and Frascells, D. (1968). Blood respiratory gases, lactate, and pyruvate during thermal stress in the chicken. *Proc. Soc. Exp. Biol. Med.* 127: 997–999.

Freeman, B.M. and Misson, B.H. (1970). pH, P_{O_2} and P_{CO_2} of blood from the foetus and neonate of *Gallus domesticus. Comp. Biochem. Physiol.* 33: 763–772.

Girard, H. (1971). Respiratory acidosis with partial metabolic compensation in chick embryo blood during normal development. *Respir. Physiol.* 13: 343–351.

Kawashiro, T. and Scheid, P. (1975). Arterial blood gases in undisturbed resting birds: Measurement in chicken and duck. *Respir. Physiol.* 23: 337–342.

Natelson, S. (1951). Routine use of ultramicro method in the clinical laboratory. *Am. J. Clin. Pathol.* 21: 1153–1172.

Piiper, J., Dejours, P., Haab, P. and Rahn, H. (1971). Concepts and basic quantities in gas exchange physiology. *Respir. Physiol.* 13: 292–304.

Tazawa, H. (1971). Measurement of respiratory parameters in blood of chicken embryo. *J. Appl. Physiol.* 31: 17–20.

Tazawa, H. and Mochizuki, M. (1977). Oxygen analysis of chicken embryo blood. *Respir. Physiol.* 31: 203–215.

Tazawa, H. (1978). Gas transfer in the chorioallantois. In: Respiratory Function in Birds, Adult and Embryonic. J. Piiper, ed., Springer-Verlag. Berlin, pp. 274–291.

Tazawa, H. (1980). Oxygen and CO_2 exchange and acid-base regulation in the avian embryo. *Am. Zool.* 20: 395–404.

Tazawa, H., Ar, A., Rahn, H. and Piiper, J. (1980). Repetitive and simultaneous sampling from the air cell and blood vessels in the chick embryo. *Respir. Physiol.* 39: 265–272.

Tazawa, H. (1982). Regulatory processes of metabolic and respiratory acid-base disturbances in embryos. *J. Appl. Physiol.: Respirat. Environ. Exercise Physiol.* 53: 1449–1454.

Tazawa, H., Visschedijk, A.H.J., Wittmann, J. and Piiper, J. (1983a). Gas exchange, blood gases and acid-base status in the chick before, during and after hatching. *Respir. Physiol.* (in press).

Tazawa, H. and Piiper, J. (unpublished). Carbon dioxide dissociation and buffering in chicken blood during development.

Tazawa, H., Visschedijk, A.H.J. and Piiper, J. (1983b). Blood gases and acid-base status in chicken embryos with naturally varying egg shell conductance. *Respir. Physiol.* (in press).

25. The energetics of development in a very large altricial bird, the brown pelican

G.A. Bartholomew and D.L. Goldstein

Department of Biology, University of California, Los Angeles, CA 90024, U.S.A.

Abstract

The brown pelican, *Pelecanus occidentalis*, is the largest altricial species for which the energetics of embryonic development has so far been reported.

The brown pelican produces a relatively large chick from a relatively small egg (80–105 g) when compared with other altricial species. A normal clutch of three eggs is equal to less than 8% of the average adult mass (3500–2800 g). External pipping occurs on day 29 and hatching occurs on day 30.5. The mass at hatching is ~72 g, of which about 10 g is yolk retained in the abdominal cavity.

Embryonic development does not follow a simple exponential curve; daily relative growth rate diminishes throughout incubation. Water vapor conductance of the egg, 16.76 mg $\cdot$ d^{-1} $\cdot$ torr^{-1} is somewhat larger than predicted on the basis of egg mass.

$\dot{V}_{O_2}$ increases steadily throughout incubation but the rate of change diminishes from about 40% $\cdot$ d^{-1} on day 8 to about 10% $\cdot$ d^{-1} on day 26. The total energy expenditure through hatching is about 121 kJ of which the process of hatching uses about 3 kJ.

The pattern of embryonic development in brown pelicans is different enough from that in other orders to be consistent with the hypothesis that avian altriciality is polyphyletic.

Introduction

The patterns of avian development form a continuous series from the extreme precociality of megapodes to the extreme altriciality of parrots and passerines. It is generally assumed that precociality is the ancestral condition in the class Aves (Portmann, 1950). If so, altriciality has probably evolved independently several times in different avian orders. In this regard the order Pelecaniformes is of particular interest, because it includes the largest of altricial birds, and also the only marine birds that are altricial.

There are no published data available on the energetics of embryonic develop-

Seymour, R.S., (ed.) Respiration and metabolism of embryonic vertebrates.
© *1984, Dr W. Junk Publishers, Dordrecht/Boston/London. ISBN-13: 978-94-009-6538-6*

ment in the order Pelecaniformes. The genus *Pelecanus* includes the largest birds in the order. The brown pelican (*P. occidentalis*), with a body mass of 3500 to 3800 g, is among the smaller members of the genus, but it is the largest altricial species for which data on the energetics of development have so far been obtained. It is of interest not only because of its size, but because its eggs are unusually small, and because its incubation period (30.5 d) is only about half as long as that of procellariiform birds of similar body mass (Pettit et al., 1982).

Material and methods

Brown pelicans occur widely in the warm temperate and tropical coastal waters of North and South America. They breed in large numbers in the Gulf of Panama (Wetmore, 1965; Montgomery, 1982).

Eggs and incubation

Eggs were obtained during February and March, 1983, from an extensive colony of brown pelicans on Isla Uraba, a steep volcanic island in the Gulf of Panama (8 °N, 78 °W). They were taken immediately to the nearby island of Taboga where we had established a temporary laboratory in a house managed by the Smithsonian Tropical Research Institute. The pelican nests were located 10 to 20 m above the ground in the outer branches of trees, usually gumbo limbo (*Bursera semaruba*). We removed entire clutches (one to three eggs), labeled them, and put them in an insulated container for transfer to Taboga where we weighed them to the nearest mg and measured their lengths and maximum breadths to 0.01 mm with calipers. The eggs were then placed in a commercial incubator at a temperature of 36–37 °C and a relative humidity of ~60%. They were turned three or four times a day by hand.

Water vapor conductance

Water vapor conductance (G_{H_2O}) was determined at 36 °C during the first week of incubation using the method of Ar et al. (1974), except that we made repeated measurements until the rate of loss in mass was constant. The data were converted to 25 °C with the formula of Paganelli et al. (1978).

Oxygen consumption

The rate of O_2 consumption ($\dot{V}_{O_2}$) during development was measured in 38 eggs at 36 °C. We used a closed respirometry system until $\dot{V}_{O_2}$ reached $10\,ml \cdot h^{-1}$, and a flow-through system thereafter. In the closed system we employed the procedures described by Bucher (1983) and corrected for egg volume and water vapor pressure with the formulae given by Hoyt et al. (1978). Egg volume was calculated with the formula $V = K_v LB^2$ (Hoyt, 1979), where V is volume in cm^3, L and B are length and maximum breadth in cm, and K_v, the volume coefficient, is 0.507. In the flow-through system CO_2 and water vapor were absorbed downstream from the pump and upstream from the flowmeter and respirometer chamber, and again before entering the O_2 analyzer (Applied Electrochemistry, S3A). $\dot{V}_{O_2}$ $(ml \cdot min^{-1})$ was calculated by the formula

$$\dot{V}_{O_2} = \dot{V}\,(F_I - F_E)/(1 - F_I) \tag{1}$$

where $\dot{V}$ is flow rate $(ml \cdot min^{-1})$ corrected to standard conditions of temperature and pressure, and F_I and F_E are initial and end fractional concentrations of O_2 in the dry, CO_2-free air passing through the system. Individual eggs were measured repeatedly over periods of 3 to 20 d. For comparative purposes we assume that 1 ml O_2 is the equivalent of 19.88 J.

Eggs and embryos

Wet mass was determined immediately after removing the embryo from the egg. The embryos were separated from their extra-embryonic membranes, blotted, and weighed to the nearest mg. In computing daily averages we interpolated values of mass for days 9, 11, 26, and 28, on which no embryos were measured.

The embryos were staged by examining them under a dissecting microscope and comparing them with the established embryonic stages of the chick of the domestic fowl (Hamburger and Hamilton, 1951; Freeman and Vince, 1974).

The wet masses of yolk, albumin, and egg shell were obtained from eggs hard-boiled during the first week of incubation. The three components were carefully separated and weighed to the nearest mg.

All eggs were measured daily for at least 5 d. Their ages were estimated to the nearest half day by superimposing the individual curves of $\dot{V}_{O_2}$ vs time on a complete curve derived from eggs of known age. This $\dot{V}_{O_2}$ curve, complete from day 1 through hatching, was compiled from two eggs of known and overlapping ages.

Results

Eggs and embryos

The eggs of the brown pelican are relatively small (Table 1). Assuming an adult mass of 3600 g, an average egg weighs only about 2.5% of adult mass and a clutch of 3 weighs less than 8% of adult mass. The proportions of fresh egg mass in yolk and in albumin were variable (Table 2). The yolk averaged 28.1% of the total mass of the egg, the same as reported by Lawrence and Schrieber (1974). This value is in the range characteristic of altricial species (Carey et al., 1980). The albumin was a viscous liquid during most of development, but by the last 10 days of incubation it had become a firm, gelatinous body that maintained its coherence when removed from the egg. The albumin body disappeared by the onset of pipping.

The mass of the embryos increased continuously from the start of incubation through hatching (Fig. 1). However, the relative growth rate (% per day increase in mass) decreased throughout development, indicating that embryonic growth does not follow a simple exponential curve (Fig. 2).

By the 12th day of incubation the pelican embryos had reached stage 25 of the

Table 1. Morphometrics of brown pelican eggs.

	Mass* (g)	Length (cm)	Breadth (cm)	Volume (cm³)
Number	38	38	38	38
Maximum	104.6	7.90	5.10	99.61
Minimum	80.6	6.81	4.65	79.36
Mean	93.8	7.48	4.88	90.27
s.d.	6.2	0.21	0.12	4.94

* Weighed at time of collection (mean time of collection = day 8 of incubation).

Table 2. Mass (g) of components of brown pelican eggs (mean mass of eggs = 87.01 g).

	Yolk	Albumin	Shell
Number	5	5	5
Maximum	42.33	67.14	11.09
Minimum	14.96	27.54	10.24
Mean	24.46	51.87	10.67
s.d.	11.16	15.42	0.39
Mean as % Total	28.11	59.61	12.26

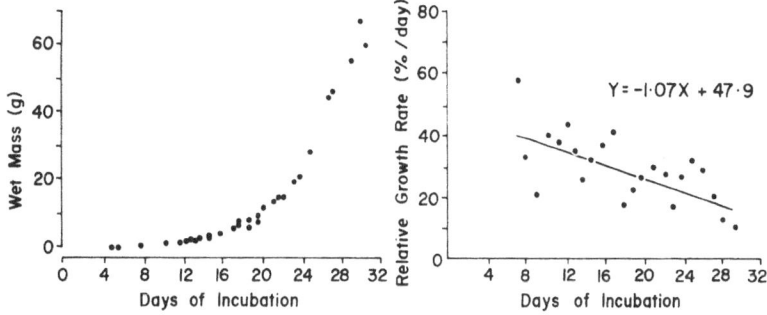

Fig. 1. Yolk-free wet mass as a function of age during embryonic development.

Fig. 2. Changes in daily relative growth rate during embryonic development. Points are based on 3-day running averages of yolk-free mass. The slope is significantly different from zero (P = 0.001).

domestic chicken, which the latter reaches during the 4th day of incubation (Table 3). By day 13 the beak of the pelican embryo had clearly started to become enlarged, and the gular pouch had begun to develop (Fig. 3). The rate of growth of the beak relative to the mass of the body was initially rapid but diminished markedly after the embryo reached about 20% of hatchling mass (Fig. 4). By day 18 the down feather follicles occurred over almost all of the body surface and all major features of external morphology were well developed. Thereafter, embryonic development primarily involved increase in size. The total length of incubation was 30.5 d. We could discern no sign of internal pipping. External pipping occurred on day 29.

The masses of the two chicks which we followed all the way through hatching were 68.9 and 75.6 g, which represented 73.8 and 74.2% of fresh egg mass, respectively. Of each chick's mass, 10.5 g was yolk retained in the abdominal cavity.

The eyelids never fused during development and the eyes were open at time of hatching. During the period of pipping the chicks vocalized steadily – an ability which is lacking in the adults, which are mute. Although the chicks are functionally naked at hatching, they possess rows of tiny needle-like feathers on the dorsal pterylae and on the trailing surfaces of the wings and tail. The grayish-purple color of the skin of the hatchlings is not apparent until the pipping stage is reached.

Water vapor conductance

Water vapor conductance for three eggs (mean mass = 94.9 g), measured on days 4, 5, and 8 of incubation, averaged 16.76 (s.d. = 1.03) mg · d^{-1} · torr^{-1} (Table 4). This value is approximately 11% greater than that predicted on the basis of egg mass (Ar et al., 1974).

351

Table 3. Development of external morphology of the brown pelican embryo compared with the domestic chicken embryo*.

Age (h)	Pelican	Chick	
	Morphological features	Closest equiv. stage	Age (h)
120	35–40 somites present; yolk stalk present; eye disc present. no pigmentation; lateral ridge in wing area.	16	51–56
228	Somites extend into tail; eyes lightly pigmented; heart distinctly chambered; legs and wings formed; elbow distinct; first indication of beak.	22	~84
252	Somites extend to tip of tail; digits demarcated by grooves. 4 on legs. 2 on wings.	25	~108
312	Yolk sac complete; 3 distinct digits on wings; upper and lower mandible separated; gular pouch beginning.	29–30	~150
336	Eyelids beginning; wing bones clearly visible; webbing between toes; egg tooth visible; single row of feather primordia present on tail; first movements observed; auditory meatus present.	30	~156
360	Femur visible; feather primordia on shoulder region.	**	**
384	Body moves jerkily; opens mouth. swallows; gular pouch well developed.	**	**
432	Eyelid covers all of eye except iris; down feather primordia over most of body; strong jerky movements; all major features of pelican hatchlings well developed.	**	**
480	Nictitating membrane present; chorioallantoic membrane covers 7/8 of shell surface.	**	**

* Stages from Hamburger and Hamilton, 1951.
** Divergence between pelican and chick too great for matching.

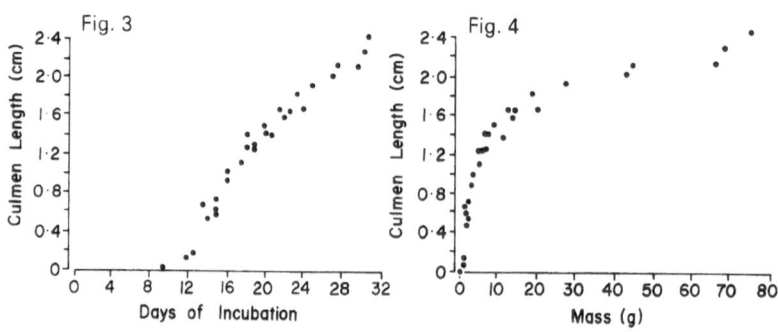

Fig. 3. Culmen length as a function of age during embryonic development.

Fig. 4. Culmen length as a function of mass during embryonic development.

Table 4. Observed and predicted values of developmental parameters for the brown pelican.

Parameter	Observed value	Predicted value	Predictor variable	Reference
G_{H2O} ($mg \cdot d^{-1} \cdot torr^{-1}$)	16.76	16.0	egg mass & length inc.	Ar and Rahn, 1978
		15.6	egg mass	Ar and Rahn, 1978
Length of inc. (d)	30.5	31.8	egg mass	Ar and Rahn, 1978
		35.8	adult mass	Rahn et al., 1975
Egg mass (g)	93.8	153.3	adult mass	Rahn et al., 1975
Cost of development*	120	90	egg mass & length inc.	Bucher, 1983
(kJ)		84.3	egg mass	Bucher, 1983

* Integration of metabolic rate from day 0 through hatching.

Oxygen consumption

$\dot{V}_{O_2}$ increases steadily throughout development (Fig. 5). The pattern of increase closely resembles that found in altricial birds of other orders (Vleck et al., 1980). As with the rate of change in mass, the rate of change in $\dot{V}_{O_2}$ diminishes with age, from about 40% $\cdot d^{-1}$ on day 8 to about 10% $\cdot d^{-1}$ on day 26, indicating a non-exponential pattern of increase (Fig. 6).

Mass-specific $\dot{V}_{O_2}$ decreases in relation to age throughout embryonic development (Fig. 7). $\dot{V}_{O_2}$ during development is a function of mass to the 0.8 power; this exponent is significantly different from 1.0, $P<0.001$, (Fig. 8).

Pre-external pipping $\dot{V}_{O_2}$ was 28 to 30 ml $O_2 \cdot h^{-1}$. We were able to measure $\dot{V}_{O_2}$ during the process of external pipping and during hatching. Integrating V_{O_2} over the time of hatching and subtracting from this value the integrated $\dot{V}_{O_2}$ of non-hatching chicks of the same age (calculated from extrapolated $\dot{V}_{O_2}$) yields the energy cost of pipping and the energy cost of hatching. There was no discernable difference between the $\dot{V}_{O_2}$ of chicks that were pipping and the extrapolated values for $\dot{V}_{O_2}$ of non-pipping birds of the same age. Hatching involved the consumption of 150 ml O_2 above the extrapolated level. Thus the cost of hatching was approximately 2.98 kJ.

The total cost of embryonic development in the brown pelican, integrated from day zero of incubation through the completion of hatching, was 6100 ml O_2, or about 121 kJ.

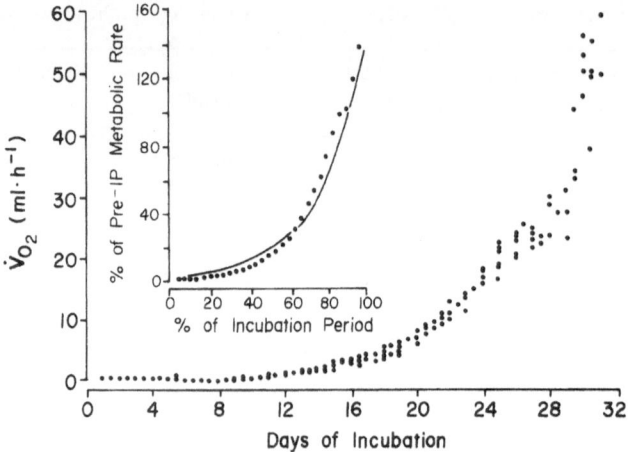

Fig. 5. Increase in O_2 consumption with age during embryonic development. N = 288 measurements on 38 individuals. Inset compares daily average metabolic rate during the development of pelicans (●) with the mean curve for altricial birds in other orders (from Fig. 3C in Vleck et al., 1980). Both the X and Y axes of the inset are normalized to allow comparison of species of different sizes.

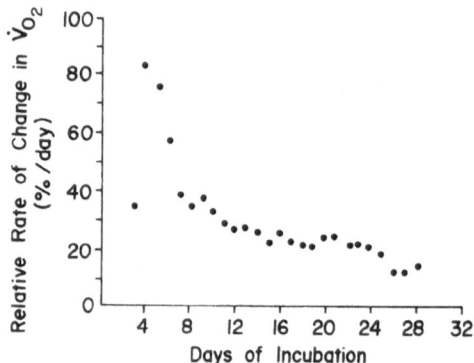

Fig. 6. Relative rate of change in $\dot{V}_{O_2}$ during embryonic development. Points are calculated from 3-day running averages.

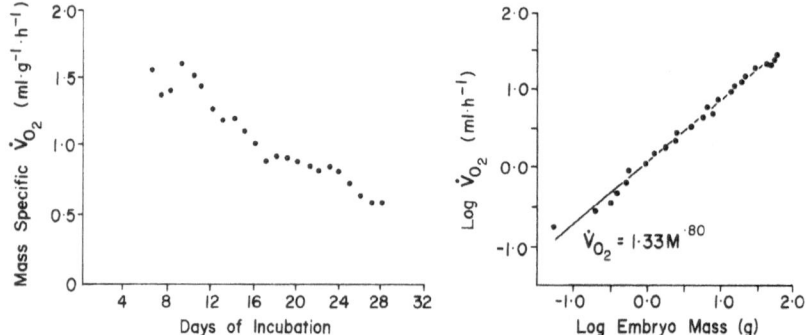

Fig. 7. The relation of daily average mass-specific $\dot{V}_{O_2}$ to age during development.

Fig. 8. The relation of the log-transformed values of $\dot{V}_{O_2}$ and mass during development.

354

Discussion

Has altriciality in pelicans been independently evolved?

For over a century the conventional wisdom has held that in birds precociality is primitive and altriciality is derived (Coues, 1872). We assume that altriciality may have evolved several times in different avian orders. If this is so, one would expect that details of embryonic development in the brown pelican should differ in several ways from those found in altricial birds of other orders such as parrots, passerines, or pigeons. In looking for contrasting developmental patterns in different taxa, one must of course isolate the differences that result from allometric factors.

Water vapor conductance of the egg and length of incubation in brown pelicans conform with allometric predictions, but a number of other developmental parameters do not (Table 4). Brown pelican eggs are only about 60% of the predicted mass, but at hatching the chicks (including retained yolk) are about 15% heavier than average relative to egg mass (see Bucher, 1983, for refs.). The amount of yolk retained at hatching in brown pelicans, about 15% by weight, is not as much as in precocial birds of the same mass (Romanoff, 1944), but it is larger than in members of the Procellariiformes (Pettit et al., 1982) and it is almost twice as great as the average value for 31 species of altricial birds (Schmekel, 1960). The total cost of development in the brown pelican (121 kJ) is 1.4 times that predicted for an altricial species either from egg mass or from mass and length of incubation (Bucher, 1983). However, the eggs of the brown pelican are twice as large as the largest eggs in the sample on which the allometric equation is based. On the basis of the available information it is difficult to be sure whether the high energy cost of development is a function of large size or is a developmental characteristic of the taxon.

The brown pelican thus produces a relatively large chick at a relatively high metabolic cost compared with other (and smaller) altricial species. This implies a larger energy content per unit fresh egg mass in pelicans than in other altricial birds, which is indeed the case. The egg of the brown pelican has the greatest energy density ($\sim$6.28 kJ $\cdot$ g^{-1} of egg mass) of the 24 species of altricial birds for which data are available (Carey et al., 1980; Lawrence and Schreiber, 1974; Bucher and Bartholomew, present volume).

On balance, we think that the special features of the altricial development of pelicans (eyes open at hatching, the high energy content of the egg, the large amount of yolk retained by the hatchling, the relation of mass-specific $\dot{V}_{O_2}$ to embryonic age and mass, and the high total cost of development) are sufficiently different from those found in the altricial development of other orders that they do not falsify the hypothesis that altriciality is of polyphyletic origin.

The cost of pipping and hatching

Of the total cost of development, about 3 kJ (2.5%) is accounted for by the activity associated with hatching. We found no incremental cost for the activity associated with earlier stages of external pipping. It is not surprising that specific costs assignable to the process of pipping were difficult to find. Avian embryos are almost continuously active during the latter stages of development; pipping may not involve increased amounts of activity, but rather the orientation of most of the embryo pelican's normal movements towards a particular goal. Similar data are not available for other species.

Ecological considerations

Birds which breed in the marine environment include all members of the orders Sphenisciformes and Procellariiformes, and some members of the Pelecaniformes, Anseriformes and Charadriiformes. Except for members of the Pelicaniformes all these marine birds produce chicks which are relatively well developed at hatching – they are covered with down and are capable of thermoregulation and/or locomotion. Thus, among marine birds, altriciality is unique to the order Pelecaniformes. Pelecaniform birds are largely confined to coastal waters, but they occur in the Arctic, temperate, tropical, and Antarctic regions. Altriciality in this order is clearly not an adaptation to any set of climatic conditions. It is equally obvious that altriciality is a viable pattern of development for a marine breeder. Our observations are limited to the first half of the altricial developmental process. For the brown pelican the principal advantages that accrue during this period appear to be (1) a relatively small total energy investment per egg and (2) rapid development of a relatively large chick. The extent of the advantages of altriciality which may accrue during the interval between hatching and fledging remains to be assayed.

Acknowledgements

This work was supported by a grant from the National Geographic Society and a grant from the U.S. National Science Foundation (DEB 81-03513 to G.A.B.), and was carried out using facilities provided by the Smithsonian Tropical Research Institute. Our research could not have been done without the advice and assistance of G.G. Montgomery and the support of the STRI Seabird Research Center.

References

Ar, A., Paganelli, C.V., Reeves, R.B., Greene, D.G. and Rahn, H. (1974). The avian egg: water vapor conductance, shell thickness, and functional pore area. *Condor* 76: 153–158.

Ar, A. and Rahn, H. (1978). Interdependence of gas conductance, incubation length, and weight of the avian egg. In: Respiratory function in birds, adult and embryonic. J. Piiper, ed., Springer-Verlag, Berlin and New York, pp. 227–236.

Bucher, T.L. (1983). Parrot eggs, embryos and nestlings: patterns and energetics of growth and development. *Physiol. Zool.* 56: 465–483.

Bucher, T.L. and Bartholomew, G.A. (1984). Patterns of growth, gas exchange, and energy utilization in parrot and other avian embryos. Present volume.

Carey, C., Rahn, H. and Parisi. P. (1980). Calories, water, lipid and yolk in avian eggs. *Condor* 82: 335–343.

Coues, E. (1872). Key to North American birds. Estes and Lauriat, Boston.

Freeman, B.H. and Vince. M.A. (1974). Development of the avian embryo. John Wiley and Sons, New York.

Hamburger, V. and Hamilton, H.L. (1951). A series of normal stages in the development of the chick embryo. *J. Morphol.* 88: 49–92.

Hoyt, D.F. (1979). Practical methods of estimating volume and fresh weight of bird eggs. *Auk* 96: 73–77.

Hoyt, D.F., Vleck, D. and Vleck, C.M. (1978). Metabolism of avian embryos: ontogeny and temperature effects in the ostrich. *Condor* 80: 265–271.

Lawrence, J.M. and Schrieber, R.W. (1974). Organic material and calories in the egg of the Brown Pelican, *Pelecanus occidentalis. Comp. Biochem. Physiol.* 47A: 435–440.

Montgomery, G.G. (1982). A Panamanian national seabird refuge on Taboga and Uraba Islands. *Revista Medica de Panama,* 10 pp.

Paganelli, C.V., Ackerman, R.A. and Rahn, H. (1978). The avian egg: In vivo conductances to oxygen, carbon dioxide and water vapor in late development. In: Respiratory function in birds, adult and embryonic. J. Piiper, ed., Springer-Verlag, Berlin and New York, pp. 212–218.

Pettit, T.N., Grant, G.S., Whittow, G.C., Rahn, H. and Paganelli, C.V. (1982). Embryonic oxygen consumption and growth of Laysan and black-footed albatross. *Am. J. Physiol.* 242: R121–R128.

Portmann, A. (1950). Le developpement postembryonaire. In: Traité de Zoologie, Vol. 15. P.P. Grassé, ed., Masson et Cie., Paris. pp. 522–535.

Rahn, H., Paganelli, C.V. and Ar, A. (1975). Relation of avian egg weight to body weight. *Auk* 92: 750–765.

Romanoff, A.L. (1944). Avian spare yolk and its assimilation. *Auk* 1944: 235–241.

Schmekel, L. (1960). Datum uber des Gewicht des Vogeldottersackes von Schlupftag bis zum Schwinden. *Rev. Suisse, Zool.* 68: 103–110.

Vleck, C.M., Vleck, D. and Hoyt, D.F. (1980). Patterns of metabolism and growth in avian embryos. *Am. Zool.* 405–416.

Wetmore, A. (1965). The birds of the Republic of Panama, Part 1. *Smithson. Misc. Collect.,* Vol. 150.

26. Analysis of variation in gas exchange, growth patterns, and energy utilization in a parrot and other avian embryos

Theresa L. Bucher and George A. Bartholomew

Department of Biology, University of California, Los Angeles, California 90024, U.S.A.

Abstract

Analysis of variation can offer insights into the factors affecting avian embryonic development that refine the understanding based on central tendencies. The magnitude of individual parameters may vary greatly between individuals as well as between species. For example, in *Agapornis roseicollis* water vapor conductance of eggs of different individuals varies by a factor of at least 7. This results in an extreme range of air cell gas tensions (P_{AO_2}, as low as 46 torr; P_{ACO_2}, as high as 90 torr) which appear to be well tolerated by the embryos. Completion of the chorioallantois occurs much later in incubation in *Agapornis* and *Pelecanus* than in the precocial chicken.

The total production efficiency of *Agapornis* embryos is 44.8%. Comparative data are not available for other birds, but this value is higher than published values for mammals.

Comparative analysis of some aspects of embryonic development such as the patterns of relative growth rate and of mass-specific energy expenditure indicate that growth phenomena are affected by a set of diverse factors that include egg mass, time, energy expenditure, phylogeny, and ecological niche. The relative importance of each of these to the cost of embryonic development can be assayed by multivariate statistics. For example, among 49 species with eggs ranging in mass from 0.9 g to 1.45 kg, no single factor adequately accounts for the observed five-fold variation in the energy expended during embryonic development per gram of egg. However, in order of importance, the three most critical variables are log egg mass, log days of incubation, and developmental type – a categorization reflecting phylogeny and ecological niche.

Introduction

After lying fallow for decades, the study of the physiology of avian eggs has recently undergone a remarkable renaissance. Few areas of comparative physiology have developed in as rapid and coherent a manner. The expansion of this

Seymour, R.S., (ed.) Respiration and metabolism of embryonic vertebrates.
© *1984, Dr W. Junk Publishers, Dordrecht/Boston/London. ISBN-13:978-94-009-6538-6*

important area of vertebrate biology and its orderly pattern of growth have in large part been the result of the innovative studies of Hermann Rahn and his associates. They presented more than just phenomenological reports. They used allometric analyses so structured as to identify major trends in the physiology of bird eggs and embryos and, most importantly, to yield predictive equations which invited testing. These papers stimulated the acquisition of a mass of new data of physiological, ecological, and evolutionary interest. This challenge continues to affect most of us at this meeting.

Intriguingly, the final truth is never available to scientists, particularly biologists. With the wealth of new information that has appeared since 1974, it has become apparent that some of the trends that Rahn and associates so astutely perceived are less inclusive than they appeared to be at first, and that single factor analysis often accounts inadequately for the observed variations.

This paper attempts to extend the insights of Rahn and his co-workers by concentrating on variation rather than on central tendencies. We think that our approach is important because many major trends in the biology of avian eggs are now apparent and because the time is ripe to target variability as a principle element in the study of ecologically and evolutionarily relevant aspects of physiology.

Variation, both within and between species, is one of the most fundamental attributes of organisms. Analysis of this variation can offer insights into the factors affecting normal development just as surely as can the delineation of the central tendencies.

We shall deal primarily with four species belonging to different orders, using our studies of parrots as a point of departure. Parrots constitute a distinct and homogeneous order. They lay relatively small eggs that require long periods of incubation and produce extremely altricial hatchlings. The prolonged incubation, the slow development of the embryos, and the slow growth of the hatchlings contrast strongly with the situation in most altricial birds and invite comparison with the ontogenetic patterns and energetics of embryonic development in other avian taxa, precocial as well as altricial.

The peach-faced lovebird, *Agapornis roseicollis*, is a small (50 g) parrot from southwestern Africa. It breeds readily in captivity and produces large clutches (3–7 eggs; mean mass per egg, 4.2 g). Its incubation period is 22.5 d, and the mean mass of its chicks at hatching is 2.52 g (Bucher, 1983). It is a convenient species for study. The characteristics of its eggs and embryonic development are sufficiently unique to be of interest in their own right and also to invite comparison with other species.

Comparative patterns of growth and gas exchange

Calculations based on egg gas conductance, rate of oxygen consumption, and gas tensions within the air cell measured throughout incubation indicate that growth of the chorioallantoic membrane in *A. roseicollis* is not complete until about day 18, 80% of the way through incubation (Bucher and Barnhart, 1984). Whether or not this is a general feature of altricial birds is not known, but recent information (Bartholomew and Goldstein, present volume) indicates that in the brown pelican (*Pelecanus occidentalis*), another and very large altricial species, the chorioallantois is not complete until at least 70% of the way through incubation. The slow development of the chorioallantois in these two altricial species contrasts strongly with the situation in the domestic chicken, the archetypal precocial species, in which growth of the chorioallantois is completed 55% of the way through its 22 d incubation period (Ackerman and Rahn, 1980).

Embryonic growth in *A. roseicollis* is not exponential; the relative growth rate (%/day, or g added per day per g of existing tissue) of wet body mass decreases throughout incubation (Fig. 1). In *A. roseicollis* relative growth rate is about 80% per day at 35% of the way through incubation, and less than 20% per day during the last quarter of incubation. *P. occidentalis* lays eggs (mean mass of about 94 g) almost 25 times larger than those of *A. roseicollis;* relative growth rate of *P. occidentalis* is about 40% per day when the embryo is 40% of the way through incubation and between 10 and 30% per day during the last quarter of incubation (Bartholomew and Goldstein, present volume). The zebra finch (*Poephila guttata*), a very small, altricial bird, lays eggs with a mean mass of 0.94 g, only one quarter as large as that of *A. roseicollis*; its relative growth rate is about 150% per day at 40% of the way through incubation and declines to 5% per day during the last quarter of incubation (recalculated from Vleck et al., 1979). Thus, the smallest of these three altricial species has the highest relative growth rate early in

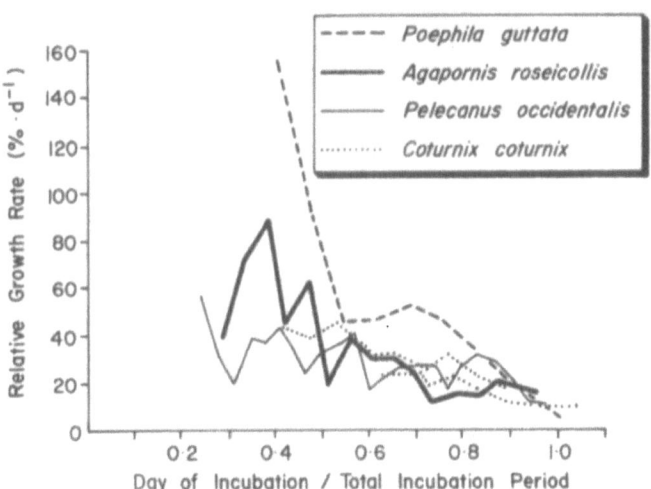

Fig. 1. Relative growth rate of embryonic wet mass as a function of day of incubation/total incubation period. Growth rates are based on 3-day running averages in this and all following figures.

incubation, and the largest species has the lowest. Towards the end of incubation the differences in relative growth rate between the three species become less obvious, if indeed they exist at all.

The coturnix quail (*Coturnix coturnix*) is a precocial species that lays eggs ranging in mass from 8 to 13 g – about 3 times the size of the eggs of *A. roseicollis*. At 40% of the way through incubation the relative growth rate of the *C. coturnix* embryo is about 45% per day. During the last quarter of incubation it declines from about 25% per day to less than 10% per day (recalculated from Vleck et al., 1979; Hoyt, unpublished data). During the last half of incubation the relative growth rate of *C. coturnix* is similar to that of the very much larger *P. occidentalis*, and these rates do not differ appreciably from that of the much smaller *A. roseicollis*.

Despite the large difference in embryo size during the last half of incubation, in three of the four species considered, relative growth rate is not inversely related to mass. This contradicts the frequently stated relationship between mass and various single measures of growth rate (see for example Calder, 1982). The patterns of relative growth rate show the same independence of mass in nestlings of four species of parrots having a 4-fold variation in adult mass even though the growth constants (Ricklefs, 1979) of the species vary inversely with mass (see Figs. 6, 7, and 8 of Bucher, 1983). The smallest species of those considered in the present paper, *P. guttata*, does have a higher relative growth rate throughout most of incubation than do the other three. We suggest that this may not be a function of its small size, but rather of phylogeny; i.e., it is a passerine, and in this order many physiological characteristics reach their extremes among birds.

In the species for which we have data the relative growth rates of embryonic dry mass are higher than those of wet mass because the percentage of water in avian embryos decreases throughout incubation (Bucher, 1983; Romanoff, 1967). Nevertheless, the patterns of growth in dry mass in *P. guttata*, *A. roseicollis*, and *C. coturnix* are similar to those of wet mass.

In *A. roseicollis* mass-specific oxygen consumption ($\dot{V}_{O_2}$) declines during the first two thirds of incubation and is relatively stable during the last third (Bucher, 1983; Fig. 2). *P. occidentalis* and *P. guttata* have the same general pattern of change in mass-specific $\dot{V}_{O_2}$ as *A. roseicollis*, although the magnitude of energy metabolism is different in each of the three species. When lengths of incubation are normalized, at any given point in incubation the smallest species has the highest mass-specific $\dot{V}_{O_2}$ and the largest species has the lowest (Fig. 2). This pattern is expected because of the least two separate consideration: (1) At the same relative time in incubation, the smaller the species, the higher its mass-specific maintenance metabolism will be – this assumes that the usual allometric factors are operating. (2) Early in incubation, smaller species have higher relative growth rates than do large ones. If gram *a* of tissue is producing more new tissue than gram *b* and if the cost of producing new tissue is constant (cost of maintenance metabolism excluded), the component of gram *a*'s mass-specific energy

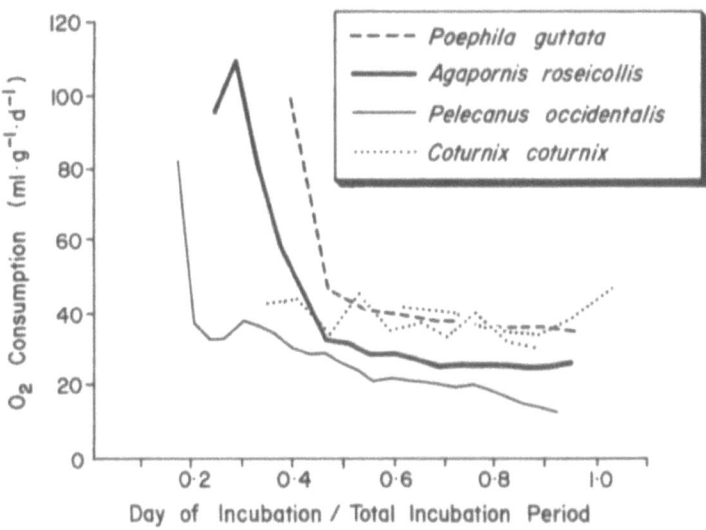

Fig. 2. Mass-specific oxygen consumption as a function of day of incubation/total incubation period.

metabolism that is assigned to growth will obviously be greater than that of gram *b*.

At the same relative age, *C. coturnix* embryos are an order of magnitude larger than those of *P. guttata* and have a lower relative growth rate except for very late in incubation. Therefore, the mass-specific metabolism of *C. coturnix* throughout its incubation period should be lower than that of *P. guttata*. However, this is not the case. Early in incubation mass-specific energy metabolism in *C. coturnix* is lower than in *P. guttata*, but during the second half of incubation it is the same (Fig. 2). This pattern can probably be accounted for on the basis of developmental type, a ranking based on the classification of Nice (1962) which reflects phylogeny and adaptation to ecological niche (see Table 1). It has previously been shown that the total cost of embryonic development is relatively higher in precocial than in altricial species (Bucher, 1983; Vleck et al., 1980).

The largest mass at which the embryos of these four species can be compared directly is about 0.5 g because that is the mass at which *P. guttata* hatches (Fig. 3). At this size, *P. occidentalis*, eventually the largest of the three altricial species, has the highest mass-specific metabolism. It is also noteworthy that between 0.5 and 2.5 g *P. occidentalis* has a higher mass-specific metabolism than does *A. roseicollis*. What accounts for these relationships? Several factors appear to be involved: (1) At a mass of 0.5 g *P. occidentalis* is at an early time in its incubation period and grows about 40%/day, whereas *A. roseicollis* and *P. guttata* are at respectively later times when they reach this mass and grow only about 25%/day and 10%/day respectively. (2) Between 0.5 and 2.5 g the relative growth rate of *P. occidentalis* always exceeds that of *A. roseicollis*. In each case the relative growth rate of *P. occidentalis* is always the highest. If mass-specific maintenance costs are

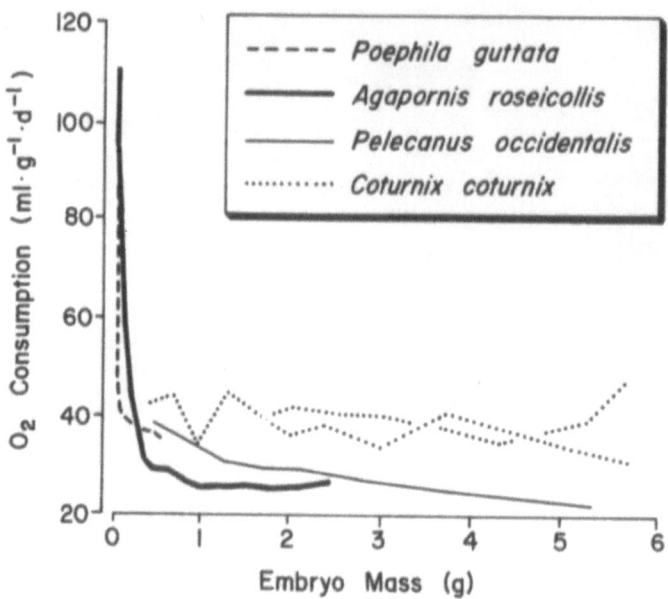

Fig. 3. Mass-specific oxygen consumption as a function of embryo mass.

the same because the species are the same size, the differences in mass-specific metabolism can be attributed to the higher relative growth rate of *P. occidentalis*. This argument does not explain why at a size of 0.5 g the mass-specific metabolism of *P. guttata* exceeds that of *A. roseicollis*. We suggest that phylogenetic factors and/or changes associated with the imminent hatching of *P. guttata* be invoked.

In *C. coturnix* $\dot{V}_{O_2}/g$ as a function of mass is consistently high (Fig. 3), just as it is as a function of relative embryonic age. Once again, this probably can be accounted for by developmental type. If as previously assumed the cost of growth is constant, growth costs cannot account for the high mass-specific metabolism of *C. coturnix* because the relative growth rate of *C. coturnix* embryos weighing between 0.5 and 2.5 g is generally lower than that of *P. occidentalis* embryos of the same mass.

Conductance and air cell gas tensions

In *A. roseicollis* the water vapor conductance of eggs laid by 'wild type' hens differs significantly from that of eggs laid by mutant color hens (Fig. 4). Nevertheless, the mean rate of oxygen consumption of eggs throughout incubation and mean mass of chicks at hatching of these two groups are not different. However, mean air cell gas tensions (P_A) before pipping in eggs of different conductances (Fig. 5) vary greatly ($P_{A_{O_2}}$ from 53 to 102 torr; $P_{A_{CO_2}}$ from 88 to 42 torr). *A. roseicollis* eggs with low gas conductances also pip earlier than those with high gas

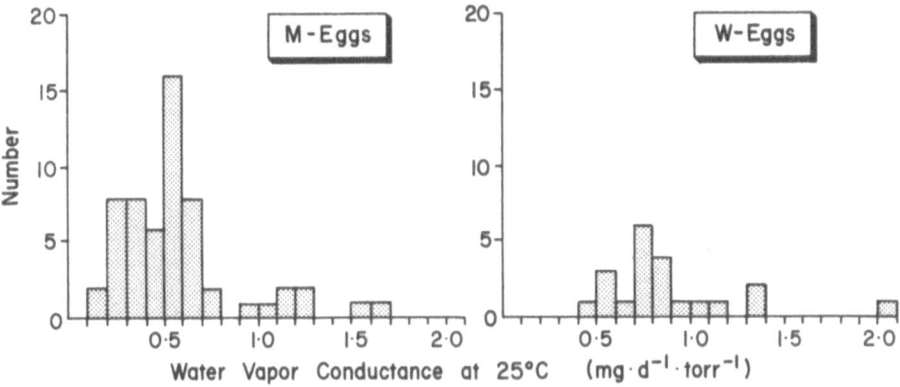

Fig. 4. Water vapor conductance of eggs laid by wild-type hens (W-eggs) and mutant colored hens (M-eggs). Adapted from Bucher and Banhart (1984).

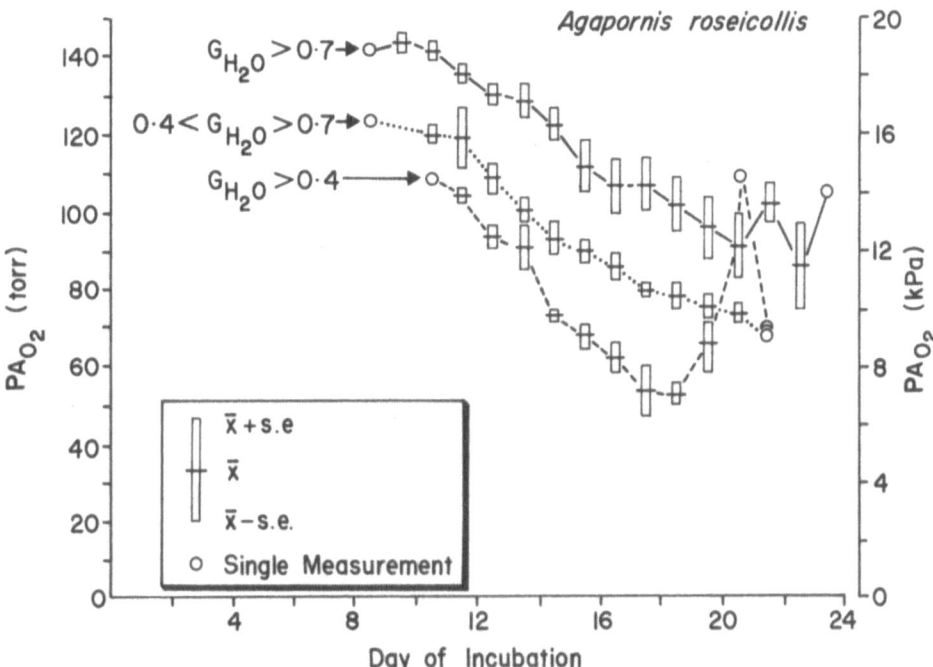

Fig. 5. Air cell P_{O_2} throughout incubation of eggs from three selected ranges of water vapor conductance. Adapted from Bucher and Barnhart (1984).

conductances (Bucher and Barnhart, 1984). The P_{O_2} difference across the shell at the pre-pipping stage of development is often as much as 100 torr, more than twice the allometrically predicted value derived by Rahn (1982). Contrary to Rahn's generalizations, the diffusing capacity (conductance) of the shell of *A. roseicollis* is not directly proportional to the overall development rate (defined by Rahn,

1982, as fresh egg mass/length of incubation), nor does the diffusing capacity of the shell limit pre-pipping $\dot{V}_{O_2}$. In this regard it is noteworthy that gas conductance of the eggs varies by a factor of at least 7 in normally developing eggs laid by hens of different color morphs. Almost as much variation exists among the eggs laid by hens of the 'wild type' color (green). All the *A. roseicollis* eggs are of about the same size. The 7-fold difference in their gas conductances should correspond to a 10-fold difference in egg mass using equation (1) of Ar and Rahn (1978).

Energy utilization

The solid content of *A. roseicollis* yolk is 46.5%, and 18.6% of the total egg content is solids. This is typical of parrot eggs, but these are high values for altricial species in general (Bucher, 1983; Carey et al., 1980). A typical *A. roseicollis* egg weighs 4.2 g, and contains 20.31 kJ chemical potential energy – 13.94 kJ of yolk, and 6.36 kJ of albumin – (Fig. 6). At the time of hatching the caloric value of the chick, including its yolk reserve and its membranes, is at least 10.01 kJ or 49.3% of the energy contained in the original egg contents. The energy content of the yolk-free hatchling is 44.8% of the chemical potential energy of the egg contents used during incubation (chemical potential energy contained in the fresh egg minus the chemical potential energy of the yolk reserve retained in the hatchling). This value is equivalent to total production efficiency, as defined by Kleiber (1961), and is higher than values (26–37%) he quotes for various mammals producing tissue from milk at the time of weaning. This high value is understandable. (1) Avian embryos operate in closed systems in which their direct costs for energy procurement are negligible, and (2) they are functionally ectotherms and theoretically should have higher production efficiencies than endotherms (Turner, 1970).

The caloric value of embryonic dry mass in *A. roseicollis*, as determined by bomb calorimetry, increases significantly during incubation from 18.84 kJ/g at day 5 to 23.45 kJ/g at hatching (Fig. 7). This pattern is qualitatively similar to that of the domestic chicken, which of course is precocial. However, the values for the domestic chicken are higher, increasing from 23.03 kJ/g to 25.96 kJ/g (Needham, 1931).

As is typical of altricial birds, the metabolic rate and growth rate of *A. roseicollis* increase throughout incubation. The caloric equivalence of the *A. roseicollis* embryo mass produced per day per kJ of metabolic expenditure increases throughout incubation also (Fig. 8, insert). It is not immediately evident why the caloric equivalence of the tissue produced per unit of energy expended should increase during development. A possible explanation can be suggested. If the cost of producing a gram of new tissue is constant (excluding the maintenance cost for the embryo), and if the mass-specific maintenance costs decline as mass increases, then the ratio of the caloric equivalence of new mass produced to the

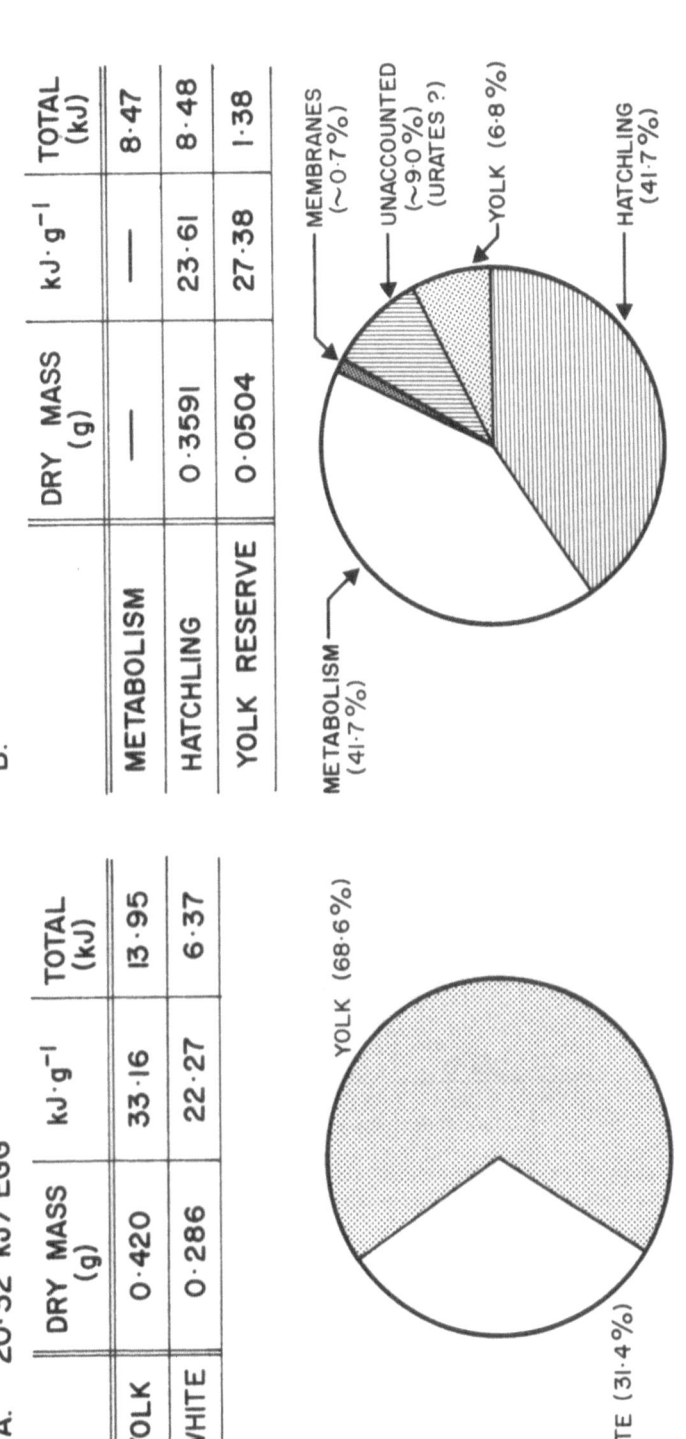

A. 20·32 kJ / EGG

	DRY MASS (g)	kJ·g⁻¹	TOTAL (kJ)
YOLK	0·420	33·16	13·95
WHITE	0·286	22·27	6·37

YOLK (68·6%)

WHITE (31·4%)

B.

	DRY MASS (g)	kJ·g⁻¹	TOTAL (kJ)
METABOLISM	—	—	8·47
HATCHLING	0·3591	23·61	8·48
YOLK RESERVE	0·0504	27·38	1·38

MEMBRANES (~0·7%)

UNACCOUNTED (~9·0%) (URATES ?)

YOLK (6·8%)

HATCHLING (41·7%)

METABOLISM (41·7%)

Fig. 6. Mean apportionment of chemical potential energy (A) of the eggs of *Agapornis roseicollis* and (B) mean chemical potential energy in the hatchlings and mean energy expended during embryonic life span.

367

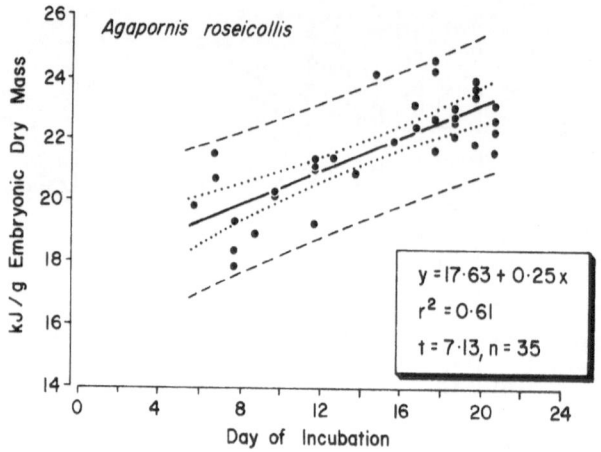

Fig. 7. Chemical potential energy per gram of embryonic dry mass as a function of embryonic age. The dotted lines indicate the 95% confidence interval for the regression and the dashed lines, the 95% confidence interval for an individual datum.

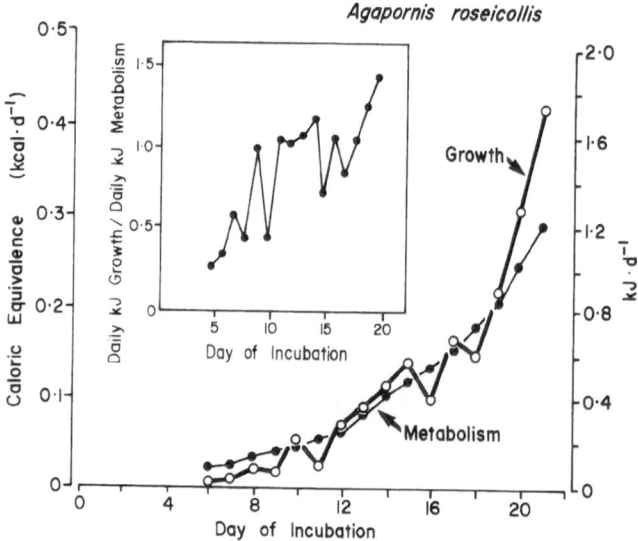

Fig. 8. Caloric equivalence of dry mass growth rate and of metabolic rate as functions of day of incubation. Insert: The ratio between the caloric equivalence of dry mass growth rate and metabolic rate as a function of day of incubation.

amount of energy expended should increase. Similar data are not yet available for other species so the generality of this trend is not known.

It has been predicted that during the embryonic life span the total amount of O_2 consumed per gram of fresh egg is a constant (approximately 0.5 kcal/g fresh egg, or 2.09 kJ/g fresh egg) which is independent of both egg mass and length of incubation (Rahn, 1982). However, as previously demonstrated by Ackerman

Table 1. Avian developmental parameters.

Ref.		a Egg mass (g)	b Inc. (days)	c Develop- mental type*	d Energy expended kJ/g of egg
1	*Poephila guttata*	0.9	14	1	2.186
1	*Troglodytes aedon*	1.5	12	1	2.512
1	*Parus major*	1.7	13	1	2.463
1	*Ploceus cucullatus*	2.8	12	1	1.794
2	*Pelecanus occidentalis*	93.8	30.5	1	1.293
1	*Columba livia*	18.5	17	1	1.576
3	*Melopsittacus undulatus*	2.58	18.4	1	2.002
3	*Bolborhynchus lineola*	3.52	21.5	1	2.139
3	*Agapornis personata*	4.0	22.5	1	1.984
3	*Agapornis roseicollis*	4.24	22.5	1	1.997
3	*Aratinga solstitialis*	8.74	27.5	1	2.063
3	*Enicognathus ferruginous*	10.29	27.2	1	2.317
4	*Aptenodytes forsteri*	47.5	69	2	1.740
1	*Oceanodroma leucorhoa*	9.3	42	2	4.502
1	*Oceanodroma furcata*	13.5	37	2	4.125
5	*Pterodroma hypoleuca*	39.2	48.7	2	2.87
6	*Puffinus pacificus*	60.04	52	2	2.70
5	*Diomedea immutabilis*	284.8	64	2	2.09
5	*Diomedea nigripes*	304.9	64	2	1.99
1	*Bulbulcus ibis*	28	23	2	1.151
1	*Casmerodius albus*	49	27	2	1.102
1	*Eudocimus albus*	51	22	2	0.895
7	*Sterna hirundo*	20.5	23	3	1.551
1	*Sterna maxima*	68	28	3	1.545
8	*Gygas alba*	23.3	35.5	3	2.22
1	*Larus argentatus*	94	27	3	2.583
1	*Apteryx australis*	403	79	5	2.128
1	*Rhea americana*	611	39	5	1.731
1	*Dromiceius novaehollandiae*	637	51	5	1.503
1	*Struthio camelus*	1450	47	5	1.194
1	*Coturnix coturnix*	12	17	5	2.303
7	*Coturnix coturnix*	10.6	17	5	2.465
6	*Gallus gallus*	60	21	5	1.749
7	*Phasianus colchicus*	33.8	24	5	2.008
7	*Calonetta leucophrys*	29.2	23	5	2.306
7	*Anas cyanoptera*	29.4	25	5	2.803
7	*Anas clypeata*	39.7	24	5	2.505
7	*Dendrocygna autumnalis*	50.9	27	5	2.187
7	*Dendrocygna bicolor*	51.2	25	5	2.624
7	*Aythya fuligula*	51.4	23	5	2.127
7	*Netta rufina*	54.3	24	5	1.908
7	*Aythya novaeseelandia*	67.2	26	5	2.127
7	*Anas platyrhynchos*	82.3	28	5	2.246
7	*Anser caerulescens*	120	23	5	1.948
1	*Anser anser*	126	28	5	1.877
7	*Anser canagicus*	131	24	5	1.988
7	*Cygnus atratus*	255.3	36	5	1.789
9	*Leipoa ocellata*	173	62	6	3.56
9	*Alectura lathami*	180	49	6	2.63

* Ranked in an order (1 = altricial, 6 = extremely precocial) based on the classification of Nice (1962), which reflects phylogeny and adaption to ecological niche.
References: (1) Vleck et al., 1980, (2) Bartholomew and Goldstein, 1984, (3) Bucher, 1983, (4) Bucher and Bartholomew, unpublished data; (5) Pettit et al., 1982; (6) Ackerman et al., 1980; (7) Hoyt and Rahn, 1980; (8) Pettit et al., 1981; (9) Vleck et al., 1984.

(1981), this is not the case in sea turtle eggs. A compilation of the data now available for birds shows that in fact this value varies 5-fold – $\bar{x} = 2.14 \pm 0.10$ s.e., $n = 49$; range, 0.89–4.50 – (Table 1). When multiple regression techniques (Afifi and Azen, 1972) are applied to these data, it becomes clear that several factors contribute to the observed variation in metabolic expenditure per gram of fresh egg.

If d = log (kJ/g of egg), in order of importance as ranked by partial correlation coefficients these factors are:

(1) log mass (a): $r_{da \cdot bc} = -0.78$, $t = -8.36$, P<.001
(2) log length of incubation (b): $r_{db \cdot ac} = 0.74$, $t = 7.34$, P<.001
(3) developmental type (c): $r_{dc \cdot ab} = 0.68$, $t = 6.20$, P<.001

The square of the partial correlation coefficient ($r_{yx \cdot z}$) indicates the percent of the variation in the variable y that is explained by variation in the variable x when a subset of variables, z, is held constant.

Multiple regression shows that the log transformed value of the cost of embryonic development per gram of egg varies significantly with log mass ($t = -8.55$), with log length of incubation ($t = 7.43$), and with developmental type ($t = 6.27$). Thus.

$$\log (kJ/g \; egg) = -0.52 - 0.28 \log mass + 0.75 \log inc + 0.06 \; dev \qquad (1)$$
$$(r_a = 0.07, \; r_b = 0.23, \; r_c = 0.32, \; r^2 = 0.62)$$

where mass is in g, inc is length of incubation in d, and dev is developmental type (Table 1).

Cost per gram of egg is not independent of either the mass, or the length of incubation. There are no simple allometric regression lines that elucidate the relationships because of the complications resulting (1) from the variability around the simple regression lines which confound such analysis and (2) from the effects of developmental type. As suggested by Hoyt (1980) multivariate statistics are required here because they deal with the variability around regression lines.

Conclusions

Examination of variability can assist us in understanding which of various possible parameters are of consequence in biological systems and which are not. It is worth reminding ourselves that 'the differences between biological individuals are real, while the mean values . . . are man-made inferences' (Mayr, 1982: p. 47). Biological variability is the raw material of natural selection, and in attempting to understand biological phenomena we must deal with variability as a central element of the system rather than ignore it.

Because most of the parameters of embryonic growth change in value continuously, simple arithmetic averages and single measurements of growth are sometimes inadequate and may mask information that can be found from comparison of patterns. Our analysis suggests a minimal list of parameters that must be considered.

Mass and time are the traditional dimensions of growth, but growth necessarily involves the expenditure of energy. Moreover, in biology natural selection can never be safely ignored. Therefore, at least two other dimensions, phylogeny as a long term consequence of natural selection and the ecological niche as an immediate expression of natural selection, must be included. Thus it seems reasonable to assert that an understanding of avian embryonic growth should encompass at least five dimensions, mass, time, energy, phylogeny, and ecological niche. The first three of these can be combined in various ways as rates and ratios. The last two should be handled individually, but they are difficult to separate.

Acknowledgements

These studies were supported in part by a grant from the United Stated National Science Foundation, DEB-81-03513 and by a U.C.L.A. Research Grant, # 1176.

References

Ackerman, R.A. (1981). Oxygen consumption by sea turtle (*Chelonia, Caretta*) eggs during development. *Physiol. Zool.* 54: 316–324.

Ackerman, R.A. and Rahn. H. (1980). In vivo O_2 and water vapor permeability of the hen's eggshell during early development. *Respir. Physiol.* 45: 1–8.

Ackerman, R.A., Whittow. G.C., Paganelli, C.V. and Pettit, T.N. (1980). Oxygen consumption, gas exchange, and growth of embryonic wedge-tailed shearwaters (*Puffinus pacificus chlororhynchus*). *Physiol. Zool.* 53: 210–221.

Afifi, A.A. and Azen, S.P. (1972). Statistical Analysis a Computer Oriented Approach. Academic Press, New York.

Ar, A. and Rahn. H. (1978). Interdependence of gas conductance, incubation length, and weight of the avian egg. In: Respiratory Function in Birds, Adult and Embryonic. J. Piiper, ed., Springer-Verlag, Berlin, pp. 227–236.

Bartholomew, G.A. and Goldstein. D.L. (1984). The energetics of development in a very large altricial bird, the brown pelican. Present volume.

Bucher, T.L. (1983). Parrot eggs, embryos, and nestlings: patterns and energetics of growth and development. *Physiol. Zool.* 56: 465–483.

Bucher, T.L. and Barnhart. C.M. (1984). Varied egg gas conductance, air cell gas tensions and development in *Agapornis roseicollis*. *Respir. Physiol.* 277–289.

Calder, W.A. (1982). The pace of growth: an allometric approach to comparative embryonic and post-embryonic growth. *J. Zool.* 198: 215–225.

Carey, C., Rahn, H. and Parisi. P. (1980). Calories, water, lipid and yolk in avian eggs. *Condor* 82: 335–343.

Hoyt, D.F. (1980). Adaptation of avian eggs to incubation period: variability around allometric regressions is correlated with time. *Am. Zool.* 20: 417–425.

Hoyt, D.F. and Rahn. H. (1980). Respiration of avian embryos – a comparative analysis. *Respir. Physiol.* 39: 255–264.

371

Kleiber, M. (1961). The Fire of Life. John Wiley & Sons, Inc., New York.

Mayr, E. (1982). The Growth of Biological Thought. Harvard Univ. Press, Cambridge, Mass.

Needham, J. (1931). Chemical Embryology. Vol. 2. Cambridge Univ. Press, Cambridge.

Nice, M.M. (1962). Development of behavior in precocial birds. Linnean Soc. New York.

Pettit, T.N., Grant, G.S., Whittow, G.C., Rahn, H. and Paganelli, C.V. (1981). Respiratory gas exchange and growth of white tern embryos. *Condor* 83: 355–361.

Pettit, T.N., Grant, G.S., Whittow, G.C., Rahn, H. and Paganelli, C.V. (1982). Respiratory gas exchange and growth of Bonin petrel embryos. *Physiol. Zool.* 55: 162–170.

Rahn, H. (1982). Comparison of embryonic development in birds and mammals: birth weight, time and cost. In: A Companion to Animal Physiology, Ch. 9. C.R. Taylor, K. Johansen and L. Bolis, eds., Cambridge Univ. Press. Cambridge, pp. 124–137.

Ricklefs, R.E. (1979). Adaptation, constraint, and compromise in avain development. *Biol. Rev.* 54: 269–290.

Romanoff, A.L. (1967). Biochemistry of the Avian Embryo. Wiley, New York.

Turner, F.B. (1970). The ecological efficiency of consumer populations. *Ecology* 51: 741–742.

Vleck, C.M., Hoyt, D.F. and Vleck, D. (1979). Metabolism of avian embryos: patterns in altricial and precocial birds. *Physiol. Zool.* 52: 363–377.

Vleck, C.M., Vleck, D. and Hoyt, D.F. (1980). Patterns of metabolism and growth in avian embryos. *Am. Zool.* 20:405–416.

Vleck, D., Vleck, C.M. and Seymour, R.S. (1984). Energetics of embryonic development in the megapode birds, mallee fowl (*Leipoa ocellata*) and brush turkey (*Alectura lathami*). *Physiol. Zool* (in press).

27. The development of the respiratory system in placental mammals

John E. Maloney

Monash University Centre for Early Human Development, Queen Victoria Medical Centre, Lonsdale Street, Melbourne, Victoria, 3000, Australia

Abstract

Various strategies are used for the supply of oxygen and substrate to developing animals either in utero or in ovo. In placental mammals the delivery of these substances to the developing tissue is by bulk transport and diffusion. The morphological development of the fetal lung depends not only on this supply but on other features which arise during pregnancy. The lung commences as an endodermal tubular outgrowth of the primitive foregut which penetrates and branches into the surrounding mesenchyme. From its beginnings the lung is fluid filled, not by amniotic fluid or an ultrafiltrate of plasma or a combination of both, but by fluid actively secreted by the developing epithelium. The volume of the potential air spaces of the lung depends in part on the balance between the tracheal outflow of this fluid and the net transepithelial rate of fluid movement. Lung volume influences the parenchymal tissue stresses which in turn alter cell proliferation and differentiation during pulmonary development. Secondly, tissue stresses are also influenced by the activity of the developing respiratory muscles. Investigation of this activity has revealed significant fetal 'breathing' movements before birth which are associated with the development of the central nervous system. Towards term the fetus is permitted to breathe during a state equivalent to rapid eye movement sleep in the adult animal. This cyclical fetal breathing activity occurs with a period of approximately 40 min in the fetal lamb. A reduction in oxygen and substrate delivery reduces fetal breathing movements and appears not to influence the cyclical nature of this activity, but leads to the development of an immature respiratory system.

Introduction

The ability of the newborn animal to leave the environment in which it developed depends in part on the degree of maturation of the cardio-respiratory system at the time of birth. The level of maturation in turn, depends upon extrinsic factors such as the delivery of oxygen and substrate to the developing embryo and fetus as

Seymour, R.S., (ed.) Respiration and metabolism of embryonic vertebrates.
© *1984, Dr W. Junk Publishers, Dordrecht/Boston/London. ISBN-13: 978-94-009-6538-6*

well as intrinsic factors which relate to the cellular timetable for cell proliferation and differentiation.

Strategies for the delivery of oxygen and substrate in developing vertebrates varies considerably. In eutherian mammals there is bulk delivery of both through the umbilical circulation while in some elasmobranchs diffusive processes govern the transport of these materials to the developing embryo (Hogarth, 1976). A close study of the variety of transport systems occurring both between and within various species must enhance our knowledge of basic mechanisms which are important for successful viviparous and oviparous development.

This review examines in particular the development of the respiratory system in placental mammals, a subject which has received considerable attention in the past decade due to the establishment of techniques which enable the longterm monitoring of the fetus in utero. This article proceeds with a brief discussion of the morphological and biochemical development of the lung, followed by a review of the functional characteristics of the developing respiratory system and the influence of a reduced oxygen and substrate delivery on development.

The morphological and biochemical development of the lung in placental mammals

Introduction

Recent interest in the morphological and biochemical development of the lung prior to birth has centered around a morphometric and morphologic description of lung growth and maturation and studies of the secretion, absorption and tracheal movement of the liquid which fills the potential air spaces of the lung. This fluid is actively secreted by the epithelial lining layer of the lung and is not as earlier supposed an ultrafiltrate of plasma or inhaled amniotic fluid. It is likely that the active production of this lung liquid is not confined to placental mammals but occurs in the developing lung of other species. Finally, studies have been designed to elucidate those factors responsible for the control of cell proliferation and differentiation in the lung. The central hypothesis is that lung parenchymal tissue stress is an important factor in the control of cell proliferation and differentiation within the lung. These tissue stresses originate from the interaction between lung parenchyma and the developing chest wall and from the distension of the lung caused by the volume of fluid in the potential air spaces.

Lung growth and development

Following birth the fluid which remains in the airless lung is replaced by air after the initiation of air breathing. This fluid has filled the developing lung since the earliest embryonic period. The relative increase in the volume of this potential air space through gestation is shown in Fig. 1 (Alcorn et al., 1981). This figure provides a convenient reference to introduce the classical morphological division of lung development: the embryonic stage prior to day 40 in the fetal lamb; the pseudoglandular stage day 40–80; the canalicular stage day 80–120; and the alveolar stage day 120-term.

The embryonic stage has been discussed in particular by Bryden et al. (1973). In placental mammals, as in other species, the lung commences as a tubular outgrowth of the ventral wall of the primitive foregut and penetrates into the surrounding mesenchyme, branching as it grows. As noted in Fig. 1 this tubular penetration is characterised by a relatively small change in future air space volume until the end of the pseudoglandular phase (Alcorn et al., 1981) by which time there has been considerable cell differentiation within the airways. The cells range in shape from columnar to cuboidal, are often ciliated within the upper airways and many demonstrate microvilli. Prominent junctional complexes join these tightly packed cells at their apical borders.

The most rapid change in potential airspace volume occurs during the next phase (day 80–120), the canalicular stage. As a proportion of the peripheral lung, the volume of the future air spaces develops from 8.1% at day 80 to 61.8% by day 108 (Maloney et al., 1980a), though significant cell differentiation at the future alveolar level does not take place until later. Mesenchymal cell number matches the growth rate of the whole lung remaining at approximately 50% of all pe-

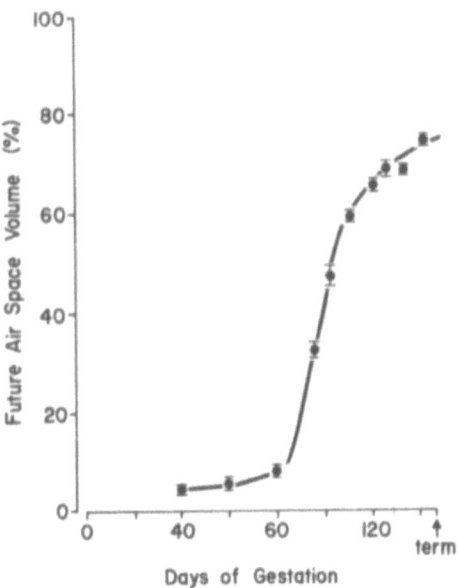

Fig. 1. The relative percentage of the lung occupied by the liquid filled potential air space through gestation in the fetal lamb (Alcorn et al., 1981).

ripheral cell types. During the later part of the canalicular stage there is some flattening of the epithelial cells and the early stages of alveolar development commence. The differentiation of the two major cell types of the mature alveoli are first evident during this phase of development. The Type I cell which covers most of the alveolar surface in the mature lung is observed by day 95. This cell provides a thin cytoplasmic diffusion barrier between the alveolus and the pulmonary capillary. By day 110 the Type II cell has appeared. This cell is responsible for the production of the low surface tension phospholipid lining layer of the alveolus which covers the alveolar surface from birth and prevents airways collapse upon the establishment of air breathing.

During the final stage of prenatal development, the relative proportion of the future airspace increases by approximately 15% while there is an increase of approximately 60% in the wet weight of the lung. Thus in this stage there is a fine control over the growth of the lung with cell proliferation and differentiation being closely linked. There is a constant relationship of 3:1 between endothelial and alveolar Type I cell numbers once alveolar formation has commenced (Alcorn et al., 1981). This stage is characterized by the development of very thin alveolar-capillary tissue barriers and an approximate doubling of the numbers of inclusion bodies in the alveolar epithelial Type II cells.

Although the phases of lung development outlined above are characteristic of placental mammals, differences in the timing of these phases occurs between species. For instance, the fetal lamb lung is more advanced than the human lung at term. There is a threefold difference in the numbers of alveoli in a terminal respiratory unit between the two species with the lamb being more mature (Alcorn et al., 1981). In the more altricial animals such as the mouse, rabbit and rat 'saccular structures' but not true alveoli are evident at birth, and alveoli develop in the neonatal period.

Secretion, absorption, and dynamics of lung liquid

For those associated with fetal lung development in placental mammals the following comment of L.B. Strang (1977a, p. 20) is of no surprise. 'The presence of lung liquid within fetal lungs is almost certainly important for lung development. It must determine the shape and volume of peripheral lung units as well as the growth and form of vascular, epithelial and connective tissue components. In its absence the air spaces would be collapsed and all these structural elements would have a different shape and size.' The site and detailed mechanism of the formation of lung liquid remain unknown though cellular pumps are needed to develop the ionic gradients noted in this fluid. The unique ionic concentrations of this fluid with the high concentrations of sodium and chloride ions, very low protein concentration and low pH confirm that this liquid is not an ultrafiltrate of plasma, amniotic fluid or a mixture of both but uniquely produced by the lung itself (Strang. 1977a). The secretion of fetal lung liquid is probably achieved by 'the active transport of Cl^- in excess of a reverse HCO_3^- flux, with Na^+ following passively down the electrical gradient set up by Cl^- movement; net water flow can be attributed to the osmotic force of NaCl' (Strang, 1977b; Olver and Walters, 1977). As birth approaches and prior to labor, there is a reduction in the secretion rate of lung liquid (Kitterman et al., 1979) and the volume of liquid in the potential air spaces is $20-30\, ml \cdot kg^{-1}$ or approximately the functional residual capacity of the lungs following birth. The volume of extravascular liquid in the lungs is reduced during labor and some experiments indicate that there is no loss of fluid during vaginal delivery (Bland et al., 1979, 1982). The residual fluid is removed from the lungs following the initiation of air breathing. The volume of fluid in the potential air spaces of the lung represents a balance between the net transepithelial exchange of fluid and the tracheal outflow of fluid into the oropharynx. The mean rate of outflow of this fluid is $5-10\, ml \cdot h^{-1}$ over the last third of gestation (Adamson et al., 1975; Goodlin and Rudolph, 1970) and superimposed on this outflow in the fetal lamb are the very small cyclical changes in tracheal flow which are the result of diaphragmatic muscle contractions prior to birth (Fig. 2). Whilst the hourly mean flow is relatively constant during this period there are marked minute to minute changes in its values.

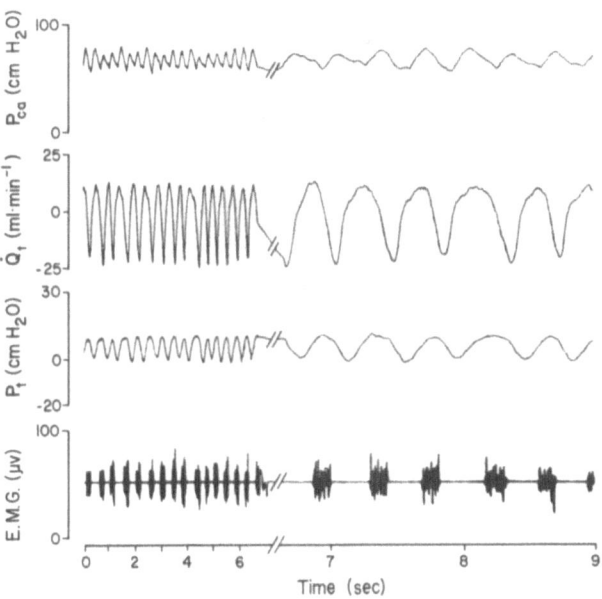

Fig. 2. A photograph of a tracing from a fetus in utero showing carotid artery pressure (P_{ca}), fetal lung liquid flow ($\dot{Q}_l$), fetal tracheal pressure (P_t), and the electrical activity in the fetal diaphragm E.M.G.

377

In the previous two sections a brief review has been given of the development of the fetal lung and the lung liquid dynamics throughout the last one-third of gestation. In the section which follows a description will be given of the development of respiratory muscle function in the thoracic wall. The purpose of this section is to indicate that disturbances in lung liquid dynamics, and disturbances in the interaction between the chest wall and the developing lung, influence pulmonary cell proliferation and differentiation. A mechanical factor which is common to these. i.e. tissue stress may be an important element in determining the ordered sequential structural development of the lung. This concept is also supported by circumstantial evidence from clinical studies in man which indicate that the disturbed activity of the diaphragm as a result of upper cervical and motor neurone problems, diaphragmatic hernia, and tracheo-oesophageal fistula alter pulmonary cell proliferation and differentiation (Wigglesworth et al., 1977).

Tables 1 and 2 summarize the results of a series of experiments in which the mechanical environment of the lung was altered for a period of approximately 4 weeks from day 100–106 until day ~133. Three experimental situations were examined: (1) lung liquid was continually drained from the lung by the application of a small negative intratracheal pressure (D); (2) lung liquid pressure was allowed to increase by tying the trachea in utero (L); and (3) the phrenic nerves were sectioned (Ph). Each experimental manipulation resulted in altered cell differentiation and proliferation. In the drained (D) and phrenectomized animals (Ph) tissue growth was inhibited; the lungs were smaller and the pattern of

Table 1. Morphometry of fetal lamb lungs. The percentage of lung as future air space, lung liquid volume. lung weight. and dry tissue weight (percent) is shown at four gestational ages and following lung liquid drainage (D). tracheal ligation (L), and bilateral phrenectomy (Ph). Drainage and phrenectomy lead to smaller lungs with reduced tissue mass, whereas ligated lungs are significantly enlarged in both volume and tissue mass. Term is ~147 d. Percentages are given as mean ± standard error of the mean.

Age (d)	Future air space volume. as a percentage of total lung volume	Lung liquid volume (ml)	Lung weight including lung liquid (g)	Dry tissue weight, as a percentage of wet weight of lung
80	8.1 ± 1.3		16	–
108	61.8 ± 0.9	31	97	8.6 ± 0.2
133	64.5 ± 1.1	45	187	8.6 ± 0.4
141	70.9 ± 1.2	55	228	7.3
~ 133 (D)	50.3 ± 1.3	16	70	10.9 ± 0.3
~ 133 (L)	72.8 ± 0.8	275	584	7.6 ± 0.3
~ 133 (Ph)	56.0 ± 1.1	13	74	10.5 ± 0.4

Table 2. Lung morphometry of the fetal lamb. T_u = undifferentiated epithelial cells; M = mesenchymal cells; and E = endothelial cells. The percentage of cells in the future gas exchanging region of the lung is shown at four gestational ages and following lung liquid drainage (D), tracheal ligation (L), and bilateral phrenectomy (Ph). Disturbances of the mechanical environment of the lung lead to altered cell differentiation and proliferation.

| Age (d) | Peripheral cell types as a percentage of cell number | | | | |
	I	II	T_u	M	E
80					
108	2 ± 1	1 ± 1	18 ± 1	58 ± 4	21 ± 2
133	11 ± 1	9 ± 1	< 1	47 ± 3	32 ± 3
141	12 ± 2	9 ± 2	0	50 ± 6	29 ± 4
~ 133 (D)	6 ± 1	19 ± 2	2 ± 1	55 ± 3	19 ± 2
~ 133 (L)	10 ± 1	5 ± 1	< 1	54 ± 3	31 ± 3
~ 133 (Ph)	9 ± 2	14 ± 2	1 ± 1	47 ± 4	30 ± 3

peripheral cell differentiation was altered with the Type II: Type I cell ratio being disturbed from the normal control situation (Tables 1 and 2). Increasing intratracheal pressure in the ligated lungs (L) reverses these observations by reducing the Type II: Type I cell ratio and increasing cell proliferation (Maloney et al., 1980a; Alcorn et al., 1980; Alcorn et al., 1977).

Thus the innervation of the diaphragm and its tone and/or its small movement influences the course of cell differentiation and proliferation in the lung. The directions of change are interesting because bilateral phrenectomy results in an alteration of the lung morphology, which is similar to that of lungs which have a chronic tracheal drainage catheter in place. In each case the lungs are small and there is an increased Type II: Type I cell ratio. Phasic activity of the diaphragm is present during lung drainage and the similarities of the lungs in the two situations suggest that this alone is not important for correct lung development. Presumably the innervated diaphragm increases the size of the thorax, thus increasing tissue stress. In chronic tracheal catheterization these tissue stresses are offset by lowering the pressure within the liquid filled future air spaces of the lung. In the ligated (L) lungs tissue stresses are increased and cell proliferation is enhanced. It would appear as if the mechanical tissue stresses can influence those nuclear processes within the cell which govern the timing of cell division and the nature of the cell type established.

Functional development of the respiratory muscles before birth

Introduction

Observations of fetal breathing movements in man and experimental animals have been the subject of many reviews which contain between them cross-references to most of the relevant literature (Preyer, 1885; Bonar et al., 1938; Duenhoelter and Pritchard, 1977; Wilds, 1978; Dawes and Henderson-Smart, 1981; Purves, 1981; Maloney, 1983). Despite a significant number of experiments demonstrating breathing before birth, intrauterine fetal breathing movements were not accepted as a normal physiological process until 1970. Barcoft's detailed observations in 1937 of the fetal lamb exposed under spinal or urethane ana-esthesia following laparotomy of the ewe supported the concept that fetuses did not spontaneously activate their respiratory muscle in utero.

A change in scientific attitude towards fetal respiratory activity came in 1970 when two abstracts (Dawes et al., 1970; Merlet et al., 1970) were published indicating fetal respiratory activity in the unanaesthetized fetal lamb in utero from day 100 (0.68 of gestation) to day 140 (0.95 term). Respiratory activity in utero has since been demonstrated in placental mammals such as the goat (Towell and Salvador, 1974), monkey (Martin et al., 1974), rabbit (Merlet-Benichou, 1973), and guinea pig (Kendall, 1977).

Fetal respiratory muscle activity

Dawes et al. (1972) first described quantitatively fetal respiratory movements using tracheal pressure as an indication of fetal respiratory muscle function. Subsequently other investigators used diaphragmatic electrical activity (Maloney et al., 1975; Bowes et al., 1981a) as an indicator of the output of the respiratory motor neurone pool. Fig. 2 illustrates measurements recorded from a fetus in utero, showing fetal carotid artery pressure P_{ca}, fetal tracheal flow $\dot{Q}_t$ and tracheal lung liquid pressure P_l, and the electrical activity of the diaphragm EMG.

Studies of fetal respiratory activity were further extended by a detailed analysis of the diaphragmatic electromyogram (Bowes et al., 1981a) in six fetal lambs from a gestational age of 101 d (Fig. 3). This study demonstrated that the maturation of fetal respiratory activity altered significantly over the last third of gestation with the emergence of a periodic modulation of respiratory rate alternating between active and quiet phases with a mean cycle length of 37 min by 130 d gestation; a threefold increase in percentage apnoea (expiratory time >10 sec) from 20% at 110 d to 50% at 140 d; and a gestational decline in average respiratory rate caused by an increasing interval of diaphragmatic electrical silence.

Fetal laryngeal muscles are also active before birth, the abductor muscles contracting in phase with the diaphragm. Adductor muscles are tonically active during high voltage activity of the cerebral cortex but there is insufficient tone to close the glottis (Harding, 1980; Harding et al., 1980). Fig. 3 thus reveals that there exists a marked alteration in the ontogenic regulation of the undisturbed fetal respiratory system.

380

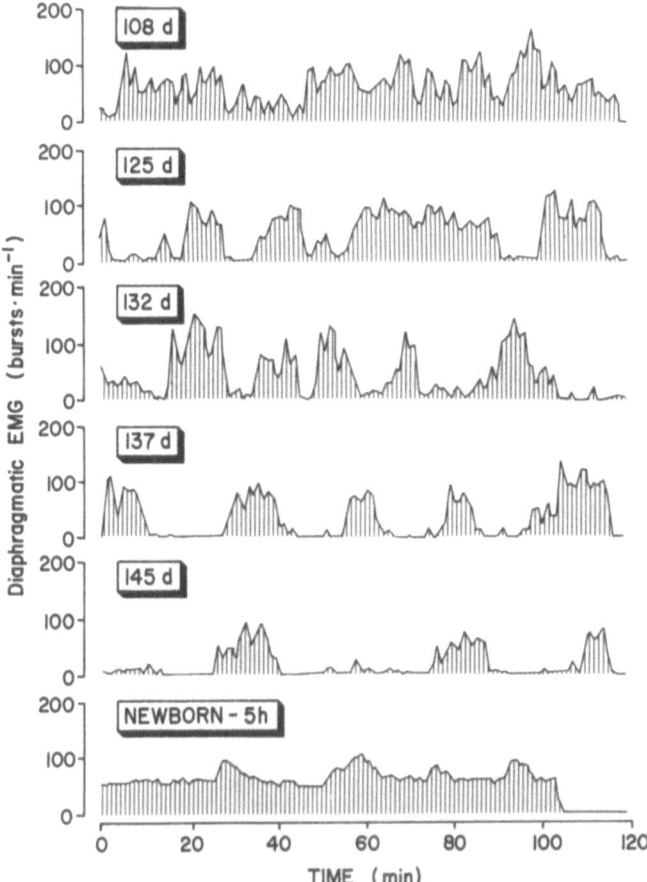

Fig. 3. The changing pattern of fetal respiratory activity throughout the last third of gestation in the fetal lamb. Note the increasing periods of diaphragmatic silence.

The central nervous system and fetal respiratory activity

A working hypothesis could be proposed that the cycles that occur in diaphragmatic activity during life in utero represent an outcome of the development of the non rapid eye movement (n-REM) sleep state or its fetal analogue. It is known that when adult unanaesthetized animals are in this state, withdrawal of their vagal and chemoreceptor afferent input significantly reduces respiratory activity (Phillipson et al., 1978; Sullivan et al., 1978). This is not so in the REM state where other central nervous system inputs are sufficient to maintain the firing of the respiratory rhythm generator when afferent input is removed. The activity state of the central nervous system in the adult animal also influence the normal regulation of respiratory patterns and the response of the respiratory system to

381

various stimuli. The central nervous system is implicated in the development of diaphragmatic silence by a recent study of breathing patterns in fetal lambs in utero after mid-brain transection (Dawes et al., 1980). In these preparations breathing is virtually continuous though respiratory rate is lower than in control animals. High voltage cortical activity is only present for short invervals of 3–4 min.

Acute observation in fetal lambs of 114–147 d gestation (Dawes et al., 1972) delivered into a warm saline bath revealed three types of behaviour which corresponded to awake, quiet and paradoxical sleep. In this study preparations were implanted with one or more sets of electrodes to measure a combination of the triad of biopotentials used to identify behavioural state in mature animals; electrocorticographic activity, electro-oculographic activity and nuchal muscle activity. In two lambs near-term, REM was associated with 'rapid irregular breathing' and in nine lambs low-voltage fast-cortical activity characteristic of paradoxical sleep in the fetal lamb was similarly associated with fetal breathing. As noted the cortical electrical activity did not become well differentiated before 128 d, an observation confirmed by others (Maloney et al., 1980b; Bowes et al., 1981a). Cyclical regulation in fetal respiratory rate alone is present between 100 and 110 d with a period similar to that of the later well-defined REM-non-REM cycles, and before the emergence of distinct ocular and cortical patterns. Ocular activity is associated with fetal breathing (Maloney et al., 1980b) by 125 d and strong visual correlations with cortical activity emerge after this time. As an aid to identifying behavioural state the third of the triad of bipotentials, the nuchal muscle electromyogram, has been found helpful is some laboratories (Ioffe et al., 1980) though not in others (Ruckebusch, 1972).

The most interesting finding in the above studies is that the modulation of fetal breathing is linked to an endogenous rhythm associated with the maturation of the central nervous system. It has been suggested that the development of non-REM sleep is associated with the maturation of cortical tissue (Jouvet-Mounier et al., 1969), an idea consistent with the gradual evolution of non-REM characteristics in the fetus and a diminution of respiratory muscle activity in late gestation in the fetal lamb. The earlier descriptions of fetal breathing in multiple pregnancies (Snyder and Rosenfeld, 1937) noted that not all fetuses breathe simultaneously supporting the concept of a rhythm endogenous to each fetus. Further studies in models such as those described are necessary to provide important insights into the spontaneous neural regulation of fetal 'breathing' in utero.

The influence of oxygen and substrate delivery on respiratory system development before birth

Introduction

The importance of understanding the control of oxygen and substrate delivery to the embryo and fetus and the consequences of sub-optimal variations in this delivery is clear when it is realised that a significant proportion of human perinatal mortality and morbidity is related to the pathophysiology of these areas. The opportunity to experimentally investigate variations in delivery in placental mammals is available by alterations in the maternal environment such as under-nutrition or acute changes in the inspired gas composition; alterations in the uterine circulation or the size of the placental exchanging surface or alterations in the umbilical circulation. Alterations in circulatory parameters obviously affect both the delivery of oxygen and substrate to the fetus and thus studies of other groups such as the eggs of birds or reptiles or the young of viviparous vertebrates provide alternative models which enable a comparative physiological analysis of the problem and separation of the interaction between the two main delivered components, oxygen and substrate.

Body and organ weights

In placental mammals an experimental reduction in placental size (Alexander, 1964) or maternal nutritional deprivation (Nitzan and Groffman, 1971) leads to a reduction in fetal body size. The various organ systems respond differently with the brain and heart usually being spared and maintaining their weight relative to that of the body. Redistribution of blood flow following the imposed stress is important in selectively distributing oxygen and metabolites to the brain, heart and kidneys during hypoxia or a graded embolisation of the uterine circulation (Creasy et al., 1973; Alexander, 1974; Cohn et al., 1974). Lung size is also reduced relatively to body size in placentally insufficient animals. While the DNA content of various organs such as the brain and the lung is not reduced in such circumstances protein content appears to be diminished (Creasy et al., 1973). Such observations indicate that cell proliferation is not altered in these circumstances, but rather that cell size is reduced. A variation in the supply of oxygen and nutrient to the fetus also alters the cell types in the diaphragm of fetal lambs, and their size (Maloney et al., 1982). The reduction in the numbers of fatigue resistant fibres may therefore influence the ability of the neonate to withstand a challenge to the respiratory system in the early neonatal period.

The function of the respiratory muscles

Acute reduction in oxygen delivery to the fetus reduces the electrical activity of the diaphragm and fetal respiratory movements (Maloney et al., 1975b; Boddy et al., 1974). This is in contradistinction to the response of the infant and adult where exposure to an hypoxic environment stimulates respiratory activity. In chronically deprived fetal animals where the fetal blood glucose concentrations were reduced to approximately 66% of the control value with a small reduction in the arterial oxygen tension of the fetus from 22.3 ± 0.3 to 19.0 ± 0.4 torr there were marked changes in respiratory muscle function in the last third of gestation (Maloney et al., 1982). Thus a reduction in oxygen and substrate delivery had altered the function of the diaphragm during development. Some recent evidence suggests that a major site of influence of a reduced oxygen supply on respiratory muscle function may reside within the brain, and not peripherally, as brain section immediately above the pons releases the respiratory system from its hypoxic suppression (Dawes et al., 1983).

Conclusions

Studies of the development of the respiratory system in placental mammals over the last thirteen years have confirmed observations over the previous 200 years that the respiratory muscles are active prior to birth and demonstrated that this activity changes its pattern throughout gestation in a manner which is correlated with the neural activity of the central nervous system. They have also indicated that the developing lung actively secretes a liquid which fills the developing future air spaces of the lung. This liquid is not 'inhaled' amniotic fluid nor an ultra-filtrate of plasma. The morphological development of the lung has been shown to be dependent on the dynamics of this lung liquid and the activity of the developing respiratory muscles. Further variations in oxygen and substrate delivery alter both the morphological and functional development of the respiratory system.

This review has not attempted to discuss the development of the homeostatic control of the respiratory system during in utero development, or the physiology of the respiratory system during labor, birth and the immediate neonatal period. A detailed study of the respiratory physiology of the perinatal period in placental mammals remains an area for investigation in the future.

Acknowledgements

I wish to acknowledge the significant contribution made by my colleagues in the improvement of our understanding of the ontogeny of the respiratory system. These scientists include Drs Blair Ritchie, Adrian Walker, T. Michael Adamson,

Glenn Bowes, Joseph Smolich, Daine Alcorn and Mal Wilkinson, and research associates Vojta Brodecky, John Cannata, and Margaret Dowling.

References

Adamson, T.M., Brodecky, V., Lambert, T.F., et al. (1975). Lung liquid production and composition in the *in utero* fetal lamb. *Aust. J. Exp. Biol. Med. Sci.* 53: 65–75.

Alcorn, D., Alexander, I.G.S., Adamson, T.M., Maloney, J.E., Ritchie, B.C. and Robinson, P.M. (1977). Morphological effects of chronic tracheal ligation and drainage in the foetal lamb lungs. *J. Anat.* 123: 649–660.

Alcorn, D., Adamson, T.M., Maloney, J.E. and Robinson, P.M. (1980). Morphological effects of either chronic bilateral phrenectomy or vagotomy in the fetal lamb lung. *J. Anat.* 130: 683–695.

Alcorn, D., Adamson, T.M., Maloney, J.E. and Robinson, P.M. (1981). A morphometric analysis of fetal lung development in the sheep. *Anat. Rec.* 201:655–667.

Alexander, G. (1964). Studies on the placenta of the sheep (*Ovis aries* L.). Effect of surgical reduction in the number of caruncles. *J. Reprod. Fertil.* 7: 307–322.

Alexander, G. (1974). Birth weight of lambs: influences and consequences. In: Size at Birth. CIBA Foundation Symposium 27, Elsevier, North Holland, pp. 215–245.

Bland, R., Bressack, M. and McMillan, D.M. (1979). Labor decreases the lung water content of newborn rabbits. *Am. J. Obstet. Gynecol.* 135: 364–367.

Bland, R.D., Hanson, T., Haberkern, C.M., Bressack, M.A., Hazinski, T.A., Usha Raj, J. and Golberg, R.B. (1982). Lung fluid balance in lambs before and after birth. *J. Appl. Physiol.: Respirat. Environ. Exercise Physiol.* 53: 992–1004.

Boddy, K., Dawes, G.S. and Fisher, R. (1974). Fetal respiratory movements, electrocortical and cardiovascular responses to hypoxia and hypercapnia in sheep. *J. Physiol.* 243: 597–618.

Bonar, B.E., Blumenfeld, C.M. and Fenning, C. (1938). Studies on fetal respiratory movements. *Am. J Dis. Child.* 55: 1–11.

Bowes, G., Adamson, T.M., Ritchie, B.C., Wilkinson, M.H. and Maloney, J.E. (1981a). Development of patterns of respiratory activity in unanaesthetised fetal sheep in utero. *J. Appl. Physiol.: Respirat.Environ. Exercise Physiol.* 50: 693–700.

Bryden, M.M., Evans, H. and Binns, W. (1973). Embryology of the sheep. III. The respiratory system, mesenteries and coelom in the fourteen to thirty-four day embryo. *Anat. Rec.* 175: 725–736.

Cohn, H.E., Sacks, E.J., Heymann, M.A. and Rudolph, A.M. (1974). Cardiovascular response to hypoxemia on acidemia in fetal lambs. *Am. J. Obstet. Gynecol.* 120: 817–824.

Creasy, R.K., Barrett, C.T., De Swiet, M., Kahanpaa, K.V. and Rudolph. A.M. (1973). Experimental intrauterine growth retardation in the sheep. *Am. J. Obstet. Gynecol.* 112: 566–573.

Dawes, G.S., Fox, H.E., Leduc, D.M., Liggins, G.C. and Richards, R.T. (1970). Respiratory movements and paradoxical sleep in the foetal lamb. *J. Physiol., (London)* 210: 47–48P.

Dawes, G.S., Fox, H.E., Leduc, D.M., Liggins, G.C. and Richards, R.T. (1972). Respiratory movements and rapid eye movement sleep in the foetal sheep. *J. Physiol. (London)* 220: 119–143.

Dawes, G.S. and Henderson- Smart, D.J. (1981). Breathing before and after birth. In: International Review of Physiology. Respiratory Physiology III: 23: 75–110, Ed. J.F. Widdicombe, University Park Press, Baltimore.

Dawes, G.S., Gardner, W.N., Johnston, B.M. and Walker, D.W. (1980). Breathing patterns in fetal lambs after mid brain transection. *J. Physiol.* 308: 298.

Dawes, G.S., Gardner, W.N., Johnson, B.M. and Walker, D.W. (1983). Breathing in fetal lambs: the effect of brain stem section. *J. Physiol.* 335: 535–553.

Duenhoelter, J.H. and Pritchard, J.A. (1977). Fetal respiration: a review. *Am. J. Obstet. Gynecol.* 129: 326–338.

Goodlin, R.C. and Rudolph, A.M. (1970). Tracheal fluid flow and function in foetuses *in utero*. *Am. J. Obstet. Gynecol.* 106: 597–606.

Harding, R. 1980). State related developmental changes in laryngeal function. *Sleep* 3: 307–322.

Harding, R., Johnson, P. and McCelland, M.E. (1980). Respiratory function of the larynx in developing sheep and the influence of sleep state. *Resp. Physiol.* 40: 165–179.

Hogarth, P.J. (1976). In: Viviparity. Edward Arnold Ltd. London, pp. 7, 8.

Ioffe, S., Jansen, A.H., Russell, B.J. and Chernick, V. (1980). Sleep, wakefulness and the mono-synaptic reflex in fetal and newborn lambs. *Pflugers Arch.* 388: 149–157.

Jouvet-Mounier, D., Astic, J. and Lacote, D. (1969). Ontogenesis of the state of sleep in rat, cat, and guinea pig during the first postnatal month. *Dev. Psychobiol.* 2: 216–239.

Kendall, J.Z. (1977). Respiratory movements in the fetal guinea pig *in utero*. *J. Appl. Physiol.: Respirat. Environ. Exercise Physiol.* 42: 661–663.

Kitterman, J.A., Bollard, P.L., Clements, J.A., Mescher, E.J. and Tooley, W.H. (1979). Tracheal fluid in fetal lambs: spontaneous decrease prior to birth. *J. Appl. Physiol.: Respirat. Environ. Exercise Physiol.* 47: 985–989.

Maloney, J.E., Adamson, T.M., Brodecky, V., Cranage, S., Lambert, T.F. and Ritchie, B.C. (1975). Diaphragmatic activity and lung liquid flow in the unanesthetized fetal sheep. *J. Appl. Physiol.* 39: 423–428.

Maloney, J.E., Adamson, T.M., Brodecky, V., Dowling, M. and Ritchie, B.C. (1975b). Modification of respiratory centre output in the unanaesthetised foetal sheep *in utero*. *J. Appl. Physiol.* 39: 552–558.

Maloney, J.E., Alcorn, D., Bowes, G. and Wilkinson, M. (1980a). Development of the future respiratory system before birth. *Sem. in Perinatol.* 4: 251–258.

Maloney, J.E., Bowes, G. and Wilkinson, M. (1980b). Fetal breathing and the development of patterns of respiration before birth. *Sleep* 3: 299–306.

Maloney, J.E., Bowes, G., Brodecky, V., Dennett Xenia, Wilkinson, M. and Walker, A. (1982). Function of the future respiratory system in the growth retarded fetal lamb. *J. Dev. Physiol.* 279–297.

Maloney, J.E. (1983). Review of experimental studies on the functional development of the respiratory system in the foetal lamb. *Aust. J. Biol. Sci.* 36: 1–14.

Martin, C.B., Murata, Y., Petrie, R.H. and Parer, J.T. (1974). Respiratory movements in fetal rhesus monkeys. *Am. J. Obstet. Gynecol.* 119: 939–948.

Merlet, C., Hoertner, J., Devilleneuve, C. and Tchobrouotsky, C. (1970). Mise en evidence de mouvements respiratoires chez la foetus d'angeau in utero au cors du dernier mois de la gestation. *Compt. Rend.* 270: 2462.

Merlet-Benichou, C. (1973). Movements respiratoires chez le foetus et installation de la respiration a la naissance. *Bull. Physio-Pathol. Respir.* 9: 1365–1387.

Nitzan, M. and Groffman, H. (1971). Hepatic gluconeogenesis and lipogenesis in experimental intrauterine growth retardation in the rat. *Am. J. Obstet. Gynecol.* 109: 623–627.

Olver, R.E. and Walters, D.V. (1977). The effect of catecholamines on lung liquid secretion. *J. Physiol.* 273: 58–59.

Phillipson, E.A., McClean, P.A., Sullivan, C.E. and Zamel, N. (1978). Interaction of metabolic and behavioural respiratory control during hypercapnia and speech. *Am. Rev. Respir. Dis.* 117: 903–909.

Preyer, W.F. (1885). Specialle Physiologie des Embryos: Unter-suchungen uber die Lebens-erscheinungen vor der Geburt. Leipzig, Grieben.

Purves, M.J. (1981). The neural control of respiration before and after birth. In: Reviews in Perinatal Medicine, Vol. 4. E.M. Scarpelli and E.V. Cosmi, eds. Raven Press, N.Y., pp. 299–336.

Ruckebusch, Y. (1972). Development of sleep and wakefulness in the foetal lamb. *Electroenceph. Clin. Neurophysiol.* 32: 119–128.

Snyder, F.F. and Rosenfeld, M. (1937). Direct observation of the intrauterine respiratory movements of the fetus and the role of carbon dioxide and oxygen in their regulation. *Am. J. Physiol.* 119: 153–166.

Strang, L.B. (1977a). In: Neonatal Respiration. Blackwell Scientific Publications, Oxford, London, Edinburgh, Melbourne.

Strang, L.B. (1977b). Growth and development of the lung: Fetal and postnatal. *Ann. Rev. Physiol.* 39: 253–276.

Sullivan, C.E., Kozar, L.F., Murphy Elaine and Phillipson, E.A. (1978). Primary role of respiratory afferents in sustaining breathing rhythm. *J. Appl. Physiol.: Respirat. Environ. Exercise Physiol.* 45: 11–17.

Towell, M.E. and Salvador, H.S. (1974). Intrauterine asphyxia and respiratory movements in the fetal goat. *Am. J. Obstet. Gynecol.* 118: 1124–1131.

Wigglesworth, J.S., Winston, R.M.L. and Bartlett, K. (1977). The influence of the central nervous system on fetal lung development. *Arch. Dis. Child.* 52: 965–976.

Wilds, P.L. (1978). Observations of intrauterine fetal breathing movements – A review. *Am. J. Obstet. Gynecol.* 131: 315–338.

28. Structural adaptations of the eggs and the fetal membranes of monotremes and marsupials for respiration and metabolic exchange

R.L. Hughes

Department of Anatomy, University of Queensland, St. Lucia, Queensland, 4067, Australia

Abstract

Microvilli, in both monotremes and marsupials, serve to increase the area of respiratory and metabolic surfaces throughout embryonic development.

Significant respiratory and metabolic exchange occurs during a prolonged period of intrauterine development in both monotremes and marsupials. Embryonic tissues, during this time, are separated from the maternal uterine endometrium by an intact monotreme egg shell or, in marsupials, by a thin shell membrane that ruptures after about two thirds of gestation has elasped. The egg shell or shell membrane both consist of a distensible highly permeable structural protein homologous with the rigid shell membranes or proteinaceous egg shells of the cleidoic eggs of sauropsid vertebrates.

The early intrauterine stages of monotreme development are characterised by meroblastic cleavage within the tiny peripheral germinal disc of the yolky egg. The resultant blastodisc proliferates and differentiates to form a bilaminar blastoderm (yolk-sac) that completely encapsulates the 4 mm diameter vitellus of yolky spheres. This embryonic stage is homologous with the marsupial bilaminar blastocyst. The monotreme embryo exhibits a fourfold increase in diameter during the terminal stages of intrauterine development when the non-vascularized yolk-sac, closely invested by the fully formed egg shell, constitute the only extraembryonic structures concerned with respiration and metabolic exchange.

The newly laid monotreme egg contains a 19–20 somite embryo at the neurula stage. A functional vascular net progressively invests the wall of the yolk-sac during the incubation period. The allantoic rudiment appears relatively late during incubation and only becomes a large vascularized vesicle during the terminal prehatching stage. The incubation period concludes with the highly vascularized vesicular allantois and the now almost completely vascularized yolk-sac lining virtually the entire inner surface of the egg shell, and occupying opposite, approximately equal hemispheres.

The vitellus of the relatively small marsupial eggs has a diameter of between 210–229 μm and exhibits indeterminate holoblastic cleavage. The blastomeres become arranged around the inner surface of the zona pellucida to form a

Seymour, R.S., (ed.) Respiration and metabolism of embryonic vertebrates.
© *1984, Dr W. Junk Publishers, Dordrecht/Boston/London. ISBN-13:978-94-009-6538-6*

unilaminar blastocyst, lacking the inner cell mass characteristic of eutherian mammals. The cell wall of the unilaminar blastocyst consists of a single layer of protoderm cells. Primitive endoderm cells, within the future animal pole of the protoderm, become arranged beneath its surface to form the bilaminar blastocyst. The wall of the fully formed bilaminar blastocyst constitutes the yolk-sac. In the majority of marsupials the vascularized and non-vascularized regions of the fully differentiated yolk-sac constitute the major extraembryonic tissues for respiration and metabolic exchange.

The marsupial allantois appears relatively late in gestation becoming an enlarged vesicle only during the terminal stages. The allantois is poorly vascularized in most marsupials and is excluded from fusion with the chorion by folds of the yolk-sac splanchnopleure. In a few marsupials the highly vascularized distal hemisphere of the vesicular allantois fuses with the chorion to form a non-invasive respiratory chorio-allantois. Only during the terminal days of gestation, in bandicoot marsupials, does the highly invasive discoidal chorio-allantois supersede the yolk-sac as the major extraembryonic organ for respiration and metabolic exchange.

Introduction

As early as 1834 Sir Richard Owen suggested that reproduction in female monotremes and marsupials might profitably be compared with that seen in certain sauropsid vertebrates.

Since the mid 1950's, an adequate perspective of marsupial embryology has accumulated through the establishment of breeding colonies of major marsupial groups. Unlike marsupials, no laboratory colonies of breeding monotremes have ever been established. Consequently almost the total body of data concerned with monotreme embryology has been derived from wild animals caught prior to the 1930's. A considerable body of unpublished information on both monotreme and marsupial embryology is now curated as the Hill Collection of the Hubrecht Laboratory at Utrecht in the Netherlands. It is some of this unpublished data, together with selected papers from the literature and material from my own personal collection that constitutes the basis for this paper. This communication seeks to provide an overview of the broad structural adaptations for respiration and metabolic exchange in monotremes and marsupials through the description of an appropriately graded series of embryological specimens.

Monotremes

Oocyte yolk synthesis

The newly ovulated monotreme egg, of both the platypus *Ornithorhynchus anatinus* and the echidna *Tachyglossus aculeatus*, is a spheroid of about 4 mm in diameter, where all but a thin peripheral zone consists of tightly packed yolk spheres. According to Flynn and Hill (1939), the yolk spheres form on both the inner and outer surfaces of a cortical zone within the oocyte cytoplasm, as illustrated in Figs. 1 and 5.

In the fully formed oocyte and early post fertilization developmental stages of the egg, the inner zone of yolk formation is reduced to a residual latebra and its associated latebral neck which is characterised by yolk spheres of greatly reduced size. The latebra complex, as shown in Fig. 4, extends from the centre of the egg to a thickened peripheral lenticulate germinal disc of relatively yolk-free cytoplasm which contains the egg nucleus and marks the future site of the blastodisc that forms as a result of meroblastic cleavage.

At the ultrastructural level, the outer peripheral region of yolk deposition in developing preovulatory oocytes is characterised by small membrane bound yolk vesicles that contain a granular product (Fig. 3). The enlargement of these yolk vesicles is accompanied by an increasing coarseness of their granular contents. The maturing yolk vesicles consist of a central zone of amorphous material and a granule free peripheral zone. The largest yolk spheres contain only amorphous material. Although the exact mechanism of yolk synthesis has not yet been determined, it would seem that the numerous microvilli of the oocyte cell membrane that interdigitate within the zona pellucida with similar processes arising from the investing follicle cells provide a greatly increased surface area which could facilitate metabolic exchange. Although the nutrients involved in the marked increase in oocyte diameter that accompanies vitellogenesis must traverse the zona pellucida, these are not resolved with the electron microscope as morphological entities. However small vesicles attached to expanding yolk vesicles suggest a possible accretional mechanism for their growth.

Mucoid coat and egg shell

The glycoprotein mucoid coat and the egg shell of monotremes are tertiary egg membranes, typical of the cleidoic eggs of sauropsid vertebrates and are both contributed to the ovulated eggs during their passage through the uterine tube or in the uterus (Hughes and Carrick, 1978).

The fully formed mucoid coat of an unsegmented early intra-uterine echidna egg varies in thickness from 17 μm to 30 μm. The mucoid coat (albumen layer) persists in the echidna as a compacted zona pellucida-albumen layer until at least

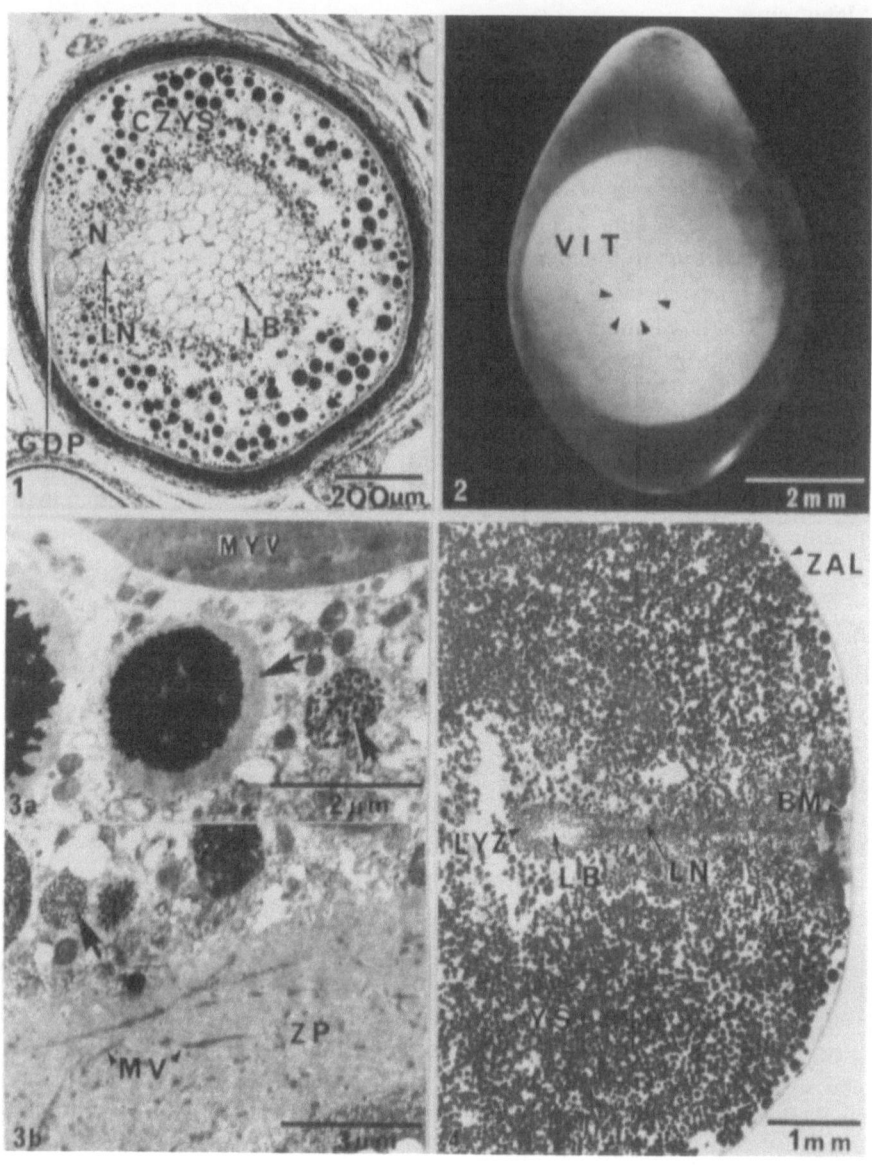

Fig. 1. Early vitellogenesis in a sectioned platypus *Ornithorhynchus anatinus* primary oocyte with a diameter of 0.86 × 0.81 mm. After Flynn and Hill (1939). CZYS, cortical zone of yolk sphere synthesis; GDP, primordium of the germinal disc; LB, latebral body; LN, latebral neck; N, nucleus of primary oocyte.

Fig. 2. The vitellus (VIT) of this early uterine echidna *Tachyglossus aculeatus* egg has a diameter of about 4.5 mm. The circular peripheral blastodisc (arrowed) of about 0.69 mm in diameter was found to contain approximately 500 cells. After Flynn and Hill (1939).

392

the establishment of the early primordia of the primitive streak (Flynn and Hill, 1947). The zona-albumen layer of monotremes may persist until the egg is laid. However this requires confirmatory histochemical studies as discussed by Hughes and Carrick (1978).

The egg shell that invests a fully developed intra-uterine platypus embryo, collected by the present author (RLH), consists of three concentrically arranged components and the following details have been previously described by Hughes and Carrick (1978). A compact inner basal layer with a thickness of $2\,\mu$m has a fine granular ultrastructure. The basal layer is in turn invested by a relatively open meshwork of rodlets with a width of approximately $20\,\mu$m. The outer matrix layer, deposited exclusively during intrauterine development, has a thickness of $49\,\mu$m and consist of loosely arranged but relatively large irregular particles (Fig. 7). It should be noted that the matrix layer is much thinner than the $130\,\mu$m to $208\,\mu$m previously reported for full-term monotreme eggs (Hill, 1933). The total thickness of the full-term platypus egg shell collected by the present author is equivalent to the overall thickness of the double shell mebrane of the domestic fowl.

The fully developed egg shell of the echidna (Fig. 6) has an essentially similar structure to that of the full-term platypus egg, thus confirming the earlier observations of Hill (1933).

Scanning electronmicroscopy of the surface of the full-term intrauterine platypus egg (Fig. 7) shows a loose arrangement of irregular shaped matrix particles that contrasts with the fibrous structure of the ovokeratinous shell membrane of the domestic fowl (Fig. 8).

In the platypus, erythrocytes (with a diameter of $5.5\,\mu$m after fixation), liberated from the endometrial blood vessels during the recovery of a full term intrauterine egg, had penetrated both the outer matrix layer as well as the rodlet layer in large numbers: however none was found within the compact basal layer (Hughes and Carrick, 1978). The egg shell of monotremes is unique among sauropsid vertebrates in that it undergoes marked stretching as a result of the transport of embryonic nutrients to the yolk-sac of the embryo during intrauterine development.

←

Fig. 3. Electron micrograph of the peripheral region of yolk deposition in a primary oocyte from a .6 mm diameter ovarian follicle of an echidna. Maturing yolk vesicles (arrowed); MV, microvilli; MYV, portion of a mature yolk vesicle: ZP, zone pellucida.

Fig. 4. Vertical section through the embryonic hemisphere of a 4.5 mm uterine echidna egg. After Flynn and Hill (1939). BM, one of the 31 cells of the blastodisc; LB, lateral body; LN, lateral neck; LYZ, small yolk spheres of the lateral zone; YS, large yolk spheres packing the cytoplasm of the vitellus; ZAL, zona-albumen layer.

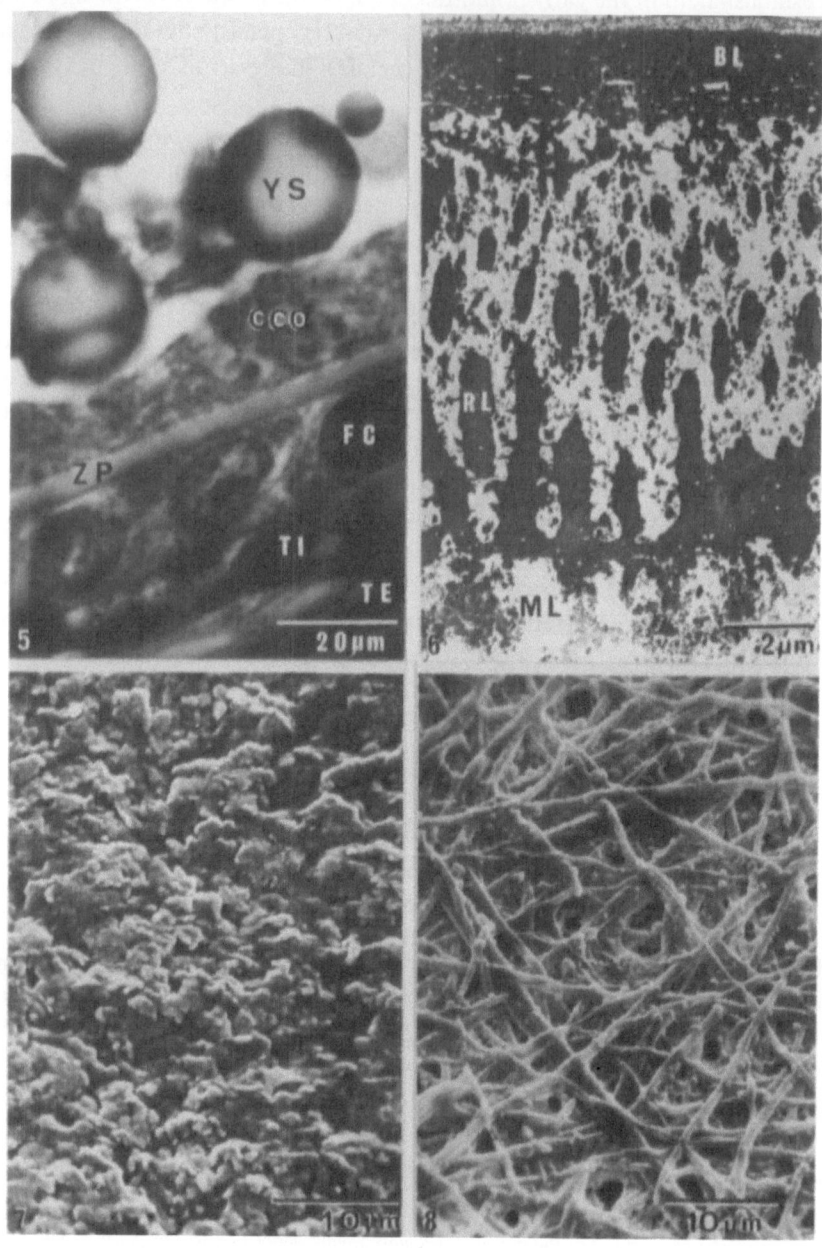

Fig. 5. Advanced stage of vitellogenesis in a 2.5 mm diameter ovarian follicle of a platypus. CCO, cortical cytoplasm of primary oocyte; FC, cuboidal layer of follicle cells; TE, theca externa; TI, theca interna; YS, yolk spheres; ZP, zona pellucia.

Fig. 6. Transmission electron micrograph showing the deeper portion of the fully formed egg shell of the echidna. BL, basal layer of shell; ML, matrix layer of shell; RL, rodlet dayer of shell.

As in non-mammalian vertebrates with a yolky cleidoic egg, monotremes exhibit a peripheral incomplete cleavage of the vitellus designated as meroblastic cleavage. This is illustrated in Fig. 9, which shows the surface view of an early platypus blastodisc located on the surface of a 4.0 mm diameter spheroidal vitellus that was recovered from a uterine egg with an overall diameter of 5.5 mm. This platypus was collected 23.8.1901. The peripheral walls of each of the eight blastomers are incompletely formed, as is typical for meroblastic cleavage of yolky eggs. The blastomeres are of unequal size and the bilateral symmetry exhibited by this early blastodisc is not universal for all early cleavage stages of monotremes (Flynn and Hill, 1939).

A later stage in the formation of the blastodisc is shown in Fig. 10. The uterine egg containing this blastodisc was obtained from an echidna collected 14.7.1933. The major and the minor axes of the slightly oval egg shell measured 6.0 mm and 4.5 mm respectively. The dimensions of the contained vitellus were 4.7 mm by 4.5 mm (Fig. 2). The circular blastodisc of about 500 surface cells has a diameter of about 0.69 mm. Centrally the blastodisc is up to five cells in thickness and at its margin is one or two cells in thickness. Here the cytoplasm of the deeper cells is usually rich in yolk-spheres. The margin of the blastodisc is surrounded by a nucleated cytoplasmic germ ring that is in continuity with the uncleaved cytoplasm of the vitellus.

According to Flynn and Hill (1942), meroblastic cleavage results in the formation of a biconvex blastodisc of up to 6 or 7 cells in thickness centrally. Specialized vitellocyte cells at the margin of the blastodisc are capable of migration and also of enclosing yolk-spheres. Peripherally the blastodisc thins and is connected with a migratory germ ring that spreads over the surface of the underlying yolk. As the diameter of the blastodisc expands behind the migratory face of the germ ring, the biconvex blastodisc diminishes in thickness to become a blastoderm of only two or three cells in maximum thickness.

At a later stage in development, the rapidly expanding blastoderm of an echidna embryo is almost circular in outline (Fig. 11), measuring about 2.7 mm in diameter and exhibiting a well defined blastodermic rim. The blastoderm at this stage is unilaminar over its entire extent and is closely associated with the under surface of the zona-albumen layer that is presumably formed by the fusion of the zona pellucida with the overlying glycoprotein mucoid coat. The blastoderm consists of prospective ectodermal cells as well as increasing numbers of primitive

←

Fig. 7. Scanning electron micrograph showing the outer surface of the matrix layer of the egg shell that invested a fully developed, 20 somite, intrauterine platypus embryo. The matrix layer consists of irregularly shaped particles.

Fig. 8. Scanning electron micrograph showing the fibrous structure of the outer surface of the shell membrane of the domestic fowl.

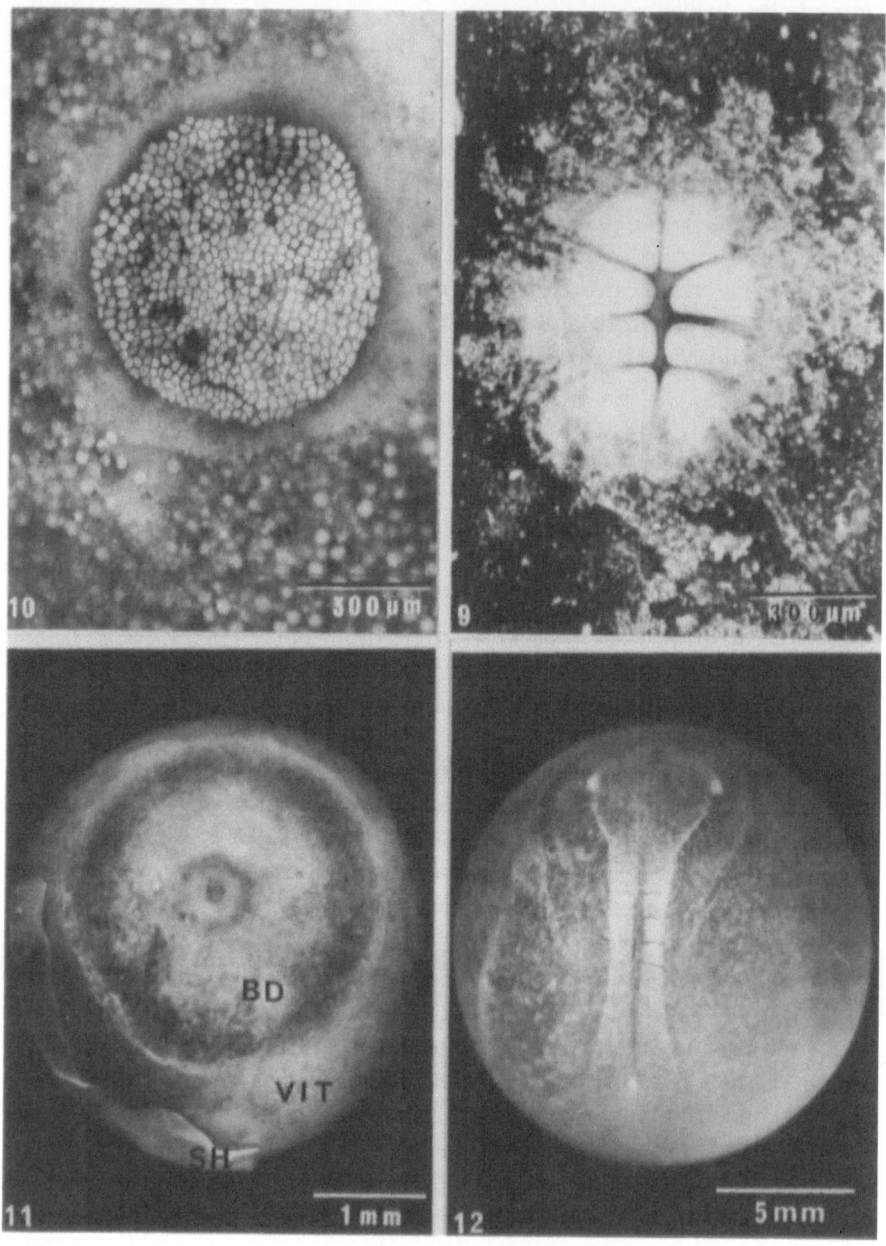

Fig. 9. Portion of the surface of the 4 mm diameter vitellus from a uterine platypus egg, showing an elipsoidal blastodisc with eight blastomeres exhibiting meroblastic cleavage. After Flynn and Hill (1939).

Fig. 10. Surface view of approximately 500 cells of the blastodisc of the uterine egg of the echidna shown in fig. 2. The circular blastodisc of about 0.69 mm in diameter consists of up to five cells in thickness centrally and this thins marginally to 1–2 cells in thickness. After Flynn and Hill (1939).

endodermal cells that have not yet commenced the formation of an inner endodermal layer. The marginal region of the expanding blastodisc consists of a small
number of vitellocytes that contain small yolk spheres in their cytoplasm, as well
as an actively expanding germ ring. The diameter of the shell of the preserved egg
shown in Fig. 11 is 5.5 mm. The vitellus within the partly dissected shell measures
about 3.85 mm in diameter and consequently exhibits no significant increase in
size when compared with the 4 mm diameter of early intrauterine eggs. At later
stages in development, the prospective ectoderm cells remain passive at the
surface of the blastoderm and give rise to a continuous layer of flattened ectoderm. The endoderm cells, segregated from the blastoderm, form cell-networks,
cell-plates, and isolated cells that gradually differentiate into a continuous layer
underlying the ectoderm cells. The expanding blastoderm eventually encloses the
entire yolk mass. The last point of closure of the blastoderm is marked by a
thickening at the future abembryonal pole of the egg in the form of a yolk navel
that develops just prior to the appearance of the primitive streak, after the
blastoderm has become bilaminar throughout. The enclosure of the yolk in a
complete cellular wall marks the establishment of the typical mammalian blastocyst, capable of actively absorbing the nutrient uterine fluid and consequently
increasing in size. In marsupial species the wall of the unilaminar blastocyst,
according to Flynn and Hill (1942), gives rise to endoderm in an almost identical
way to monotremes. However more recent studies have indicated that the
marsupial blastocyst expands before the bilaminar stage is attained.

From the early primitive streak stage and onwards, the yolk-sac endoderm comes
to consist of plate-like polygonal cells, with cytoplasm crowded with discrete yolk-
spheres of large size (Flynn and Hill, 1947). The maximum development of this type
of endoderm occurs in the extra-embryonal region of the yolk-sac.

In a presomite echidna embryo collected 7.8.1933, the fully developed primitive streak had a length of 6.9 mm. The now partly trilaminar yolk-sac completely
encloses the yolk mass that has expanded, with the result that the overall
diameter of the intrauterine egg, before preservation, now measures 9.0 mm.
Gastrulation in the platypus has been investigated by Wilson and Hill (1908).

A near terminal stage in the expansion of the yolk-sac, as a result of the
absorption of uterine endometrial fluids, is represented by an early somite stage
echidna embryo obtained from an intrauterine egg with a diameter of 11.7 mm

←——

Fig. 11. Uterine egg of an echidna showing the blastoderm in the process of investing the vitellus. BD,
circular unilaminar blastoderm of about 2.7 mm in diameter; SH, portion of the dissected egg shell;
VIT, spheroidal vitellus of 3.85 mm in diameter. Photograph from the Hill collection, Hubrecht
Laboratory.

Fig. 12. Surface view of a flat five somite uterine echidna embryo curving from a distance of about
13.7 mm around the surface of the egg (yolk-sac) that has now expanded to a diameter of 11.7 mm
before preservation. The diameter of the preserved yolk-sac measured 15 mm. Photograph from the
Hill collection, Hubrecht Laboratory.

before preservation. This egg was located within the right uterus of an echidna collected on 30.7.1930. The flat embryo curved around the surface of the yolk-sac for a distance of about 13.7 mm and was located within an oval embryonal area with major and minor axes of about 22 × 11 mm. A relatively narrow pear-shaped crescentric region of translucent yolk-sac, within the wider embryonal area, surrounded the brain plate and somite region of the embryo proper (Fig. 12).

The embryo proper consisted of a flattened spatulate brain plate with a length and width of 4.7 mm and 4.0 mm respectively. Two small lateral protrusions, a little in front of the middle of the brain plate, constitute the rudiments of the future trigeminal brain plates. The brain plate narrows in its caudal extremity, so that at the level of the first of the five pairs of prominent paraxial somites the width of the embryo is about 2.2 mm. A prominent primitive streak of 5.6 mm in length is located distal to the mesoderm that follows the last pair of somites (Fig. 12).

The subterminal stage of intrauterine development is illustrated by an echidna embryo recovered from a preserved spherical intrauterine egg of 14 mm in diameter, collected 8.8.1930 (Fig. 13). This eleven somite neurula stage had an overall greatest length of about 17.5 mm. The greatest width of the embryo was 12 mm measured across the trigeminal head plates. The width of the embryo tapered distally from about 6–7 mm at the mid somite region to about 2 mm at the caudal extremity. Facial and prefacial nerve rudiments were reported to be present. The yolk-sac contained loose nutritive material. Hill considered that the primordia of the pericardial canals were differentiating. It is important to note that, although this is a near terminal stage of intrauterine development, the non-vascularized yolk-sac is the only extraembryonic fetal membrane. The expansion

Fig. 13. Cephalic portion of subterminal intrauterine echidna embryo located on the surface of a 14 mm diameter spheroidal, non-vascularized yolk-sac. This eleven somite neurula stage had an overall greatest length of about 17.5 mm. S, somite; THP, trigeminal head plate. Photograph from the Hill collection, Hubrecht Laboratory.

Fig. 14. Prefetal branchial cleft stage echidna embryo from a pouch egg with a diameter of 15 mm before preservation and an estimated postlaying age of approximately 6 d. The embryo is viewed from the inner surface of the yolk-sac and shows folds of the combined proamnion and amnion investing its forelimbs and head region. Al, allantois; FL, forelimbs; VYS, vascularized yolk-sac. Photograph from the Hill collection, Hubrecht Laboratory.

Fig. 15. The fully formed extraembryonic membranes of the echidna. Redrawn from Fig. 41 of Semon (1894). AL, allantois; SH, egg shell; YS, yolk-sac. The amnion and the chorion are not shown.

Fig. 16. A fully developed pre-hatching platypus fetus with a greatest length of 9 mm, obtained 31 August 1898 from a laid egg with a diameter of 15 mm. The upturned snout constitutes the oscaruncle. A sharp recurved median egg tooth of 0.29 mm in length, situated in the upper jaw is further described by Hill and De Beer (1949) page 513. Portions of the highly vascularized extraembryonic membranes attached to the umbilical region include a flattened allantois of 10 × 6.5 mm that is fused with the chorion. Photograph from the Hill collection, Hubrecht Laboratory.

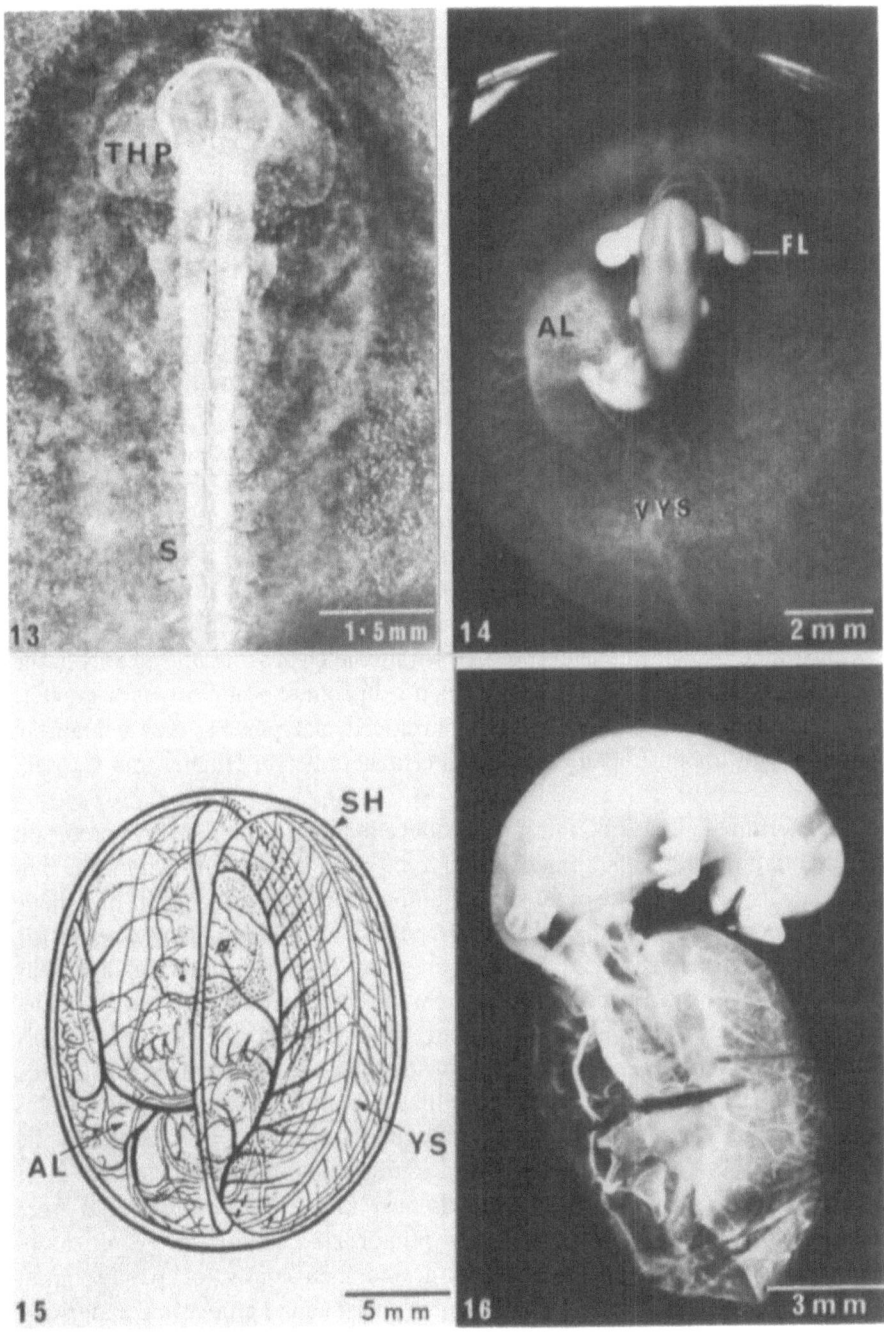

in diameter from 4 mm at the earliest stage of intrauterine development to this 14 mm diameter embryo has resulted from the transport and absorption, via an intact shell, of nutrient fluids derived from the progesterone stimulated uterine endometrium.

The terminal stage of intrauterine development is characterised by no apparent further growth in the diameter of the yolk-sac and the persistance of a flat embryo at the somite neurula stage. The fundamental ground plan for future fetal development has now been established. The well formed primordia of the blood vascular system together with early specialisations of the yolk-sac constitute important events in relation to embryonic metabolism and gaseous exchange during the approaching incubation period of the laid egg (Wilson and Hill, 1908).

The shape of the fresh monotreme egg at the end of intrauterine development varies from spheroidal to ovoidal, with major and minor axes of about 17 × 14 mm in echidna and 18 × 14 mm for platypus. The preserved eggs exhibit shrinkage, with a reduction of as much as one sixth of the length of the major axes of the shell. In both echidna and platypus, the egg shell encloses a relatively flat 19 somite embryo at the neurula stage. The embryo curves for a distance of some 19 mm around the surface of the partly trilaminar yolk-sac. A degree of variation occurs in the stage of development attained at the end of the intrauterine period, as illustrated by an echidna egg located within the cloaca and presumably in the process of being laid. This egg contained a 16 somite embryo with a greatest overall length of 14.5 mm. The fully developed platypus egg may contain an embryo with about 20 pairs of somites (Hughes, 1974b; Hughes and Carrick, 1978).

The terminal period of intrauterine development is marked by important differentiation within the primary embryonic germ layers, both within the embryo and its yolk-sac. The outer surface of the entire yolk-sac consists of a single layer of ectoderm cells that are mostly polygonal in shape. However, in the immediate extraembryonic region they are cuboidal. The ectodermal layer may be more than one cell thick in the region of the embryo and the medullary plate. The endoderm in the region of the embryo consists of a single layer of closely packed flattened cells. In the region of the yolk-sac immediately outside the embryonal area, large endoderm cells with cytoplasm distended by yolk spheres are increasingly numerous. This type of endoderm cell will eventually line the entire yolk-sac. The only intraembryonic differentiation of the endoderm layer consists of two bands of cuboidal cells near the heart primordia that were considered to represent the pharyngeal primordia. The intraembryonic mesoderm at the terminal stage of intrauterine development consists of up to 19 pairs of paraxial somites situated on either side of the axial neural groove and underlying notochord. The somites contain a flattened myocoele cavity. Immediately lateral to the somites, the intermediate mesoderm exhibits a region where the primordia of embryonic mesonephric kidney have developed. Still farther laterally, the mesoderm has divided into somatic and splanchnic layers, separated by an as yet

400

undeveloped primitive intraembryonic coelom. Two conspicuous lateral head plates, situated on either side of the future forebrain and midbrain areas, were originally described by Hill and Martin (1895) as consisting of mesoderm. However, according to Wilson and Hill (1908) and in Hill's later extensive note books these are considered to be trigemial ganglionic neural plates. The three future primary brain vesicles are present as yet not at the end of the intrauterine period. The hindbrain component is characterised by at least eight pairs of segmentally arranged neuromeres. The thickened triangular plates acustico-facial ganglionic of ectoderm are situated opposite the second and third pairs of hindbrain neuromeres. The hindbrain gradually narrows into the precursors of the spinal cord that extends caudally in the midline between the somites to the distal extremity of the embryo, where a well formed but contracting primitive streak is located. The notochord arises from the thickened cephalic end of the primitive streak and extends beneath the developing spinal cord in a cephalic direction, to a point beneath the middle portion of the forebrain.

The rudiments of a non-functional circulatory system are present as a pair of endothelial heart tubes situated within the lateral margin of the mesoderm located on either side of the hindbrain. Each endothelial heart tube is reduced in diameter distally to form the primordia of the sinus venosus and the omphalomesenteric vein. Blood islands are present in both the embryonic area as well as in the immediately adjacent extraembryonic mesoderm that exhibits a marked expansion lateral to the posterior region of the embryo, where mesoderm extends from a quarter to a halfway around to yolk-sac. The blood islands are in an early stage of the formation of a blood vascular net. However the lumina of these primitive vessels are discontinuous and do not contain fetal blood corpuscles and as yet are not connected with the heart primordia.

The shelled intrauterine egg with its contained embryo is now sufficiently differentiated to relinquish its reliance on the uterus for metabolic and gaseous exchange. The embryo is about to embark on an almost autonomous existence, with the yolk-sac that closely underlies the shell initially constituting the sole organ for both nutrition and gaseous exchange.

Development during the incubation period

On the 7th September 1900, Hill received a consignment of echidnas in a box. Eight days later he recovered a damaged laid egg from the box. Externally the fresh spheroidal egg measured 15×14.5 mm and contained a strongly torted 29–30 somite prefetal branchial cleft stage. with a greatest length of 7.8 mm. A small allantois was present; however a description of the development of the organ systems showed this embryo to be slightly less developed than either of the prefetal branchial cleft echidna and platypus stages shortly to be described. Hill's opinion that the egg was probably recovered a day or two after laying should now be interpreted as an underestimate. In the light of all the material collected since 1900, a more appropriate postlaying age is suggested to range between 3 to 6 d. The incubation period for the echidna is 10–10.5 d (Griffiths et al., 1969).

A prefetal branchial cleft echidna embryo was obtained on 28.7.1933 from a pouch egg with a diameter, before preservation, of 15 mm (Fig. 14). Although the fetal length of the embryo proper was about 10 mm, this was reduced by a marked degree of torsion to give a greatest length of 5.5 mm. The well developed forelimbs of this embryo are represented as ventrolaterally directed paddles consisting of a shaft and a manus plate. The hindlimb rudiments were present as limb buds. The figure shows extensive portions of the proamnion and the amnion investing the head and forelimbs of the embryo (Fig. 14). Approximately three quarters of the embryonal pole of the yolk-sac was well vascularized and was served by two prominent vitelline veins that merge just distal to the cervical flexure and then pass in front of the forelimbs, before entering the embryo. The vitelline artery divides into two branches that encircle the yolk-sac, before overlapping for a short distance on the opposite side of the yolk-sac. These branches of the vitelline artery are the homologues of the vessels that form the marsupial sinus terminalis and serve to separate the non-vascularized abembryonal quarter of this echidna yolk-sac from the vascularized area. A vascularized flattened ovoidal allantoic vesicle of 4 mm by 3 mm is closely associated with the caudal umbilical region of this echidna embryo.

According to Hill's unpublished notes, the prefetal branchial cleft stage of the echidna embryo just described is very similar to a prefetal branchial cleft stage platypus embryo from a single laid egg with major and minor axes of 17 mm and 14.5 mm respectively. This platypus embryo was also extremely torted with a greatest length of 6.5 mm measured from the cervical flexure to the caudal extremity.

The details of its development can be summarised as follows: The head of the embryo is sunken into the yolk-sac and is covered by a proamnion as far caudally as the forelimbs. An amniotic fold also invests the caudal 0.6 mm of the embryo. The primordia of the face consist of nasal pits surrounded by medial and lateral nasal swellings. Deep lacrimal grooves are present. Eye primordia have developed. Maxillary and mandibular processes of the first pharyngeal arch have developed. The second and the third pharyngeal arches are prominent, while the fourth and fifth arches are just visible. The forelimbs are differentiated into a shaft and a manus plate. The hindlimb buds are present. The respiratory primordia consist of two primary bronchi with well developed distal lung primordia. The midgut opens into the yolk-sac behind the heart primordia. Structures arising from the alimentary canal include a gall bladder, dorsal pancreatic rudiment together with the right and the left ventral pancreatic rudiments. The hindgut and the anal plate are well formed. Bowman's capsules are a component of the well formed mesonephric kidneys, which are served by as yet solid mesonephric ducts that enter the cloaca. The following cranial nerves have differentiated: V, VII,

VIII, IX, X, XI, XII. The primordia of the intraembryonic blood vascular system is well developed and includes atrial, ventricular and truncal components of the heart. Five pairs of aortic arches are present and two pulmonary vessels reportedly join a sixth aortic arch. Segmental arteries of the dorsal aorta supply the spinal cord and the body wall. The seventh pair of segmental arteries vascularize the forelimbs. A coeliaco-mesenteric branch of the dorsal aorta is present. Aortic sinusoids occur in the region of the mesonephros. A large omphalomesenteric artery is present. Twelve pairs of arteries arise from the dorsal aorta to enter the wall of the yolk-sac. A pair of aortic arteries vescularize the hindgut. The dorsal aortae end in a sinus located in the anal part of the hindgut.

Components of the venous system include the anterior and the posterior cardinal veins that enter the sinus venosus by way of the common cardinal veins. Blood returns from the head by way of two inferior jugular (internal jugular) veins. The posterior cardinal veins drain the mesonephros. Two vitelline veins enter the liver, with the larger right vitelline vein forming the ductus venosus. The smaller left vitelline vein unites with the right, but has a small separate branch that enters the sinus venosus.

The greater part of the yolk-sac with its contained mass of free yolk was highly vascularized in a manner very similar to that previously described for the prefetal pharangeal cleft echidna embryo. However there is a considerably greater area of non-vascularized yolk-sac that occupies somewhat less than one half of the abembryonal pole. The rudimentary allantois, measuring 1.4 mm by 0.8 mm, was highly vascularized by five small arteries and two umbilical veins.

The stage of development attained in this platypus embryo is reminiscent of that found in human development at approximately the end of the 4th week of gestation.

To put the major events of the prefetal incubation period of monotreme development into some perspective, it should be noted that the flat somite stage embryo of the early period of incubation now exhibits prefetal development where the primordia of the major organs have appeared. The prefetal period of incubation has also been characterised by the formation of a functional vascular net within the greater portion of the wall of the yolk-sac. This must constitute the major source of fetal nutrition and virtually the sole organ for gaseous exchange. It is inconceivable that the tiny vascularized allantoic rudiment could be of importance in gaseous exchange until the onset of fetal development has commenced. In this connection, the differentiation of the monotreme allantois remarkably parallels that found in marsupials, where the allantoic rudiment also first appears at the prefetal branchial cleft stage and is also destined in some species to attain a relatively short functional respiratory life during the terminal stages of fetal development (Sharman, 1961).

The fetal membranes in the two existing Families: Tachyglossidae and Ornitho-rhynchidae are essentially similar in their arrangement. The extraembryonic membranes practically fill the interior of the shell, which measures in echidna 16.5×13 mm and in the playpus: 16.5×14 mm; 18×14 mm; 17×14 mm in three eggs taken at random from the 'Hill Collection'.

The fully formed extraembryonic membranes, according to Hill, are charac-terised by a fetus enclosed by an amnion and lying within the extraembryonic coelom between two highly vascularized vesicular extraembryonic membranes, the allantois and the yolk-sac (Fig. 15; echidna redrawn from Fig. 41 of Semon 1894). A relatively narrow signet ring-shaped zone of pure chorion encircles the fetus and separates the lesser right chorio-allantoic hemisphere from the greater left chorio-vitelline hemisphere.

The relatively flat inner wall of the yolk-sac that constitutes the yolk-sac splanchnopleure is not markedly invaginated by either the fetus or the allantois. A very thin short yolk-sac stalk connects the apex of the herniated intestinal loop of the fetus with the yolk-sac splanchnopleure.

According to the unpublished notes of Hill, the yolk-sac is lined by large yolk-laden endoderm cells and is filled by fluid in which yolk granules are dispersed. It is important to note that, although the formation of the yolk-sac resembles that found in the sauropsida, its marked increase in size is primarily due to the absorption of nutritive fluid, during the presumed month long (28 d) period of intrauterine development (Broom, 1895) that can be causally linked with a functional corpus luteum and a progesterone stimulated uterine endometrium (Hughes and Carrick, 1978). The progesterone stimulated uterine endometrium, rather than the ovary, as the major source of embryonic nutrients constitutes a fundamental departure from the sauropsid pattern of embryonic metabolism and is a feature that remarkably parallels marsupial development.

The single vitelline artery that passes from the yolk stalk onto the yolk-sac splanchnopleure was presumed to be derived from two fused elements. The vitelline artery gives off small branches in the yolk-sac splanchnopleure, before continuing for a short distance as a single vessel into the yolk-sac omphalopleure (chorio-vitelline area), where it divides into two main branches that closely parallel each other to almost encircle the entire lower wall of the yolk-sac. At intervals these two arterial branches unite, throughout the remainder of their course enclosing a very narrow strip of non-vascularized yolk-sac. At an earlier stage of development in a platypus embryo with a greatest length of 6.5 mm, Hill notes that the two major branches of the vitelline artery bounded a conspicuous non-vascularized circular area on the lower pole of the yolk-sac. In embryos with a greatest length of over 8.0 mm, the non-vascular area of the yolk-sac becomes markedly reduced. These two major branches of the vitelline artery are homolo-gous with similar vessels that give rise to the sinus terminalis in the marsupial yolk-sac. However, in monotremes, almost the entire yolk-sac becomes vascularized and, although a very small non-vascularized area remains, a definitive sinus terminalis fails to develop. The major arteriolar net of the fully formed monotreme yolk-sac omphalopleure arises as small branches from the upper side of the two main arterial trunks. The fine venous vessels drain into two vitelline veins in the lateral wall of the yolk-sac that pass separately into the yolk-sac splanchnopleure, before uniting to form a single trunk that enters the body of the fetus immediately in front of the yolk stalk.

The capillaries of the richly vascularized yolk-sac must obviously be involved in the transport of the nutritive contents of the yolk-sac to the embryo. Hill also suggested that the large vascularized area of the yolk-sac omphalopleure that immediately underlies more than half the surface area of the shell might reasonably be presumed to be involved with gaseous exchange.

The allantois is a flattened, richly vascularized, vesicular structure connected to the fetus by a short stalk that contains the major allantoic blood vessels and the allantoic canal. The outer wall of the allantois is fused with the chorion to form an extensive richly vascularized allanto-chorion, which is closely applied to the inner surface of most of the right hemisphere of the shell. The monotreme allantois was believed by Hill to subserve a purely respiratory function. The thinner inner wall of the allantoic vesicle is concave and forms a cap-like investment on the right side of the fetus, as well as carrying the main vascular branches of the allantoic vesicle. These comprise a pair of allantoic arteries and a pair of allantoic veins.

In the late stages of development of both the platypus and echidna, a relatively narrow sero-amniotic connection links the amnion enclosing the fetus with the adjacent area of pure chorion. This connecting partition extends from in front of the fetal head along its right side as far back as the anterior margin of the forelimb.

Newly hatched monotreme

After an unknown period of incubation in the platypus and 10 to 10.5 d in the echidna (Griffiths et al., 1969), the fetus hatches out as a relatively helpless little creature of some 15–16 mm in greatest length, with a state of development somewhat resembling a new born marsupial. The characteristic features of the newly hatched include both a nasal oscaruncle and egg tooth, as well as remnants of the yolk-sac in the umbilical region.

In a newly hatched platypus mammary fetus with a greatest length of 16.8 mm and a head length of 6.0 mm, believed to be not more than 2 d old, the prerostral region at the distal extremity of the nose is upturned to form the oscaruncle. This bluntly conical structure has a height of 1.7 mm measured from the margin of the upper lip to the tip of the caruncle. Internally the oseous skeletal foundation of the caruncle is derived from the fusion of two upturned inferior lamellae of the premaxillae, to form a nodule of bone that is capped externally by a cornified epidermis; further details have been reported by Hill and De Beer (1949). In addition to the oscaruncle, an egg tooth is also present and consists of a thick basal portion that provides a remarkably strong support. Distally the egg tooth tapers and bends slightly inwards as a curved pointed lancet-like structure. The egg tooth exhibits a translucent horn-like appearance and, according to Hill and De Beer (1949), consists of a pulp region, vascularized by small capillaries, successively invested by two inner layers of dentine and an outer film-like layer of presumptive enamel. The oscaruncle and the egg tooth are well shown in the fully developed prehatched platypus fetus (Fig. 16).

Marsupials

Microvilli

The early growth phase of the ovarian primary oocyte in the marsupial *Trichosurus vulpecula* is characterised, as is typical for mammals, by the appearance of numerous microvilli arising from the cell membrane that interdigitate, via the developing zona pellucida, with other microvilli that emanate from the investing follicle cells (Shorey, 1968). The resultant increase in the surface area at the interface of the membranes of these interacting cells coincides with oocyte growth and presumably an increase in metabolic activity. In ovulated *Trichosurus* eggs, an inner less dense region of the zona pellucida continues to be traversed by numerous microvilli of up to 1.9 μm in length, arising from the surface of the oocyte (Figs. 17, 18); the microvilli of follicular cell origin having now disappeared

Fig. 17. Electron micrograph of portion of an unfertilized uterine egg of the brushtail possum *Trichosurus vulpecula*. The zona pellucida has an overall thickness of about 4 μm. IZP, inner less dense region of the zona pellucida traversed by numerous microvilli that arise from the cell membrane of the oocyte. OZP, outer dense region of the zona pellucida; MC, mucoid coat of strongly acidic glycoprotein; VIT, vitellus of the secondary oocyte.

Fig. 18. Electron micrograph of an unfertilized uterine secondary oocyte of *Trichosurus vulpecula* showing numerous microvilli arising from its surface and extending into the inner region of the zona pellucida. The longitudinally sectioned microvillus (arrowed) has a length of 1.9 μm.

Fig. 19. Electron micrograph showing a transverse section through the surface of a diapausing unilaminar blastocyst of the tammar wallaby *Macropus eugenii*. IMZ, inner subzonal microvillous region; OZP, outer dense region of zona pellucida; MC, mucoid coat; PRO, protoderm cellular wall of the blastocyst; SM, shell membrane.

Fig. 20. Electron micrograph showing portion of the medullary plate of a large trilaminar blastocyst of *Trichosurus vulpecula* at the primitive streak stage and with an overall diameter of 9.8 mm. The mucoid coat and the zona pellucida have now disappeared and a granular layer, GL of 0.9 μm in thickness separates the cells of the medullary plate, MPC, from the intact shell membrane; SM, of 1.2 μm in thickness.

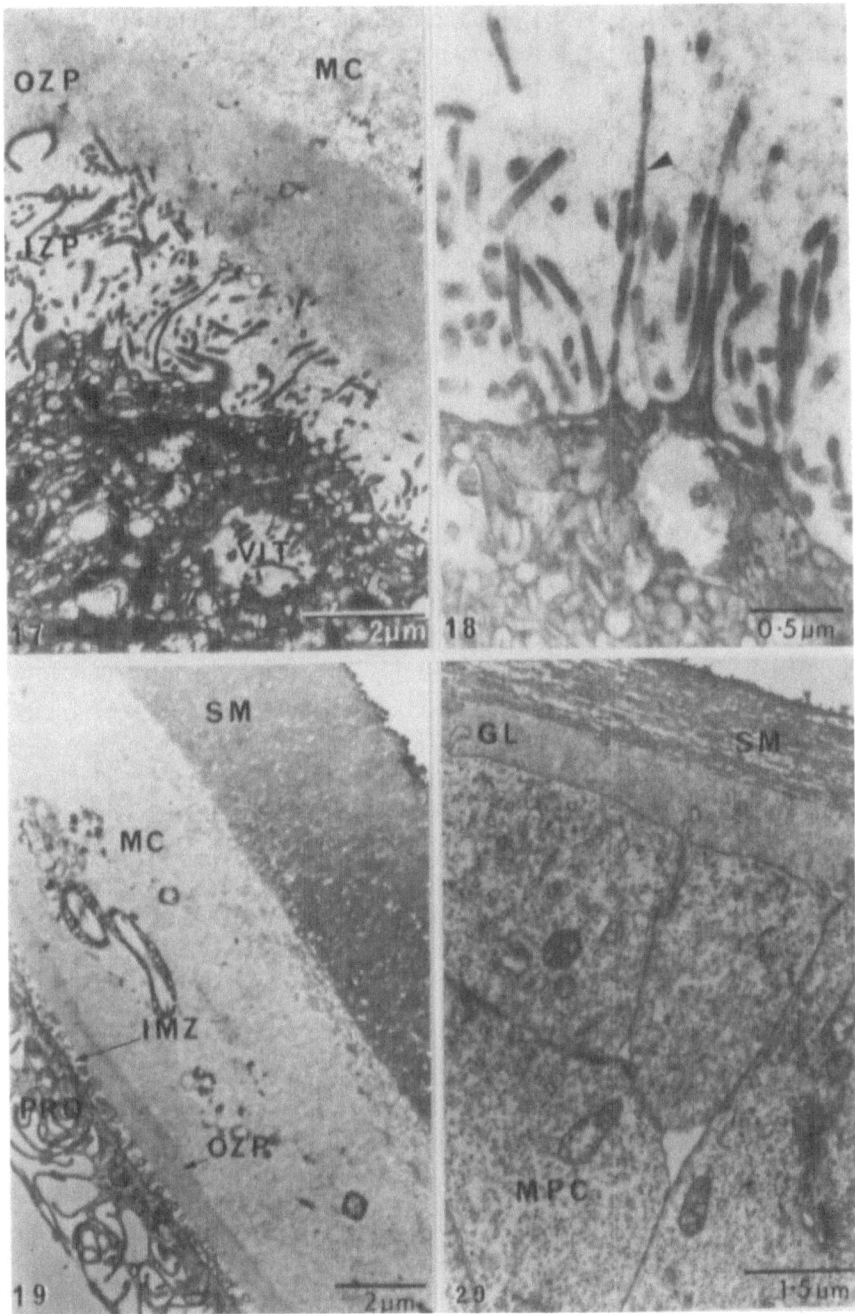

407

from the outer dense zone of the zona pellucida as a result of preovulatory maturation (Fig. 17; Shorey, 1968). Microvilli persist as a feature of the outer cell membrane of blastomeres in the late 31 to 32 cell pre-blastocyst cleavage stage in both *Perameles nasuta* and *Isoodon macrourus*, as evidenced by the text figures of Lyne and Hollis (1977). Microvilli of about 0.5 μm in length emanate from the outer surface of the protoderm cells that comprise the wall of the unilaminar diapausing blastocyst of *Macropus eugenii* (Fig. 19). In the fully formed bilaminar blastocyst of *Isoodon*, microvilli were more numerous on the outer surface of the non-embryonic ectoderm than they were on the outer surface of embryonic ectoderm (medullary plate – Hollis and Lyne, 1977). In a spheroidal 9.8 mm trilaminar blastocyst of *Trichosurus* at the late primitive streak stage, the ectoderm cells in the region of the medullary plate lack microvilli. However, at the margins of these cells, outwardly projecting processes of about 0.6 μm in length enter an overlying dispersed granular layer (Fig. 20). The outer surface of the ectoderm cells of *Isoodon* at the primitive streak stage continues to feature microvilli (Lyne and Hollis, 1977).

In *Macropus rufogriseus*, microvilli are a feature of the outer surface of the protoderm cells of the unilaminar blastocyst at both the diapausing stage and at 4 d after the removal of pouch young (RPY). These microvilli persist as a feature of the ectodermal cells of the bilaminar blastocyst at 10 d RPY and are particularly well developed in the ectoderm of the bilaminar yolk-sac at 14 d RPY, as evidenced by the text figures of Walker (1983). The near terminal stages of gestation in *Macropus rufogriseus* at 26 d RPY is characterised by cells of the uterine epithelium forming extensive microvillous contact with the hypertrophied ectodermal cells of the fetal bilaminar yolk-sac. These fetal ectodermal cells were said to be active in synthesis and nutrient transport and to exhibit a well developed outer microvillous border that interdigitates with the immediately adjacent maternal microvilli. Microvilli are also present on the outer surface of ectodermal cells in the vascularized region of the yolk-sac. However at no stage during gestation do junctional complexes develop between fetal and maternal tissues (Walker, 1983).

Egg membranes

The definitive ovum comprise a vitellus of ovarian origin with a diameter of 229 μm and 210 μm for *Trichosurus* and *Sarcophilus harrisii* respectively (Hughes 1974a, 1982). The vitellus is invested by four egg membranes (not to be confused with fetal membranes), which are classified according to their mode of origin by Boyd and Hamilton (1952).

The primary egg membranes are products of the definitive egg itself and include the cell membrane of the ovum (definitive vitelline membrane) together with its outwardly projecting microvilli that are described in a previous section of this paper and will not be considered further.

The secondary egg membranes arise in the ovary as products of the follicle cells and include the acellular matrix of the zona pellucida that is located between the interdigitating microvilli of the cell membrane of the developing oocyte and its investing follicle cells. The zona pellucida of marsupials consists of a weakly acidic glycoprotein (Hughes, 1974a). In *Trichosurus* the zona pellucida arises from both the vitelline membrane of the egg and the cell membranes of the investing membrana granulosa cells (Shorey, 1968). The fully formed zona pellucida of marsupials has a thickness of less than 6 μm (Hughes, 1974b, 1977). The fully formed zona pellucida of 3.9 μm in thickness is shown in an unfertilized uterine *Trichosurus* egg to be more or less equally divided between an outer dense region and an inner less dense region traversed by microvilli arising from the egg (Fig. 17). In the diapausing unilaminar blastocyst of *Macropus eugenii*, the inner zona pellucida matrix, formerly adjacent to the egg, has disappeared to leave free microvilli occupying a sub-zona space of about 0.3 μm in thickness, beneath the greatly thinned 0.6 μm outer dense component of the zona pellucida (Fig. 19). A similar arrangement also occurs in the diapausing blastocyst of *Macropus rufogriseus*, where the free microvilli of the protoderm cells lie beneath a thinned dense outer microvillous free component of the zona pellucida (Walker and Hughes, 1981; Walker, 1983). The zona pellucida usually fragments shortly after the expansion of the unilaminar blastocyst (Hughes, 1977). In *Trichosurus* the zona pellucida is not found in bilaminar blastocysts (Hughes, 1974a). The zona pellucida of *Trichosurus* is readily permeable to molecules as large as ferritin (mol. wt. 460,000) and consequently the zona is permeable to virtually all nutrients and wastes (Hughes and Shorey, 1973).

The zone pellucida of the platypus has a maximum thickness of about 11 μm at about the onset of vitellogenesis. However it thins to less than 0.5 μm in the greatly expanded preovulatory oocyte (Fig. 5).

The tertiary egg membranes arise from secretions of the female genital tract that invest the egg during its oviducal passage. In both monotremes and marsupials, these consist of an inner mucoid coat of strongly acidic glycoprotein, contributed to the egg during its passage through the uterine. tube. An outer tertiary egg membrane is a shell in monotremes and a shell membrane in marsupials. In both cases it consists of structural protein that exhibits strong cross linkage by disulphide bonds in marsupials and can consequently be designated as an ovokeratin (Needham, 1931). There is some evidence to suggest that the monotreme egg shell may exhibit a different composition than that reported for marsupials (Hughes 1974a, 1977).

In marsupials the mucoid coat ranges in thickness from between 15 μm to 150 μm and rarely persists beyond the bilaminar blastocyst stage (Hughes, 1974b). The mucoid coat investing an unfertilized uterine egg of *Trichosurus*, and the mucoid coat of a diapausing unilaminar blastocyst of *Macropus eugenii* are shown in Fig. 17 and Fig. 19 respectively. The mucoid coat of *Trichosurus* is readily permeable to ferritin and, like the zona pellucida, permits the free passage of nutrients and wastes (Hughes and Shorey, 1973).

The shell membrane investing the early cleaving egg of *Trichosurus*, the unilaminar blastocyst of *Macropus eugenii*, and a trilaminar primitive streak stage of *Trichosurus* are shown in Fig. 21, Fig. 19, and Fig. 20 respectively. The fully formed shell membrane of marsupials attains a maximum thickness ranging from a little under 7 μm to about 10 μm at the unilaminar blastocyst stage (Hughes, 1977). In *Trichosurus* the shell membrane precursors are exclusively produced by the uterine endometrial glands, so that in this species it is not homologous in origin with the inner component of the three layered monotreme egg shell that is a product of the distinctive tubal glands occupying the lower third of the uterine tube (Hughes, 1974a). Deposition of the marsupial shell membrane is not evident after the unilaminar blastocyst stage is attained. However in monotremes the shell continues to thicken by the addition of uterine secretions to its outer matrix layer, as the developing intrauterine egg continues to expand (Hughes, 1974a; Hill, 1933). Thus it is the outer matrix layer of the monotreme egg shell (not the inner component as reported in the earlier literature) which is homologous with the shell membrane of *Trichosurus* and presumably other marsupials.

The monotreme egg shell is obviously permeable to nutrients, as its diameter exhibits more than a fourfold increase during the intrauterine development, due to the absorption of uterine fluids.

The shell membrane of marsupials thins with the expansion of the blastocyst, bringing embryonic and maternal tissues even closer together. The marsupial

⟶

Fig. 21. An unpreserved cleaving uterine egg obtained from *Trichosurus vulpecula* 2 d after oestrus. The overall diameter of the cell mass is 197 μm and consists of 22 to 25 blastomeres of unequal size, that surround a presumptive blastocele. The cell mass is enclosed by a thin zona pellucida and this is in turn invested by a laminated mucoid coat, MC, of 56 μm in thickness that contains sperm remnants (arrowed). A shell membrane, SM, of 4.7 μm in thickness invests the mucoid coat. Shell platelets, SP, are attached to the outer surface of the shell membrane.

Fig. 22. Portion of a transversely sectioned pregnant uterus of *Trichosurus vulpecula*. The young fetus had an overall length of 9 mm and the oval yolk-sac had major and minor axes of 19 mm and 11 mm respectively. The zona pellucida and the mucoid coat are absent, the shell membrane, SM, of 0.4 μm in thickness although ruptured in places is extensively intact and is not applied to either the bilaminar yolk-sac, BL YS, or to the luminal epithelium of the uterine endometrium, UT EP.

Fig. 23. Portion of the transversely sectioned uterine endometrium of the full term pregnant koala *Phascolarctos cinereus* shown in Fig. 28. Giant fetal ectodermal cells of the nonvascularized bilaminar yolk-sac, BL YS ECT, have retracted, after preservation, from their loose interdigitation with the uterine luminal epithelium, UT EP.

Fig. 24. Portion of the transversely sectioned uterus of a full term koala *Phascolarctos cinereus* in a region of the nonvascularized yolk-sac near the sinus terminalis. An enlarged ectodermal cell of the bilaminar yolk-sac, BL YS ECT, has penetrated (arrowed) the luminal epithelium of the uterus, UT EP.

410

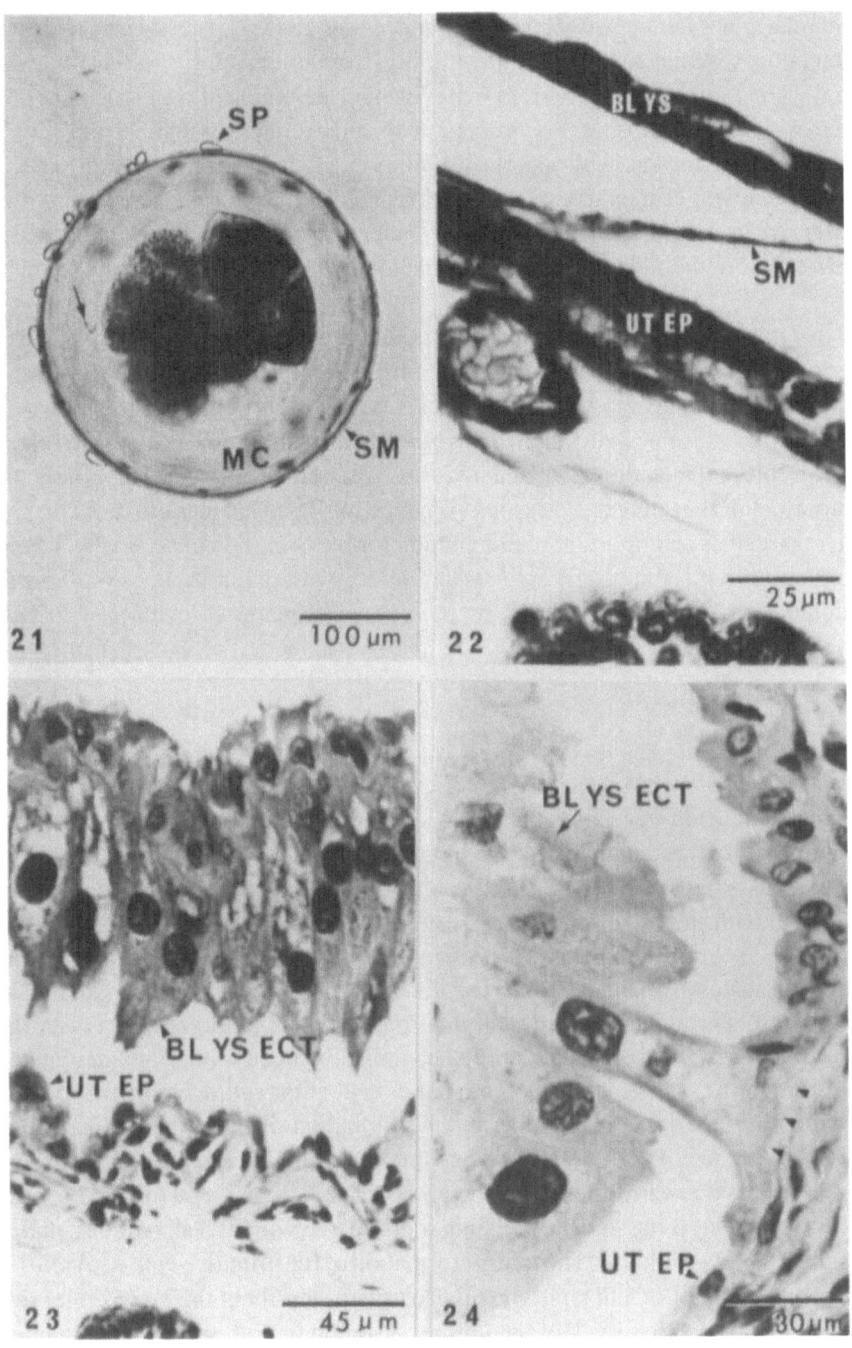

411

shell membrane remains unruptured for about two thirds of gestation, to about the time when fetal development commences and the yolk-sac has established a functional vascular supply (Hughes, 1974b). Even after rupture, extensive areas of shell membrane may persist, as was evident in the region of the bilaminar yolk-sac of an early fetus of *Trichosurus* with an overall length of 9 mm and a vascularized yolk-sac with major and minor axes of 19 mm and 11 mm respectively. In this specimen, the zona pellucida and the mucoid coat were absent, but extensive areas of the slightly ruptured shell membrane of only 0.4 μm in thickness separate the fetal ectoderm from the luminal epithelium of the uterus (Fig. 22).

In *Trichosurus* the dense granular lattice of the shell membrane is permeable to peroxidase (mol. wt. 40,000), but some filtration of ferritin occurs (Hughes and Shorey, 1973). This would suggest a free passage of nutrients and wastes.

As in monotremes and marsupials, the shell membrane of the avian egg is permeable and includes nutrients such as glucose (Hughes, 1977). This is not only important in the context of gaseous exchange for the avian embryo, but also can be regarded as an important preadaptation for the evolution of viviparity, where some presumptive ancestral embryo developing within a shelled egg was able to supplement its nutrient reserves by the uptake of maternal oviducal nutrients through the shell.

Yolk-sac

In marsupial embryogenesis, the yolk-sac is without question the extraembryonic membrane of preeminence for respiration and metabolism. Cleavage of the marsupial egg results in the formation of a group of blastomeres of unequal size, together with the extrusion of accessory egg cytoplasm lacking nuclear material (Fig. 21). Within a few days, the irregularly arranged blastomeres become pressed against the inner surface of the zona pellucida to form a unilaminar blastocyst. During subsequent cell divisions, the size differential between the blastomeres is apparently equalized. The unilaminar cell wall is then designated as protoderm, because of the potential of its cells at the future embryonal pole to form either ectoderm or endoderm. The formation of the primitive marsupial yolk-sac involves the internalization of these presumptive endodermal cells and their dispersal towards the future abembryonal pole, to give rise to a two layered blastocyst. This is the so-called bilaminar blastocyst, with its wall now designated as the primitive yolk-sac. The time of formation of the primitive yolk-sac varies in different marsupials, but typically this occurs between about the 2 d on either side of the end of the first week of the normal gestation period, while the unilaminar blastocyst is expanding. This process has recently been described at the ultrastructural level for *Isoodon* and *Perameles* (Hollis and Lyne, 1977). These authors also provide a review of the earlier light microscope studies of bilaminar blastocyst formation for a variety of marsupials. The newly formed primitive yolk-sac continues to expand as the wall of the bilaminar blastocyst, until at least about the middle of gestation, when a small discoidal area of ectoderm at the future embryonal pole thickens to form the medullary plate. In *Didelphis marsupialis*, the wall of the bilaminar blastocyst outside the medullary plate thins during the expansion of the blastocyst and this can reasonably be presumed to be as a result of internal pressure (McCrady, 1938). The primitive streak, that is soon to develop within the central region of the medullary plate, confers bilateral symmetry on the future embryo, but more importantly gives rise to mesoderm, the third of the primary embryonic germ layers. In *Antechinus stuartii* this trilaminar blastocyst stage is attained as late as day 22 of the 27 d period of gestation (Selwood, 1981). The formation of the trilaminar blastocyst marks the beginning of acceleration of all aspects of embryonic development. The blastocyst now exhibits marked expansion as mesoderm migrates away from the primitive streak, beneath the ectoderm, towards its ultimate equatorial destination in the wall of the yolk-sac. The early portion of the second half of gestation is marked by the establishment of the definitive prefetal primordia, including a neural tube and associated lateral somite blocks, as well as blood islands within the extra-embryonic mesoderm of the yolk-sac. A trilaminar blastocyst at such a stage is shown in Fig. 25. This is the stage of development attained by *Macropus eugenii* on day 18 of the 28 d gestation period (Berger, 1970). At this time the blastocyst of *Macropus eugenii* is invested by an intact shell membrane and an analysis of the yolk-sac fluids indicates that, although the main nutrients are absorbed uterine secretions, the composition of the yolk-sac fluids is in fact different from the uterine secretions, with three times the level of free amino acids. The proteins of the yolk-sac fluids are also of embryonic origin, with only pre-albumin proteins present at this stage (Renfree, 1977). With the subsequent rupture of the shell membrane and the development of a vascularized yolk-sac, the composition of the yolk-sac fluids alter in a manner that suggests less dependence on uterine secretions. Proteins are now believed to be synthesized by the fetus; however some may enter the yolk-sac by selective transfer through its placental wall, although their concentration is some thirty times lower than the levels in maternal serum. This is a period of rapidly accelerating fetal growth and glucose levels within the yolk-sac have now increased to levels of from two to four times higher than those for maternal serum (Renfree, 1970). The high concentrations of material within the yolk-sac fluids of *Macropus eugenii* were considered to assist in the maintenance of turgidity within the yolk-sac, as well as providing a nutrient store (Renfree, 1970).

During the terminal stages of marsupial pregnancy, the yolk-sac is approximately equally divided by a prominent ring vessel, the sinus terminalis, into a highly vascularized embryonal pole and non-vascularized bilaminar portion of the yolk-sac (Fig. 26). During these terminal stages, attachments frequently develop between the luminal epithelium of the uterus and one or other area of the

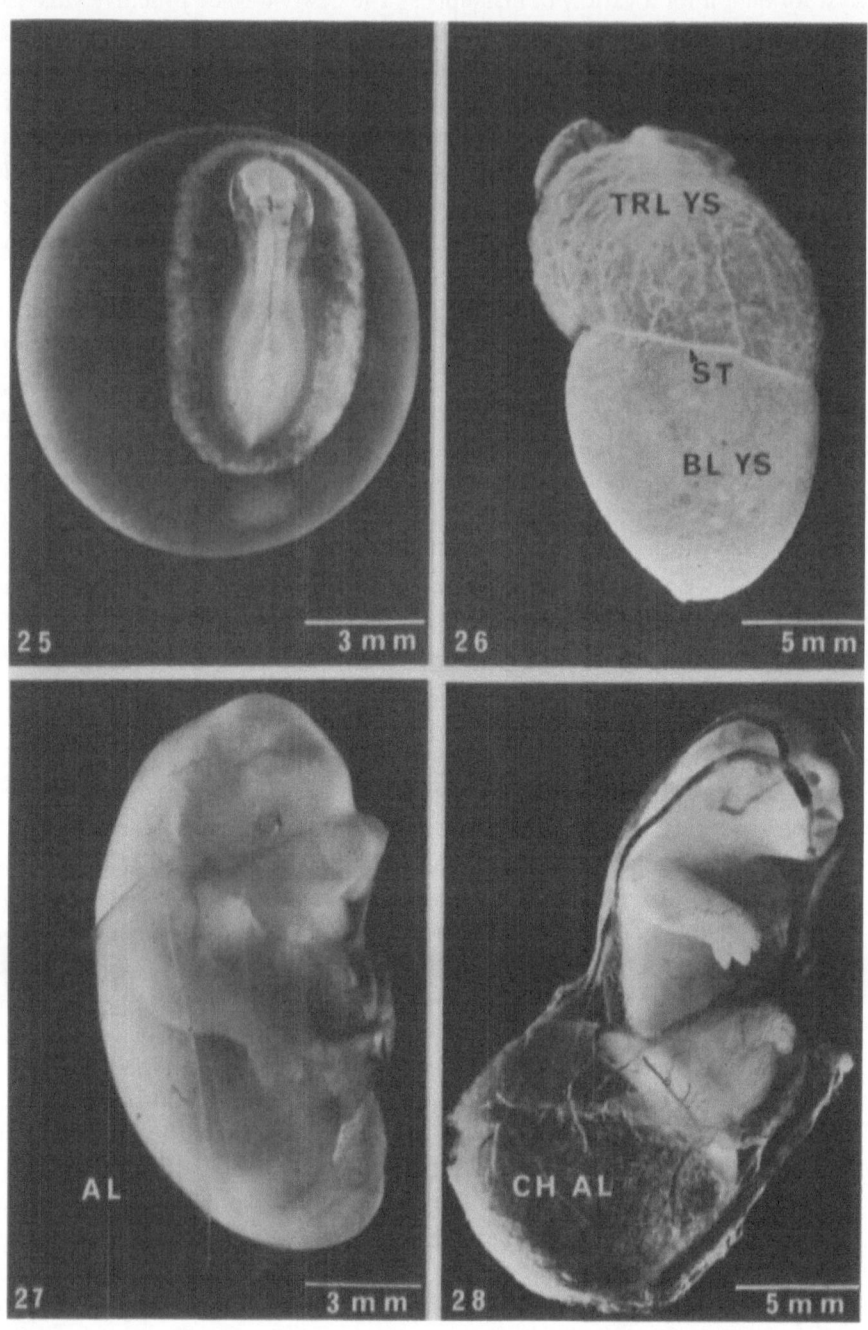

Fig. 25. Embryonic vesicle of the tammar wallaby *Macropus eugenii* on day 18 of the 28 d gestation period. The embryonic vesicle has an overall diameter of 10 mm and the overall length of the definitive somite stage embryo, including the primitive streak is 7 mm.

Fig. 26. The fully formed yolk-sac of the rat-kangaroo *Potorous tridactylus* completely encloses an advanced fetus with an overall length of 8.7 mm. The yolk-sac has a major and a minor axis of 18 mm and 12 mm respectively. The prominent equatorial sinus terminalis, ST, separates the vascularized trilaminar portion of the yolk-sac, TRL YS, from the bilaminar nonvascularized region, BL YS.

Fig. 27. A near full term fetus of the ringtail possum *Pseudocheirus peregrinus* with a head length of 6.5 mm and a greatest overall length of approximately 13.5 mm. The yolk-sac has been removed to show the fetus closely invested by its amnion. The relatively poorly vascularized allantois, AL, of 8 mm in diameter was not attached to the chorion.

Fig. 28. A full term fetus of the koala *Phascolarctos cinereus* with an overall length of 17 mm. The oval yolk-sac with major and minor axes of 34.5 mm and 20 mm respectively has been removed to show a highly vascularized allantois of 12 mm in diameter fused with a noninvasive oval region of the chorion, CH AL.

yolk-sac (Hughes, 1974b); in this paper it was reported that the yolk-sac ectoderm of both *Potorous tridactylus* and *Macropus rufogriseus* (= *Protemnodon rufogrisea*) were not invas ive. However recent studies in both of these marsupials show the presence of a slightly invasive yolk-sac ectoderm (Walker, 1983; Shaw and Rose, 1979). Such a slightly invasive yolk-sac placenta of marsupials is possibly more widespread than previously reported. In its simplest form this consists of interdigitation of fetal and maternal microvilli.

Attachments between fetal and maternal elements during the terminal stages of pregnancy in the koala *Phascolarctos cinereus* involve giant bilaminar yolk-sac cells, near the sinus terminalis, that interdigitate so weakly with maternal elements that the union is easily disrupted by techniques of tissue preparation (Fig. 23). However occasional giant fetal ectodermal cells penetrate the luminal epithelium of the uterus (Fig. 24).

Allantois

The allantois first appears during the terminal third of intrauterine development as a small vesicular rudiment connected to the umbilical region of the prefetal or early fetal stages (Fig. 30). The allantois remains relatively small until towards the end of the terminal week of gestation, when its volume may approach about two thirds of the fetal dimensions (Figs. 27, 28).

The fully formed marsupial allantois exhibits three grades of complexity.

Type one allantois

The chorion overlying the allantoic area fails to form invasive ectodermal villi and the ectodermal area of the chorion is obliterated by the encroachment of the yolk-sac. At term the allantois may be a large vesicle; however, it is poorly vascularized. The allantois does not reach the chorion and is invested by folds of the yolk-sac splanchnopleure. This is the most common type of marsupial allantois and is found in *Pseudocheirus peregrinus* (Fig. 27) (Hughes et al., 1965). The vascularized and the non-vascularized regions of the yolk-sac are the sole organs for nutritive and respiratory exchange. However the allantois may store waste.

Type two allantois

The allantoic vesicle forms during the branchial cleft prefetal stage. Villi fail to develop within the immediately adjacent area of pure chorion. The small allantoic vesicle fuses with the overlying chorion at a relatively early stage of development to become a large richly vascularized vesicle, with obvious respiratory potential, by the terminal stages of pregnancy. Approximately two thirds of the superficial hemisphere of the allantoic vesicle fuses with the adjacent chorion to form the non-invasive chorio-allantois. The remainder of the allantoic vesicle is less vascularized and is invested by folds of the yolk-sac splanchnopleure. The stalk of the allantois is poorly developed. This is the type of allantois found in both *Vomabatus ursinus* and *Phascolarctos cinereus* (Fig. 28). The yolk-sac remains the major organ for nutritive and respiratory exchange. However the allantois is considered to have a respiratory function and also to store wastes.

Type three allantois

The allantoic vesicle forms, during the branchial cleft prefetal stage, immediately adjacent to an area of pure chorion that develops invasive ectodermal villi (Fig. 29). At this stage the allantois is not fused with the chorionic area (Fig. 30). At a later stage the allantois fuses with the chorion to form a discoidal placenta with a long stout umbilical cord (Figs. 31, 32). Both a vascular and non-vascular yolk-sac placenta develops; however at term this is subservient to the discoidal chorio-allantoic placenta in regard to respiration, metabolic exchange, and disposal of wastes. This type of placenta involves the close apposition of the endothelial linings of both fetal and maternal blood vessels within the uterine endometrium and this is achieved by highly invasive fetal tissues. This type of placenta characterises the Peramelidae (bandicoots) and was thoroughly described by Flynn (1923) and at the ultrastructural level by Padykula and Taylor (1977).

Concluding comments

In considering the striking parallel morphological adaptations for respiration and metabolic exchange in the embryos of monotremes and marsupials, it would seem reasonable to suggest that each group could have been derived from an ancestor with an even larger energy rich egg, and a diminished uterine dependence. Consequently it is tempting to postulate a monotreme-like ancestor for marsupials and a more sauropsid-like ancestor for the monotremes.

The cleidoic egg of monotremes, with its 4 mm diameter vitellus, largely consists of yolk-spheres and is exceedingly large by mammalian standards. However in relation to oviparous submammalian vertebrates its energy reserves are grossly deficient and totally inadequate to support the development of an independent organism. This deficit in nutrients is made good in part by the passage uterine endometrial substances to the embryo through the highly permeable intrauterine egg shell. The fourfold expansion of the egg within the uterine endometrium indicates that by far the major volume of nutrients that sustain the embryo after the egg is laid are derived from the progesterone primed uterine endometrium, rather than of ovarian origin. This constitutes a major departure from the mode of development in sauropsid vertebrates. The rather small and relatively helpless, undeveloped offspring that hatches from the monotreme egg is brought to full independence during a period of maternal care, when the energy for development is derived from milk suckled from nippleless mammary glands.

The late appearance and relatively limited extent of the extraembryonic coelom of both monotremes and marsupials results in the greater portion of the yolk-sac remaining in direct contact with the peripheral trophoblastic ectoderm for the entire period of intrauterine development. In monotremes this results in the yolk-sac being the only extraembryonic structure concerned with respiration and metabolic exchange during intrauterine development. The non-vascularized region of the yolk-sac of both monotremes and marsupials are important organs of metabolic exchange during intrauterine development.

The allantois of the majority of marsupials does not fuse with the chorion and almost the entire trophoblast is directly underlain by either a vascularized or a non-vascularized region of the yolk-sac, which is in these species virtually the exclusive extraembryonic organ for respiration and metabolic exchange. In the remaining marsupial species the allantois, which appears late in all marsupial intrauterine development, fuses with and vascularizes the trophoblast (chorionic region). However during the terminal days of gestation in the bandicoot marsupials, the vasculature of the highly invasive chorio-allantois greatly exceeds that of the yolk-sac. This indicates that the dominance of the yolk-sac has been supplanted by the chorio-allantois as the major extraembryonic structure for respiration and metabolic exchange.

The new-born marsupial is relatively small and undifferentiated, and this might reasonably be presumed to bear some relationship with limitations for respiration

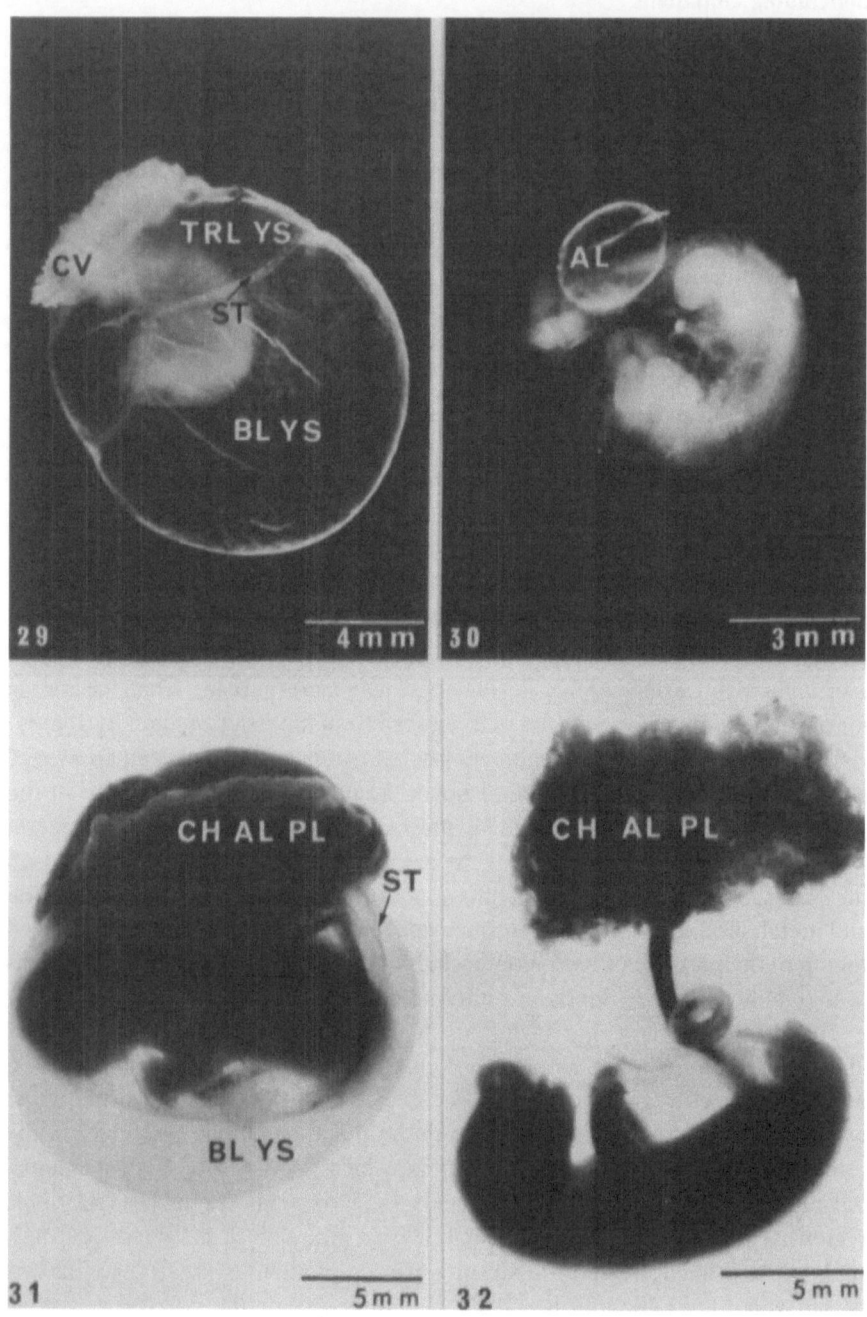

Fig. 29. An early implantation stage of the long nosed bandicoot *Perameles nasuta*. The prefetal embryo with an overall length of 6.5 mm is completely invested by extraembryonic membranes with a major and a minor axis of 13.5 mm and 12 mm respectively. Three regions can be distinguished within the peripheral extraembryonic membranes. A nonvascularized portion of the bilmainar yolk-sac, BL YS, occupies about two thirds of the abembryonal region. A sinus terminalis, ST, marks the limits of the vascularized portion of the trilaminar yolk-sac, TRL YS. A small thickened discoidal area of true chorion with a diameter of about 6.3 mm is firmly united with the uterine endometrium by highly invasive chorionic villi, CV.

Fig. 30. A strongly torted prefetal branchial cleft stage bandicoot embryo, revealed after the removal of the extraembryonic membranes of the specimen shown in Fig. 29. The small vesicular allantoic vesicle, AL, of 2.5 × 2.1 mm had not yet fused with the highly invasive chorionic area.

Fig. 31. A full term fetus of the long nosed bandicoot *Perameles nasuta* enclosed within its fetal membranes. The limiting extraembryonic membranes have a major and minor axis of 17 × 11.5 mm respectively. The nonvascularized bilaminar portion of the yolk-sac, BL YS, now occupies slightly less than half the surface area of the extraembryonic membranes and the sinus terminalis, ST, marks its boundary with the immediately adjacent vascularized trilaminar region of the yolk-sac. The allantois has now fused with the chorion in a discordal area of 11.5 × 10 mm that has invaded the uterine endometrium. CH AL PL, portion of the undissected uterine wall immediately continuous with the discoidal chorio-allantoic placenta.

Fig. 32. A further stage in the dissection of the bandicoot specimen shown in Fig. 31. The fetus has an overall length of 13.8 mm. The umbilical cord with its single loop connects the fetus with the discoidal chorio-allantoic placenta, CH AL PL.

and metabolic exchange, particularly during the terminal stages of intrauterine development. A variety of causal contributing factors might be suggested: The fully developed vascularized uterine endometrium consists of a relatively small volume of tissue; The shell membrane, although permeable, separates embryonic and maternal tissues for about two thirds of gestation; Organogenesis, including the vascularization of the yolk-sac, is delayed until the terminal stages of gestation; The vascularized extraembryonic membranes are not sufficiently invasive for the provision of the necessary nutrients or respiratory exchange for the development of a large fetal mass; The marsupial fetus is unable to sufficiently regulate the maternal endocrine system to prolong intrauterine development beyond the limits of uterine reserves that result from the postovulatory uterine proliferative phase.

The allantoic vesicle of monotremes develops during the latter stages of the incubation period of the laid egg. As the allantois fuses with the trophoblastic ectoderm (chorion) the extraembryonic coelom markedly enlarges. In this way the yolk-sac is displaced from about half the area of the trophoblast that immediately underlies the egg shell by a highly vascularized and presumably respiratory allantois. The trophoblast of the opposing hemisphere also immediately underlies the egg shell and is intimately fused with the now almost totally vascularized yolk-sac, which might reasonably be presumed to serve a respiratory function as well as the sole extraembryonic structure for the vascular transport nutrients to the embryo.

Acknowledgement

I wish to thank Dr Elze C. Boterenbrood, curator of the Hill Collection, at the Hubrecht Laboratory, Netherlands for introducing me to this valuable material.

References

Berger, P.J. (1970). The Reproductive Biology of the Tammar Wallaby *Macropus eugenii*, Desmarest (Marsupialia). Ph.D. Thesis. Tulane University, New Orleans, U.S.A.

Boyd, J.D. and Hamilton, W.J. (1952). Cleavage, Early Development and Implantation of the Egg. In: Marshalls' Physiology of Reproduction, 3rd Edn. Vol. 2. A.S. Parkes, ed., Longmans Green and Co., London, pp. 1–110.

Broom, R. (1895). Notes on the period of gestation in echidna. *Proc. Linn. Soc. N.S.W.* 10: 567–577.

Flynn, T.T. (1923). The yolk-sac and allantoic placenta in *Perameles*. *Q.J. Micros. Sci.* 67: 123–182.

Flynn, T.T. and Hill, J.P. (1939). The development of the Monotremata. Part. 4. Growth of the ovarian ovum, maturation, fertilization and early cleavage. *Trans. Zool. Soc. London* 24: 445–622.

Flynn, T.T. and Hill, J.P. (1942). The later stages of cleavage and the formation of the primary germ-layers in the monotremata. *Proc. Zool. Soc. London* Series A3: 233–253.

Flynn, T.T. and Hill, J.P. (1947). The development of the Monotremata. Part 6. The later stages of cleavage and the formation of the primary germ-layers. *Trans. Zool. Soc. London* 26: 1–151.

Griffiths, M., McIntosh, D.L. and Coles, R.E.A. (1969). The mammary gland of the echidna, *Tachyglossus aculeatus* with observations on the incubation of the egg and on the newly-hatched young. *J. Zool. London* 158: 371–386.

Hill, J.P. (1933). The development of the Monotremata. Part 2. The structure of the egg-shell. *Trans. Zool. Soc. London* 21: 443–476.

Hill, J.P. and De Beer, G.R. (1949). The development and structure of the egg-tooth and caruncle in the monotremes and on the occurrence of vestiges of the egg-tooth and caruncle in marsupials. *Trans. Zool. Soc. London* 26: 503–544.

Hill, J.P. and Martin, C.J. (1895). On a platypus embryo from the intrauterine egg. *Proc. Linn. Soc. N.S.W.* 10(2nd Ser.): 43–74.

Hollis, D.E. and Lyne, A.G. (1977). Endoderm formation in blastocysts of the marsupials *Isoodon macrourus* and *Perameles nasuta*. *Aust. J. Zool.* 25: 207–223.

Hughes, R.L. (1974a). The Tertiary Egg Membranes of the Marsupial *Trichosurus vulpecula*. Ph. D. Thesis, University of N.S.W.

Hughes, R.L. (1974b). Morphological studies on implantation in marsupials. *J. Reprod. Fertil.* 39: 173–186.

Hughes, R.L. (1977). Egg Membranes and Ovarian Function During Pregnancy in Monotremes and Marsupials. In: Reproduction and Evolution. J.H. Calaby and C.H. Tyndale-Biscoe, eds., Australian Academy of Science, Canberra, pp. 281–291.

Hughes, R.L. (1982). Reproduction in the Tasmanian Devil *Sarcophilus harrisii* (Dasyuridae, Marsupialia). In: Carnivorous Marsupials. M. Archer, ed., Publ. Royal Zoological Society of N.S.W., Surrey Beatty, Sydney, pp. 49–63.

Hughes, R.L. and Carrick, F.N. (1978). Reproduction in Female Monotremes. In: Monotreme Biology. Vol. 20. M.L. Augee, ed., *Aust. Zool.*, Royal Zoological Society of N.S.W., pp. 233–253.

Hughes, R.L., Thomson, J.A. and Owen, W.H. (1965). Reproduction in natural populations of the Australian ringtail possum, *Pseudocheirus peregrinus* (Marsupialia: Phalangeridae), in Victoria. *Aust. J. Zool.* 13: 383–406.

Hughes, R.L. and Shorey, C.D. (1973). Observations on the permeability properties of the egg membranes of the marsupial *Trichosurus vulpecula*. *J. Reprod. Fertil.* 32: 25–32.

Lyne, A.G. and Hollis, D.E. (1977). The Early Development of Marsupials, with Special Reference to Bandicoots. In: Reproduction and Evolution. J.H. Calaby and C.H. Tyndale-Biscoe, eds., Australian Academy of Science, Canberra, pp. 293–302.

McCrady, E. (1938). The Embryology of the Opossum. *Amer. Anat. Mem.* Wistar Institute, Philadelphia 16: 1–233.

Needham, J. (1931). The Unfertilized Egg as a Physico-Chemical System. In: Chemical Embryology, Vol. 1. Cambridge University Press, London, pp. 232–367.

Owen, R. (1834). On the generation of the marsupial animals, with a description of the impregnated uterus of the kangaroo. *Philos. Trans. R. Soc. London* B124: 333–364.

Padykula, H.A. and Taylor, M.J. (1977). Uniqueness of the Bandicoot Chorioallantoic Placenta (Marsupialia: Perameliadae) Cytological and Evolutionary Interpretations. In: Reproduction and Evolution. J.H. Calaby and C.H. Tyndale-Biscoe, eds., Australian Academy of Science, Canberra, pp. 303–324.

Renfree, M.B. (1970). Protein, amino acids and glucose in the yolk-sac fluids and maternal blood sera of the tammar wallaby, *Macropus eugenii* (Desmarest). *J. Reprod. Fertil.* 22: 483–492.

Renfree, M.B. (1977). Feto-placental Influences in Marsupial Gestation. In: Reproduction and Evolution. J.H. Calaby and C.H. Tyndale-Biscoe, eds., Australian Academy of Science, Canberra, pp. 325–331.

Selwood, L. (1981). Delayed Embryonic Development in the Dasyurid marsupial, *Antechinus stuartii*. In: Embryonic Diapause in Mammals. A.P.F. Flint, M.B. Renfree and B.J. Weir, eds., *J. Reprod. Fertil. Supp.* 29: 79–82.

Semon, R. (1894). Die Embryonalhüllen der Monotremen und Marsupialier. *Denkschr. Med. Naturwiss. Ges. Jena.* 5: 19–58.

Sharman, G.B. (1961). The embryonic membranes and placentation in five genera of diprotodont marsupials. *Proc. Zool. Soc. London* 137: 197–220.

Shaw, G. and Rose, R.W. (1979). Delayed gestation in the potoroo. *Potorous tridactylus* (Kerr). *Aust. J. Zool.* 27: 901–912.

Shorey, C.D. (1968). An Electron Microscope Study of Pituitary-Ovarian Changes During Oogenesis and Embryonic Development in *Trichosurus vulpecula*. M.Sc. Thesis, University of N.S.W.

Walker, M.T. (1983). Gestational Hormones and Embryonic Maintenance in the Marsupial *Macropus rufogriseus*. Ph.D. Thesis, University of Queensland.

Walker, M.T. and Hughes, R.L. (1981). Ultrastructural Changes after Diapause in the Uterine Glands, Corpus Luteum and Blastocyst of the Red-necked Wallaby *Macropus rufogriseus banksianus*. In: Embryonic Diapause in Mammals. A.P.F. Flint, M.B. Renfree and B.J. Weir, eds., *J. Reprod. Fertil. Suppl.* 29: 151–158.

Wilson, J.T. and Hill, J.P. (1908). Observations on the development of *Ornithorhynchus*. *Philos. Trans. R. Soc. London* Ser. B 199: 31–168.

29. Gas transfer by the neonate in the pouch of the tammar wallaby, *Macropus eugenii*

David Randall, Bren Gannon, Sue Runciman and
R.V. Baudinette

*Department of Human Morphology, School of Medicine, The Flinders
University of South Australia, Bedford Park, S.A., 5042, Australia*

Abstract

The gas exchange characteristics of, and gaseous environment within, the wallaby pouch containing a single neonate (joey) (age 3–33 d) was determined in conscious animals from pouch gas samples obtained via an indwelling catheter. Pouch gas tensions were constant, with O_2 approximately 3 torr below, and CO_2 4 torr above, ambient values. For a surgically closed and sealed pouch without a joey, the equilibrium values of O_2 and CO_2 were approximately 70 and 40 torr, respectively. For joeys 3–33 d old, O_2 uptake and CO_2 excretion rates were approx. 0.60 and 0.52 ml $\cdot$ g^{-1} $\cdot$ h^{-1}, and the respiratory exchange ratio averaged 0.94. Of the O_2 from the atmosphere entering the pouch containing a 3-d joey, only 15% was consumed by the joey, the remainder being removed by the pouch wall; similarly, 85% of CO_2 leaving the pouch was added to the pouch gas from the pouch wall. Histological studies confirmed that the maternal skin lining the pouch was well vascularized and the joey lung highly vascularized, but that the joey skin was poorly vascularized. Gas exchange in neonatal joeys must be almost totally pulmonary, given the markedly poorer skin vascularity and longer diffusion distances from skin capillaries across its epithelium, compared to that of the capillary network of the lung.

Introduction

Marsupial reproduction includes a short gestation period with birth of the young at an early stage of embryological development and subsequent growth of the young firmly attached to a nipple within the pouch. CO_2 levels are reportedly high and O_2 levels low in pouch gas, and the pouch is considered poorly ventilated (Bailey and Dunnett, 1960; Farber and Tenney, 1971). The neonates in the pouch are hairless and small; the lungs, however, are considered well developed and probably represent the major, and perhaps only, site for gas transfer (Krause and Leeson, 1975; Johnson and Nicholas, 1979). The youngest marsupial neonate investigated, a 5-d Virginia opossum (*Didelphis virginiana*), showed ventilatory

Seymour, R.S., (ed.) Respiration and metabolism of embryonic vertebrates.
© *1984, Dr W. Junk Publishers, Dordrecht/Boston/London. ISBN-13:978-94-009-6538-6*

responses to hypoxia, hypercapnia and anoxia, and a 7-d neonate showed a reflex ventilatory response to lung inflation and exercise (Farber et al., 1972; Farber, 1972; Farber and Marlow, 1977). Thus, marsupials are of particular interest because, compared with their eutherian counterparts, the young must operate outside the uterus at a very early stage of development. This is only possible if the lungs and pulmonary circulation are functional for gas transfer in the neonate. Yet we know little of the development of the lungs or of the environment in which the animal lives, namely the marsupial pouch.

The object of this study was to investigate gas transfer by the neonate (joey) in the pouch of the tammar wallaby, *Macropus eugenii*. The aims were (1) to describe the nature of the lung and skin vascular beds of the neonate and of the pouch circulation in the mother, using conventional histology of tissue samples, and (2) to measure gas tensions in the pouch gas and determine O_2 uptake and CO_2 production by both the neonate and the pouch wall.

Material and methods

The experiments were carried out on 14 adult female tammar wallabies (*Macropus eugenii*) at Flinders Chase National Park, Kangaroo Island, South Australia.

The wallabies were caught by netting at night on the farming property Brookland Park with the aid of the owner, Mr. P. Davies during the period 20.2.82 to 10.3.82. The mean mass of all animals was 4.25 kg, but there were two distinct classes, the first group (2.69 ± 0.14 kg, $\bar{x} \pm$ s.e., N = 5) were probably one year old, whereas the others (5.07 ± 0.22 kg, N = 9) were 2 or more years old. No differences were observed in the other data collected from these two groups, so the results were grouped together.

Adult females were anaesthetized for surgery by an intra-muscular injection of 10 to 20 mg · kg^{-1} ketamine (Ketalar; Parke, Davis & Co, Caringbah, N.S.W.), followed by spontaneous inhalation of 1.5 to 2% halothane (Fluothane, I.C.I., Villawood, N.S.W.), using compressed air as the carrier gas. Animals were killed for dissection by an intraperitoneal injection of 720 mg pentabarbitone (Nembutal, Ceva Chemicals, Hornsby, N.S.W.) following an intramuscular injection of 10 mg · kg^{-1} ketamine.

The pouch was cannulated with a 1 m length of PE 160 polyethylene tubing (Intramedic, Clay Adams, New Jersey). The pouch end of the cannula was heat flared and was sewn to the pouch wall; the cannula was then threaded under the skin and securely tied where it emerged at the back of the neck. The cannula was air filled and could be used to flush or sample gas in the pouch while the wallaby rested in a wire cage ($0.6 \times 0.65 \times 0.5$ m) which was draped with jute sacks.

Our protocol was to implant the cannula and allow the wallaby to recover overnight in the cage. With the joey inside, the pouch was then flushed three times with 60 ml of N_2 and changes in P_{O_2} and P_{CO_2} in the pouch gas were

424

measured. The joey was then removed and the pouch was sealed, using a series of stitches and silicone glue. The wallaby was allowed another overnight recovery. The pouch was then flushed three times with 60 ml of either N_2 or air, following removal of any gas in the pouch. The pouch gas volume at the start of measurements was 60 ml. Changes in P_{O_2} and P_{CO_2} of pouch gas were then measured following this filling of the sealed pouch with either air or N_2. The cannula had a dead space volume of 1.2 ml and the gas sample was only 1 ml. To avoid dead space problems, 2 ml samples were pumped in and out of the pouch 10 times before a final 1 ml sample was taken. The cannula protruded from between the sacks covering the cage and the wallaby appeared undisturbed by the sampling procedure.

O_2 and CO_2 levels in the pouch gas were measured with a blood gas analyser (Model PHM71b, Radiometer, Copenhagen) and appropriate electrodes (E5036-0, P_{CO_2} and E5046-0, P_{O_2}), thermostated at 37 °C and colibrated with gas mixing pumps (Wosthöff, Bochum, West Germany).

The joey was removed from the pouch before it was sealed, and was aged from leg, head and foot measurements (Murphy and Smith, 1970). O_2 uptake ($\dot{V}_{O_2}$) and CO_2 production ($\dot{V}_{CO_2}$) of six joeys were determined by placing them singly into a sealed, 100% humidified, air filled chamber at 37 °C and following changes in P_{O_2} and P_{CO_2} with time. The size of the joey was always less than 5% of the chamber volume.

For histological studies, the pouch was fixed in 2% paraformaldehyde/0.5% glutaraldehyde in neutral buffered normal saline by installation with or without prior perfusion, and then dissected free and stored in a 10% formol saline solution. The joeys were killed by an intraperitoneally injected overdose of Nembutal; at apnoea, a coarse ligature was tied tightly around the neck to occlude the trachea, and the thorax and abdomen then opened by a mid-line incision. Several of the larger joeys were first perfusion-fixed by cannulation of the pulmonary trunk via the infundibulum of the right ventricle; all joeys were then immersed in the formaldehyde/glutaraldehyde fixative for storage.

Samples of adult and joey lung and skin, and for several of the smaller joeys the whole thoracic region, were embedded in either wax or acrylic resin, sectioned at 5 μm (wax) or 1 μm (resin) and stained with either hematoxylin and eosin, or with Verhoff and Van Gieson stains. Details of equipment used and exact preparative procedures are published elsewhere (Gannon, 1983).

Results

Kangaroo Island tammar wallabies breed from January to July and begin breeding as one-year-olds (Murphy and Smith, 1970; Andrewartha and Barker, 1969). A single joey is found in the pouch, leaving when between 200 and 250 d old and about a kilogram in mass. The present study is based on adult female tammar wallabies and joeys less than 34 d old.

Fig. 1. Photomicrographs of histological sections of the skin of the tammar wallaby *Macropus eugenii.*
   A. Section of the skin of the thoracic region of a 3-d joey. Note the location of the most superficial capillaries (arrows) fairly deep in the dermis underlying the epithelium (e), the keratinized surface (k) of which is in contact with pouch air (a). Bar = 100 μm.
   B. Higher magnification of 3-d joey skin. The most superficial capillary (arrow) is separated from pouch air (a) by a substantial distance comprised by dermis (d) and non-keratinized (e) and keratinized epithelium (k). Bar = 50 μm.
   C. Skin lining the ventral surface of the pouch. The capillaries of the dermis (arrows) closest to the pouch air (a) are indicated. Note the plexus of larger blood vessels (b) deep in the dermis, and the profile of a sweat gland (s). Perfusion fixed specimen. Bar = 100μm.
   D. Higher magnification of pouch lining skin. Note the most superficial capillaries (arrows) in the dermis, separated from the pouch air by the stratified squamous epithelium; the keratin layer has been detached and lost during processing in this region; perfusion fixed specimen. Bar = 50 μm.
   E. Striated muscle in ventral pouch wall. Note degree of capillarity cf. Figs. B.D.G.; perfusion fixed specimen. Bar = 50 μm.
   F. Section of surface skin overlying the ventral surface of the pouch. Note sections of hair follicles (arrows) with adjacent sebaceous glands. The dermis is not prominently vascularized. Bar = 100 μm.
   G. Higher magnification of the surface skin. The keratinized (k) and non-keratinized epithelium (e) bounding the air surface (a) is underlain by a dermis in which no capillaries are seen in this area. Bar = 50 μm.

Grossly, the pouch is a large, compliant, skin lined bag on the lower ventral surface of the abdomen. The size of the pouch is fairly independent of body size, with a deflated volume of about 25 ml, and expandable to about 250–300 ml in animals containing a 1 g joey. The temperature of the pouch is $36.5 \pm 0.4\,^{\circ}$C ($\bar{x} \pm$ s.e.). The pouch wall is supplied by a branch of the femoral artery running directly to mammary glands on each side of the pouch. A vein drains the pouch, on each side running parallel to the artery and emptying into the femoral vein just distal to the inguinal ligament. There is a minor blood supply to the base of the pouch, chiefly to prominent paired lymph nodes. The epithelium of the pouch wall is stratified squamous and keratinized, of total thickness approximately 25–35 μm with the keratin layer approximately 5 μ thick (Fig. 1C, D). The underlying dermis (300–600 μm thick) is densely collagenous, containing cells of sebaceous glands. The dermis of the pouch wall is notably better vascularized than that of the skin overlying the pouch, (Fig. 1F, G) chiefly by larger blood vessels (50–100 μm luminal diameter in perfusion fixed specimens) located in the deepest level of the dermis (typically about 300 μm below the pouch epithelium). Profiles of coiled sweat glands are also located at this level. Larger arteries and veins (up to 700 μm and 2000 μm luminal diameter respectively) lie deeper in an adipose layer immediately adjacent to the abdominal/pouch wall muscle. The capillarity of the dermis of the pouch is markedly lower than that of the abdominal or pouch wall muscle (Fig. 1E), but somewhat higher than that of the dermis of the surface skin (Fig. 1G). The most superficial capillaries are located typically 75–100 μm below the pouch epithelium (i.e. some 100–130 μm from the pouch lumen).

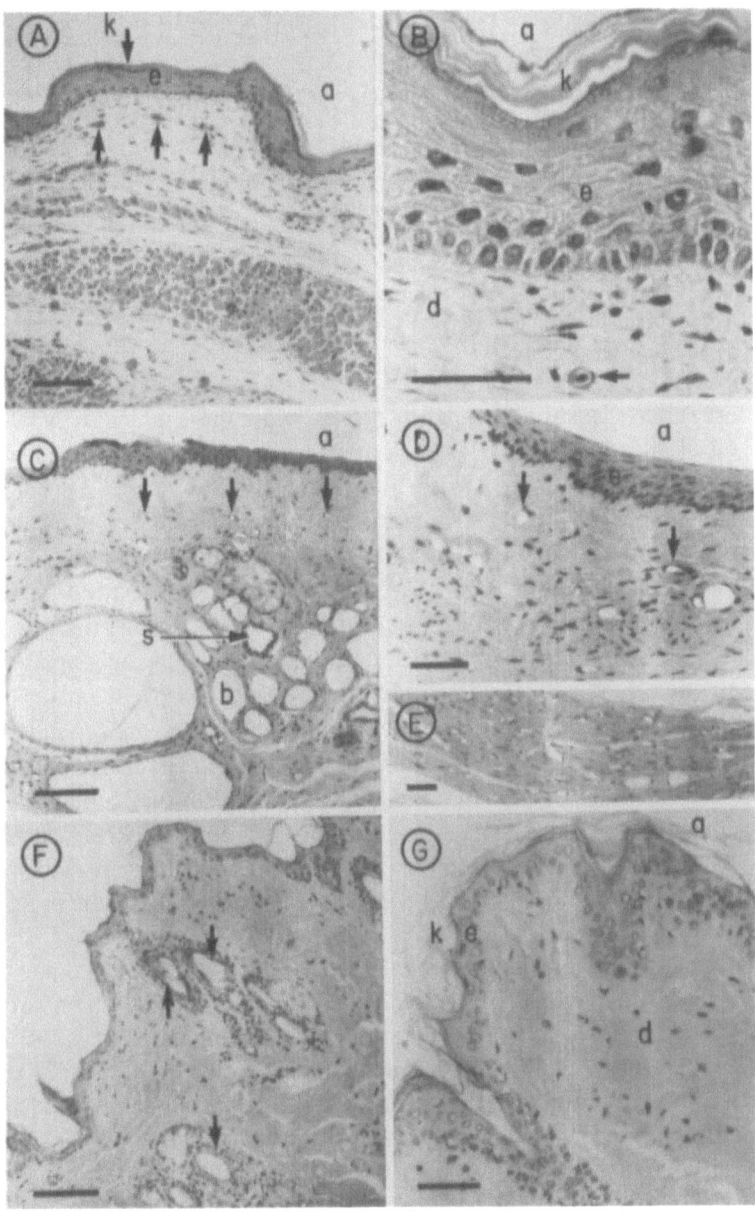

The joeys used in this study were either about 4 or 26 d old. All were hairless and blind, but attached firmly to a nipple in the pouch. Histological study of neonatal (<3 d) joeys showed that the skin epithelium is stratified, squamous, and keratinized, with a total thickness of 50–70 μm, including some 10–15 μm of keratinized layer (Fig. 1A). The dermis, some 100–200 μm thick, is of rather loose mesenchymal connective tissue which contains occasional small blood vessels.

Fig. 2. Photomicrographs of histological sections of lungs of tammar wallaby *Macropus eugenii*.
  A. Transverse section of the thorax of a 3-d joey. Note large terminal air sacs and conducting airways in the lung. Bar = 1 mm.
  B. Higher magnification of 3-d joey lung. Note dual subepithelial capillary networks (c) in the septa (s) separating adjacent terminal air sacs (tas). Bar = $100\,\mu$m.
  C. Section of 33-d joey lung. Note terminal air sacs and larger conducting airways. Bar = $500\,\mu$m.
  D. Higher magnification of 33-d joey lung. Note septa separating adjacent terminal air sacs (tas) are thinner than in the 3-d joey lung. Bar = $100\,\mu$m.
  E. Section of adult wallaby lung. Note the large number of small thin-walled alveoli. Bar = $50\,\mu$m.
  F. Higher magnification of adult lung. Note ciliated epithelium (e) of bronchiole (b), with bronchial smooth muscle (m) and microvessels of bronchial circulation (c), and only a single pulmonary capillary network in the septum (s) separating adjacent alveoli (a). Bar = $100\,\mu$m.

The most superficial capillaries lie 10–$50\,\mu$m below the epithelium (60–120μm below the skin surface (Fig. 1B)). Erythrocytes are nucleated in these neonatal joeys.

The lung parenchyma of the neonatal (<3 d) joeys is characterized by large air spaces of 150–$350\,\mu$m diameter (Fig. 2A) and is clearly at a terminal air sac rather than alveolar stage of development (Krause and Leeson, 1973; Moore, 1982). Each air sac is encircled with an almost continuous sheet of capillaries filled with red cells (Fig. 2B); the air-blood barrier is not measurable by light microscopy, but is clearly less than $1\,\mu$m. Each air sac is surrounded by its own separate capillary bed; the capillary beds underlying adjacent air sacs are separated by a loose connective tissue layer 10–$20\,\mu$m thick (Fig. 2B). The larger conducting airways are identifiable by their virtually avascular walls and cuboidal to low cuboidal simple epithelium which ranges from 15–$5\,\mu$m in thickness; large thin walled pulmonary blood vessels (up to 150μm luminal diameter) are evident near the lung hila. No specific staining of collagen or elastin is observable in the lung parenchyma at this age.

At 33 d, the lung is qualitatively similar to that observed in the neonates (Fig. 2C); however the erythrocytes are no longer nucleated and some cartilage nodules have begun to develop around the largest conducting airways. Although a vast increase in the number of terminal air sacs is apparent, most of the air spaces are still around $250\,\mu$m diameter; however significant numbers of smaller air chambers around $50\,\mu$m diameter are also developing at this age. The individual air spaces are still each surrounded by a unique capillary network and the septum separating adjacent air spaces thus containes two capillary networks (Fig. 2D). Stainable collagen is present around the conducting airways, larger vessels, in the pleura, and sparsely in the parenchyma; occasional fine elastic fibres are evident in the parenchyma.

In sections of the adult lung, fixed at residual volume at apnoea, most alveoli are 50–$100\,\mu$m in diameter (Fig. 2E). The most apparent differences between the adult and the 33-d joey are (1) the presence of only a single capillary layer in the tissue septum separating adjacent air spaces, with resultant loss of the connective

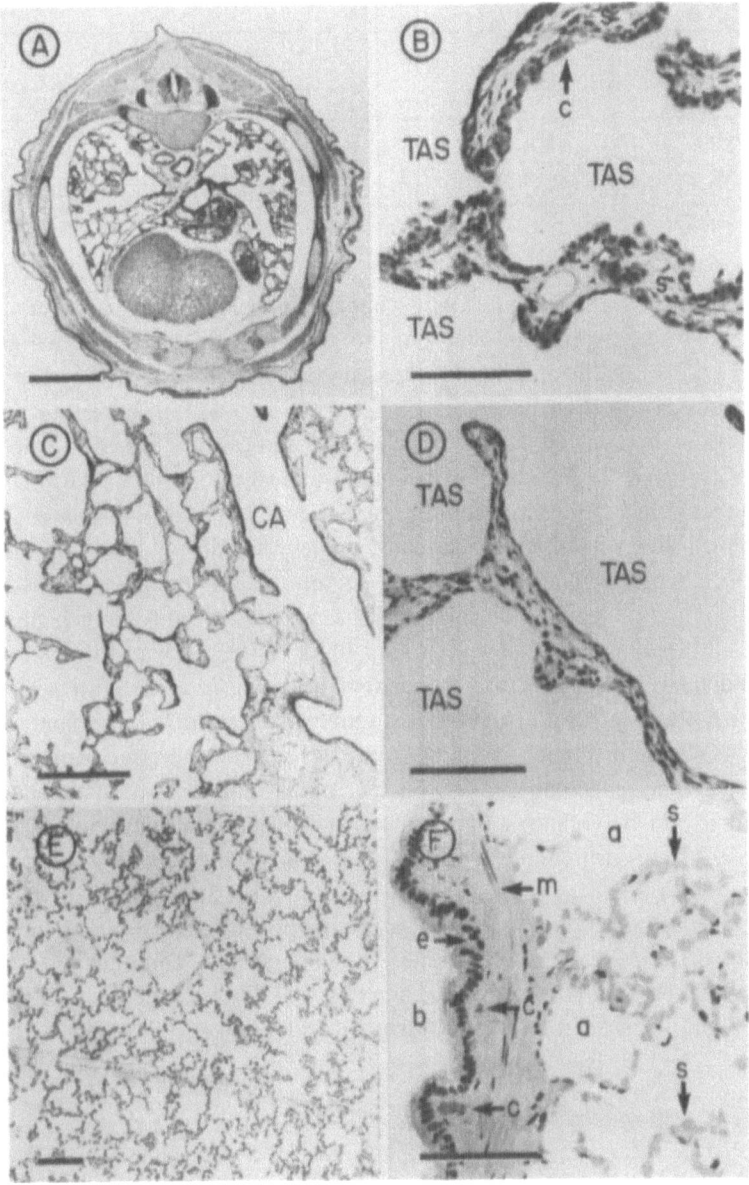

tissue layer between adjacent alveoli and thinning of the interalveolar septum (Fig. 2F); (2) the clearly evident increase in the ratio of air space to tissue volume, and (3) the marked development of muscle, ciliated epithelium and bronchial circulation of the conducting airways (Fig. 2F).

Gas samples from a normal open pouch containing a young joey have high P_{O_2} and low P_{CO_2} values (Table 1); however, if the animal is curled up while restricted

Table 1. O_2 and CO_2 levels in gas from a normal, intact pouch (values $\bar{x} \pm$ s.e.).

| | Animal in cage | | Animal in sack | |
	Pouch gas	Air	Pouch gas	Air in sack
P_{O_2} (torr)	143.4 ± 0.6	146.2 ± 0.6	134.5 ± 2.1	143.0 ± 0.7
P_{CO_2} (torr)	4.3 ± 0.5	<0.8	13.4 ± 3.5	2.7 ± 0.5

within a jute sack, rather than a cage, then pouch gas P_{O_2} is reduced and P_{CO_2} is increased somewhat (Table 1). Gas exchange between the pouch and ambient air is extensive because, if the open pouch is flushed with N_2, the pouch gas P_{O_2} rapidly rises to a value of around 140 torr, a value typical of gas in the open pouch with a joey in situ (Fig. 5).

If the joey is removed and the pouch sealed and filled with N_2 or air, then the P_{CO_2} levels rise in the pouch gas to an equilibrium value of 39.4 ± 3.1 torr (Fig. 4). The increase in P_{CO_2} in more rapid if the pouch is filled with air than N_2. If the pouch is flushed with air and then sealed, O_2 levels in the pouch fall; however, if the pouch is first flushed with N_2, some O_2 diffuses into the pouch, presumably from the maternal circulation (Fig. 3). If gas is sampled from a pouch closed for over 24 h, then the P_{CO_2} is always around 40 torr but P_{O_2} is very variable, on one occasion being as low as 7 torr, the mean value being 70.1 ± 15.2 torr ($\bar{x} \pm$ s.e.).

Clearly the pouch wall (together with any bacterial flora it may contain) can remove O_2 from the pouch gas and CO_2 can diffuse from maternal tissues into the pouch gas. The calculated rates of pouch O_2 uptake and CO_2 production based on changes in pouch CO_2 and O_2 between 5 and 60 min after flushing with air (Figs. 3, 4), are 2.85 ml $O_2 \cdot h^{-1}$ and 2.40 ml $CO_2 \cdot h^{-1}$, respectively; the respiratory exchange ratio ($\dot{V}_{CO_2}/\dot{V}_{O_2}$) is 0.84; these values probably apply under the high P_{O_2}, low P_{CO_2} conditions normally found within the pouch. If the pouch is filled with N_2, then O_2 as well as CO_2 accumulates in the pouch gas. The rate of O_2 transfer under these conditions is much less than that for CO_2, presumably reflecting the much higher transfer rates of CO_2 through tissue.

Table 2. O_2 uptake and CO_2 production by six wallaby joeys (values $\bar{x} \pm$ s.e.).

Body mass (g)	0.86 ± 0.08	6.71
Age (d)	4 ± 0.3	33
O_2 uptake		
$\quad \dot{V}_{O_2}$ ml $\cdot$ g$^{-1} \cdot$ h^{-1}	0.60 ± 0.13	0.67
CO_2 excretion		
$\quad \dot{V}_{CO_2}$ ml $\cdot$ g$^{-1} \cdot$ h^{-1}	0.52 ± 0.07	0.60
Gas exchange ratio		
$\quad RE = \dot{V}_{CO_2}/\dot{V}_{O_2}$	0.94 ± 0.12	0.9
N	5	1

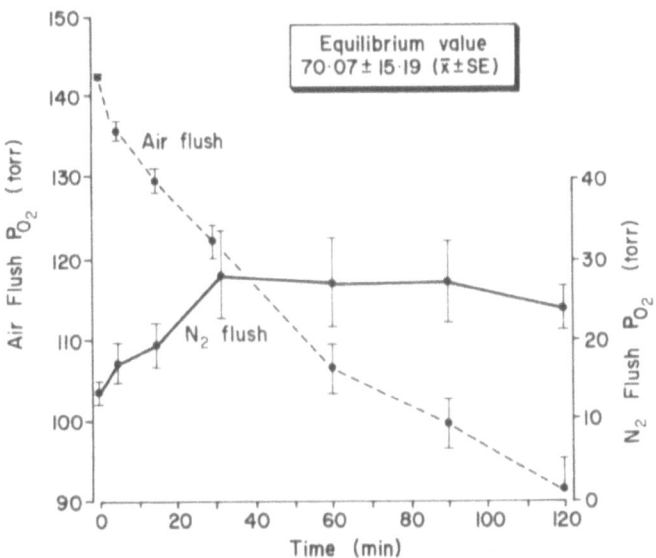

Fig. 3. Changes in P_{O_2} in gas contained in a sealed pouch flushed with either N_2 or air at zero time. The equilibrium value is that obtained at 24 h after flushing.

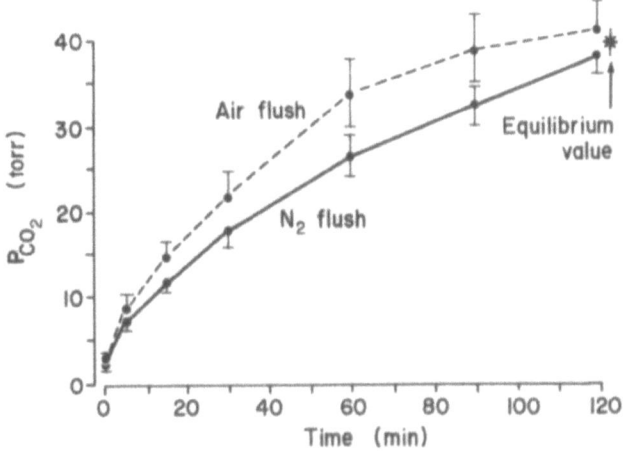

Fig. 4. Changes in P_{CO_2} in gas in a sealed pouch flushed with either N_2 or air at zero time. The equilibrium value is that obtained at 24 h after flushing.

The O_2 conductance from air to pouch gas can be determined by dividing the total O_2 uptake of the pouch ($\dot{V}_{O_2}$ of joey plus $\dot{V}_{O_2}$ of pouch) by the mean P_{O_2} difference between pouch gas and ambient air (Table 1). Similar calculations can be made for the CO_2 conductance (Table 3). Thus, if the open pouch is flushed with N_2, the pouch gas P_{O_2} rapidly rises to a value of around 140 torr, a value

Table 3. The conductance (G) for O_2 and CO_2 of the wallaby pouch opening containing a 4-d joey calculated by dividing the total O_2 ($\dot{V}_{O_2\ tot}$ = O_2 uptake by joey plus uptake by pouch) and total CO_2 transfer ($\dot{V}_{CO_2\ tot}$) by the mean difference in P_{O_2} (ΔP_{O_2}) or P_{CO_2} (ΔP_{CO_2}) between pouch gas and ambient air.

$\dot{V}_{O_2\ tot}$	$= 3.36\,\text{ml } O_2 \cdot h^{-1}$
$\dot{V}_{CO_2\ tot}$	$= 2.86\,\text{ml } CO_2 \cdot h^{-1}$
Conductance of pouch opening for O_2 (G_{O_2})	
	$= \dfrac{\dot{V}_{O_2\ tot}}{\Delta P_{O_2}}$
	$= 1.12\,\text{ml} \cdot \text{torr}^{-1} \cdot h^{-1}$
Conductance of pouch opening for CO_2 (G_{CO_2})	
	$= \dfrac{\dot{V}_{CO_2\ tot}}{\Delta P_{CO_2}}$
	$= 0.88\,\text{ml} \cdot \text{torr}^{-1} \cdot h^{-1}$

typical of gas in the open pouch with a joey in situ (Fig. 5).

Oxygen uptake by the joey is shown in Table 2. Five out of six joeys measured were between 2 and 5 d old, one was 33 d old. This much older joey had the same mass-specific O_2 and CO_2 exchange rates as the much smaller joeys; there was also no difference in the gas exchange ratio.

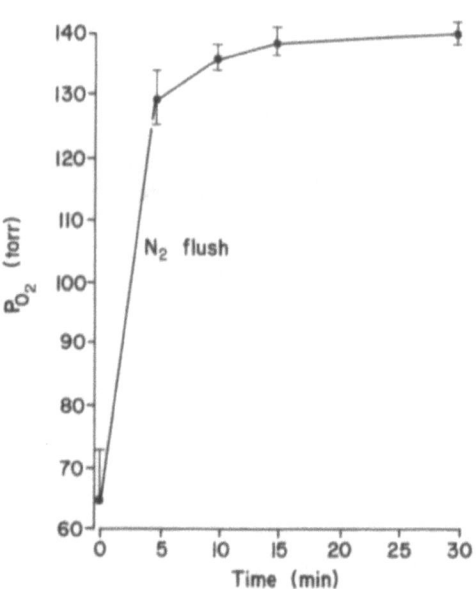

Fig. 5. Changes in P_{O_2} of gas contained in an open pouch with a joey in situ. The pouch was flushed with N_2 at zero time.

Discussion

The 4-d joey consumes only 15% of the O_2 utilized from the pouch gas (Fig. 6). The other 85% is consumed by either the pouch wall or its contained bacterial flora or else taken up into maternal blood. Some preliminary observations suggest that bacterial $\dot{V}_{O_2}$ is significant because total $\dot{V}_{O_2}$ is reduced if the sealed pouch is disinfected before closure. Even in this situation there is still a rapid reduction of O_2 levels in gas in the disinfected sealed pouch, O_2 apparently either being consumed by the pouch wall or passing into maternal blood. What is clear is that the mother and bacterial flora in the pouch, along with the joey, remove O_2 and add CO_2 to the pouch gas.

The $\dot{V}_{O_2}$ for the joey of about $0.6 \, ml \cdot g^{-1} \cdot h^{-1}$ is within the range reported for small mammals (Altman and Dittmer, 1974). It is higher than the value of $0.29 \, ml \cdot g^{-1} \cdot h^{-1}$ reported for adult wallabies (Altman and Dittmer, 1974) but this is to be expected, for the joey is active and growing rapidly. The recorded value for $\dot{V}_{O_2}$ may not be representative of that for the joey in the pouch, since, in our experiments, the joey was removed from the nipple and was continually moving around in the chamber in which $\dot{V}_{O_2}$ was measured. Joeys move around when the pouch is open, but we do not know if this movement continues when the joey is not exposed. Thus our $\dot{V}_{O_2}$ values are for a wallaby neonate that is exposed but not feeding, but is very active. Whether $\dot{V}_{O_2}$ is higher or lower when the neonate is in the pouch, we do not know.

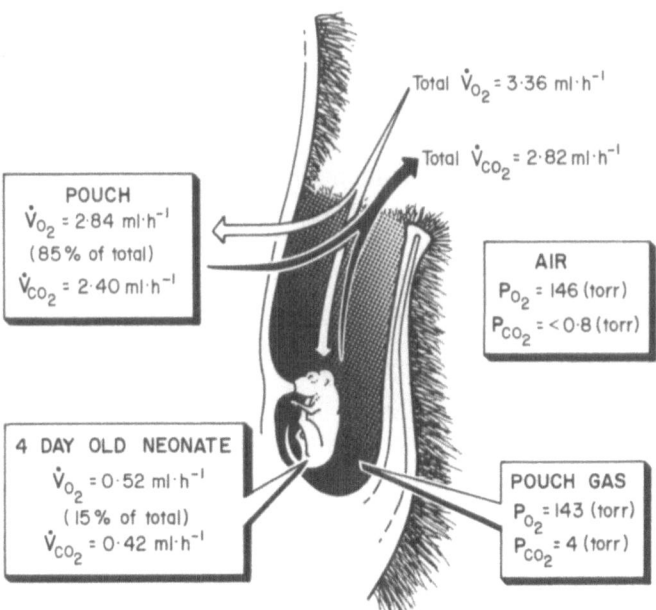

Total $\dot{V}_{O_2}$ = 3·36 ml·h^{-1}

Total $\dot{V}_{CO_2}$ = 2·82 ml·h^{-1}

POUCH
$\dot{V}_{O_2}$ = 2·84 ml·h^{-1}
(85% of total)
$\dot{V}_{CO_2}$ = 2·40 ml·h^{-1}

AIR
P_{O_2} = 146 (torr)
P_{CO_2} = < 0·8 (torr)

4 DAY OLD NEONATE
$\dot{V}_{O_2}$ = 0·52 ml·h^{-1}
(15% of total)
$\dot{V}_{CO_2}$ = 0·42 ml·h^{-1}

POUCH GAS
P_{O_2} = 143 (torr)
P_{CO_2} = 4 (torr)

Fig. 6. A schematic diagram illustrating O_2 and CO_2 transfer between ambient air, pouch gas, the pouch wall and the neonate (ca. 4-d) of the tammar wallaby *Macropus eugenii*.

From the histological study, it is apparent that the lungs must be the principal gas exchange surface of the neonatal wallaby joey, as in the neonatal Virginia opossum where the lung is also well developed (Krause and Leeson, 1973, 1975). Both because of their relative infrequency, and also because of their much greater distance from the gas surface, the skin capillaries can play virtually no part in the overall gas exchange of the wallaby neonate. Preliminary direct measurements of the skin/lung partitioning of O_2 uptake indicate that O_2 removal across the skin of an 11-d joey accounts for less than 0.1% of total (L. Matwiejczyk, pers. comm.). The skin erythema commonly observed in wallaby joeys (and other marsupial neonates) therefore has little to do with gas exchange, despite frequent suggestions to the contrary (Richardson and Russell, 1969); it may be that the newborn skin is relatively more transparent and thus permits visualization of the circulation in the most superficial muscle layer, which is only some 150–200 μm below the surface. The nucleated red cells of the neonate are a common feature of marsupial pouch young, also being observed in neonatal kangaroos and the Virginia opossum (Block, 1964; Richardson and Russell, 1969).

While as yet we have insufficient data to accurately describe the postnatal development of the wallaby lung through to the adult, it is apparent that the initial development of alveoli is occurring by about 30 d, and that differentiation of the conducting airways is commencing by this time; it seems likely that the postnatal lung development will be similar qualitatively to that observed in the Virginia opossum (Krause and Leeson, 1973, 1975).

Oxygen levels are higher and CO_2 levels lower in the pouch gas of the tammar wallaby than those reported for the brush-tailed possum and the Virginia opossum (Bailey and Dunnet, 1960; Farber and Tenney, 1971). Our values appear to be the most extensive pouch values recorded for young about 4 d old and the differences between species may simply reflect the age of the neonate. It is possible that, as the joey grows and metabolism increases, the pouch gas P_{CO_2} rises and P_{O_2} falls to levels similar to that observed by others in the possum and opossum with older young (Bailey and Dunnet, 1960; Farber and Tenney, 1971).

The outer wall of the pouch is thick and covered with hair, and, most likely, all the O_2 and CO_2 transferred between ambient air and pouch gas occurs through the pouch opening. When the joey is small this opening is constricted and covered with hair. In this study the pouch was at 37 °C. The pouch gas was probably 100% saturated with water vapor because the inside surface of the pouch and the joey skin was slightly moist to the touch. The ambient temperature varied but the maximum was in the region of 20 °C, the temperature falling to below 10 °C during the night in the unheated laboratory housing the animals. The air was dry so the ambient water vapor pressure, although not measured, could have been as low at 5 to 15 torr. Thus there was a large pressure gradient for water vapor between pouch gas and ambient air, perhaps of the order of 20 or 30 torr. The water vapor pressure gradient was, therefore, an order of magnitude greater than that for O_2 and CO_2. There is undoubtedly heat loss from the pouch, probably with a sharp temperature gradient at the mouth of the pouch (Bird et al., 1960). The net movement of water vapor and CO_2 from the pouch as well as the thermal gradient from the pouch tends to retard O_2 diffusion into the pouch. As a result, given that the O_2 and CO_2 partial pressure differences between ambient air and pouch gas are similar, there is probably some convective exchange in addition to diffusion, to maintain O_2 transfer between air and pouch gas. Exactly how this occurs is not clear, as the pouch opening is constricted and covered with hair and we observed no large or rapid changes in pouch temperature or gas tensions indicative of the episodic exchange of large volumes of pouch gas with ambient air.

As the joey increases in size in relation to pouch volume, movements of the joey and a less constricted pouch opening probably facilitate convective air movements, culminating in the joey intermittently pushing his head out through the pouch opening during the latter stage of pouch dwelling.

Acknowledgements

This work was supported by grants from the Flinders University Research Committee, and from the Australian Research Grants Scheme. We are grateful to the Zoology Department, University of Adelaide for use of their Field Station, to Peter and Louise Davis of Brookland Park, Kangaroo Island, for assistance with catching and handling wallabies, and Ms Kathy Roberts and Julie Dodwell for typing of the manuscript.

References

Altman, P.L. and Dittmer, D.S. (1974). Biology Data Book Vol. III. Federation of American Societies for Experimental Biology, Bethesda Maryland.

Andrewatha, H.G. and Barker, S. (1969). Introduction to the study of the ecology of the Kangaroo Island Wallaby, *Protemnodon eugenii* (Desmarest) within Flinders Chase, Kangaroo Island, South Australia. *Trans. R. Soc. South Aust.* 93: 127–133.

Bailey, S.W. and Dunnett, G.M. (1960). The gaseous environment of the pouch young of the brush-tailed possum, *Trichosaurus vulpecula. C.S.I.R.O. Wildl. Res.* 5: 149–157.

Bird, R.B., Stewart, W.E. and Lightfoot, D.N. (1960). *Transport Phenomena.* Wiley, N.Y., 780 pp.

Block, M. (1964). The blood forming tissue and the blood of the newborn opossum (*Didelphis virginiana*). 1. Normal development through to about the one hundredth day of life. *Ergeb. Anat. Entwicklungsgesch* 37: 239–245.

Farber, J.P. (1972). Development of pulmonary reflexes and pattern of breathing in the Virginia opossum. *Respir. Physiol.* 14: 278–286.

Farber, J.P., Holtgren, H.N. and Tenney, S.M. (1972). The development of the chemical control of breathing in the Virginia opossum. *Respir. Physiol.* 14: 267–277.

Farber, J.P. and Marlow, T.A. (1977). Ventilatory effects accompanying spontaneous movement in the sucking opossum. *Respir. Physiol.* 31: 241–250.

Farber, J.P. and Tenney, S.M. (1971). The pouch gas of the Virginia opossum (*Didelphis virginiana*). *Respir. Physiol.* 11: 335–345.

Gannon, B.J., Randall, D.J., Browning, J., Lester, R.J.G. and Rogers, L.G. (1983). The microvascular organization of the gas exchange organs of the Australian lungfish *Neoceratodus forsteri* (Krefft). *Aust. J. Zool.* 31: 651–676.

Johnson, R.G. and Nicholas, T.E. (1979). The pattern of pulmonary morphological development and surfactant phospholipid composition in the marsupial *Megaleia rufa*. *Proc. Aust. Physiol. Pharmacol. Soc.* 10: 50 p.

Krause, W.L. and Leeson, C.R. (1973). The postnatal development of the respiratory system of the opossum. 1. Light and scanning electron microscopy. *Am. J. Anat.* 137: 337–356.

Krause, W.J. and Leeson, C.R. (1975). Postnatal development of the respiratory system of the opossum. II. Electron microscopy of the epithelium and pleura. *Acta. Anat.* 92: 28–44.

Moore, K.L. (1982). The developing human: Clinically oriented embryology. 3rd ed. W.B. Saunders, Philadelphia.

Murphy, C.R. and Smith, J.R. (1970). Age determination of pouch young and juvenile Kangaroo Island wallabies. *Trans. R. Soc. South Aust.* 94: 15–20.

Richardson, B.J. and Russell, E.M. (1969). Changes with age in the proportion of nucleated red blood cell types and in the type of haemoglobin in kangaroo pouch young. *Aust. J. Exp. Biol. Med. Sci.* 47: 573–580.

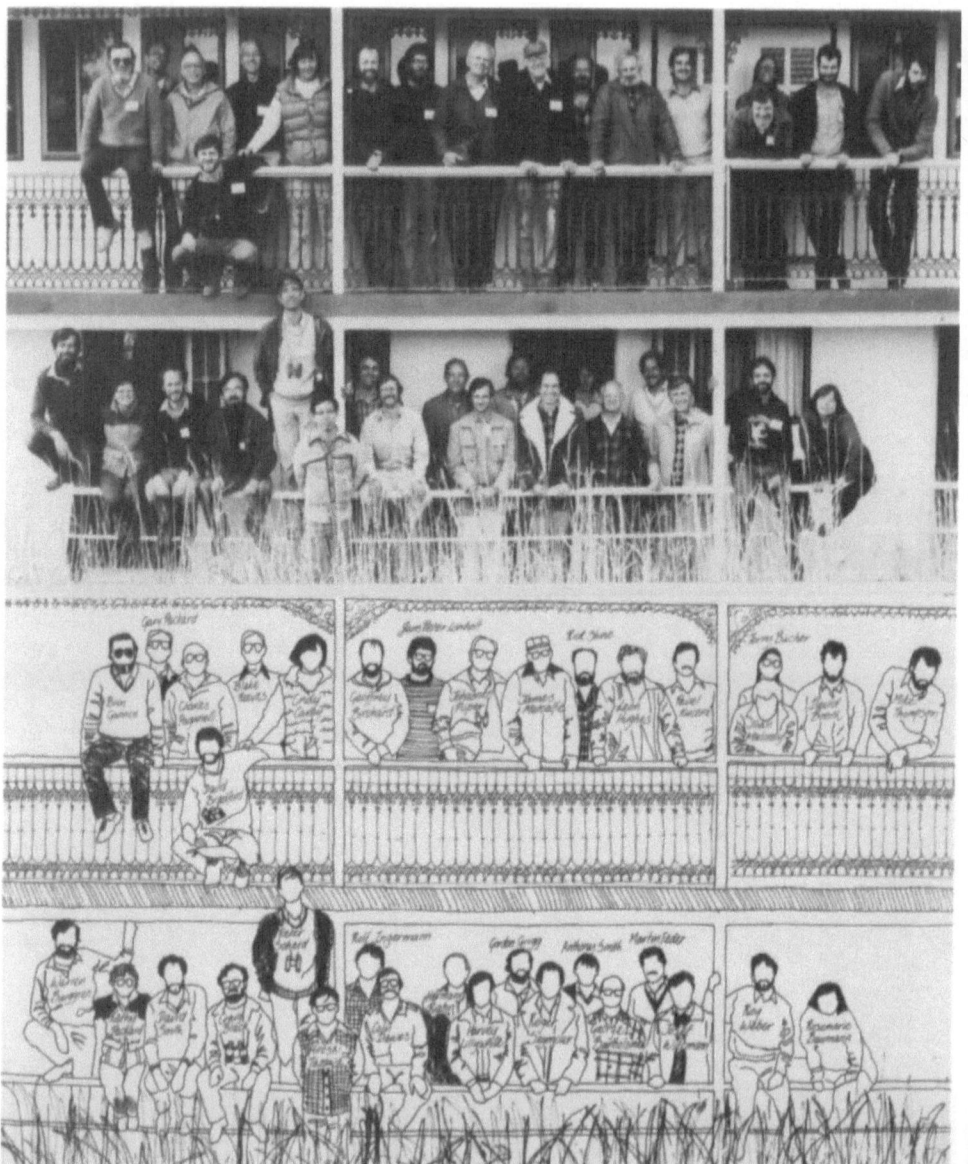

438

List of participants

George A. Bartholomew,
Department of Biology,
University of California,
LOS ANGELES, CA. 90024,
U.S.A.

Rosemarie Baumann,
Physiologisches Institut,
Medizinische Hochschule Hannover,
3000 HANNOVER,
Karl-Wiechert-Allee 9,
FEDERAL REPUBLIC OF GERMANY.

Geoffrey F. Birchard,
Department of Physiology,
Dartmouth Medical School,
HANOVER, NH 0355,
U.S.A.

Craig P. Black,
Department of Biology,
University of Toledo,
TOLEDO, Ohio 43606,
U.S.A.

David Booth,
Department of Zoology,
University of Adelaide,
G.P.O. Box 498,
ADELAIDE, S.A. 5001.
AUSTRALIA.

David F. Bradford,
Department of Zoology,
University of Adelaide,
G.P.O. Box 498,
ADELAIDE, S.A. 5000.
AUSTRALIA.

Theresa L. Bucher,
Department of Biology,
University of California,
LOS ANGELES,
California 90024,
U.S.A.

Warren W. Burggren,
Department of Zoology,
University of Massachusetts,
Amherst, MA 01003,
U.S.A.

Cynthia Carey,
Department of Environmental, Population and
Organismic Biology,
University of Colorado,
BOULDER, Colorado 80309,
U.S.A.

Colin M. Dawes,
Department of Physiology,
Royal Veterinary College,
Royal College Street,
LONDON, NW1 0TU,
ENGLAND.

Martin E. Feder,
Department of Anatomy,
The University of Chicago,
1025 East 57th Street,
CHICAGO, Illinois 60637,
U.S.A.

Bren Gannon,
School of Medicine,
Flinders University,
Sturt Road,
BEDFORD PARK, S.A. 5042.
AUSTRALIA.

439

Gordon C. Grigg,
School of Biological Sciences,
Zoology Building,
The University of Sydney,
SYDNEY, N.S.W. 2006.
AUSTRALIA.

R. Leon Hughes,
Department of Anatomy,
University of Queensland,
ST. LUCIA, Qld. 4067.
AUSTRALIA.

Rolf L. Ingermann,
Obstetrics and Gynecology Research – L458,
Oregon Health Sciences, University,
3181 S.W. Sam Jackson Park Road,
PORTLAND, Oregon 97201,
U.S.A.

Pavel Kucera,
Institute of Physiology,
University of Lausanne,
7, Rue du Bugnon,
1011 Lausanne CHUV,
SWITZERLAND.

Harvey B. Lillywhite,
Department of Zoology,
University of Florida,
GAINESVILLE, Florida 32611,
U.S.A.

Jens Peter Lomholt,
Department of Zoophysiology,
University of Aarhus,
DK-8000 AARHUS C,
DENMARK.

John Maloney,
Centre for Early Human Development,
Monash University,
Queen Victoria Medical Centre,
172 Lonsdale Street,
MELBOURNE, VIC. 3000
AUSTRALIA.

James Metcalfe,
Department of Medicine,
University of Oregon,
Health Sciences Center,

PORTLAND, Oregon 97201,
U.S.A.

Gary C. Packard,
Department of Zoology and Entomology,
Colorado State University,
FORT COLLINS, Colorado 80523,
U.S.A.

Mary J. Packard,
Department of Zoology and Entomology,
Colorado State University,
FORT COLLINS, Colorado 80523,
U.S.A.

Charles V. Paganelli,
Department of Physiology,
Schools of Medicine and Dentistry,
120 Sherman Hall,
State University of New York at Buffalo,
BUFFALO, New York 14214,
U.S.A.

Johannes Piiper,
Abteilung Physiologie,
Max-Planck-Institut für experimentelle
Medizin,
Hermann-Rein-Strasse 3,
D-3400 GÖTTINGEN,
FEDERAL REPUBLIC OF GERMANY.

Hermann Rahn,
Department of Physiology,
Schools of Medicine and Dentistry,
120 Sherman Hall,
State University of New York at Buffalo,
BUFFALO, New York 14214,
U.S.A.

R. Blake Reeves,
Department of Physiology,
Schools of Medicine and Dentistry,
120 Sherman Hall,
State University of New York at Buffalo,
BUFFALO, New York 14214,
U.S.A.

Roger S. Seymour,
Department of Zoology,
University of Adelaide,
G.P.O. Box 498,
ADELAIDE, S.A. 5001.
AUSTRALIA.

440

Peter Scheid,
Institut für Physiologie,
Ruhr-Universität Bochum,
Universitätstrasse 150,
D-4630 BOCHUM,
FEDERAL REPUBLIC OF GERMANY.

Richard Shine,
School of Biological Siences,
Zoology Building,
The University of Sydney,
SYDNEY, N.S.W. 2006
AUSTRALIA.

Anthony Smith,
Research School of Biological Sciences,
Australian National University,
CANBERRA CITY, A.C.T. 2601,
AUSTRALIA.

David Smith,
Natural History Unit,
Australian Broadcasting Corporation,
Box 9994 G.P.O.,
MELBOURNE, 3001, VIC,
AUSTRALIA.

Hiroshi Tazawa,
Department of Physiology,
Muroran Institute of Technology,
27-1 Mizumoto,
MURORAN,
JAPAN.

Michael B. Thompson,
Department of Zoology,
University of Adelaide,
G.P.O. Box 498,
ADELAIDE, S.A. 5001,
AUSTRALIA.

Roy E. Weber,
Institute of Biology,
Odense Universitet,
Campusvej 55,
DK 5230 Odense M,
DENMARK.

Josef Wittmann,
Institut für Physiologie,
Tierärztliche Fakultät,
Ludwig-Maximilians-Universität,
Veterinärstrasse 13,
D-8000 München 22,
FEDERAL REPUBLIC OF GERMANY.

Index